Thomas Brock
Michael Groteklaes
Peter Mischke

European Coatings Handbook

2nd revised edition

Brock, Groteklaes, Mischke: European Coatings Handbook

ISBN 978-3-86630-849-7

Cover: Evonik Tego Chemie GmbH

Thomas Brock, Michael Groteklaes, Peter Mischke
European Coatings Handbook, 2nd Edition
Hannover: Vincentz Network, 2010
European Coatings Tech Files
ISBN 3-86630-849-3
ISBN 978-3-86630-849-7

Please ask for our book catalogue
Vincentz Network, Plathnerstr. 4c, 30175 Hanover, Germany
T (202) 684-6630, F (202) 380-9129
E-mail: books@american-coatings.com, www.american-coatings.com

Layout: Maxbauer & Maxbauer, Hannover, Germany

ISBN 3-86630-849-3
ISBN 978-3-86630-849-7

European Coatings Tech Files

Thomas Brock
Michael Groteklaes
Peter Mischke

European Coatings Handbook

2nd revised edition

Brock, Groteklaes, Mischke: European Coatings Handbook

ISBN 978-3-86630-849-7

Foreword

Anyone working in the coatings sector, whether in manufacturing or processing, knows - or will soon observe if they are new to the business - that an extremely broad knowledge base is a prerequisite for mastering this unique protective and finishing material. Coating chemistry in its widest sense, and especially polymer science, is of central importance. However, today's coatings specialist also requires a knowledge of process engineering in relation to the use of production or application equipment, an understanding of materials science in regard to substrate materials and more generally in terms of the quality of the paint system, and finally a familiarity with environmental and safety aspects.

Very few teaching institutions are able to offer a training programme that is specially designed to cover such an extensive field of knowledge. The Niederrhein University of Applied Scienes in Krefeld, Germany, is one of these - an institution with a long tradition and good reputation, whose name comes up repeatedly in discussions with leading figures in the coatings sector. A good many of them proudly and gratefully acknowledge that the framework for their career was built in Krefeld.

For this reason the publishers and editors are extremely grateful to the current teaching faculty, represented here by Peter Mischke, Michael Groteklaes and Thomas Brock, for deciding to make a large part of the Krefeld curriculum available to practitioners in the field. The authors have produced a contemporary handbook of coating technology. Each was responsible for around a third of the content, based on his own specialist subject areas and written in roughly the above sequence. The work merits the title of handbook for two reasons: firstly because of its solid theoretical basis, augmented by "in-depth" explanations (shown on a grey background) where necessary, and secondly because of its consistently relevant use of practical references to exemplify its themes. These features are underpinned by a constant awareness of emerging developments in the coatings sector, which remains as dynamic as ever.

The book covers the principles of raw materials, manufacture, application and testing of coatings; as a handbook, however, its principal aim is to illustrate and to create connections. Naturally only the essential themes could be addressed within the stated limits of the book. It does not wish or claim to be complete; the authors felt that it was more important to explain the foundations and principles as clearly as possible. For this reason also, the book does not contain all of the material taught to budding coating engineers at Krefeld; this would far exceed the scope of a single-volume handbook.

This work is intended to fill a gap in the current specialist literature: as an accompanying handbook it is intended on the one hand to provide a trainee or student with the basic knowledge to form a solid foundation for a closer study of coating technology; on the other hand it is designed to help people from other disciplines - scientists, engineers, business people - to find out more about this subject which, in its fascinating diversity, is difficult to assimilate. Experienced coating specialists may use it to refresh or to extend their knowledge. It may also enable them to take a glance over the "garden fence" into neighbouring disciplines, into the raw materials used every day or into the application and usage of coating materials.

The authors hope that also the second edition of the book will meet the expectations of the reader and stimulate him (or her) to take it out of the shelves very often.

No book is ever perfect. - There will certainly be specialists amongst our readers who can offer changes or improvements to particular topics; the authors will be grateful for any constructive suggestions!

Krefeld, August 2009
Peter Mischke, Michael Groteklaes and Thomas Brock

Brock, Groteklaes, Mischke: European Coatings Handbook

ISBN 978-3-86630-849-7

Contents

Brock, Groteklaes, Mischke: European Coatings Handbook

ISBN 978-3-86630-849-7

1 Introduction

1.1 Historical perspective

The earliest known use of paint dates back around 30,000 years. People used mixtures of coloured earth, soot, grease and other natural substances to ornament their bodies and to decorate their homes and places of worship, one such example being the cave paintings discovered in southern France and northern Spain.

In ancient times ...

The advanced civilisations of the Egyptians (from 4000 years BC), Greeks and Romans used sophisticated painting techniques to decorate or to identify vessels, statues, tools and buildings. Raw materials included vegetable gums, starches, hide glue, milk (products), beeswax, charcoal and various minerals. Natural dyes such as indigo, purple and madder were used to dye textiles, fibres, wood, paper and leather.

In contrast to the decorative or colour-giving use of paints described so far, the art of lacquerwork was developed in China from around 2000 years before Christ, to produce smooth and glossy surfaces. The lacquers were based on the sap of the Chinese rhus tree and, in addition to their decorative effect, they also had a protective function. Raw materials such as balsams and resins, vermilion and ultramarine, came predominantly from India. The word "lacquer" itself stems from the term "Laksha", from the pre-Christian, sacred Indian language Sanskrit, and originally referred to shellac, a resin produced by special insects ("lac insects") from the sap of an Indian fig tree.

Seafaring brought with it another important area of application for coatings. The fourth century before Christ saw a wave of migration spreading from Asia Minor as far as England and Scandinavia - some of it by land and some by sea. The wooden ships that carried the migrants were made watertight with mixtures of non-drying (non-curing) oils and tree resins or rock asphalt.

Leaping further forward in time, around the year AD 1100 the German goldsmith and monk *Roger von Helmarshausen (Theophilus)* described the manufacture of a coating by boiling linseed oil with molten amber. This process, known as paint boiling, continued to develop and by the 17th century there were numerous recipes for coatings made from a variety of natural resins, linseed oil and spirit.

In modern times ...

In the 18th century the Industrial Revolution brought about a dramatic rise in the demand for paints and coatings. In particular, the increasing numbers of goods and buildings produced from rust-prone iron needed to be treated to protect them against weathering. Furthermore, countries with a strong seafaring economy required large quantities of marine paints. The first paint factories, which appeared in England in 1790, grew out of the larger paint workshops. They were followed by factories in Holland and later in Germany and other countries.

With the exception of a few synthetic pigments already produced on an industrial scale (Berlin blue, cobalt blue, mineral green, chromium yellow), the raw materials for coatings were all of natural origin even in the 19th century. A distinction was made between "volatile paints", "varnishes" and "long-oil paints". This last group were manufactured by boiling resins with drying oils in "brewing kettles", adding pigments if required. The addition of pigments became increasingly mechanised - first using cone mills then, from the early 20th century, cylinder mills. One weak point of these products was their extended drying time; it could take several weeks to paint an entire coach or car.

Brock, Groteklaes, Mischke: European Coatings Handbook

ISBN 978-3-86630-849-7

In the 20th century…

Huge innovations took place after the turn of the century. In terms of coating technology, the following advances were particularly important:

- the development of synthetic polymer chemistry
- the invention of the production line by *Henry Ford* (1913) and the mass production of cars arising from it.

In response to the demand for faster coating technologies, the spraying of coatings based on cellulose nitrate (nitrocellulose) was introduced.

In 1907 the first entirely synthetic resins, phenol-formaldehyde condensates ("Bakelite") were launched on the market. These were followed in rapid succession by vinyl resins, urea resins and, from the 1930s onwards, alkyd resins, acrylic resins, polyurethanes and melamine resins. Epoxy resins were introduced in the late 1940s. Titanium dioxide established itself as the leading white pigment when it went into mass production in 1919.

These developments in *coating chemistry* were paralleled (finally) by advances in *coating technology*. The various methods of brush application and spraying were supplemented by electrodeposition, electrostatic coating and powder coating techniques. Ambient air drying was joined by infrared and radiation drying methods (UV, electron beam), and the automation of coating processes continued to advance. It is also worth mentioning environmental technologies for the control of air and water pollution and for waste reduction.

Measuring techniques for coatings can be regarded as the pillar supporting modern coating technology. The reproducible quantifiability of flow properties, optical characteristics, drying behaviour, adhesion, anti-corrosion action and many other properties of coating materials and/or coatings is the precondition for selective product development and the practical usage of products. Many companies now sell instruments for performing the various measuring techniques - most of them governed by standards - for coatings and related products.

At the start of the 21st century, there do not appear to have been any clear-cut revolutionary innovations in the coatings sector, but rather a great many individual developments have been aimed at improving and attaining highly specific functional properties and effects. The primary development goal of the last two decades, namely enhanced environmental compatibility of products, now seems to be taking a backseat (see section 1.2), aside from a general tendency to substitute renewable materials for mineral-oil-based products, wherever possible, in the long term.

To recap and to summarise, we can see that:

The production and use of paints and coatings has developed from a prehistoric art form via an empirical craft into the multi-disciplinary, highly complex coating technology of today.

1.2 The economic importance of paints and coatings

The coatings industry is a medium-sized sector, albeit with a growing tendency towards internationalisation. In 2005 some 250 paint factories in Germany employed around 20,500 people. In 2007 they produced over 2.4 million tonnes of coatings, paints and thinners, with an overall worth of approximately 5 thousand million euro. In terms of value this equates to about 1 % of Germany's commodities production. The tonnage produced is the equivalent of 4000 fully laden trains.

The economic benefit of these products only becomes clear if we look at their applications. The overwhelming majority of coatings and paints have, in addition to their optical or aesthetic function, a protective and hence value-retaining function. 2.0 million tonnes of paint could cover and therefore protect against corrosion, weathering and/or mechanical damage an area of around 3,000 km^2 (roughly the size of Oxfordshire) with a 200 μm thick (dry) coat.

The overall quantity of coatings produced covers a vast range of product types:

- emulsion paints and renderings: approx. 45 %
- solvent-based coatings: approx. 21 %
- powder coatings: approx. 3 %
- electrodeposition coatings and water-thinnable industrial coatings approx. 2 %
- other miscellaneous coatings: approx. 29 %

In the breakdown above, the proportion of powder coatings seems too low, since one part by weight of powder coating is equivalent to two to three parts of wet paint.

We will end with a look at the three environmentally friendly coating classes: water-based coatings, powder coatings and high solid coatings (see Figure 1.1). Annual production of high solid coatings and especially powder coatings increased steadily during the 1990s, whereas water-based volumes stagnated. The current overall trend is negative, with production of water-based coatings seemingly having collapsed altogether. In view of manufacturers' efforts to be able to supply water-based products, this observation comes somewhat as a surprise. It would also appear that more and more attention has been paid since 2000 to coating properties, such as new optical and haptic effects, enhanced scratch resistance and easy-to-clean properties, all grouped together under the fashionable term "performance", frequently in combination with the buzzword "nanotechnology". For the sake of completeness we should add that there are other low-solvent or solvent-free products, e.g. radiation-curing coatings, solvent-free two-component systems and paint-like emulsion coatings.

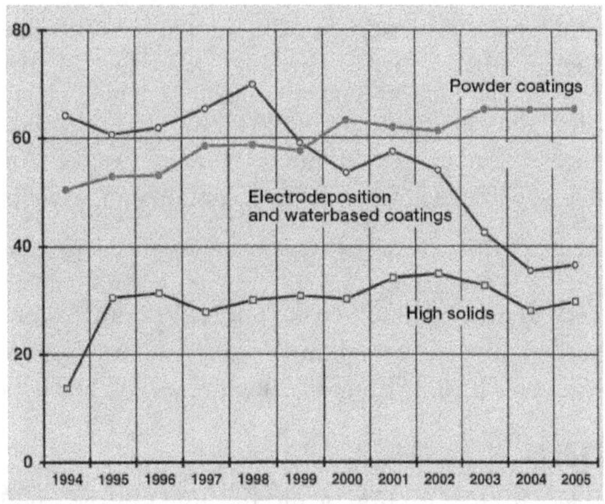

Figure 1.1: Evolution in production of environmentally friendly coatings 1995-2006 in Germany. (source "besser lackieren!" Jahrbuch 2008, Vincentz Network 2007)

1.3 *Classification and material structure of coatings*

The title of EN ISO 4618 is "Paints and varnishes - Terms and definitions". In Germany, this standard has been supplemented by DIN 55945 (2007-03) "Paints and varnishes - Additional terms and definitions to EN ISO 4618".

A **paint** is defined in the standard as a product in liquid or paste form that is applied primarily by brushing, rolling or spraying.

A product based on organic binders (modified natural substances, synthetic resins), which when applied to a substrate produces a cohesive, virtually water-impermeable (non-absorbent), protective and possibly decorative film, is called a **coating material**. The coating itself is properly termed a **coating system**; it comprises the **coating film**.

Products not covered by this definition include (polymer) emulsion paints, silicon emulsion paints and distempers. Printing inks are naturally also excluded.

A **coating powder** is a powder coating which produces a film after it has been applied to and fused onto the substrate. (Since coating powders are not in liquid or paste form, they do not by definition belong to the category of paints). The term "paint" in the trade sense refers to a pigmented coating ("paint") or alternatively to a pigmented varnish ("gloss paint").

<table>
<tr><th colspan="2">Coating material</th></tr>
<tr><td>non-volatile matter</td><td>volatile matter</td></tr>
<tr><td>pigments
fillers
film-formers
non-volatile additives</td><td>solvents or dispersants
volatile additives
(any elimination products from stoving)</td></tr>
</table>

Other materials covered by the standard include fillers, synthetic resin renderings and floor coating compounds.

All coating materials are based on the structure shown in the table below. Not every coating material necessarily contains all the components listed. (A clear varnish does not contain any pigments or fillers, whilst a powder coating contains no solvents).

Pigments

Very finely dispersed colouring and/or corrosion-inhibiting powder that is practically insoluble in the application medium.
Examples include titanium dioxide, carbon black, pearlescent pigments, zinc phosphate.

Fillers

Powders that are practically insoluble in the application medium and which impart or improve particular technological properties and give the coating material greater volume (body).
Examples include chalk, talcum, cellulose fibres.

Film formers

Macromolecular or macromolecule-forming substances responsible for film formation.
Examples include chlorine rubber, alkyd resin, polyester/polyisocyanate blends (two-component systems), polyester acrylate (radiation-curable).

Additives

Substances that are generally added in small quantities and which have particular chemical or technological effects.
Examples include hardening accelerators (catalysts), thickeners, dispersants, flow control agents, flatting agents, preservatives.

Solvents

Liquids or blends of liquids that are able to dissolve the film former(s).
Examples include butyl acetate, butyl glycol, white spirit, water.
Also known as *thinning agents* or *thinners* when used to adjust processing characteristics (viscosity).

Dispersants

Liquids that do not dissolve the film former(s) but instead hold them in a fine, microheterogeneous dispersion (or emulsion).
Examples include water and, in non-aqueous dispersions, hydrocarbons.

One further term, which is frequently used incorrectly, is “binder”. According to the standards, the **binder** is the non-volatile part of the coating material, excluding pigments and fillers but including non-volatile additives such as plasticisers and driers. In common usage, however, binder is frequently used to mean film former.

1.4 *Technology of paints and coatings ("Coating technology")*

In broad terms the whole teaching of coatings (and paints) - as is the case in this book - can be regarded as "coating technology". In more precise terms, however, coating technology - as opposed to coating chemistry - refers to the process technology of the manufacture and processing of coatings and paints, where processing can be subdivided into the processes of application (spraying, dipping, brushing, etc.) and of drying or curing (air drying, stoving, radiation curing).

A typical problem in coating technology, which is representative of many others, is the balance in a coating between spreading and dripping. After a coating has been applied it should normally form a uniform surface. In order to achieve this, any irregularities arising from its application, such as brush strokes, build-up of droplets from spraying or roller marks, should even out naturally if the coating is still sufficiently flowable, i.e. not too dry. On the other hand, however, the flowable coating must not drip when applied to vertical surfaces, since this would lead to "running", "sagging" and other unattractive forms of curtaining. We can see that two conflicting properties are expected of the coating, and these can only be balanced by means of skilful formulation and adjustment of application conditions.

The following parameters are specifically involved in this problem:

- roughness of the substrate
- form and degree of the initial unevenness of the wet film
- evaporation behaviour of the solvent or solvent blend
- change in viscosity during evaporation
- rheological behaviour (Newtonian, pseudoplastic, thixotropic)
- surface tension (size and uniformity)
- slope of the surface in question.

This example not only illustrates the complexity of coating technology but also shows that the development of coatings and paints requires the properties of the material to be precisely adjusted to the particular technological conditions existing in individual coating workshops.

Sources and references for Chapter 1

[1] G. Benzing et al.: Pigmente und Farbstoffe für die Lackindustrie. 2nd edn., Expert-Verlag, Ehningen 1992

[2] H. Biegel: Industrielacke (Die Bibliothek der Technik, vol. 39). Verlag moderne Industrie, Landsberg/Lech 1990

[3] Brockhaus Enzyklopädie, vol. 22, 19th edn., F. A. Brockhaus GmbH, Mannheim 1993

[4] www.colour-europe.de/textMenue_II_VDL-Statistik.html (August 2008)

[5] www.colouring.de/Lackindustrie/... (August 2008)

[6] www.destatis.de/... (Statistisches Bundesamt, August 2008)

[7] DIN 55945 (2007-03)

[8] EN ISO 4618 (2007)

[9] H. Kittel (Ed.): Lehrbuch der Lacke und Beschichtungen. Vol. 1/Part 1, Verlag W. A. Colomb, Stuttgart - Berlin 1971

[10] D. Ondratschek (Ed.): besser Lackieren! Jahrbuch 2008. Vincentz Network, Hanover 2007

2 Raw materials for coatings

2.1 Film formers

Film formers, which are frequently also referred to imprecisely as binders (→1.3), are polymers or oligomers (prepolymers) that are generally organo-chemical in nature and which polymerise as the coating cures. The role of the film former is to form a cohesive coating or paint film on a given substrate and - where relevant - to hold together or to embed the other non-volatile components of the coating, particularly the pigments and fillers. The film former thus constitutes the basis for the coating material in question.

Depending on their origin, film formers can be categorised into

- natural substances
- modified natural substances
- synthetic substances.

The importance of the product types increases through the above sequence. Unmodified natural substances are used in very few coatings now and never as the sole film former. With the exception of "bio-coatings" or "natural coatings", natural film formers are now used primarily in certain printing inks. Before exploring the chemistry and properties of individual film formers, it is important to establish a grounding in polymer science, which is covered in the section below.

2.1.1 General polymer science

2.1.1.1 Basic concepts

The following section introduces the basic concepts of polymer chemistry that are relevant to the field of coating film formers.

A **monomer**

is a substance consisting of small, reactive molecules that can be converted to a polymer through what is known as a polymerisation reaction (see below).

Examples:

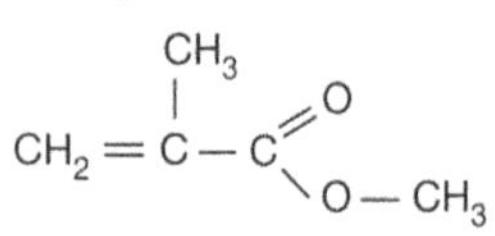

$CH_2 = CH - CH = CH_2$

methyl methacrylate (methacrylic acid methyl ester)

1,3-butadiene

HO — R — OH
a diol

HOOC — R' — COOH
a dicarboxylic acid

(R, R' = molecule skeleton, not otherwise specified)

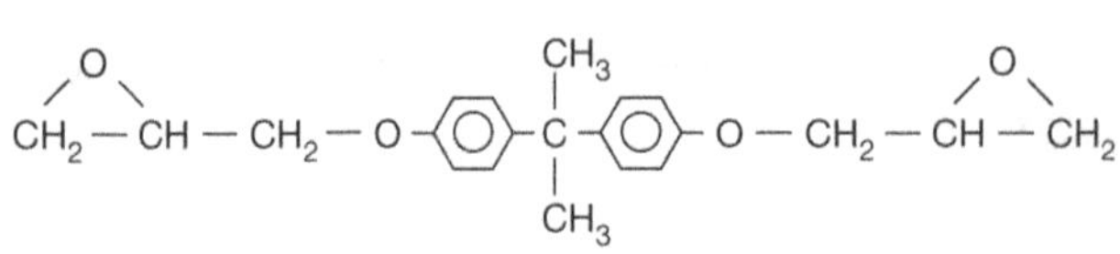

bisphenol-A-diglycidyl ether (BADGE)

Brock, Groteklaes, Mischke: European Coatings Handbook

ISBN 978-3-86630-849-7

$H_2N-CH_2-CH_2-NH_2$

ethylene diamine

A **polymer** (macromolecular substance)

is a substance consisting of (very) long molecules (polymer or macromolecules) or extended molecule networks. Individual polymer molecules may have a molecular weight ranging from a few thousand to several million g/mol.

Examples:

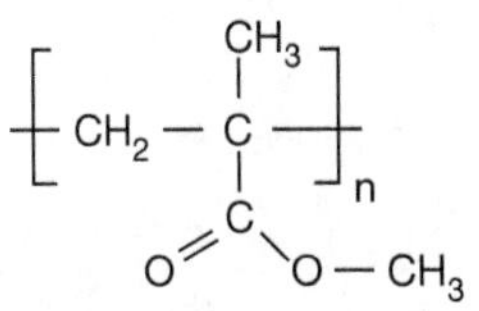

polymethyl methacrylate

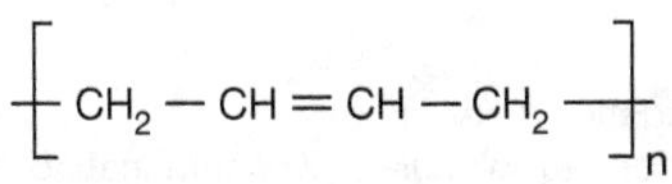

1,4-polybutadiene

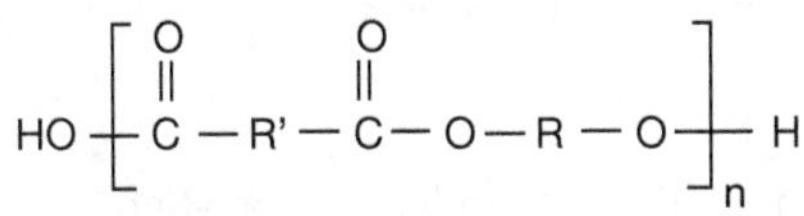

linear polyester

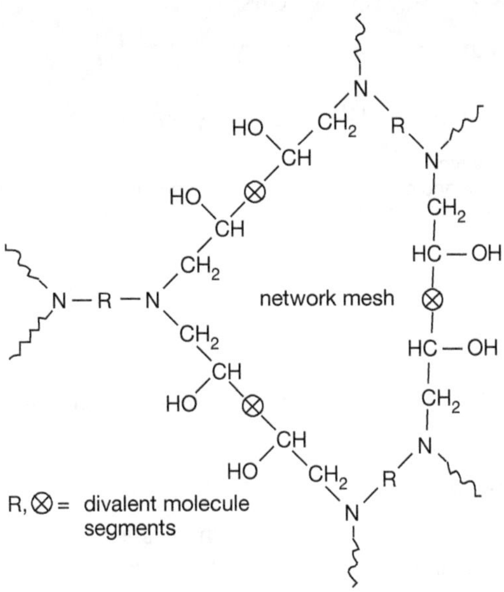

amine-cured epoxy resin
(section from three-dimensional network)

A relatively low-molecular polymer, up to a molecular weight of around 2000 g/mol, is classed as an **oligomer**.

Polymerisation is a chemical reaction in which one or more (different) monomers are converted into a polymer.

There are three basic types of polymerisation reactions, as briefly described below.

- **Addition polymerisation as a chain reaction**

A polymer molecule is produced after a starting reaction, typically within a maximum of a few seconds, through the chemical bonding of numerous monomer molecules with no separation of by-products. This process can be formulated as shown below in relation to a radical polymerisation:

$$R^{\cdot} \xrightarrow[(1)]{+M} R-M^{\cdot} \underbrace{\xrightarrow{+M} \xrightarrow{+M} \xrightarrow{+M}}_{(2)} \ldots R\!-\!\!\left(M\right)_{n}\!\!-M^{\cdot} \xrightarrow[(3)]{+X^{\cdot}} R\!-\!\!\left(M\right)_{n+1}\!\!-X$$

initiator fragment (radical) — initiator radical — (2) — macro-radical — polymer molecule (polymer)

(1) Initiation reaction (2) Propagation (very rapid)
(3) Chain termination M = monomer molecule $X^{\cdot}$ = terminating radical

In chemical kinetics terms, this is a chain reaction in that one propagation step inexorably draws the next immediately after it.

Acrylic resins and polymer dispersions are examples of polymerisation products.

- **Condensation polymerisation**

The monomer molecules react relatively slowly in discrete, mutually independent propagation steps to form the polymer, causing small molecules (mainly water) to be separated. The macromolecules are formed successively over a long period of time, generally several hours.

Schematic reaction equation (for bifunctional monomers):

$$n\,X-M^1-X + n\,Y-M^2-Y \longrightarrow \longrightarrow \longrightarrow \ldots X\!-\!\!\left(M^1-M^2\right)_{n}\!\!-Y + (2n-1)\,XY$$

monomer molecule 1 — monomer molecule 2 — (slowly) — polymer molecule (polycondensate) — small, separated molecule

or (more rarely)

$$n\,X-M-Y \longrightarrow \longrightarrow \longrightarrow \ldots X\!-\!\!\left(M\right)_{n}\!\!-Y + (n-1)\,XY$$

The synthesis of a polyester (see above) or a melamine resin proceeds as a condensation polymerisation with separation of water.

- **Addition polymerisation as a stepwise reaction**

The reaction proceeds in approximately the same way as a condensation polymerisation, but with no separation of molecules. Schematic reaction equation (for bifunctional monomers):

$$n\,M^1 + n\,M^2 \longrightarrow \longrightarrow \longrightarrow \ldots M^2\!-\!\!\left(M^1-M^2\right)_{n-1}\!\!-M^1$$

(moderately fast) — polymer molecule (polyadduct)

M^1, M^2 = bifunctional monomer molecules

The crosslinking of an epoxy resin with an amine (see above) and the formation of a polyurethane from a polyol and a polyisocyanate are examples of stepwise addition polymerisation reactions.

The products of polymerisation reactions are known as *polymers*, *polycondensates* or *polyadducts*, according to the reaction type.

Monomer unit or *basic unit* is the name given to a section of a polymer molecule produced from a monomer molecule. (The monomer changes on transition into the polymer).

The term *structural element* (*structural unit, constitutional repeating unit*) refers to the smallest possible chain section of a polymer molecule; when arranged in series - up to the end groups - these constitute the complete polymer molecule. Only simple polymer molecules with very regular structures, such as homopolymers (see below), contain a structural element. Most synthetic coating film formers are random copolymers (see below) and therefore by definition do not have any structural elements.

Examples:

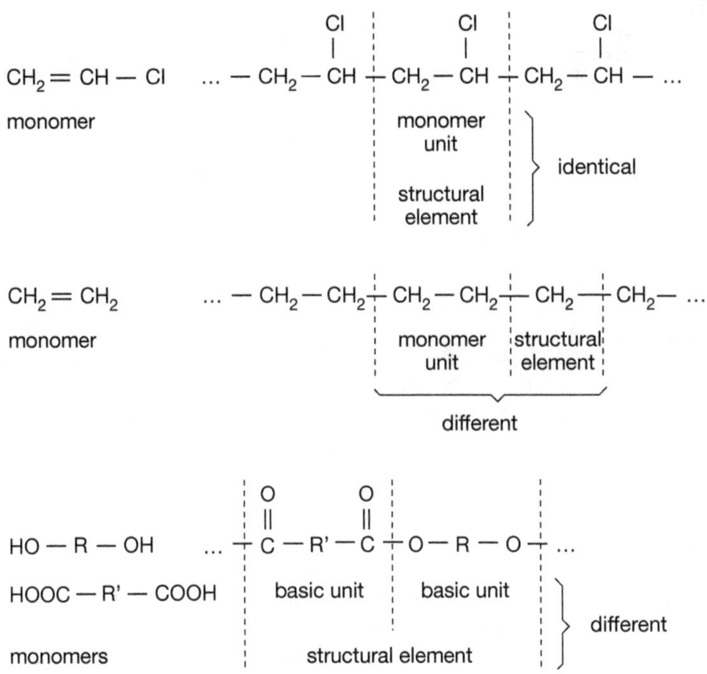

Linear, branched, crosslinked polymers

As Figure 2.1 illustrates, *linear polymers* consist exclusively of chain-like, unbranched molecules. *Branched polymers* consist of branched molecular chains; where possible, one should distinguish between the main chain and side chains. *Crosslinked polymers* consist of three-dimensional molecular networks. The average mesh width of the network can also be expressed by the term *crosslink density* (→ 2.1.1.4).

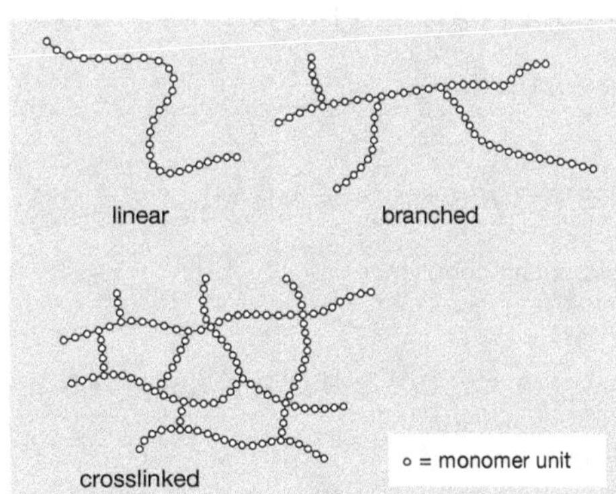

Figure 2.1: Linear, branched and crosslinked molecules/polymers

According to these structures, polymers can be divided into the following three types:

- *Thermoplasts*

These are linear or branched, soften at elevated temperatures, and are soluble in suitable solvents[1)]

- *Elastomers*

Loosely crosslinked[2)], rubbery-elastic (not plastic), insoluble in solvents, but readily swellable

- *Thermosets*

These are closely crosslinked, scarcely softening at elevated temperatures, insoluble in solvents but slightly swellable

In the case of polymers of dienes, i.e. of molecules with two conjugated double bonds, a distinction must be made between *cis-* and *trans-*polymers and between *1,2-* and *1,4-*polymers as shown by way of example in the formulae below for 1,3-butadiene:

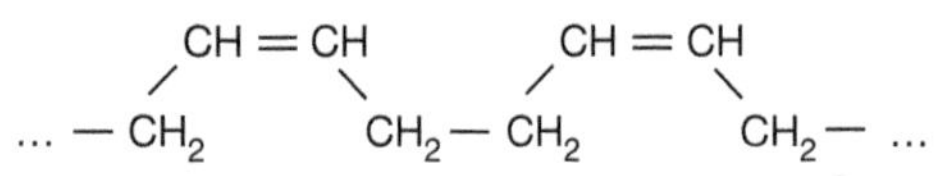

cis-1,4-polybutadiene

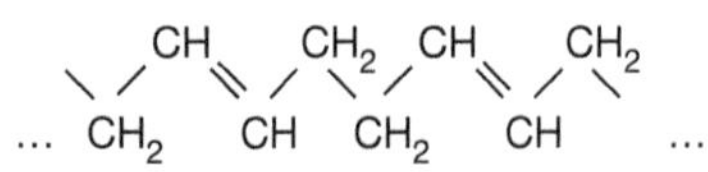

trans-1,4-polybutadiene

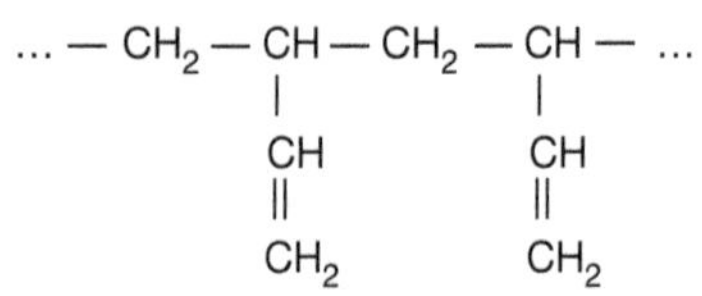

1,2-polybutadiene

Homopolymers, copolymers

Polymerisation of a single monomer produces what is known as a *homopolymer*. When two or more monomers are involved, we refer to a *copolymer*. The term *terpolymer* is also used to refer to a polymer produced from three monomers. Depending on the sequence of the various monomer units in a copolymer, we can distinguish between a *random* copolymer, an *alternating* copolymer, a *block* copolymer and a *graft* copolymer:

A A A B A A B B B B B A B A A A B B A   A, B = monomer units

random copolymer

A B A B A B A B A B A B A B A B A B A

alternating copolymer

A A A A A B B B B B B B A A A A B B B

block copolymer

---

1) Apart from extreme exceptions such as PTFE

2) Uncrosslinked thermoplastic elastomers are a relatively new type of polymer.

```
A A A A A A A A A A A A A A A A A A A
    B       B             B
    B       B             B
    B       B             B
            B             B
            B
```

graft copolymer

The structural features of polymer molecules described so far are grouped together under the collective term of *primary structures*. *Secondary* and *aggregate structures* develop because molecules form in various ways in a space and then congregate below one another.

2.1.1.2 Degree of polymerisation, molecular weight, molecular weight distribution

Only an average value can be given for the size of the molecules of an engineering polymer, since polymerisation reactions lead to a random distribution of molecule sizes. The following two dimensions are conventionally used:

- Average *degree of polymerisation* ($\overline{P}$):

Average number of monomer units (basic units) per polymer molecule

- Average molecular weight ($\overline{M}$):

Average molecular weight of a polymer molecule

These two quantities are linked by the molecular weight of the monomer unit (in the case of homopolymers) or the average molecular weight of a monomer unit M_{mono} (in the case of copolymers):

$$\overline{M} = \overline{P} \cdot M_{mono}$$

It is more usual to quote the average molecular weight than the average degree of polymerisation.

There are a number of differently defined average values for the molecular weight of a polymer. The two most important are explained below.

Number average:

$$\overline{M}_n = \sum \left(\frac{N_i}{\Sigma N_i} \right) \cdot M_i = \frac{\Sigma N_i M_i}{\Sigma N_i}$$

where N_i is the number of molecules having molecular weight M_i and ΣN_i is the total number of molecules in the quantity of polymer under consideration.

The number average - mathematically expressed - is the arithmetic mean of the molecular weight.

Weight average:

$$\overline{M}_w = \sum \left(\frac{N_i M_i}{\Sigma N_i M_i} \right) \cdot M_i = \frac{\Sigma N_i M_i^2}{\Sigma N_i M_i}$$

Since weight average considers not the number but the mass of the molecules, large molecules have a greater influence on the calculation of the average value than do the same number of small molecules. As a consequence, the weight average works out higher than the number average.

The greater the relative difference between the number average and the weight average (known as *non-uniformity*, U), the broader the molecular weight distribution.

$$U = \frac{\overline{M}_w - \overline{M}_n}{\overline{M}_n} = \frac{\overline{M}_w}{\overline{M}_n} - 1$$

(The quotient $D = \overline{M}_w / \overline{M}_n$ (*polydispersity*) is frequently used in place of U.)

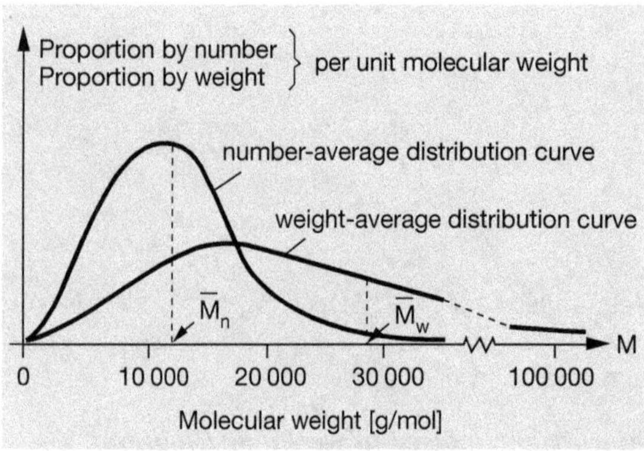

Figure 2.2: Molecular weight distribution curves for a coating resin (schematic view, curves smoothed)

In practice, molecular weight distribution curves are not symmetrical but distorted, as shown diagrammatically in Figure 2.2. In polymerisation reactions in particular, the molecular weight distribution can frequently be extremely irregular, which is due to the fact that there is no temporal or spatial consistency in the reaction conditions for typical industrial polymerisations. As a consequence of this, the overall distribution is the product of an overlay of many narrower individual distributions.

Three methods are conventionally used in industry to determine average molecular weight:

- **Viscometry**

Viscometry utilises the relation between the molecular weight of a polymer and the viscosity of the solution in a suitable solvent; the viscosity average of the molecular weight is obtained (or other quantities such as the K-value).

- **Vapour phase osmometry (vapour pressure osmometry)**

This is a thermodynamic (microcalorimetric) method; it determines the number average of the molecular weight.

- **Gel permeation chromatography (GPC)**

GPC has established itself as the standard method. A sample of the polymer in dissolved form is allowed to migrate through a columnar gel bed of a swollen, microporous substance. The larger the molecules, the less frequently or less deeply they enter the pores and hence the more quickly they pass through the column. Various evaluation methods enable one to calculate the number average, the weight average or the molecular weight distribution.

2.1.1.3 Secondary and aggregate structures of polymers

Uncrosslinked polymer molecules, both in the undissolved state and in solution, generally take the form of coils extended to a greater or lesser degree (see Figure 2.3). The reason for this is firstly that the many atoms of a polymer chain linked together through single bonds can be twisted in virtually any direction in respect of one another, and secondly that the bonds are angled, i.e. they are not aligned with one another. In an extreme case a zigzag chain is theoretically possible, although the probability of all the bonds orienting themselves in the appropriate pattern by chance is virtually nil. (As a comparison, gas molecules like-wise fill an available space uniformly and do not voluntarily arrange themselves into a chain.)

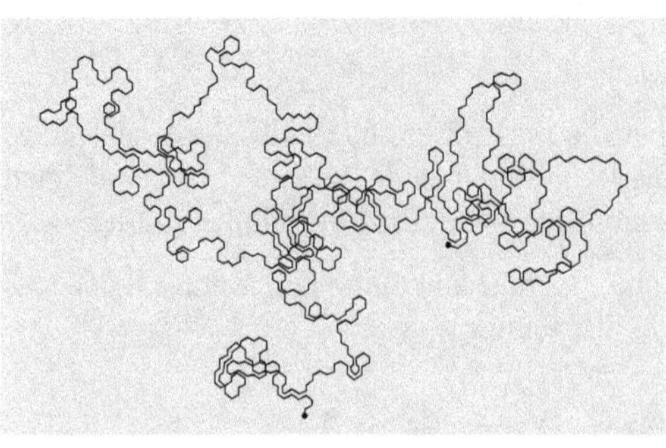

Figure 2.3: Random coil of a linear polymer molecule (macromolecule) according to [3]

Where many such polymer coils occur together, they may either lie adjacent to one another whilst remaining largely separate (cell structure, see Figure 2.4 (a)), or they may interpenetrate with one another, forming a kind of molecular felt (network structure, see Figure 2.4 (b)). For coating films the latter is preferable because of its superior mechanical properties.

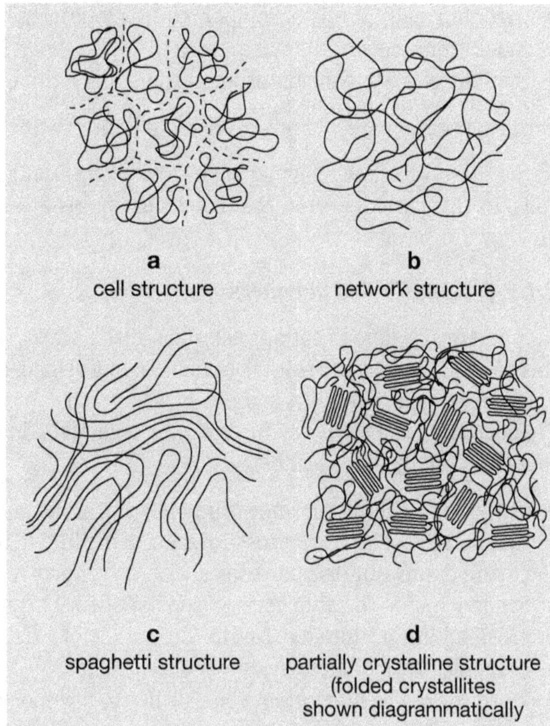

Figure 2.4: Aggregate structures of polymers

In addition to these random conditions, the intra- and intermolecular forces of attraction[1] (van der Waals' forces and hydrogen bridges) between polymer segments must be taken into account. A distinction is normally made between the following basic types of intermolecular interactions:

- *Dispersion forces*:
Weak; exist between all atoms and molecules (cause: temporary asymmetries in the charge distributions within the atoms or molecules).

- *Polar forces*:
Moderately strong; exist between polar bonds (permanent dipoles) or ions and polarisable bonds (induced dipoles).

- *Hydrogen bridges*:
Strong; form primarily between OH or NH bonds and free pairs of electrons of O or N atoms.

(The order of magnitude of normal chemical bonds (primary valency bonds) is ten times greater than that of hydrogen bridges).

The following general rule applies: the stronger the intermolecular forces of attraction, the stronger the mutual coherence between different polymer molecules through the layering of molecule segments. One of the mechanical effects of intermolecular interactions is increased tensile strength.

If in addition to their mutual attraction the molecules or segments are arranged in a regular structure, bundles or concentrations of molecule segments may be produced; the braid of molecules has something of the appearance of a heap of spaghetti (see Figure 2.4 (c)). These bundles can sometimes have a role to play in coating technology. They have a positive impact in polyurethane films, for example, where the stacked urethane groups are responsible for the material's good abrasion resistance. A negative example is the poor solubility of polyesters containing too high a proportion of very symmetrical basic units (e.g. terephthalic acid).

The most extreme example of molecular bundling is where many molecule segments of different molecules form crystallites (small crystals of around 10 nm in size). The crystallites in turn arrange themselves into what are known as superstructures (textures). Figure 2.4 (d) shows in diagrammatic form the structure of a partially crystalline polymer with "fold crystallites" – regular bundles of folded chain segments. The appearance of *partial crystallinity* requires a very regular polymer structure (e.g. tacticity). It plays a major role in the plastics sector and has a certain importance in adhesives technology. In coating film formers, however, partial crystallinity is undesirable since:

- partially crystalline polymers are poorly soluble in coating solvents;

1) Intramolecular: within a molecule, Intermolecular: between different molecules

- partial crystallinity is accompanied by cloudiness (the crystallites have a different refractive index from the amorphous areas);
- the flow of coatings made from partially crystalline polymers would be impaired.

For these reasons we will not cover them in any further detail.

In the dissolved state, polymer molecules again occur as isolated or as varyingly interpenetrated coils. In this case, however, the molecular chains are surrounded by adhering solvent molecules, i.e. they are solvated.

2.1.1.4 Crosslinked polymers

In liquid or powdered coating materials, film formers occur as uncrosslinked, discrete molecules; this means that they are soluble or fusible. In finished, cured coating films polymers normally need to be crosslinked, since this is the only way to achieve the best mechanical and chemical properties. Chemically crosslinked structures can be obtained - in molecular terms - by a number of different routes, as briefly described below.

- First route: The film former originates in the form of linear polymers. In the chemical curing reaction the chains are crosslinked either directly or using short bridges. One example is the curing of unsaturated polyesters with styrene as reactive diluent and peroxide as initiator.
- Second route: The film former consists of highly branched polymer molecules. Continuous crosslinking is initiated by the formation of relatively few intermolecular chemical bonds. Example: the oxidative drying of alkyd resins.
- Third route: The film former consists of two low-molecular (oligomeric) components, which form a macromolecular crosslinked substance on curing. Example: the formation of polyurethane from low-molecular polyester polyol and polyisocyanate surface coating resin.

The most important general characteristic of crosslinked polymers is the *crosslink density* ν. This refers to the number of *network chain segments* - expressed in mol - per unit volume of the polymer, where a network chain segment is the chain unit stretching from one crosslink point to the next. Instead of crosslink density the average molecular weight $\overline{M}_c$ of a network chain segment can also be quoted. Both quantities are connected via the density ρ of the polymer:

$$\overline{M}_c = \frac{m_c}{n_c} = \frac{m_c/V_c}{n_c/V_c} = \frac{m_p/V_p}{n_c/V_p} = \frac{\rho}{\nu}$$

n_c: number of moles of network chain segments
$m_c = m_P$: mass of the network chain segments or the polymer
$V_c = V_P$: volume of the network chain segments or the polymer

The greater the crosslink density, the greater the hardness and the chemical (solvent) resistance of the polymer; its elasticity or flexibility is reduced, however.

Crosslink density should not be used when "degree of crosslinking" is meant. The latter is used variously to mean the gel content of the polymer, the degree of crosslinking conversion (curing conversion) and the crosslink density.

A crosslinked polymer can still contain molecules which are not bound to the network. This sol content (w_S) can be extracted from the polymer sample with a suitable solvent. The non-soluble, bound fraction is called the gel content (w_G). Naturally, both fractions must add up to 100%.

Crosslinking in the narrow sense always involves the chemical joining together of individual molecules to form three-dimensional networks. However, in the extended sense, crosslinking can also be the outcome of the interaction of weaker, physical secondary bonding forces, i.e. Van der Waals forces and hydrogen bonds, as well as steric effects, called entanglements. These network bonds can generally be broken by high shearing forces or rapid shearing and/or heating. They play a major role in organic rheological additives (thickeners etc).

2.1.1.5 General information about polymer solutions

If a non-crystalline ("amorphous") and uncrosslinked polymer, e.g. the film former in a physically drying coating (i.e. by evaporation of the solvent alone), is added to a solvent, the solvent molecules slowly diffuse into the substance and solvate (encapsulate) the polymer molecules. This causes the volume of the substance to increase and the mechanical strength of the polymer to decrease, since the intermolecular forces of attraction responsible for coherence are gradually replaced by the forces of attraction between the polymer chains and the solvent molecules. This process is known as *swelling*.

If the dissolving power of the solvent is strong enough, swelling continues until a polymer solution is obtained. The dissolution of a (non-crystalline) polymer thus proceeds smoothly as a continuous swelling with no distinct change of phase; there is no clear boundary between the "swollen" and "dissolved" states. Conversely, when a polymer solution is evaporated, the polymer is not precipitated as a solvent-free bottom product - like a salt, for example; instead there is a continuous transition to the solvent-free polymer.

One exception to this rule occurs when the dissolving power of a blend of a highly dissolving and a poorly dissolving solvent deteriorates in respect of the dissolved polymer - either through further addition of the "non-solvent" (also referred to as "extender") or through the preferential evaporation of the "solvent". Such an instance can lead to the *precipitation* or dissolution of the swollen polymer, i.e. to a discontinuous or multi-phase system. Such undesirable precipitation phenomena can occur in the film as the paint is drying, for example, if the solvent composition has not been correctly adjusted.

In a polymer solution the molecules occur as diffuse "gel coils" interpenetrated with solvent. The solvating part of the solvent adhering strongly to the polymer chain due to intermolecular forces of attraction is known as *bonded solvent*; the rest of the solvent in the solution is called *free solvent*. With high molecular weights the requirement for solvating (bonded) solvent can be considerable. Thus even a 5% solution of polymethyl methacrylate with an average molecular weight $\overline{M}_w$) of 500,000 g/mol in acetone still contains no free solvent. The gel coils can reach diameters of up to 100 nm.

For this reason polymer solutions represent a special category of colloid solutions, known as *molecular colloids*[1)]. Technically speaking, however, true (molecularly disperse) polymer solutions are not considered to be colloids.

The distinction between free and bonded solvent is important in coating technology: bonded solvent evaporates from a coating film much more slowly than free solvent and is subject to the phenomenon known as *solvent retention* (retention in the film).

2.1.1.6 Solubility and solubility parameters

Attempts to predict the solubility of polymers (and other substances) in particular solvents on the basis of physico-chemical material properties have led to the term "solubility parameter". We understand this to mean the following:

First of all we assume that a particular quantity of a substance - e.g. a solvent - is completely evaporated. In this process the amount of energy used is that necessary to separate completely all molecules from one another - in opposition to their intermolecular forces of attraction. This amount of energy is known as the cohesive energy E_c, which is the same as the evaporation energy $\Delta_v U$. If we divide this by the volume of the substance V, we obtain the cohesive energy density. The square root of the cohesive energy density is the *solubility parameter* δ as defined by *Hildebrandt*.

$$\delta = \sqrt{\frac{E_c}{V}}$$

1) Particle sizes in colloidal systems are by definition between 10 and 100 nm.

Table 2.1: Examples of the failure in the one-dimensional solubility parameter system

Polymer	Solvent	$\Delta \delta$ [$(J/cm^3)^{1/2}$]	Solubility
polyvinyl chloride	chloroform	1,6	poor
polyethylene terephthalate (crystalline)	acetone	1,8	poor
polystyrene	n-heptane	2,9	moderate
polyvinyl acetate	methanol	9,8	good
polymethyl phenyl siloxane	methanol	11,2	good

Table 2.2: Solubility parameters for selected solvents and polymers (in $(J/cm^3)^{1/2}$)

Substance	δ_D	δ_P	δ_H	δ
n-hexane	14.9	0	0	14.9
toluene	18.0	1.4	2.0	18.2
methyl isobutyl ketone	15.3	6.1	4.1	17.0
butyl acetate	15.8	3.7	6.3	17.4
isobutanol	15.3	5.7	15.8	22.7
water	14.3	16.3	42.6	47.8
hydrocarbon resin	17.6	1.2	3.6	18.0
long oil alkyl resin	20.4	3.4	4.6	21.2
polymethyl methacrylate	8.6	10.5	7.5	22.6
hexamethoxymethyl melamine (HMMM)	20.4	8.5	10.6	24.5
epoxy resin	20.4	12.0	11.5	26.3

It is (theoretically) the case that two substances are homogeneously miscible if their solubility parameters are approximately the same (see more detailed explanation on page 31). For polymer solutions, the difference in solubility parameters above which complete solubility is no longer achieved is around 6 $(J/cm^3)^{1/2}$.

Three-dimensional system

Unfortunately the one-dimensional solubility parameter system described above can often lead to false predictions (see Table 2.1).

For this reason *Hansen* devised a refined three-dimensional version of the one-dimensional system. In this three-dimensional system there is one parameter for the dispersion forces (δ_D), one for the polar forces (δ_P) and one for the hydrogen bridges (δ_H), which are combined to make one overall parameter d using the following formula:

$$\delta = \sqrt{\delta_D^2 + \delta_P^2 + \delta_H^2}$$

Table 2.2 shows the *Hansen parameters* for a number of selected solvents and polymers. From the parameter values given in the table we can see the following:

In the case of solvents, the values for the dispersion parameter vary very little; the diffuse π electron system for the aromatic toluene does have a rather higher value. As we would expect, the values for the polarity parameter increase from n-hexane up to water; ketones are higher than esters. The tendency to form hydrogen bridges is naturally lowest in the case of hydrocarbons, moderate in the case of the polar-aprotonic esters and ketones and highest in the case of the protonic alcohols and, particularly, water.

The parameter values for polymers (coating film formers) are less easy to interpret than those for solvents. Polarity, polarisability (displaceability of bonding electrons by adjacent dipolar molecules or ions) and proton donor effect (OH or NH bonds) are of key importance here too, however.

It is useful to consider the Hansen parameters as three $\vec{\delta}_D$, $\vec{\delta}_P$ and $\vec{\delta}_H$ arranged vertically one above the other; according to the formula above, d is given by the vectorial sum of these three vectors (see Figure 2.5). The difference in the solubility parameters of two substances, e.g. of a solvent (S) and a polymer (P), is obtained according to the rules of vector algebra by the following equation:

$$\Delta\delta = \sqrt{[\delta_D(L) - \delta_D(P)]^2 + [\delta_P(L) - \delta_P(P)]^2 + [\delta_H(L) - \delta_H(P)]^2}$$

Example:
Hansen parameter in $(J/cm^3)^{1/2}$ for polyvinyl chloride: $\delta_D = 18.7$; $\delta_P = 10.0$; $\delta_H = 3.1$
Hansen parameter in $(J/cm^3)^{1/2}$ for chloroform: $\delta_D = 17.8$; $\delta_P = 2.5$; $\delta_H = 6.1$

$$\Delta\delta = \sqrt{(18.7 - 17.8)^2 + (10.0 - 2.5)^2 + (3.1 - 6.1)^2} = 8.1\ (J/cm^3)^{\frac{1}{2}}$$

We can see that in this case the three-dimensional system, unlike its one-dimensional counterpart (see Table 2.1), delivers the correct prediction - "poorly soluble".

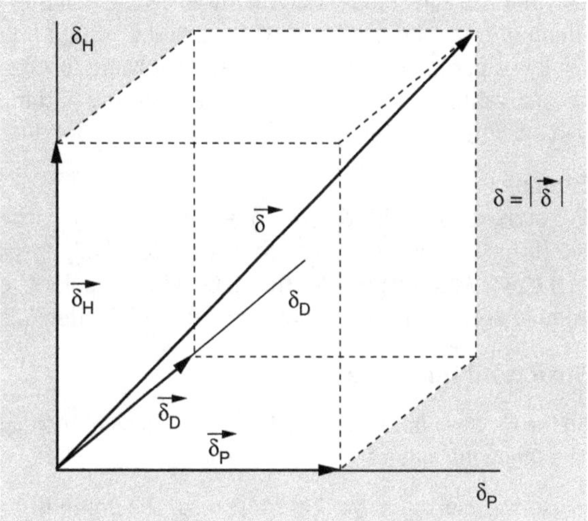

Figure 2.5: Solubility parameters as vectors in the three-dimensional system according to Hansen

Solubility parameter diagrams

Unfortunately, even the three-dimensional system does not always produce the correct predictions of solubility, however. This is particularly the case if a fixed, generally applicable difference in the values for the overall parameter is given for the soluble/insoluble boundary. In such a case, empirically determined solubility parameter diagrams are more reliable. A *solubility parameter diagram* for a polymer or coating film former is a simply cohesive region in the three-dimensional solubility parameter zone with the following characteristic: all solvents (or blends of solvents), whose solubility parameter triplet forms a point within the region will dissolve the polymer; all solvents whose parameter points lie outside the region will not dissolve the polymer. Figure 2.6 shows a two-dimensional version of a solubility parameter diagram.

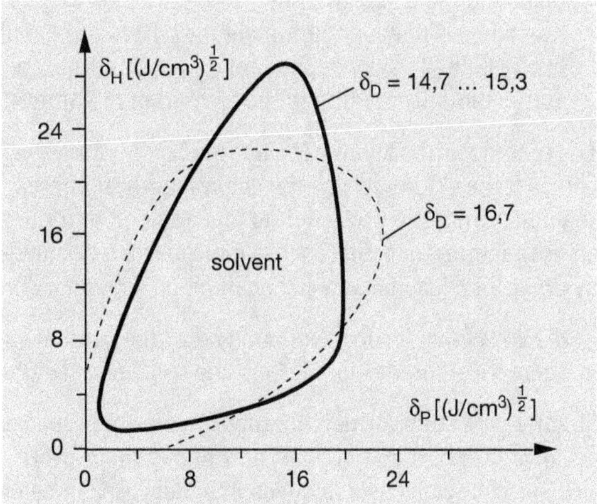

Figure 2.6: Solubility parameter diagram for a melamine resin (third parameter δ_D kept constant)

Solvent blends, latent solvents

Determining the solubility parameters for solvent blends represents a particular problem. Even for binary blends there are no simple calculation formulae which at the same time offer physico-chemical

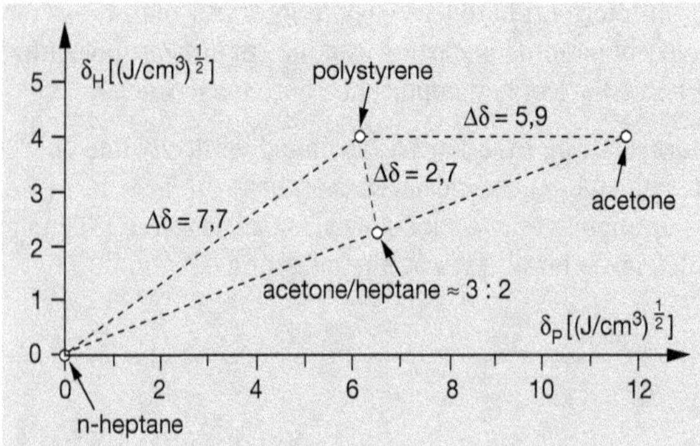

Figure 2.7: Solubility parameters δ_P and δ_H for polystyrene, n-heptane, acetone and n-heptane/acetone (δ_D not shown for clarity)

accuracy. Simple mean calculations involving the volume content Ø of components A and B according to the following formulae

$$\delta_D = Ø_A \cdot \delta_D(A) + Ø_B \cdot \delta_D(B)$$

$$\delta_P = Ø_A \cdot \delta_P(A) + Ø_B \cdot \delta_P(B)$$

$$\delta_H = Ø_A \cdot \delta_H(A) + Ø_B \cdot \delta_H(B)$$

do at least allow rough predictions to be made in regard to the change in dissolving power of one solvent following the addition of a second in respect of a given polymer.

The three-dimensional solubility parameter system can be used to provide a simple explanation of the mode of action of a "latent solvent", for example, as follows. *Latent solvents* for a particular polymer are solvents that do not dissolve the polymer by themselves but become a solvent on addition of a second solvent - which may even be a non-solvent. This is explained by the fact that too low a parameter value in the latent solvent, for example, is balanced out by too high a value in the activating component, so that the resulting parameter value moves close to the corresponding value for the polymer.

Example:
Polystyrene (δ_D = 17.5; δ_P = 6.1; δ_H = 4.0 $(J/cm^3)^{1/2}$) dissolves in neither n-heptane (δ_D = 15.1; δ_P = 0; δ_H = 0 $(J/cm^3)^{1/2}$), nor in acetone (δ_D = 15.6; δ_P = 11.7; d_H = 4.1 $(J/cm^3)^{1/2}$), yet it does dissolve in a blend of the two. Acetone provides the blend with the polarity and the hydrogen-bridging tendency that are completely absent from n-heptane. Figure 2.7 offers a simplified representation of this relation.

Structural influences

So far we have not considered the influences of molecule size and polymer structure on solubility. The following rules apply:

- As the molecular weight increases, the solubility and hence the swellability of polymers reduce.
- Branched polymers are generally more readily soluble than linear (unbranched) polymers of the same molecular weight.
- Crosslinked polymers are insoluble. They are capable of swelling, however, their swelling capacity being heavily dependent on the swelling agent used, crosslink density and temperature, and potentially leading to the formation of voluminous gels.

The reasons are as follows: the intermolecular forces of attraction and the number of entanglements between the chains rise as the chain length increases. Branching prevents the molecules or chain segments from lying adjacent to one another, and the linear chain sections of branched molecules are also shorter than unbranched molecules of the same molecular weight. Molecules linked together by crosslinks (chemical bonds) cannot be separated by solvents into discrete particles.

At the end of this section we should recall the familiar rule that "Like dissolves in like", which derives its theoretical interpretation from the concept of solubility parameters. According to this rule:

Slightly polar film formers dissolve primarily in hydrocarbons and esters, moderately polar types dissolve in esters and ketones and highly polar-protic (OH-containing) types dissolve in similarly OH-containing alcohols and also in some cases in esters and ketones. The best solvents ("true solvents") can to a certain degree, to be determined by experiment, be added to weaker solvents or even to non-solvents ("diluents"). (See section 2.1.1.9 for solubility in water).

The *Gibbs-Helmholtz equation* formulated for the dissolution process (d) is as follows:

$$\Delta_d G = \Delta_d H - T \cdot \Delta_d S$$

where $\Delta_d G$ represents free enthalpy of solution, $\Delta_d H$ enthalpy of solution, $\Delta_d S$ entropy of solution and T absolute temperature.

The lower the value of $\Delta_d G$, the greater the driving force of the dissolution process. If we leave aside $\Delta_d S$ as an imprecisely known but constant quantity, then the consequence is that the lower the value for $\Delta_d H$, the lower the value for $\Delta_d G$.

Let us consider the dissolution of a substance B in a solvent A. During this process, intermolecular interactions between molecules A and similar interactions between molecules B change into interactions between A and B as follows:

$$A\text{-}A + B\text{-}B \rightarrow 2\ A\text{-}B$$

The changes in the corresponding interactive energies (cohesive energies) emerge as heat tonality $\Delta_d H$:

$$\frac{\Delta_d H}{V_d} = \text{const.} \cdot \left[\left(\frac{E_c}{V}\right)_A + \left(\frac{E_c}{V}\right)_B - 2 \cdot \left(\frac{E_c}{V}\right)_{AB} \right]$$

where V_d is the volume of the solution and (E_c/V) the cohesive energy density. $(E_c/V)_{AB}$ can be calculated in approximate terms as the geometric mean:

$$\left(\frac{E_c}{V}\right)_{AB} = \sqrt{\left(\frac{E_c}{V}\right)_A \cdot \left(\frac{E_c}{V}\right)_B} = \sqrt{\left(\frac{E_c}{V}\right)_A} \cdot \sqrt{\left(\frac{E_c}{V}\right)_B}$$

This gives

$$\frac{\Delta_d H}{V_d} = \text{const.} \cdot \left[\sqrt{\left(\frac{E_c}{V}\right)_A} - \sqrt{\left(\frac{E_c}{V}\right)_B} \right]^2$$

$$= \text{const.} \cdot (\delta_A - \delta_B)^2$$

Since the factor in front of the brackets is positive, then the closer together δ_A and δ_B are, the lower the value for $\Delta_d H$ and hence also for $\Delta_d G$, which is what we set out to demonstrate.

2.1.1.7 Incompatibilities

By *incompatibility* of (dissolved) polymers we mean the phenomenon whereby two polymers that are completely dissolved in the same solvent, mutually precipitate to form two solvent-containing polymer phases when the solutions are combined. This precipitation is normally first recognised by a cloudiness.

The cause of the incompatibility may be even slight differences in molecular structure. For example, polystyrene is incompatible with poly-α-methyl styrene and polymethyl acrylate is incompatible with polyethyl acrylate.

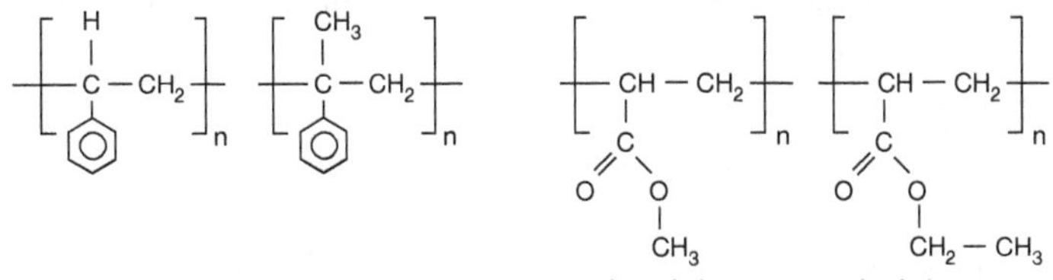

polystyrene | poly-α-methyl-styrene | polymethyl acrylate | polyethyl acrylate

Copolymers of the same monomers in somewhat different proportions may also be incompatible with each other. Apart from the molecular structure, the degree of incompatibility is also governed by the solvent, the molecular weight (the higher this is, the greater the incompatibility), the concentrations or proportion of the two polymers and the temperature.

Incompatibility between different film formers and potentially between other substances in coatings is generally undesirable and must be avoided by skilful formulation. A more practicable method at times of successfully incorporating various incompatible film formers into a coating is to "*boil*" them first, e.g. natural resins (colophony, copal) with drying oils, or phenolic resins with drying oils or epoxy resins. The incompatible components bond together chemically in such a way that their subsequent separation into two phases is prevented and at the same time their compatibility with other components of the coating is improved.

Incompatibilities between polymers are much more noticeable in the undissolved state than in solution, since here the polymer molecules are in direct contact with one another. This incompatibility can be utilised in the area of polymer materials, namely in the form of special block or graft copolymers and "*interpenetrating networks*" (IPN). The basic principle here is that the separation processes lead to submicroscopic areas (domains) with differing properties, which combine to give the overall behaviour of the material. Despite ongoing research, these multi-phase technologies are not yet widely used in the coatings sector, however.

2.1.1.8 Viscosity of polymer solutions

By the (dynamic) viscosity η - expressed in Pa·s (*Pascal seconds*) - of a flowable substance we mean its thickness (ropiness). It can be determined by the time in which a certain volume of liquid at a given pressure difference flows through a capillary of a particular length and diameter.

The following definitions are required to discuss the viscosity of polymer solutions:

- Relative viscosity $\eta_{rel} = \frac{\eta}{\eta_0}$ (η_0 = viscosity of pure solvent)
- Specific viscosity $\eta_{sp} = \frac{\eta - \eta_0}{\eta_0} = \frac{\eta}{\eta_0} - 1$
- Reduced viscosity $\eta_{red} = \frac{\eta_{sp}}{c}$ (c = concentration, g = polymer/ml solution)
- Intrinsic viscosity $[\eta] = \lim_{c \to 0} \eta_{red}$ (reduced viscosity for negligibly low concentration or "infinite dilution")

Through infinite dilution according to the definition of $[\eta]$, the distances between the polymer molecules in the solution become very large, with the result that the molecules do not influence one another.

The reduced viscosity is also commonly called the *Staudinger function.*

Intrinsic viscosity and *Staudinger index* mean the same as "limiting viscosity number". The latter can be obtained directly from the relative viscosity, instead of from the reduced viscosity:

$$[\eta] = \lim_{c \to 0} \frac{\ln \eta_{rel}}{c}$$

Material influences

The relation between molecular weight and intrinsic viscosity is described by the *Staudinger-Mark-Houwink equation* (abbreviated to SMH equation):

$$[\eta] = k_\eta \cdot \overline{M}_v^a$$

where k_η and a are polymer-, solvent- and temperature-dependent constants and $\overline{M}_v$ is the *viscosity average* of the molecular weight.

Depending on the polymer, solvent and temperature, the factor k_η lies in the range from 10^{-4} to $6 \cdot 10^{-2}$; in the case of linear molecules a is between 0.5 and 1 and is a measure for what is known as coil expansion. The coil diameter of a given polymer molecule increases as solvation (quality of the solvent) and temperature increase, expressed by a rising value for a. The state where a = 0.5 is known as the θ state (theta state). The θ state contains coils that have not been thoroughly washed through, i.e. coils filled exclusively with bonded solvent with precisely the diameter that they would have in the force-free state in vacuo.

The central quantity in the SMH equation is the viscosity $\overline{M}_v$, the molecular weight. $\overline{M}_v$ can be described pragmatically as follows: $\overline{M}_v$ is a mean of the molecular weight formed precisely to satisfy the SMH equation. It can now be demonstrated that $\overline{M}_v$ lies between $\overline{M}_n$ and $\overline{M}_w$, which means that the narrower the molecular weight distribution, the closer it moves to $\overline{M}_w$ (see more detailed explanation).

Although of great theoretical interest, the SMH equation is important only for determining the molecular weight by viscometry. If k_η and a are known from calibration measurements, then [η] and hence $\overline{M}_v$ can be calculated by viscometry.

The exact formula for the viscosity average is:

$$\overline{M}_v = \left(\frac{\Sigma N_i M_i^{1+a}}{\Sigma N_i M_i} \right)^{\frac{1}{a}}$$

a is identical to the exponent from the SMH equation.

When a = -1 $\overline{M}_v$ moves to $\overline{M}_n$, when a = +1 it moves to $\overline{M}_w$ (→ 2.1.1.2). Since a is between 0.5 and 1 and is thus much closer to +1 than to -1, $\overline{M}_v$ lies between $\overline{M}_n$ and $\overline{M}_w$, close to $\overline{M}_w$. For a uniform molecular (monodisperse) polymer of molecular weight M, all values of M_i in the mean value formulae are identical to one another, from which it follows mathematically that $\overline{M}_n = \overline{M}_w = \overline{M}$. The narrower the molecular weight distribution, the closer $\overline{M}_n$ and $\overline{M}_w$ move together, and $\overline{M}_v$ lying between them moves closer to $\overline{M}_w$.

The SMH equation has no direct application in industrial binder solutions or coatings, since the film former concentrations here are high and the film formers are frequently of a relatively low molecularity (oligomeric), i.e. they are not really macromolecular substances. As an example of the dependence of solution viscosity on film former concentration, Figure 2.8 shows the dependence on concentration of the reduced viscosity of a high solids blend of film formers in various solvents. Up to a concentration of around 0.3 g/ml the *Martin equation* applies:

$$\eta_{red} = \frac{\eta_{sp}}{c} = [\eta] \cdot \exp(K_M \cdot [\eta] \cdot c)$$

or, expressed as a logarithm:

$$\ln\left(\frac{\eta_{sp}}{c}\right) = \ln[\eta] + K_M \cdot [\eta] \cdot c$$

(K_M = Martin constant).

If we plot $\ln(\eta_{sp}/c)$ against c we obtain a straight line.

If in the Martin equation we substitute η_{sp} with $\eta/\eta_0 - 1$, [η] with $k_\eta \cdot M_v^a$ and solve for η, then

$$\eta = \eta_0 [k_\eta \cdot \overline{M}_v^a \cdot c \cdot \exp(K_M \cdot k_\eta \cdot \overline{M}_v^a \cdot c) + 1]$$

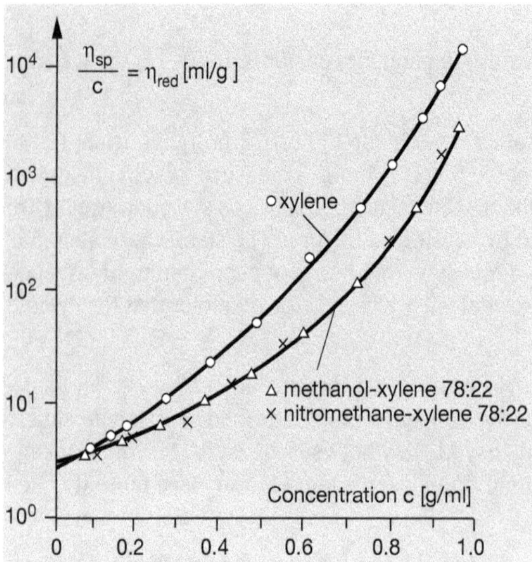

Figure 2.8: Dependence on concentration of the reduced viscosity of a blend of high solid alkyd resin and melamine resin (HMMM) in various solvents

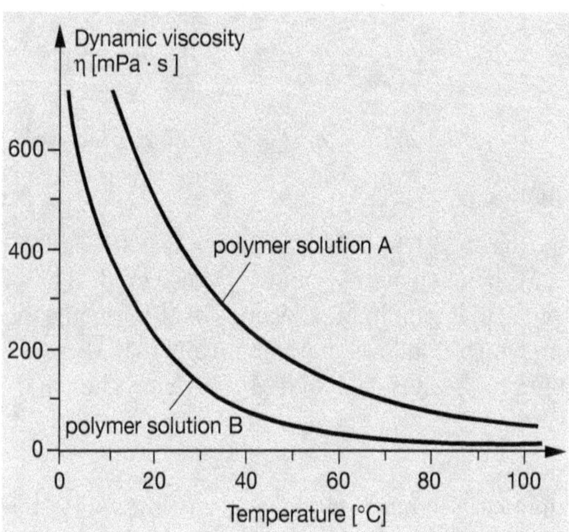

Figure 2.9: Relation between solvent viscosity and temperature

A low solution viscosity for a given film former concentration is thus obtained if the inherent viscosity η_0 of the solvent, the constants k_η and K_M and the viscosity average M_v are low.

With regard to the influence of the solvent or solvent blend on viscosity, experiments show that, in addition to inherent viscosity, solubility parameters are also influential. For the polar and hydrogen-bridging film former in the example (high solids alkyd resin/HMMM) this means that the greater the value for δ_P and/or δ_H in the solvent, the lower the solution viscosity. This is because the formation of viscosity-raising molecule associations (skeleton structures) due to dipole-dipole attraction and hydrogen bridges is prevented by solvents of the specified type.

A film former which is designed to produce a lower solution viscosity - or a higher solids coating - must therefore exhibit, in addition to as low a molecular weight as possible, a narrow molecular weight distribution, since the narrower the molecular weight distribution, the lower the viscosity average of the molecular weight.

Figure 2.8 shows that the relatively simple relationship described above (Martin equation) no longer applies to higher concentrations of polymers or film formers such as are found in "real" coatings. (The graph is concave rather than linear). The increase in solvent viscosity is greater than that described by the Martin equation. Nevertheless, the trends for all of the regularities mentioned above still apply at higher concentrations.

Temperature influence

The viscosity of a polymer solution changes dramatically under the influence of temperature. This correlation is described - with sufficient accuracy for most instances - by the *Arrhenius equation* for viscous flow:

$$\eta = \eta_\infty \cdot \exp(E_\eta/RT)$$

where η_∞ = viscosity that is formally obtained for infinitely high temperature, E_η = molar activation energy for viscous flow, R = general gas constant and T = absolute temperature. Figure 2.9 shows the graph of this function.

The temperature dependence of solution viscosity is important in many areas of coating technology, e.g. in viscometry, in the hot spraying of coatings and in coating flow. For example, a coating with a viscosity of 500 mPa·s at 20 °C has a viscosity of 100 mPa·s at 70 °C, which means that at 70 °C it can be used with a standard pneumatic spray gun.

2.1.1.9 *Aqueous systems*

Polymers in water may be either fully dissolved (molecularly disperse) or in coarsely dispersed states. In order to be soluble in water, polymer molecules must possess ionic groups like carboxylate or ammonium groups or, alter-natively, a considerable number of non-ionic, hydrophilic groups or segments such as hydroxyl, carbonyl, amino, amide groups and/or polyether chains.

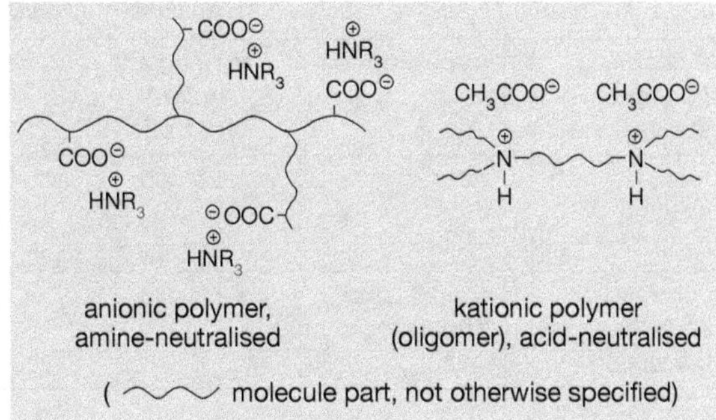

Figure 2.10: Examples of ionically dissolved film former

If the molecules are not hydrophilic enough to form molecularly disperse (true) solutions, several of them may associate to form larger, colloidal aggregates or *secondary dispersions*, which are virtually clear. The less hydrophilic the molecules are, the larger the disperse particles become. At some point, emulsifiers (surfactants), and maybe *protective colloids*, need to be added to provide stability. Such secondary dispersions generally range from being more or less hazy to opaque. If the disperse phase, e.g. an oligomeric resin, is liquid, the dispersion is an *emulsion. Microemulsions* are very fine-particle, virtually clear, stable emulsions which form when a "self-emulsifying" oil phase in water is stirred gently.

In the case of ionic polymers (polyelectrolytes) we differentiate between *anionic, cationic* and *zwitterionic* types. The anionic variety are mostly polycarboxylic acids and are neutralised with amines, the cationic film formers are usually polyamines to which simple acids (acetic acid, for example) are added. Figure 2.10 provides a schematic representation of the chemical structures of these ionic systems.

(Primary) dispersions

Primary dispersions (polymer dispersions) should really be distinguished from the disperse states described above that are formed by the dispersion or dissolution of polymeric or oligomeric substances. Primary dispersions are formed in principle by the process of emulsion polymerisation and consist of compact polymer particles (latex particles) of a high molecular weight which are finely dispersed in water. The polymer dispersions contain surfactants and possibly also protective colloids[1] to stabilise them against coagulation (agglomeration of the particles).

Aqueous-disperse states

Figure 2.11 shows the aqueous-disperse states of polymeric substances arranged in order of particle size. The typical properties of the disperse systems of greatest techno-logical importance are listed in Table 2.3 (page 38).

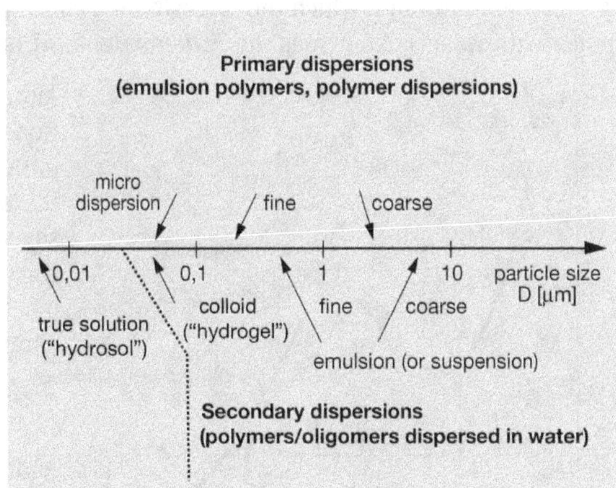

Figure 2.11: Disperse states of polymers in water

[1] Protective colloids are molecular colloids of macromolecular substances dissolved in water which stabilise other colloidally disperse particles.

Table 2.3: Typical properties of aqueous-disperse polymer systems in water

System (disperse state)	Appearance	Molecular weight [g/mol]	Particle size [μm]	Particle type	Additive
true solution	clear	< 20 000	< 0.01	ionic molecule	neutralising agent (amine/acid)
colloidal secondary dispersion	almost clear ("opalescent")	< 50 000	< 0.1	coil formed from numerous molecules	less neutralising agent
secondary dispension or emulsion	almost clear to milky	< 50 000 (or < 10 000)	> 0.1	granule	emulsifier
primary dispersion (polymer dispersion)	translucent	> 100 000	0.03 to 5	almost spherical polymer particles	emulsifier, protective colloid if necessary

Examples of the systems listed in the table include

- acidic solvent acrylic resins for water-thinnable industrial stoving coatings; water-soluble after neutralisation with amine (clear, true solutions)
- water-thinnable self-emulsifying alkyd resins with low amine requirement (colloids)
- emulsions of liquid epoxy resins for heavy-duty corrosion protection
- Styrene-acrylate copolymer dispersions for emulsion paints and coatings (polymer dispersions).

Table 2.3 does not include *cosolvents*, i.e. organic solvents which are frequently retained in the solutions or dispersions to stabilise the dispersion state and/or to obtain particular coating properties.

Viscosity anomalies

The viscosity or flow behaviour of aqueous systems differs in some ways from that of solvent-based solutions. Let us briefly consider just two phenomena.

The maximum through which the viscosity of a concentrated polyelectrolyte solution passes when diluted with water is known as the "*water mountain*" (see Figure 2.12).

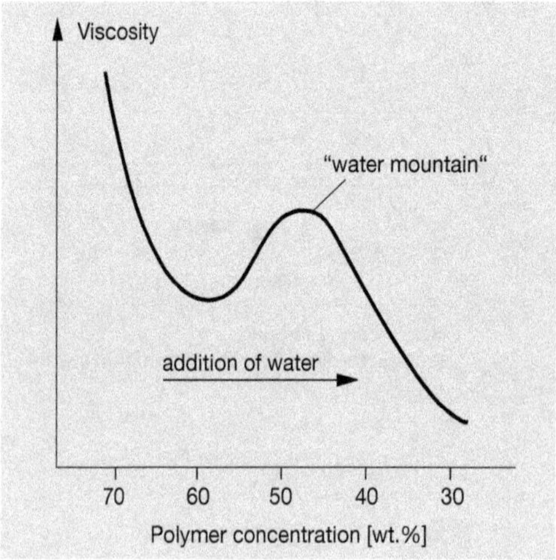

Figure 2.12: Rise in viscosity ("water mountain") on dilution of a hydrosol with water

More or less all strongly ionic true polymer solutions display this viscosity anomaly, which is undesirable in coatings. As far as the causes of the water mountain are concerned, let us simply say that at the viscosity maximum, extended coils and molecule associates exist. The following measures can be taken to flatten or completely eliminate the water mountain:

- reducing the number of ionic groups in the molecule; alternatively, increasing the hydrophilia, by incorporating polyethylene oxide chains, for example
- switching from the true solution to the colloidal secondary dispersion with lower amine content and thus fewer ionic groups
- adding water-miscible cosolvents (butyl glycol, n-butanol and the like).

The first two points have already been incorporated into modern coating film formers, e.g. third generation water-thinnable alkyd resins. The addition of cosolvents should be regarded as an "emergency solution" only, since we need to keep the content of organic solvents as low as possible for environmental and occupational health reasons.

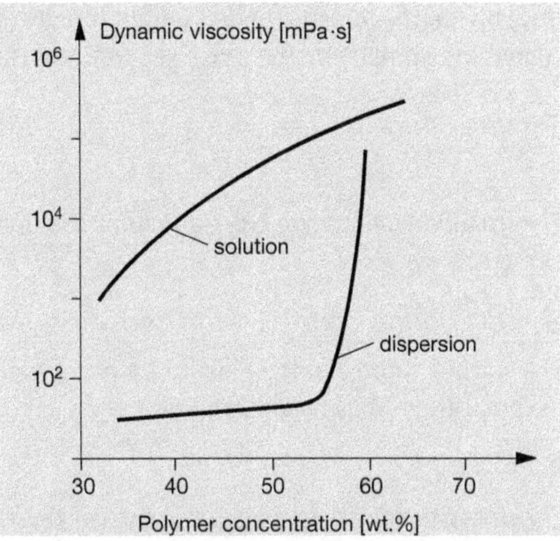

Figure 2.13: Comparison of the dependence on concentration of the viscosity of a polymer dispersion and of a polymer solution

Figure 2.13 illustrates the viscosity behaviour typical of many polymer dispersions that are free from protective colloids or thickeners. Up to moderate concentrations the viscosity is very low and almost constant, but then it suddenly rises sharply above a certain higher solids content because the particles prevent one another from flowing as the distance between them becomes too small. Since the macromolecules of polymer dispersions are contained within the compact latex particles rather than lying free in the water phase, their molecular weight can have no influence on the viscosity. This fact is of enormous value in the coatings sector because it allows the formulation of paints and coatings with high-molecular film formers which in true dissolved form would lead to unacceptably high viscosities (see Figure 2.13).

2.1.1.10 Mechanical behaviour of polymers - viscoelasticity

Let us start by describing some very simple but highly instructive experiments:

Take a jelly baby, stretch it without tearing it and then let it go. You will see that the jelly baby contracts again - rapidly to begin with, then more slowly until it returns to its original length. If you repeat the experiment with a piece of insulating tape made from plasticised PVC, you will see a similar behaviour, except that the contraction process takes longer and the tape never quite returns to its original length. Next repeat the experiment with an elastic band. We know the outcome: the elastic band very quickly returns to its original length. Finally, if you stretched a strip of dental filling compound (e.g. "Plastic Fermit"), you would see that this scarcely contracts at all but virtually retains its length.

Macroscopic model representations

To enable us to illustrate, systematise and roughly calculate this very diverse behaviour of polymers, which naturally include coating film formers, we can use as models combinations of helical springs and oil brakes (oil-filled cylinders from which the oil is expressed when force is applied to the piston), which most closely reproduce the given polymer behaviour. Since this is an interaction between viscous (oil) and elastic elements (springs), we call the resulting general mechanical behaviour "*viscoelastic*".

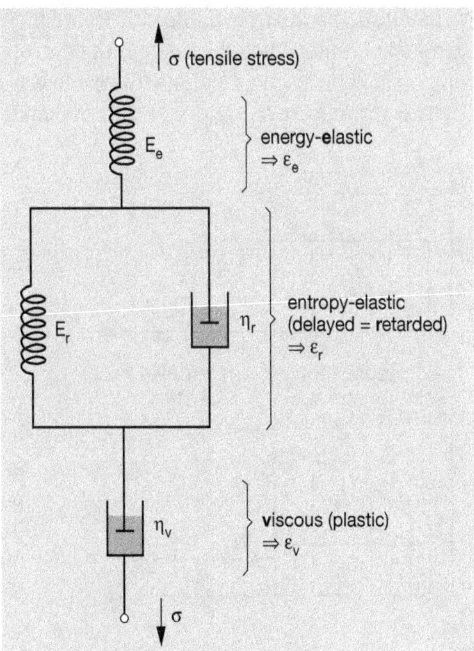

Figure 2.14: Burger's model of viscoelastic behaviour (see text for details)

Figure 2.14 shows the *Burger's model* of viscoelastic behaviour.

If we divide the force F applied longitudinally to the material specimen (bar, strip of coating film or something similar) by the cross-section A of the specimen, we obtain the tensile stress σ:

$$\sigma = \frac{F}{A}$$

The (relative) elongation ε is defined as the increase in length Δl divided by the original length l_0:

$$\varepsilon = \frac{\Delta l}{l_0}$$

The springs represent the elastic components in the polymer. Provided that the elongations are not too great then *Hooke's law* applies:

$$\sigma = E \cdot \varepsilon$$

The proportionality constant E is known as the elastic modulus. The more rigid the spring, the higher this is. The cylinders represent the plastic or viscous behaviour of the material. Here η represents the stretching viscosity of the polymer:

$$\eta = \frac{\sigma}{\dot{\varepsilon}} \quad \text{where} \quad \dot{\varepsilon} = \frac{d\varepsilon}{dt}$$

Figure 2.15 shows qualitatively how the elongation of the individual elements of the model and hence of the whole arrangement behaves over time, firstly under the application of a constant tensile stress and secondly after the spontaneous removal of the tensile stress.

The behaviour of the various polymer materials in the elongation experiments described above can be readily explained by the Burger's model or partial models as follows:

In the case of the jelly baby the cylinder η_v is missing; the spring E_e is very rigid and scarcely stretches, so this too can be discarded. What remains is the braked (entropy) elasticity. The PVC tape behaves in a similar way to the jelly baby, except that in this case the cylinder η_v, which is responsible for the permanent elongation, is present; the braking cylinder η_r contains a thicker oil. The behaviour of the elastic band is reproduced if we take away the cylinder η_v and make the oil in cylinder η_r extremely liquid, which means that the entropy-elastic elongation returns very quickly. The filling compound behaves in essentially the same way as the cylinder η_v filled with moderately viscous oil. The two springs are so rigid that they make virtually no contribution to the elongation.

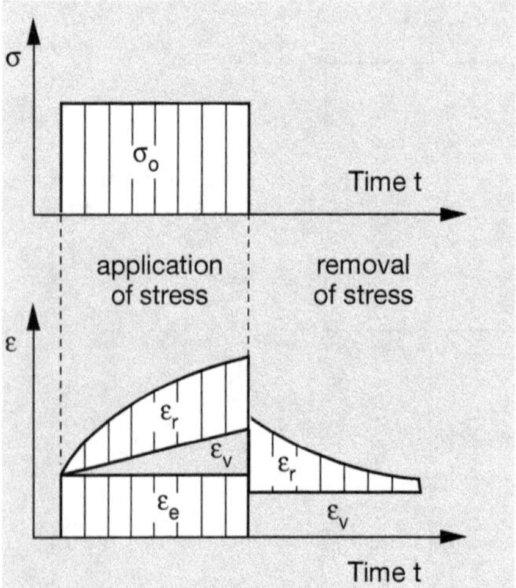

Figure 2.15: Dependence of elongation (ε) on time in the Burger's model (see Figure 2.14 for symbols)

Molecular processes

Yet how do we explain the polymer behaviour on a molecular level? To understand this question we need to consider the elements of the Burger-Kelvin model separately.

Entropy-elastic element

Imagine a tangle of interpenetrated coils of linear polymer molecules, which cohere in some places through intermolecular forces of attraction, entanglements or possibly chemical bonds. Thermal excitation means that varying lengths of molecule segments – depending on molecular structure and temperature – vibrate between the contact points (*micro-Brownian motion*).

If we now apply a moderate tensile stress to this system, the segment vibrations are inhibited and the partial parallel alignment of the molecules causes the substance to stretch. In this elongated state the molecular order is somewhat higher than before, corresponding to a reduced entropy[1] in the system. On removal of the constraint caused by the external stress, the polymer section returns - contracting more or less quickly - to its original high-entropy state.

A weakly crosslinked rubber (soft rubber) very clearly demonstrates this entropy-elastic behaviour: the vibrating segments extend from one crosslink to the next and the vibrations are very strong. The slipping of complete chains, which would lead to flow and hence to permanent elongation, is prevented by the continuous crosslinking of the substance. Figure 2.16 illustrates the entropy-elastic behaviour of an elastomer.

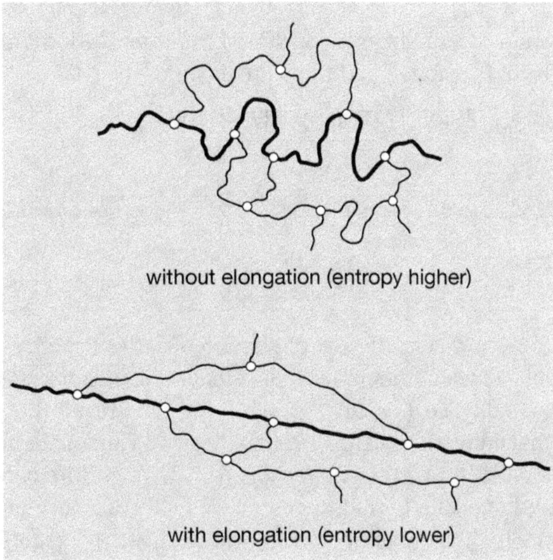

Figure 2.16: Molecular cause of the entropy-elastic behaviour of an elastomer

Purely viscous (plastic) element

Plastic behaviour in the form of what is known as flow occurs because when a stress is applied to the polymer material, entire molecules slip past one another (*macro-Brownian motion*). The higher the temperature (hence the term "thermoplastic") and the weaker the intermolecular bonding forces, the more readily this slipping occurs. The higher the molecular weight, the less readily does flow occur and the more resistant to strain and to other mechanical forces is the polymer. Crosslinked polymers (thermosets, elastomers) cannot flow and hence cannot undergo plastic deformation.

Energy-elastic element

This contribution to elongation is maintained by the change in the length and angle of chemical bonds. It occurs in all materials and is restricted to small elongations. Whilst hard solids such as steel or ceramics (provided the tensile stresses are not too great), effectively display only energy elasticity, in soft materials the energy-elastic elongation is more or less subsumed by the significantly greater entropy-elastic or plastic elongation. The term "energy-elastic" derives from the fact that the work required to produce elongation is stored as internal energy in the body and is released again when the stress is reduced. A further characteristic of energy-elastic elongation is that it disappears very rapidly ($< 10^{-6}$ s) on removal of the stress.

In thermodynamic terms entropy elasticity can be explained as follows:
From the definition of enthalpy $H = U + p \cdot V$ it follows that the total differential is

$$dH = dU + p \cdot dV + V \cdot dp$$

and under constant pressure

$$(dH)_p = (dU)_p + p \cdot dV\ .$$

With the first theorem $dU = dQ + dW$, we obtain

$$(dH)_p = (dQ)_p + (dW)_p + p \cdot dV.$$

1) The thermodynamic quantity (more accurately, state function) entropy, S, is a measure of the randomness of a system; the more irregular the internal structure and/or the higher the temperature of a substance, the greater the value of S.

For a reversible process, dQ = T · dS; the work dW can be divided into the elongation work F · dl (F = force) and the volume work -p · dV:

$$(dH)_p = T \cdot (dS)_p + \underbrace{F \cdot dl - p \cdot dV}_{dW} + p \cdot dV$$

If we solve this for F, then

$$F = \left(\frac{\partial H}{\partial l}\right)_p - T \cdot \left(\frac{\partial S}{\partial l}\right)_p .$$

If we now consider the polymer molecules to be free from forces between one another, like the molecules of the perfect gas, elongation at constant temperature does not cause any change in enthalpy, i.e. $(\partial H/\partial l)_p = 0$, with the same result that $F = -T \cdot (\partial S/\partial l)_p$. Since the entropy falls as the elongation increases, $(\partial S/\partial l)_p < 0$. The tensile force F is thus positive, and the higher the temperature T the greater it is. This temperature behaviour is characteristic of *entropy* elasticity and opposite to that of *energy* elasticity, e.g. a steel wire. We can also express this as follows: the higher the temperature, the stronger the segment vibrations which counteract the forced (one-dimensional) elongation of the body.

2.1.1.11 Measuring viscoelasticity

Tensile test

Continuing our earlier observations on viscoelasticity, let us now look at polymer behaviour under the tensile test. This method of determination involves extending for example a polymer film (free coating film) at a constant rate – generally until it tears. Figure 2.17 illustrates the stress-strain behaviour of a thermoplastic and a crosslinked coating film, a densely crosslinked thermoset material and a loosely crosslinked polymer (elastomer). The discussion that follows is restricted to the moderately densely crosslinked coating film.

First phase

Initially the tensile stress increases with a practically linear progression. In this area, known as the linear viscoelastic (entropy-elastic) zone, only vibrating molecule segments are stretched; there is no chain slippage or breaking of bonds in the polymer. On removal of the tension, the specimen would return to its original length.

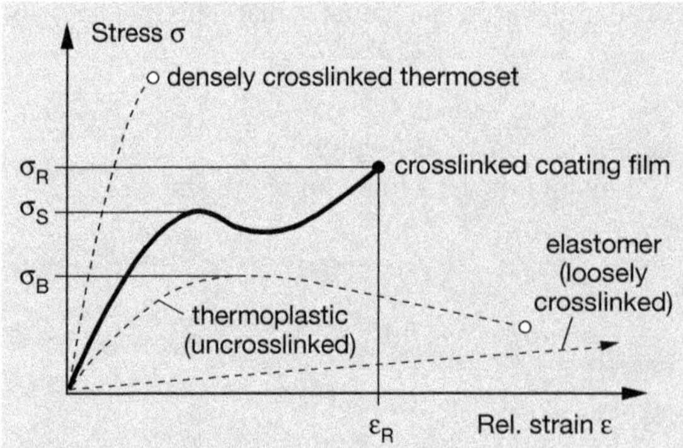

Figure 2.17: Stress-strain behaviour of varyingly densely crosslinked polymers and an uncrosslinked polymer ε_R elongation at break, σ_R tear strength (yield stress), σ_S tensile stress at yield, σ_B tensile strength

Second phase

A stress maximum is produced (*tensile stress at yield* σ_s). In this zone the stress is so great that some of the intermolecular bonds (secondary bonds, secondary valencies) are overcome and form again at different points; the chain segments are increasingly aligned in parallel. In other words, the system avoids the external force by structural rearrangements (transposition processes). This process is known as *cold stretching*. On removal of the stress, some elongation would now remain.

Third phase

Finally the elongation reaches a point above which no further stretching is possible; if the tensile stress is further increased, the chain segments rupture at *elongation at break* ε_R, causing the specimen to tear (break). (If it were not crosslinked, the polymer would flow on removal of the tensile stress).

The greatest tensile stress occurring during the tensile test is known as *tensile strength* σ_B. As the example of the thermoplastic shows, this does not always correspond with the elongation at break.

The work applied per volume element of the specimen until it breaks or tears, also called fracture energy or energy at break, is determined by the area below the curve and hence by the following integral:

$$W_R = \int_0^{\varepsilon_R} \sigma \cdot d\varepsilon$$

W_R is a measure of the *toughness* of the material; this important coating property can also be determined by measurement.

Creep, relaxation

Two more terms from polymer mechanics should be briefly mentioned here. "*Creep*" refers to the development of an irreversible (residual) elongation of the polymer specimen on application of a constant tensile stress for a defined period of time. Whilst creep in a crosslinked polymer - if it occurs at all - is limited, in thermoplastics it can continue unchecked as entire molecule chains slip off one another, i.e. by flow.

"Stress relaxation" is the phenomenon whereby the tensile stress $\sigma(t)$ produced by a spontaneously forced elongation ε_0 decreases from its initial value. The general equation for the relaxation following a decreasing exponential function is as follows:

$$\sigma(t) = E(t) \cdot \varepsilon_o.$$

E(t) is known as the "stress relaxation module".

The relaxation time τ is taken to be the time needed for the initial stress $\delta(0)$ to fall to the e-th part, i.e. to $0.368 \cdot \delta(0)$. The molecular cause of relaxation is the same as that of creep: slippage of chain segments (transposition processes) or of entire chains (flow).

Dynamic-mechanical analysis – elastic, storage and loss modulus

Dynamic-mechanical analysis (DMA), also known as "mechanical spectroscopy", is often used to conduct an accurate physical investigation or characterisation of the viscoelastic behaviour of specimens such as coating films. This technique involves applying a periodic sinusoidal elongation to the specimen and analysing the resulting stress vibration. Two quantities are obtained from this investigation: the *storage modulus* E' and the *loss modulus* E''. From these can be calculated the *dynamic (complex) elastic modulus* E* and the *loss factor* tan δ. The following correlations apply:

$$E^* = E' + i\,E'' \text{ where } i = \sqrt{-1} \text{ (imaginäry Unit)}$$

$$|E^*| = \sqrt{(E')^2 + (E'')^2}$$

$$\tan\delta = \frac{E''}{E'}$$

(δ is the phase displacement between the given elongation vibration and the resulting stress vibration).

As with a metal spring, the elongation work stored (which can be recovered on relaxation) is proportional to E'; this means that E' is a measure of the purely elastic properties of the specimen. The work expended to achieve viscous or plastic deformation of the specimen is proportional to E''. This is lost by conversion to heat. E'' is therefore a measure of the viscous behaviour of the body.

In the illustration of the spring/cylinder model of the polymer (see Figure 2.14), the interaction between all the springs is represented by the storage modulus and the interaction between all the cylinders by the loss modulus.

In a different version of the dynamic-mechanical analysis method, the specimen is not extended longitudinally but instead is twisted around its longitudinal axis to give the *shear moduli* G*, G' and G'' instead of the elastic modulus. Both are equally suitable for material characterisation. In the rheological investigation of fluid systems using oscillating shear, shear moduli are likewise obtained as characteristic values for viscoelastic behaviour.

To conclude our deliberations on viscoelasticity we should emphasise again that the mechanical behaviour of a polymer is highly dependent on deformation rate as well as temperature. One practical example of this is the differing behaviour of a coating system under "*Erichsen indentation*" (slow buckling of a coated metal specimen) on the one hand and "*impact indentation*" (rapid buckling) on the other hand. In the first case the coating film has time during elongation to reduce the resulting stresses somewhat by means of molecular transposition processes or flow. In the second case this is not really possible. The result is that Erichsen indentation values are generally significantly higher than impact indentation values.

2.1.1.12 Temperature dependency of polymer behaviour – glass transition temperature

If we heat an amorphous thermoplastic from a very low temperature, it goes through a number of different states or property zones (see Figure 2.18). The reason for this behaviour is the increasingly vigorous molecular movement as the temperature increases.

Molecular movement processes, glass transition temperature

At a sufficiently low starting temperature, the only movement in the cold molecule is the rotation and vibration of atoms and small groups of atoms. The polymer chains are almost motionless within the braid of molecules; they are "frozen". The polymer is in what is known as the vitreous state and is hard to brittle, like glass. The elastic modulus is high and there is no plastic deformation.

When we heat it we can detect a certain softening above a particular temperature. Now segments are starting to vibrate, leading to entropy elasticity and plasticity, i.e. permanent deformability or thermoformability. The behaviour of the polymer is viscoelastic. As the temperature continues to increase, the vibrating segments grow longer and the number of contact points between the chains reduces. The behaviour of the polymer becomes rubber-elastic. Finally the coherence between the chains becomes so slight that entire molecules slip off one another, forming a viscous, more or less elastic melt. Further heating - to around 250 °C or higher - leads to the decomposition temperature at which molecular fragmentation (cracking) in the form of depolymerisation and/or numerous other decomposition reactions occurs. A polymer cannot boil or evaporate.

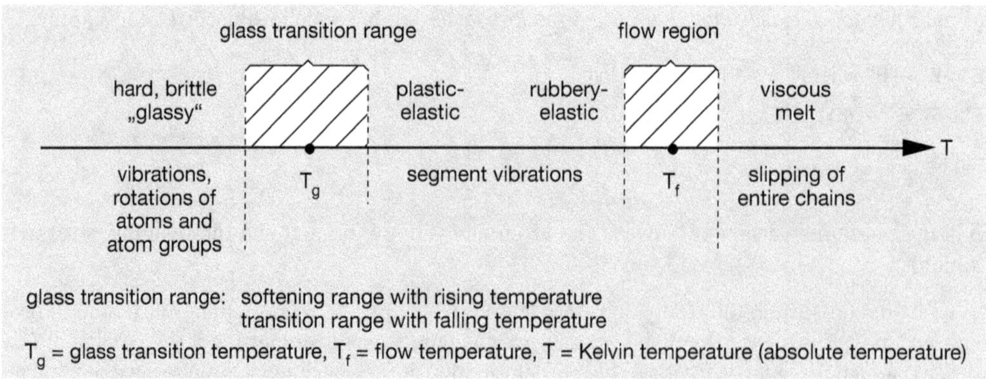

Figure 2.18: Thermal states and transitions of an amorphous thermoplastic

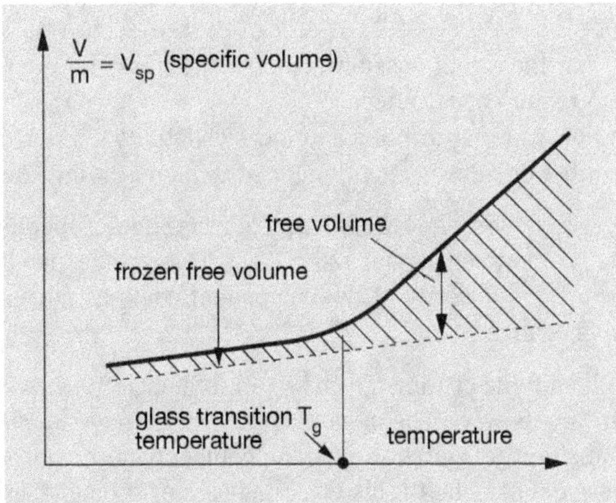

Figure 2.19: Dependence of specific volume, glass transition temperature and free volume on temperature

The temperature range for the transition from the hard-brittle to the viscoelastic, soft state or vice versa is known as the *softening* or *transition range* or more generally the *glass transition range*. The *glass transition temperature* (glass temperature, glass point) T_g is assigned to this region, although in physico-chemical terms this cannot be clearly defined or, consequently, identified. The glass transition is not a second order phase transition. For whereas the latter involves two different structures for the material and one clear transition temperature, glass transition is a sliding process which is not accompanied by molecular reorientations. Despite this, the glass transition can be recognised physically in a number of ways, e.g. by the increase in specific volume V_{sp} (see Figure 2.19).

The transition to the viscous melt occurs in what is known as the *flow region*; the temperature assigned to this is called the *flow temperature* (flow point) T_f. Figure 2.18 provides an overview of the various states of an amorphous thermoplastic.

As can be seen from Figure 2.19, above the glass transition region the specific volume of the polymer increases more quickly as the temperature increases than it does below this region. In mathematical terms this means that the differential quotient dV_{sp}/dT increases at T_g. This more rapid increase in specific volume is caused by the intensification in segment vibrations, which form what can be termed molecular cavities in the polymer. The total volume of these cavities is called the *free volume*, which is added to the actual molecular volume.

As well as specific volume, many other polymer properties can be used to determine T_g. Measurements used in industry include the change in indentation hardness (*thermomechanical analysis*, TMA), the changes in elastic or shear modulus using dynamic-mechanical measurements (→ 2.1.1.11) or the change in specific heat capacity using *differential scanning calorimetry* (DSC).

Further to what has already been said about the glass transition temperature, we should add that crosslinked polymers likewise exhibit a glass transition. However, the higher the crosslink density, the less pronounced this is.

Consequences for coating performance

So what impact does the glass transition point T_g have on the properties of a coating film? If T_g is below the temperature at which the coating is used, the film former will be in a softened state and will have free volume, i.e. a kind of microporosity. The effects of this include:

- softness (vulnerability to damage, possibly defective blocking resistance, poor grindability, but also good flexibility)
- reduced barrier effect against water vapour, SO_2, CO_2 and other substances
- increased water absorption and ability to swell with various agents
- reduced solvent retention.

If T_g is above the working temperature, it leads to

- hardness (but possibly also to brittleness and internal stresses)
- greater barrier effect
- low water absorption and low swellability
- greater solvent retention (depending on drying conditions).

The T_g values for coating films are generally - depending on chemical structure and curing conditions - between 10 and 125 °C. The lower range includes, for example, fresh long oil alkyd paints, whilst the majority of two-component coatings and hard stoving coatings lie in the middle and upper range.

Polymer and coating chemists can influence the glass transition point of a film former or coating in various ways. The monomers in film formers are classified by, amongst other things, the glass transition temperature of their homopolymers and/or their hardness in the polycondensate or polyadduct state. In this way the polymer chemist can influence the glass transition temperature of a film former to a certain degree using appropriate combinations of monomers with different T_g values or hardnesses. *Fox*'s formula holds true for polymers:

$$\frac{1}{T_g} = \frac{w_A}{T_{gA}} + \frac{w_B}{T_{gB}} + \frac{w_C}{T_{gC}} + \dots$$

where T_g is the glass transition temperature (in K) of the copolymer, T_{gA}, T_{gB}, T_{gC} are the glass transition temperatures of the homopolymers of monomers A, B, C and w_A, w_B, w_C represent the proportions by weight of monomers A, B, C in the polymer.

Fox's formula can also be applied to homogeneous blends of different film formers. The glass transition temperature for the blend can then be determined from the T_g values and proportions by weight of the individual components. Even the influence of solvent(s) on the T_g can be estimated by using the freezing temperature of a solvent without crystallisation as the T_g.

Finally let us summarise the influence of the chemical-structural molecular properties on the glass transition temperature:

- T_g rises with
  - increasing polarity and increasing hydrogen bridge content
  - increasing chemical-structural order (symmetry)
  - increasing quantity of short, rigid groups (methyl, phenyl, etc.) on the main chain
  - increasing proportion of inflexible co-monomers
  - increasing molecular weight
  - increasing crosslink density.
- T_g falls with
  - increasing content of flexible co-monomers (internal plasticisation)
  - increasing number of branches or longer side chains.

2.1.2 Natural film formers (natural substances)

2.1.2.1 Natural resins

The most important natural resin in coating chemistry is colophony. Other well-known resins include copal, dammar and shellac. (The following observations are limited to colophony and shellac).

Colophony is extracted from pine trees. The balsam obtained by slitting the trunk is separated by distillation into oil of turpentine and light-coloured colophony (*gum rosin*). The darker *wood rosin* is extracted from the rootstock; it is frequently lightened by bleaching. Sulphate wood rosin is a third type of colophony; this is isolated from the tall oil produced during the sulphate method of cellulose extraction. Colophony is a ruby red to pale yellow, brittle substance that melts at around 80 °C.

The chemical composition of colophony consists primarily of abietic acid ($C_{20}H_{30}O_2$, M = 302.5 g/mol):

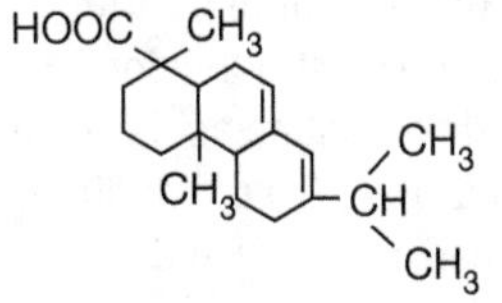

abietic acid

together with related (double-bond isomeric) resin acids, and thus has a high acid value - around 150 to 180 mg/g. This is defined in the same way for all chemical substances and blends of substances, as follows:

The acid value (AV) is the amount in mg of potassium hydroxide required to neutralise the free acid contained in 1 g of substance (unit: mg/g).

The high acid value of colophony is the main cause of its incompatibility with most other coating raw materials.

Together with the carboxyl group (-COOH), the conjugated double bond system is an important chemical feature of colophony ("conjugated double bonds": → 2.1.2.2)

A number of the properties of colophony are undesirable for applications in paints and coatings: poor oxidation and yellowing resistance, high acid value and consequently poor water resistance, compatibility and solubility, tendency to crystallise, tackiness. For these reasons it is almost only ever used in chemically modified form (→ 2.1.3.1).

Shellac is the conversion product of the sap of certain trees native to India and neighbouring countries produced by the lac insect; hence it is a resin of animal origin. It is ruby red to yellowish in colour, depending on the degree to which it has been refined, it melts at 65 to 80 °C and is readily soluble in alcohol and moderately to poorly soluble in other coating solvents. The chemical composition of shellac consists of free and esterified aliphatic and aromatic hydroxyl acids and a certain amount of wax; the acid value is around 70 mg/g. Shellac is used in unmodified form in wood varnishes, inks, India inks, polishes, flexographic printing inks and many other products, either alone or in combination with other film formers. It imparts good adhesion, fuel resistance and polishing ability to coatings. Shellac can be made water-soluble in combination with alkalis, borax, amines or ammonia.

2.1.2.2 Oils, oxidative drying

Like natural resins, oils are rarely used in unmodified form in coatings nowadays. Nevertheless, they form the basis for conversion products or synthetics that are more suitable for use in coatings and for that reason they merit a brief discussion.

The relevant oils for coating chemistry are the *"fatty oils"*, i.e. *triglycerides* of fatty acids. Each of the three hydroxyl groups (-OH) is esterified with a fatty acid molecule:

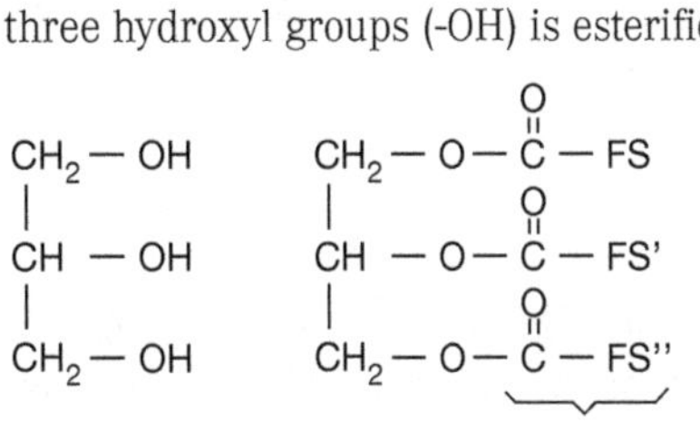

triglyceride
(fatty oil)
M ≈ 880 g/mol

glycerol fatty acid groups

Most of the oils are obtained from oleaginous plant seeds or oil plants by pressing and extracting, after which the oils are refined (purified). Fish oils and train oils are also used to a limited degree.

Fatty acid types

The fatty acids bonded in the oils are relatively long-chain, aliphatic, unbranched monocarboxylic acids; they may be saturated or varyingly highly unsaturated (containing double bonds). In terms of paint properties the number of double bonds is the most important feature. The subdivision into *isolate and conjugate acids* is also of great significance. Isolated double bonds or isolate acids occur when the double bonds in the polyunsaturated fatty acids are all separated by at least two single bonds. If there is only one single bond between each double bond then we speak of conjugated double bonds or conjugate acids. The simplified chemical structural formulae for the most important fatty acids are listed below. (Cis-trans isomerism, which is also considered in the formulae, is of only limited importance in coating chemistry).

Dehydrated castor oil or its saponification product, dehydrated castor oil fatty acid, contains approximately 30 % conjugated dehydrated castor acid and the corresponding isolate acids (double bonds in the 9 and 12 position). Dehydrated castor oil is obtained by thermocatalytic dehydration[1] of castor oil.

lauric acid (C_{12})

stearic acid (C_{18})

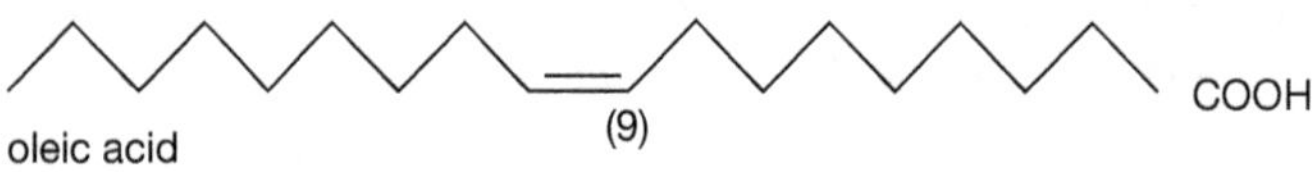

oleic acid

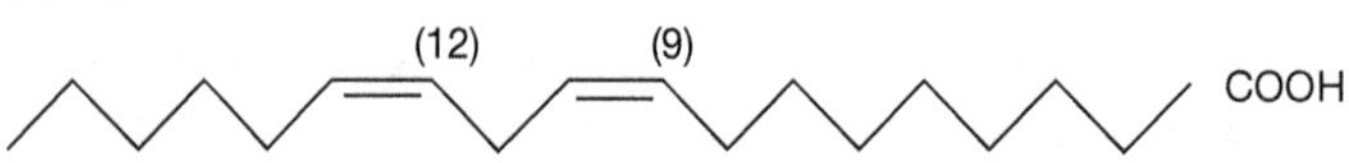

linoleic acid

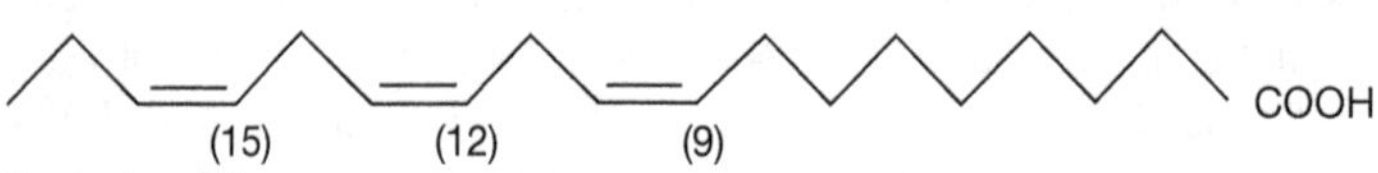

linolenic acid

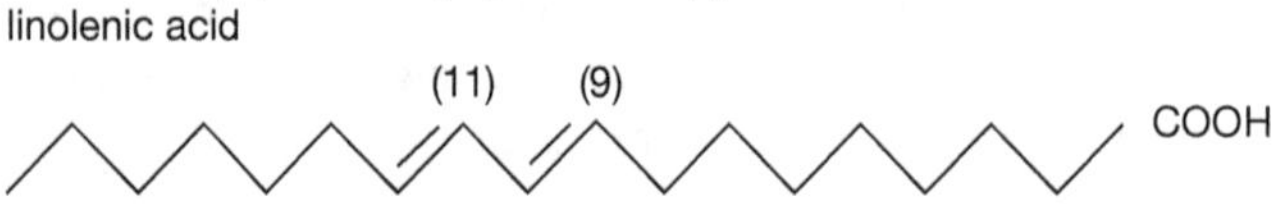

dehydrated castor acid

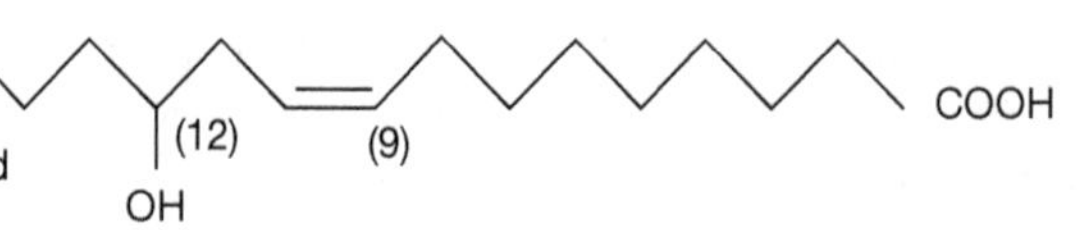

ricinoleic acid

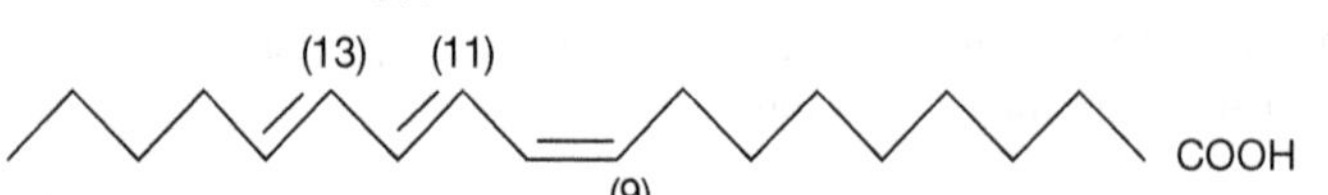

elaeostearic acid

Oxidative drying

The "*drying*" of oils containing polyunsaturated fatty acids is of enormous importance in coating technology. This term refers to the process of oxidative polymerisation accompanied by crosslinking, which - depending on the number of double bonds per molecule - causes the oil to change into a soft and tacky to hornlike and hard substance. In the first instance we speak of *semidrying* oils, in

1) Dehydration is the removal of water from a molecule; it proceeds here as an acid-catalysed β elimination.

the second of *drying* oils. Since drying takes place very slowly, it is accelerated by the addition of small quantities of *"driers"*. These are organic salts ("soaps") of particular metals such as cobalt, manganese, calcium, zirconium, cerium, etc. In this way the drying time for a linseed oil film, for example, can be reduced from 120 to around 3 hours. A dried oil is known as a *boiled oil.*

The classification of oils into *"isolate and conjugate oils"* according to the fatty acids they contain is important in relation to the speed of oxidative drying. Conjugate oils dry much more rapidly than isolate oils (see more detailed explanation).

The iodine value is a measure of the degree of chemical unsaturation of a substance. One mole of isolated double bonds adds 253.8 g (1 mole) of iodine. The iodine value is defined as follows:

*The **iodine value** (IV) is the mass of iodine in g bonded by 100 g of substance (oil). (Unit: g/100 g).*

There are a number of standard methods for determining the iodine value, but space prevents us from discussing these in further detail here.

One definition of the drying character of oils, whilst admittedly not universally applicable, is based on their iodine value:

- non-drying oils: IV < 100
- semidrying oils: IV = 100 to 150
- drying oils: IV > 150

Oxidative drying is classed as an autoxidation reaction, whereby organic substances are slowly oxidised at (or close to) room temperature by atmospheric oxygen. The underlying cause of autoxidation is the biradical structure of oxygen: each oxygen molecule has two relatively high-energy single electrons by means of which it attacks other suitable substances. In the case of isolate oils, this attack initially produces mainly hydroperoxides which trigger the crosslinking polymerisation reaction - i.e. the actual film forming process - only after decomposition:

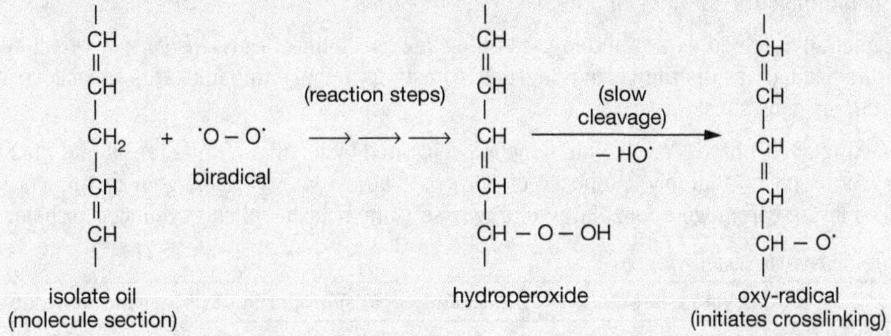

Whilst in isolate oils high oxygen absorption is observed at the start of the drying process, with the double bond content of the oil remaining largely unchanged, oxygen absorption in conjugate oils is noticeably lower and film-forming significantly faster. This is because the polymerisation of conjugate oils proceeds to a great extent with no preliminary hydroperoxide decomposition, forming C-C bonds. The reduced oxygen absorption means that, in addition to accelerated film forming, conjugate oils demonstrate a lower tendency to become yellow and brittle as compared with isolate oils. They also form fewer volatile decomposition products as they dry (with a consequent reduction in odour).

The mechanism of driers (more accurately: primary driers) consists in accelerating peroxide decomposition according to the following reaction:

$$M^{n\oplus} + \overset{(R')}{R - O - OH} \longrightarrow M^{(n+1)\oplus} + R - O^{\cdot} + \overset{(R')}{OH^{\ominus}}$$

metal cation (from drier) — hydroperoxide (peroxide) — oxy-radical (initiates crosslinking)

R, R' = fatty acid groups from oil

(In subsequent reactions the metal is further reduced and hence reactivated.)

Principal oils

In conclusion, Table 2.4 provides a summary of the oils that are most frequently used in coating chemistry. In addition, numerous other oils not listed here are of limited importance.

Note: *Tall oil,* which is not included in Table 2.4, is an oily, acid raw product derived from conifers during the production of cellulose via the "sulphate process", from which *tall oil distillate* or (the even purer) *tall oil fatty acid,* essentially a blend of linoleic acid and oleic acid, is obtained by distillation. (Another distillation product is "sulphate turpentine oil").

2.1.2.3 Bitumen, asphalt, pitch

Of these three substances only natural asphalt is a truly natural product; bitumen and pitch are obtained by distillation or heat treatment of natural products (petroleum, coal).

Bitumen is strictly the residue from petroleum distillation. It is a dark, semisolid to hard and brittle, thermoplastic material. In chemical terms it is a colloidal dispersion of hard asphaltenes in viscous malthenes, which in turn consist of large numbers of higher-molecular hydrocarbons as well as oxygen-, nitrogen- and sulphur-containing substances. Depending on production conditions, *distillation bitumen, high-vacuum bitumen* or *oxidation bitumen* (*blown bitumen*) is obtained. Oxidation bitumen has the broadest plasticity range (< 0 to 90 °C or higher), but is inferior to distillation and high-vacuum bitumen in respect of all other coating properties.

Asphalt is generally defined as a bitumen containing mineral matter. Only *gilsonite,* a very low-mineral natural asphalt (asphaltite), is of relevance in coatings and printing inks. It is a black, very hard to brittle, resinous substance.

Pitch: The coking of coal produces coal tar, which is separated by distillation into tar oil and pitch. In chemical terms pitch is largely composed of polycyclic aromatics[1]. One member of this class of substances is the carcinogen benzo[a]pyrene, whose content in tar pitch is reduced to below

Table 2.4: Important oils in coating chemistry

Oil	Drying character	Main fatty acid(s)/proportion in %	Iodine value*
linseed oil	drying	linolenic acid/50	178
tung oil	(quick) drying	elaeostearic acid/77	163
dehydrated castor oil	drying	dehydrated castor acid/30	160
fish oils	drying	highly unsaturated/9 to 28	100 to 180
soya bean oil	semi-drying	linoleic acid/54	127
safflower oil	semi-drying	linoleic acid/70	145
sunflower oil	semi-drying	linoleic acid/51, oleic acid/42	130
castor oil	non-drying	ricinoleic acid/92	84
coconut oil	non-drying	lauric acid/48	9

*) typical value

[1] Hydrocarbon molecules consisting of several adjacent aromatic (benzene) rings

0.005 % by secondary treatment processes. In spite of this, drinking water, foodstuffs and animal feed must not be allowed to come into contact with pitch coatings, and the use of coating materials containing tar pitch is declining rapidly.

Coatings based on bitumen, asphalt and pitch are referred to collectively as "*black coatings*". They are normally filled and combined with other film formers (oils, alkyd resins, chlorine rubber, etc.) in order to improve their properties. Asphalt coatings can be coloured to a limited extent using coloured pigments. Pitch or coal pitch (containing tar oil) can be combined with polyaminoamides or polyols and used as two-component systems with epoxy resin or polyisocyanate. This eliminates thermoplasticity and thus improves their heat resistance. Typical areas of application for black coatings include heavy-duty corrosion protection (physical protection), building protection (concrete coating), roof coatings, vehicle underseals, etc. Gilsonite is also used in printing inks.

2.1.3 Modified natural substances

Natural substances can be modified in order either to enhance or optimise their coating properties - as with natural resins and oils - or to render the substance suitable for use as a coating film former in the first place - as with cellulose or natural rubber.

2.1.3.1 Modified natural resins

The observations below are restricted to modifications of colophony.

As we mentioned in section 2.1.2.1, the use of colophony in coatings is limited by a number of weaknesses. To overcome these weaknesses as far as possible, various modifications can be made with a view to

- lowering the acid value
- reducing the content of reactive double bonds
- increasing the molecular weight.

The commonest modifications are briefly described below.

The *thermocatalytic* treatment of colophony essentially involves a dimerisation accompanied by decarboxylation reactions (decomposition of carboxyl groups with separation of CO_2), isomerisations and dehydrogenations. The "*polymerised colophony*" produced by this treatment has noticeably superior properties and is also suitable for further modification by the other treatments conventionally used with colophony (see below).

Salification of colophony with oxides or hydroxides of zinc, magnesium and/or calcium produces *zinc, magnesium* or *calcium resins* (*resinates*), which are characterised by a much lower acid value and improved compatibility and solubility in non-polar solvents (hydrocarbons). These products are used predominantly in the printing inks sector.

Esterification of the resin molecules with polyfunctional alcohols (polyols) such as glycerine or pentaerythritol, resulting in a dramatic lowering of the acid value and significant molecule enlargement, represents an alternative to salification. Depending on their range of properties, *colophony esters* (resin esters) are used either as *hard resins*[1] to improve the gloss, hardness and physical drying of paints and coatings or as "*tackifier* resins" in adhesives and related products.

Maleinisation is another important modification of colophony. This process involves the reaction of *maleic anhydride* (as dienophile) with laevopimaric acid (as diene) at around 190 °C using the Diels-Adler reaction. The laevopimaric acid is delivered continuously by isomerisation from abietic acid and other resin acids:

[1] Modified natural resins or synthetic resins whose softening range is above 90 °C and which produce hard, glossy coating films are known as "hard resins".

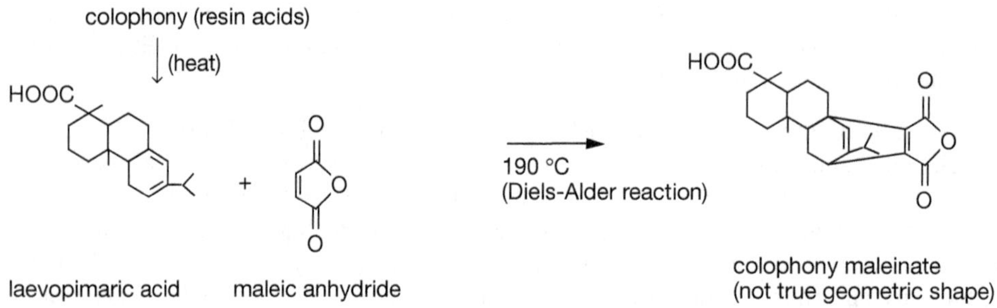

This produces an extremely acid *colophony maleinate resin*, which in most cases, with the exception of certain printing ink applications, is then more or less completely esterified with glycerol or pentaerythritol. The coating applications for which this product is suitable are largely the same as those for the non-maleinised resin ester - although its properties are superior. One peculiarity of partially esterified maleinate resins is that after neutralisation with amine their higher acid value makes them water-soluble.

Finally we should briefly mention *colophony-modified phenolic resins*. Hard resins known as "*synthetic copals*", which are generally added to quick-drying printing inks (e.g. offset printing inks) and coatings, are produced by "boiling" reactive phenolic resins (resols) with colophony to form *albertolic acids*, which are then esterified.

2.1.3.2 Modified oils

Fatty oils (triglycerides) have a relatively low molecular weight (approx. 880 g/mol), which is reflected *inter alia* in their low viscosity. The number of oxidative or other bonds required to produce a three-dimensional network is therefore relatively high and the film forming time correspondingly long. One purpose of oil modification is therefore to increase molecular weight. Other objectives may include:

- accelerated oxidative drying
- improved surface drying (through physical drying)
- improved water resistance and hygrostability
- chemical derivatisation (i.e. incorporation of functional groups)
- increased polarity.

When drying oils are heated to temperatures above 260 °C in the absence of air, this triggers complex chemical reactions which lead to oligomerisation (i.e. increase in molecular weight) and hence also to increased viscosity. In the filmed state, the *stand oils* formed by these processes, e.g. linseed standoil or tung oil standoil, demonstrate an improved resistance to moisture and weathering as well as reduced yellowing.

Passing air through hot drying oil produces a similar effect to the formation of stand oils. The tedious process of oxidative drying (molecular enlargement) has already been partially completed in the *blown oils*, which means that the subsequent film forming of the coating proceeds correspondingly faster. (Stand oils, blown oils and other chemically thickened oils are collectively known as "*bodied oils*").

The thermocatalytic isomerisation of isolate oils or isolate acids produces oils or fatty acid blends with a significantly higher content of conjugate acids. In particular, the *isomerised oils* demonstrate accelerated oxidative drying.

Styrenated oils are characterised by physical drying and consequently accelerated surface drying together with improved water resistance. These are produced as shown below by the radical grafting of styrene onto the double bond systems in the oils:

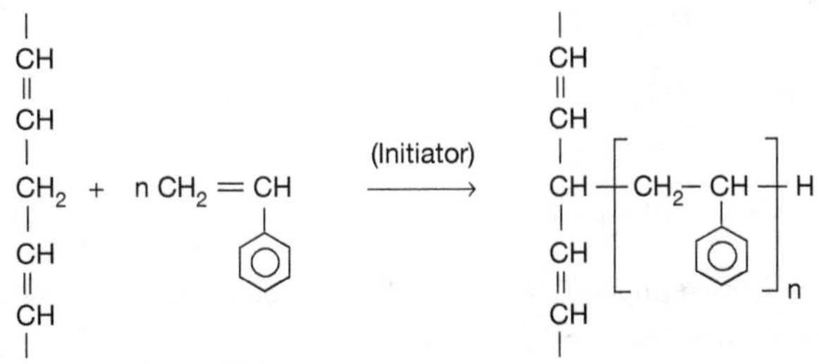

The polystyrene component transfers its hardness and hydrophobia to the respective oil. *Acrylisation* with (meth)acrylates such as methyl methacrylate, which is performed in the same way as styrenisation, produces similar effects but with vastly improved properties.

Adducts of drying oils and cyclopentadiene are known as "cyclo-oils".

We should also briefly mention the following oil conversion products:

- *Maleinate oils*: Adducts of maleic anhydride and drying oils. (More or less completely esterified with polyfunctional alcohols and/or neutralised with amine and thus rendered water-thinnable)
- *Urethane oils*: Reaction products of oils with polyfunctional alcohols and (subsequently) diisocyanates
- *Factised oils*: Oils reacted with sulphur or disulphur dichloride.

2.1.3.3 Cellulose derivatives

Although cellulose is widely available as a key skeleton-forming component of plants, its total insolubility in all common solvents means that it cannot be used directly as a coating film former. In chemical terms cellulose is poly-β-D-glucose, i.e. a polysaccharide. As is clear from the formula below, the basic unit of cellulose, the glucose unit, has three hydroxyl groups.

β-D-glucose unit
(basic unit of cellulose)

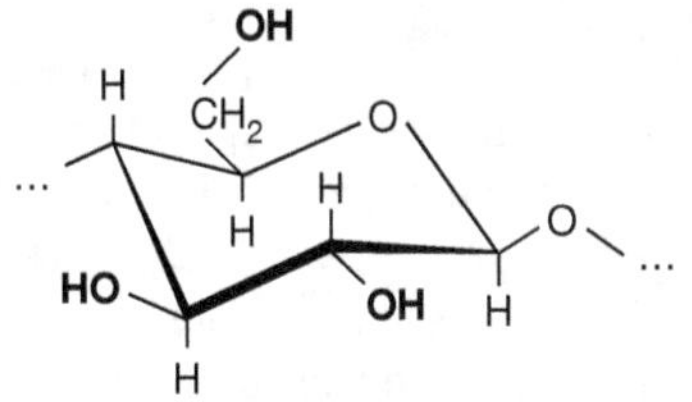

The insolubility of cellulose is attributable to its extremely high molecular weight of several 100,000 to over 10^6 g/mol, its large number of hydrogen bridges and its partial crystallinity. Any chemical modification to render cellulose suitable for use as a coating film former must therefore achieve

- a marked reduction in the average molecular weight and
- a significant decrease in the number of free hydroxyl groups.

This can be achieved by

- esterification with nitric acid or
- esterification with organic acids or
- etherification with alkyl groups

followed by molecular decomposition by means of acid hydrolysis of the ether bridges between the glucose units.

Cellulose nitrate ("nitrocellulose")

When cellulose is reacted with so-called nitrating acid (blend of nitric acid, sulphuric acid and water) a majority of the hydroxyl groups are converted into nitrate groups:

$$\text{cell} - \text{OH} \quad + \quad \text{HNO}_3 \xrightarrow[-\text{H}_2\text{O}]{(\text{H}_2\text{SO}_4)} \text{cell} - \overbrace{\text{ONO}_2}^{\text{NO}_3}$$

cellulose OH group · nitric acid · cellulose nitrate group

At the same time the molecular weight is reduced due to hydrolysis (acetal decomposition), which can be selectively continued in a second, separate stage of treatment.

Cellulose nitrate (CN) - which traditionally is also known as "nitrocellulose" (NC), collodion cotton or nitro-cotton - is sold not in pure, dry form but moistened with around 35 % ethanol, propan-2-ol, butanol or water because of its high flammability and devastating combustibility. Solutions, pastes, plasticiser-containing chips and (since the late 1980s) aqueous products (emulsions) are also commercially available.

Warning:
When storing or working with cellulose nitrate, statutory regulations and the manufacturer's safety instructions must be complied with!

Cellulose nitrate is classified according to the degree of esterification - expressed as the nitrogen content - or to the resulting (preferred) solubility.

- 10.6 to 11.2 % N: *A grades* (alcohol-soluble)
- 11.8 to 12.2 % N: *E grades* (ester-soluble)

Between these two grades lie the less common M-(AM-) grades with 11.3 to 11.7 % N. (Fully esterified cellulose ("gun-cotton") has a nitrogen content of 14.14 %.)

A second distinguishing feature is the average molecular weight, expressed by the viscosity of a defined solvent form. Unfortunately the various manufacturers refer to the viscosity or molecular weight by different, non-standardised index systems. Nevertheless, it is usually possible to match the commercial grades to the "*standard grades*" according to ISO 14 466 (formerly DIN 53 179) using the relevant product data sheets or comparative tables.

The coating properties and areas of application of the various types of cellulose nitrate are dependent above all on their molecular weight. Whereas the high-molecular grades (M > 100,000 g/mol) are suitable for use as sole film formers for coatings forming extremely quick-drying, flexible, solid but thin films, grades of average molecular weight are used in combination with other film formers and plasticisers (phthalates, phosphates, castor oil, etc.) to make quick-drying coatings, for furniture and nails in particular. Low-molecular grades of cellulose nitrate can be found in quick-drying fillers, stoppers and printing inks, for example.

The positive properties of cellulose nitrate, such as very rapid drying, good film hardness and virtually universal compatibility with other raw materials, are counteracted by some negative features, however. These include, in addition to its particular flammability, a tendency to become yellow when exposed to light (separating off traces of nitrous gases) and a certain sensitivity to alkalis.

Cellulose esters of organic acids

Efforts to find non-flammable alternatives to cellulose nitrate have led to the use of esters of lower organic acids. *Cellulose acetate* (CA) (ester of acetic acid) is of little interest in the coating sector because of its very limited solubility. Conventional coating raw materials include the mixed esters *cellulose acetopropionate* (CAP) (mixed ester of acetic and propionic acid) and especially *cellulose acetobutyrate* (CAB) (mixed ester of acetic and butyric acid). Like cellulose nitrate, the esters also contain non-esterified hydroxyl groups which can react with other binder components. The properties of mixed esters are determined by the content of acetyl, propionyl or butyryl and hydroxyl groups and by the average molecular weight.

Cellulose acetobutyrate is used as the main component or as an additive in many coating systems - both physically and reactively curing types. The addition of CAB accelerates physical drying, improves grindability (of primers and fillers) and flow and reduces cratering. In the wood coatings sector the improvement of "cold check" values (resistance to temperature change) is also important, and in conventional automotive metallic base paints CAB encourages the parallel alignment of the aluminium flakes. CAB and the other organic cellulose esters are not prone to yellowing when exposed to light, and their heat resistance is also significantly better than that of cellulose nitrate.

Attempts to develop water-dispersible CAB grades for water-borne coatings such as aqueous metallic base paints have not yet resulted in any commercial products.

Cellulose ethers

Cellulose ethers are produced by the (incomplete) etherification of cellulose - e.g. by reaction of hydroxyl groups with alkyl chlorides in the presence of caustic soda solution.

Of the various commercial cellulose ethers, only *ethyl cellulose* (EC) is soluble in organic solvents. It is a flexible thermoplastic with good resistance to heat, water and various chemicals (including alkalis). The main areas of application for ethyl cellulose are physically drying coatings on flexible substrates (paper, leather, films, textiles) and printing inks. It is compatible with many other film formers.

Methyl, hydroxyethyl, hydroxypropyl and *carboxymethyl cellulose* (the last in the form of sodium salt) are water-soluble substances that are used as thickeners and protective colloids in emulsion paints and as binders for distempers, glues and pastes.

2.1.3.4 Modified natural rubber

Natural rubber is produced from the latex (milky sap) obtained by slitting the trunk of the rubber tree. In chemical terms natural rubber is a very high-molecular cis-1,4-polyisoprene:

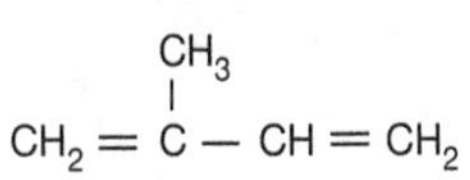

isoprene

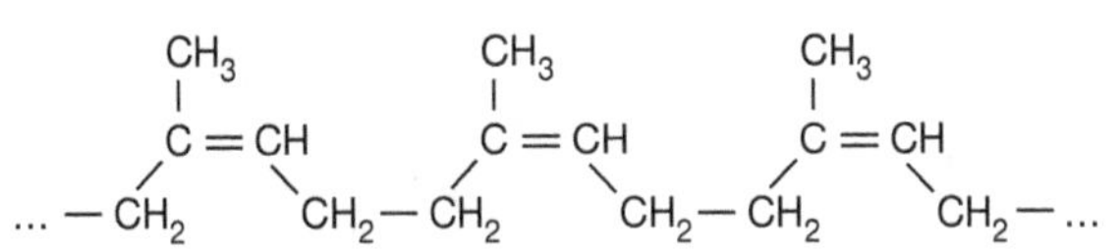

cis-1,4-polyisoprene (natural rubber)

The natural product, which is characterised by a high content of double bonds, can be decomposed by oxidation, is brittle and is insoluble in coating solvents; it can be modified in various ways according to its intended use.

Cyclorubber

Cyclorubber is formed by the action of certain reagents on natural rubber at high temperature. This triggers a cyclisation reaction which, with a considerable reduction in the number of double bonds and the average molecular weight (to 3000 to 10,000 g/mol), leads primarily to a ladder structure consisting of saturated six-membered rings aligned next to one another. The external evidence of this reaction is constituted by its transformation into a solid, resinous substance that melts at 120 to 140 °C and is readily soluble in non-polar solutions.

The non-saponifiable nature and good chemical resistance of cyclorubber explain its suitability as a film former for weather-resistant and chemical-resistant industrial coatings, anti-corrosive primers and other similar coatings. Plasticisers or additional softer film formers can be added to improve its flexibility. Naturally any additives must also display adequate chemical resistance.

Chlorine rubber

Chlorination is another means of achieving the degradation and chemical stabilisation of rubber molecules. Nowadays, however, largely because of variations in the properties of crude rubber, chlorinated natural rubber has been replaced entirely by chlorinated synthetic polymers in the coatings sector. (For further information see section 2.1.4.10 under "Chlorinated polymers").

2.1.4 Synthetic film formers

2.1.4.1 Saturated polyesters (SP)

General introduction – chemistry and properties

Saturated polyesters (SP) are produced by the polyesterification of bifunctional or higher-functional alcohols[1] with polyfunctional (predominantly bifunctional) saturated aliphatic or cycloaliphatic or aromatic carboxylic acids or the corresponding anhydrides. The formulae for some of the most frequently used monomers for saturated polyesters are given below:

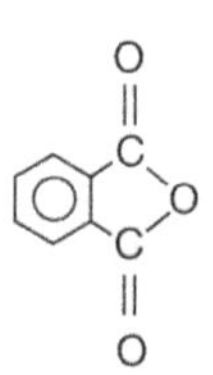

phthalic anhydride
(aromatic)

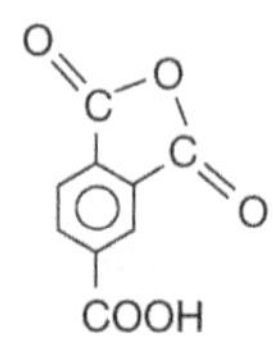

trimellitic anhydride
(aromatic)

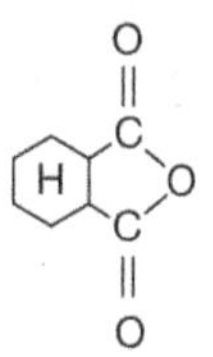

hexahydrophthalic
anhydride
(aromatic)

$HOOC-(CH_2)_4-COOH$

adipic acid
(aliphatic)

$HO-CH_2-CH_2-OH$

ethylene glycol

$HO-(CH_2)_6-OH$

1,6-hexanediol

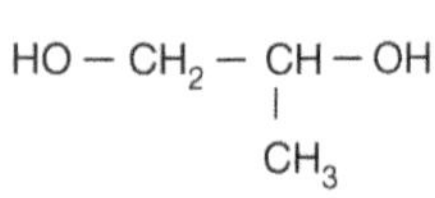

1,2-propylene glycol

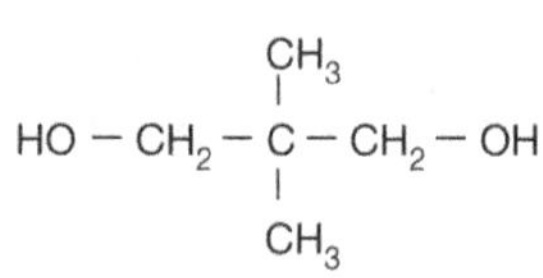

neopentyl glycol

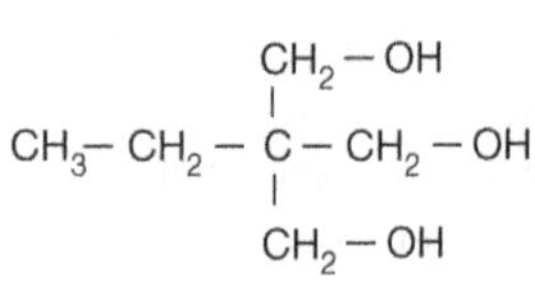

trimethylol propane

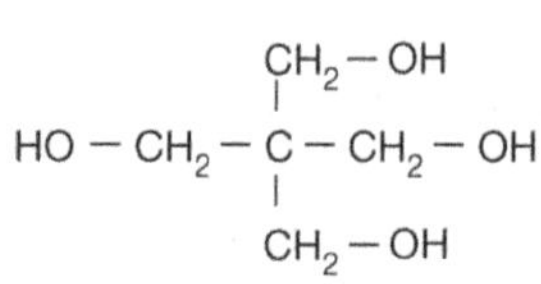

pentaerythritol

[1] Polyfunctional alcohols (molecules with several alcohol hydroxyl groups) are also known as "polyvalent" alcohols, polyols (diols, triols, tetrols, etc.) or in technical jargon – which is chemically inaccurate – as "glycols". (Strictly speaking, only 1,2-diols can be described as "glycols").

The use of exclusively bifunctional monomers produces linear polyesters. If trifunctional or higher-functional monomers are added, branched polyesters are formed. The desired average molecular weight can be controlled by the conversion (degree of conversion) of COOH- and OH-groups, i.e. the reaction period and reaction conditions, and by the COOH/OH ratio in the starting mixture (see more detailed explanation); this is somewhere between 1000 and 30,000 g/mol. The COOH/OH ratio also governs whether a *polyester polyol* (hydroxy-functional) or an *acid polyester* (carboxy-functional) is produced.

In the relatively simple case of a polycondensate or polyadduct produced exclusively from bifunctional monomers (A–A and B–B), the following dependence on the conversion conv and the molar (numerical) ratio r of the reactive groups on the average degree of polymerisation P_n as defined by *Carothers* applies:

$$\overline{P}_n = \frac{r+1}{2\,r\,(1-\text{conv})+1-r}$$

$$\text{where} \quad r = \left(\frac{n_A}{n_B}\right)_{t=0} = \frac{\text{mol minority groups (A)}}{\text{mol excess groups (B)}} \Bigg\} \text{ before reaction}$$

("working ratio")

In order to interpret or discuss this equation, it is simply a matter of being aware that not every incorporated group (A or B) is a molecule end or end group. The fewer end groups that are present, the fewer molecules exist, i.e. the greater the size of the average molecule.

Let us now consider

- Case 1: conv increases → 1-conv decreases → denominator reduces → P_n increases. Explanation: the greater the value for conv, the fewer end groups (molecule ends) there are.
- Case 2: conv = 1 (100 %), i.e. all A groups are reacted (this is never achieved in practice!)

$$\Rightarrow \overline{P}_n = \frac{1+r}{1-r} \quad \text{und} \quad \frac{d\overline{P}_n}{dr} = \frac{2}{(1-r)^2} > 0$$

In summary $\overline{P}_n$ increases with r (extreme case: $\overline{P}_n \to \infty$ for $r \to 1$).

Expressed in words: For a given reaction (1 in this case), the average degree of polymerisation increases sharply when the molar ratio of the reactive groups approaches 1:1.

Explanation: The smaller the numerical excess of a group, namely B, the fewer B groups remain as end groups when all of the A groups have been reacted.

Together with chemical composition and functionality, average molecular weight, molecular weight distribution and degree of crosslinking, the glass transition temperature, which in mechanical terms is reflected in the hardness, is of particular importance for coating applications. As a simplified summary, the following relation between monomer structure and polyester hardness applies:

- aromatic and cycloaliphatic dicarboxylic acids → hard
- longer-chain aliphatic dicarboxylic acids → soft
- short-chain and cycloaliphatic diols (and trimethylol propane, pentaerithrytol) → hard
- longer-chain aliphatic diols (and glycerol) → soft.

This is an example of the following rule of thumb which is very useful for developing synthetic resins and coatings:

The chemical and physical properties of a monomer are reproduced in the polymer, albeit to a lesser extent.

Thus, glycerol makes polyester softer while trimethylol propane, which is also trifunctional, imparts hardness. Explanation: glycerol is a liquid, while TMP does not melt below 56 °C.

Two further examples of the rule: amines are basic, and so are the epoxy resin films cured with them. Urea is readily soluble in water, and melamine sparingly so. Films cured with urea resin are therefore less water-resistant than those cured with melamine resin.

Linear homopolyesters produced from only one dicarboxylic acid and one diol (or from only one hydroxycarboxylic acid or its lactone in total) are more or less crystalline. For polymer materials (plastics), some hot-melt compounds and thermoplastic powder coatings a certain amount of partial crystallinity is definitely desirable.

By contrast, copolyesters for coating applications consisting of very irregularly structured molecules are practically amorphous and are thus sufficiently soluble and compatible.

Coating applications

Some relatively soft, long-chain, linear to slightly branched polyesters are physically drying only; when combined with harder film formers they form *polymeric plasticisers* ("*soft resins*").

However, the vast majority of polyesters are chemically crosslinked, specifically
as a two-component system with polyisocyanate at room temperature up to around 80 °C (forced):

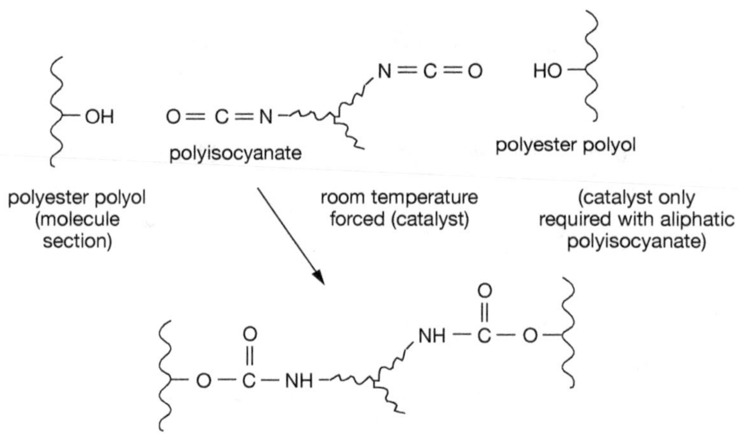

as a one-component system with amino resins, principally with melamine resin:

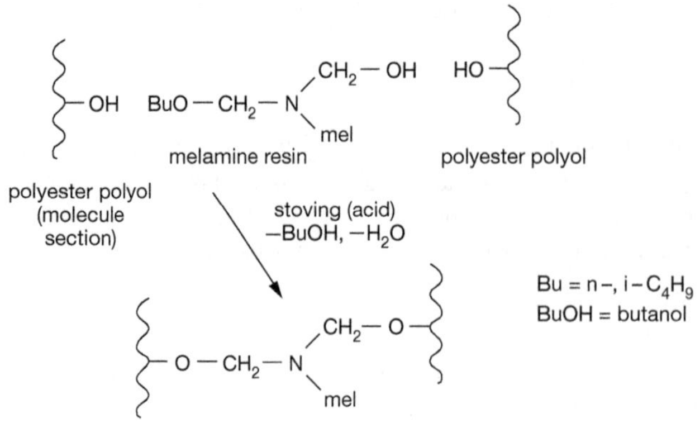

or again as a one-component system, this time with blocked polyisocyanate:

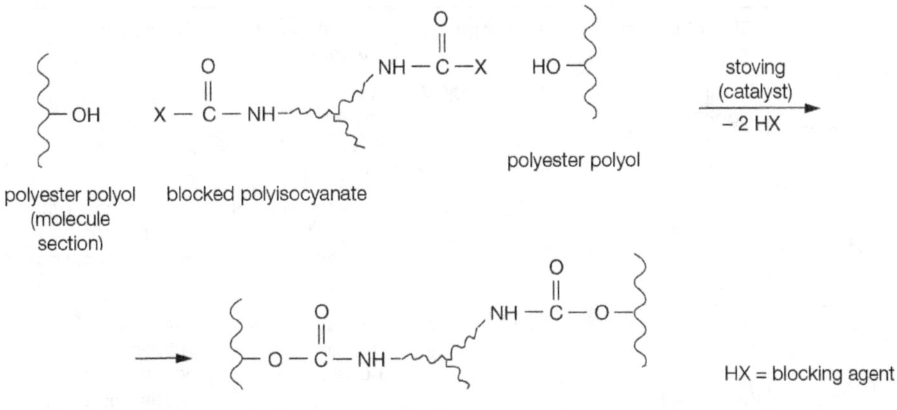

Acid polyester solid resins are used as film former components in powder coatings. They are cured either by stoving with conventional epoxy resin for indoor applications or with special glycidyl or epoxy compounds, and primarily with 2-hydroxyalkyl amides ("Primid") for outdoor applications. Triglycidyl isocyanurate (TGIC), labelled with "T" for toxic, was the standard crosslinking agent for exterior acid polyester powder coatings, but has largely disappeared from the market (especially in Germany).

Epoxy resin reacts with acid polyester as follows:

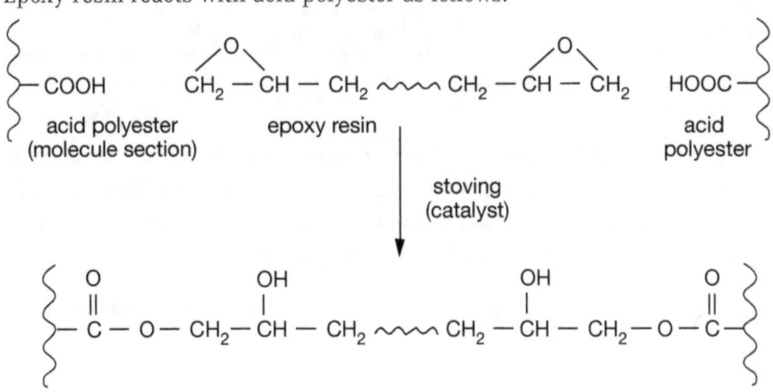

(The reaction between carboxyl and epoxy groups ("epoxy-carboxy reaction") is also used outside the area of coating powders for synthesis and curing processes).

Recent years have seen the advent of "super-durable polyesters" in powder coatings technology. These are based on isophthalic acid (1,3-benzenedicarboxylic acid) instead of phthalic acid anhydride and terephthalic acid (1,4-benzenedicarboxylic acid) as acid monomer. They are reputed to be up to four times as water-resistant as conventional polyesters, e.g. as regards gloss retention, and thus offer the same performance as pure acrylics (see section 2.1.4.5).

Table 2.5 (page 60) provides a summary of the molecular structure, crosslinking partners and areas of application of saturated polyesters. In conclusion we should mention that polyesters can also be chemically modified with non-polyester components such as diisocyanates, epoxy and silicone resins.

Water-thinnable[1)] polyesters

Water-thinnable polyesters are generally low-molecular, hydroxy-functional, branched polycarboxylates. They are normally available commercially either dissolved in water-miscible solvents

1) In this book all film formers and coating materials which can be dissolved or dispersed (emulsified, suspended) by the addition of water or which - as with polymer dispersions - are supplied as a dispersion in water due to the manufacturing process (see also ISO 4618) are described as "water-thinnable".

Table 2.5: Classification of saturated coating polyesters

Structure (schematic)	Classification	$\overline{M}_n$ range	Crosslinking partners	Applications
HO~~~~~~~~OH	linear high-molecular, hydroxy-functional	10,000 to 30,000	none, melamine/benzo-guanamine resins	coil/can coating, packaging, printing inks
HO~~~~~OH	linear, low-molecular, hydroxy-functional	1,500 to 6,000	melamine resins blocked polyisocyanates	coil/can coating, automotive/ industrial coatings
HO~~~<(OH)(OH)	branched, low-molecular hydroxy-functional	1,000 to 4,000	melamine resins free/blocked polyisocyanates	automotive/ industrial coatings, polyurethane coating powders
HO~~~<(COOH)(OH)	branched, low-molecular hydroxy and carboxy functional	1,000 to 4,000	HMMM resins, free/blocked polyisocynates	water-thinnable automotive/ industrial coatings, masonry paints
HOOC~~~<(COOH)(COOH)	branched, low-molecular, carboxy-functional (i.e. acid)	1,000 to 4,000	epoxy resins, other epoxy crosslinkers, 2-hydroxy-alkylamides	powder coatings

without neutralisation (e.g. to 80 % in butyl glycol) or *pre-neutralised* and dissolved or emulsified in a mixture of water and water-miscible solvent (*co-solvent*), (e.g. to 50 % in isobutanol/butyl glycol). Film forming is usually achieved by stoving with water-thinnable melamine resins or blocked polyisocyanates, but also as a two-component system with free polyisocyanate.

Polyesters can be modified with diisocyanates and other components to give aqueous polyurethane dispersions.

Many water-borne coatings are based on a mixture of water-soluble film formers (hydrosol), e.g. water-soluble polyester, and a polymer dispersion. These combinations are known as *hybrid systems*.

2.1.4.2 Unsaturated polyesters (UP)

General introduction to their chemistry

If diols are esterified with a mixture of maleic anhydride and other dicarboxylic acids or dicarboxylic anhydrides, an *unsaturated polyester* (UP) is obtained. (In special cases fumaric acid (trans-butenedioic acid) or other low, unsaturated dicarboxylic acids are used in place of maleic acid (cis-butenedioic acid). The cis double bonds of maleic acid incorporated primarily in the polyester chain are mostly converted to trans double bonds by thermal cis/trans isomerisation at the elevated condensation polymerisation temperatures of up to 200 °C.)

The double bonds contained in the polyester molecule chains are now capable of radical copolymerisation with monomeric (meth)acrylates, allyl compounds and other unsaturated monomers and in particular with styrene. During copolymerisation, the relatively short polyester chains M_n between 1000 and 3000 g/mol) are crosslinked by short bridges consisting on average of two styrene units, resulting in a densely crosslinked thermoset:

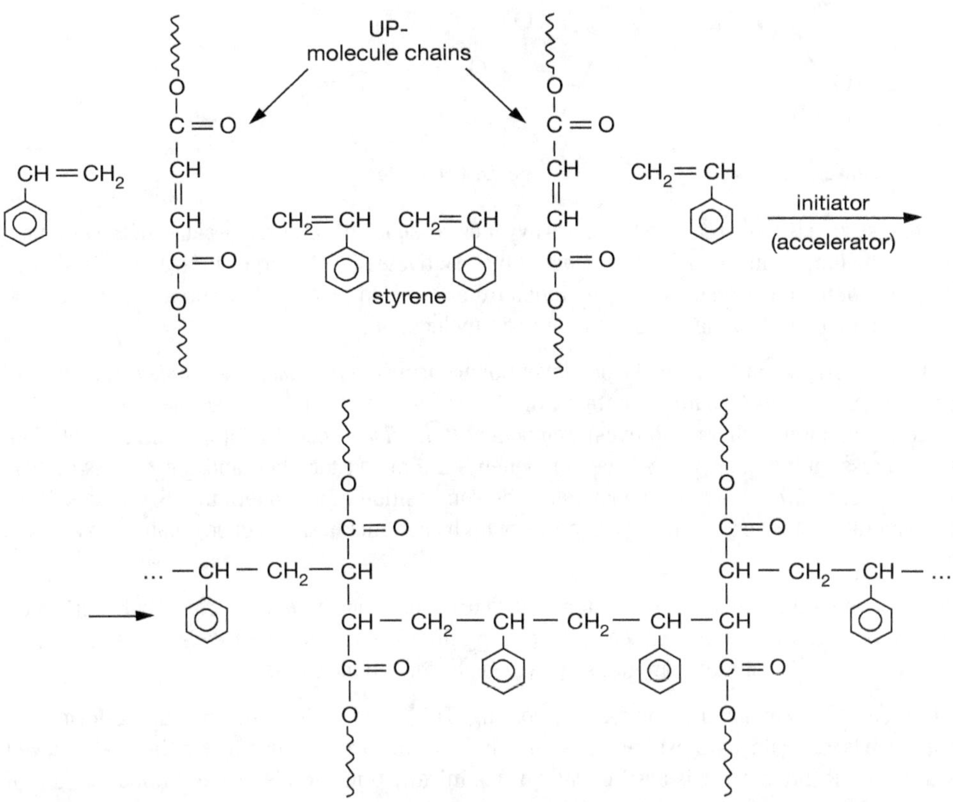

The styrene or other monomer is described as a "*reactive thinner*" here, since on the one hand it serves as a solvent and thinner whilst, on the other, it does not evaporate during processing, instead becoming part of the thermoset coating or moulding following the chemical curing reaction. UP systems are therefore classed as environmentally friendly products.

Virtually all commercial unsaturated polyesters include styrene - usually in a weight ratio of two parts resin to one part styrene. Also available however are resins dissolved in inert, i.e. non-reactive solvents (esters) and self-emulsifying 100 % resins suitable for water-thinnable systems. (Users of styrene-containing materials should be aware of the harmful effect on health of styrene vapour; MAC = 20 ppm[1)].)

Curing, aspects of relevance to coating technology

The copolymerisation of resin and reactive thinner, which in this instance constitutes the curing reaction, proceeds

- either conventionally by initiation with organic peroxide or hydroperoxide (methylethyl ketone peroxide, cyclohexanone peroxide, (di)benzoyl peroxide, etc.)
- or photochemically by initiation with a photoinitiator in ultraviolet light (UV).

The conventional curing of UP resins can proceed either as a *hot cure* or - following the addition of accelerators - at room temperature or even lower (*cold cure*). UP-based coatings are generally cold cured. Conventional combinations of initiator and accelerator include

- ketone (hydro)peroxide + cobalt octoate for UP coatings[2)]
- benzoyl peroxide + tertiary aromatic amine for highly filled systems, e.g. UP fillers.

1) MAC stands for "maximum allowable concentration"; it is usually expressed in ppm (parts per million) i.e. cm^3 of vapour or gas in 1 m^3 of air.

2) As with the oxidative drying of oils and alkyd resins, cobalt accelerates the cleavage of the peroxide bond by reduction.

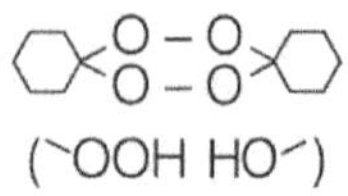

cyclohexanone(hydro)peroxide

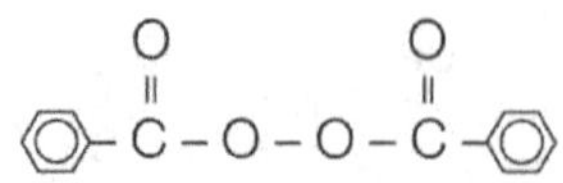

benzoyl peroxide

The second system is more reactive but starts to yellow as soon as it is cured because of the aromatic amine. On the other hand, cobalt octoate would be deactivated by the large surface area of filler in a filling compound. Filler resins can be obtained from the resin manufacturer pre-accelerated with amine. The amine binds preferentially to the resin molecules.

UP coating materials that are suitable for conventional curing are *two-component systems*: *component A* (*resin component, stock coating*) contains the UP resin and possibly the accelerator and other components (pigments/fillers, additives); *component B* (*hardener*) consists of the initiator solution. Chemical crosslinking begins when the components are mixed together, and the *processing time* (*residence time, potlife*) or *gel time* depends on the composition and temperature. Since no solvent has to evaporate from the film during curing, very high coat thicknesses can be obtained in a single operation.

Due to the action of atmospheric oxygen during the curing of a conventional UP resin, polymerisation is *inhibited* (suppressed) at the surface of the coating film. The result is that the surface of the film is tacky. This effect is countered in industry by two alternative measures:

1. A little paraffin is mixed into the resin or coating. This paraffin floats during curing, forming a protective layer against atmospheric oxygen. The matting layer of paraffin can then be removed by abrasion if the matting is aesthetically or technically undesirable. An additional benefit of the paraffin layer is that it reduces styrene evaporation.
2. Resins are formulated whose surface undergoes oxidative post-curing and thus (slowly) loses its tackiness. These resins are known as "*gloss polyesters*"; they generally contain allyl ether groups (CH_2=CH-CH_2-O-), which both copolymerise with the maleic acid double bonds and cause oxidative crosslinking (curing) with atmospheric oxygen at the film surface.

Radiation curing can be used as an alternative to or in combination with peroxide curing ("*double-cure process*"). The one-component, usually unpigmented coating material contains a few per cent of a *photoinitiator*, generally an a-cleaver, which decomposes into reactive radicals under the action of intensive UV radiation with wavelengths of 300 to 400 nm.

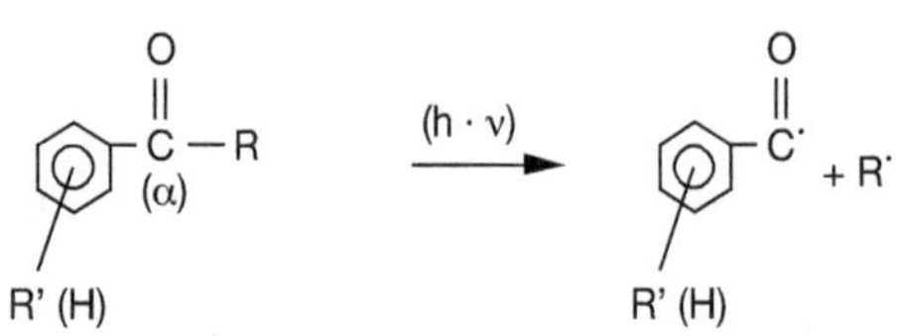

photoinitiator (here: α-cleaver) — active radicals

R : various radicals (depending on pr oduct type)
h · ν : quantum of UV radiation (photon)

These large numbers of abruptly forming radicals cause the film to cure completely within a matter of seconds. Using particular combinations of lights and photoinitiators, even pigmented coatings can be cured by UV radiation provided that the coating is not too thick.

The main area of application of UP coatings is in varnishes for wood and for furniture. The adhesion of UP coating systems on non-breathing substrates, particularly metal surfaces, is often inadequate - largely because of the stresses caused by "surface shrink". Fillers and stoppers adhere well to metal because the high filler content greatly reduces polymerisation-induced volume shrinkage. The coatings are characterised by hardness, scratch resistance, high gloss through to wet look (where sanded/polished or not matt), resistance to chemicals and high and low temperatures; they display only moderate weather resistance.

2.1.4.3 Radiation-curing acrylates

Double bonds of acrylic acid polymerise much more quickly in UP resins than do maleic or fumaric acid double bonds. *Radiation-curing acrylates* (*unsaturated acrylates*) can be produced by various relatively simple reactions of free acrylic acid or suitable derivatives of acrylic acid such as hydroxyethyl acrylate with reactive resinous parent substances to obtain *polyester, polyether, epoxy* and *urethane acrylates* → e.g.:

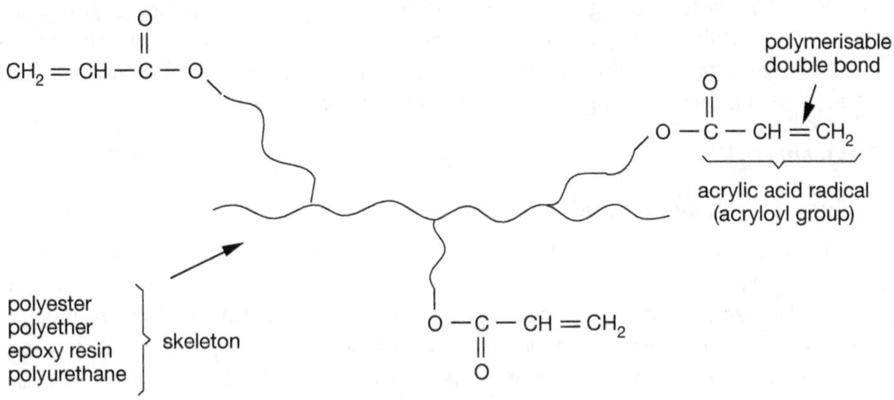

radiation-curable acrylate resin

Further, acrylated resin bases are encountered in specialty applications, e.g. melamine and silicone resins. (It should be stressed that these acrylate resins must not be confused with polyacrylates, i.e. solution acrylic resins and acrylate dispersions. The first group contain acrylic double bonds, whereas in the second group the acrylic double bonds are converted to unreactive single bonds during resin synthesis.)

Radiation-curing acrylates, like UP resins, are generally too viscous to be used without further treatment. They are therefore adjusted to working consistency by the resin or coating manufacturer using monomeric liquid acrylates, which act as reactive thinners (hexanediol diacrylate, ethylene glycol diacrylate, trimethylol propane triacrylate, etc.), and possibly also using an inert solvent such as methylisobutyl ketone. Since low-molecular acrylates are a skin irritant and may evaporate from the coating producing an odour if not completely incorporated into the resin skeleton by polymerisation, manufacturers seek to minimise the content of these substances or to leave them out altogether. The latter is possible in the case of aqueous resin emulsions or dispersions for example, although these do require a more complex processing technology (evaporation of water, difficulty of grinding due to the presence of fibres which stand up as the wood is painted).

In principle the UV radiation of acrylates proceeds in the same way as the curing of UP resins, albeit significantly more quickly. This means that belt speeds (speeds of transit through UV zones) of up to around 50 m/min can be achieved.

Where the numbers of items to be coated are high and/or where thick, pigmented coats are involved, *electron beam curing* (EBC) may be both technically superior to and more cost-effective than UV radiation. Here polymerisation (curing) is achieved in fractions of a second by an intense beam of electrons without the presence of a photoinitiator. Due to the high kinetic energy of the electrons (e.g. 150 keV (*kilo-electronvolts*)), the thickness of the coating and the presence of pigmentation have no effect on curing. The curing plant is extremely expensive, however, and thick lead casings and other radiation protection measures are needed because of the X-ray radiation generated during curing.

As with UP coatings the main application area of the radiation curable acrylates is wood, wood basic material and coating of furniture. In the meanwhile, the use of light stabilisers (see section 1.4.4) in conjunction with polyurethane acrylates and an optimised combination of lamp and photoinitiator has made it possible to increase the weather resistance of radiation-curable coatings to the extent that they are now being employed as clearcoats in automotive refinishing and OEM finishing.

UV-film former molecules which possess a sufficient number of other functional groups (usually hydroxyl) in addition to their acrylic double bonds are amenable to dual cure, i.e. conventional curing and radiation curing. Dual cure offers the advantage of ensuring that the coating cures sufficiently at those points on the article where little or no UV radiation strikes. Water-thinnable binders and paints are available for this technology, too.

A further new technology is UV curing of powder coatings. A powder coating based on a solid binder containing, e.g., acrylic groups, and applied to the article is heated by infrared radiation to 110 to 120 °C, whereupon it melts, flows and then is immediately crosslinked by UV radiation. This approach separates the melting/levelling process from the curing process.

2.1.4.4 Alkyd resins (AK)

General introduction – chemistry and properties

Alkyd resins can be defined in brief as polyesters modified with fatty acids or fatty oils or with higher synthetic carboxylic acids. The molecules consist of a polyester backbone, which may be scarcely to moderately branched depending on the raw material selected, from which fatty acid groups project as side chains; excess (free) hydroxyl and residual carboxyl groups are also present. The average molecular weight (number average) is normally between around 2000 and 5000 g/mol; in high solid grades it is even lower.

The structure of an alkyd resin consisting solely of oil, additional glycerol and ortho-phthalic acid is represented in simplified form and by way of example by the following formula:

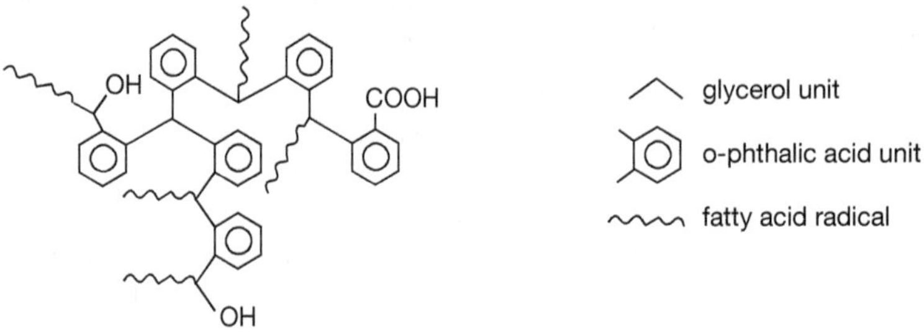

The term "**alkyd**" is derived from the combination of "**al**cohol" and "a**cid**".

From the start of their industrial production in 1930, alkyd resins rapidly developed into the most important type of synthetic resin for coating chemistry. Even today they still account for over 40 % of the world's production of synthetic coating resins (excluding polymer dispersions).

The huge success of alkyd resins can be attributed – in short – to an ideal combination of polyester and oil properties. The polyester component is responsible for physical (surface) drying and weather resistance (gloss retention, freedom from yellowing, etc.), the oil component for the suppleness of the films (internal plasticisation) and above all for the capability of oxidative crosslinking. The strengths of alkyd resins, such as

- self-curing at room temperature as a one-component system
- very broad compatibility and solubility spectrum
- virtually unlimited variability of properties by appropriate choice of raw material and synthesis conditions

- good pigment wetting
- attractive flow properties, leading to good spreadability of paints, for example
- relatively low cost

are nevertheless countered by more or less pronounced weaknesses, depending on composition, such as

- poor chemical resistance (particularly to alkalis)
- gradual embrittlement and yellowing (even away from light)
- relatively rapid loss of gloss (with chalking)
- relatively slow drying (especially where there is a high proportion of oil or fatty acid).

The main raw materials used in the synthesis of alkyd resins are

- fatty oil (for "*oil alkyds*")
  or a mixture of fatty oils (for "*mixed oil alkyds*")
  or free fatty acid(s) (for "*fatty acid alkyds*")

or synthetic monocarboxylic acid(s) (e.g. isooctane acid, "Versatic" acids[1])
and

- polyester raw materials (see summary on page 56).

In the case of modified alkyd resins (see page 64), special "non-polyester" components such as styrene, acrylate, diisocyanate, epoxy resin, silicone and many others are also used.

Coating chemistry generally classifies (unmodified) alkyd resins on the basis of their oil content or their fatty acid content calculated as triglyceride content, and of the type of oil or fatty acid ("linseed oil alkyd", "soya bean oil alkyd", "peanut fatty acid alkyd", etc.). Classification according to the *oil content* (triglyceride content) of the resin is based on the following nomenclature:

- less than 40 % oil: "*short oil alkyd (resin)*"
- 40 to 60 % oil: "*medium oil alkyd (resin)*"
- over 60 to 70 % oil: "*long oil alkyd (resin)*"
- over 70 to 85 % oil: "*very long oil alkyd resin*".

With regard to the oil type or the type of fatty acid incorporated, the first step is to distinguish between two categories: drying or non-drying. It is important to remember that a medium or long oil alkyd resin of a semidrying oil such as soya bean oil is completely drying, since the oxidative film forming of the oil component is accompanied by the physical drying of the polyester component, and there are generally more fatty acid groups in a resin molecule than in an oil molecule.

Isophthalic acid alkyds are of particular importance as a special grade of unmodified alkyd resins (see below).

Coating applications

Long oil alkyds always dry by oxidation. The high oil content promotes good flow, high flexibility and easy manual processing, but also leads to relatively slow drying. If conjugate oil(s) or acid(s) are additionally used in the resin synthesis, rather faster drying resins are produced. Long oil alkyds, most of them based on soya bean oil and oils of a related composition, are principally used as sole film former for decorator's paints, DIY paints and house paints. Linseed oil alkyds which because of their high content of linoleic acid are more prone to yellowing, are particularly suitable for anti-corrosive coatings and printing inks.

Medium oil alkyd resins may cure by oxidation or may equally be externally or non-crosslinking. Their universal compatibility is frequently utilised in combinations with a variety of other film formers, e.g. with cellulose nitrate to accelerate drying, with hard resins to increase hardness, gloss and

1) Tradename for synthetic monocarboxylic acids with three alkyl radicals at the C atom adjacent to the carboxyl group (α position), i.e. they are densely crosslinked.

fullness, or with chlorine-containing polymers to improve water and chemical resistance. Melamine resins, which react with the free hydroxyl groups in the alkyd resins, are generally used as curing agent resins in stoving coatings. Medium oil alkyds are principally used in industrial coatings (for machines, commercial vehicles, domestic appliances and many others), primers and undercoats.

Much of what has been said in regard to medium oil alkyds also applies to short oil alkyds, except that their fatty acid content is too low to allow them to crosslink independently by oxidation. Curing is generally achieved as a stoving cure with melamine resin. Alternatively crosslinking can be performed in a two-component system with urea resin and catalytic acid or for more heavy-duty coatings with polyisocyanates at room temperature or in forced conditions (up to around 80 °C). Alkyd resins based on non-drying oils such as coconut or peanut oil or non-drying fatty acids can be added to improve yellowing resistance, provided that no oxidative post-curing or supplementary curing is required.

In the (predominantly long or very long oil) *isophthalic acid alkyd resins*, the usual ortho-phthalic acid is partially or completely replaced by isophthalic acid (benzene-1,3-dicarboxylic acid). This results in increased mechanical strength and improved water, chemical and weather resistance.

"*Cardura® resins*" obtained by the incorporation of "Versatic" acids (see footnote on page 64) in the form of glycidyl esters (glycidyl "versatates", tradename "Cardura", see Figure 2.20 for formula), are particularly high-quality stoving alkyd resins.

Modified alkyd resins

The already wide variety of unmodified alkyd resins discussed above can be expanded even further by the chemical incorporation of "non-alkyd resin" components. A thorough analysis of all of the major modifications would exceed the scope of this textbook, so we will simply list the commonest modifications, together with the properties that are most improved:

- Resin-modified alkyd resins
  (Modification with colophony or "albertolic acid")
  → Surface drying, bond strength
- Styrenated alkyd resins
  → Surface drying, water resistance, chemical resistance
- Acrylated alkyd resins
  → Surface drying, weather resistance
- Epoxy (resin) modified alkyd resins
  → Bond strength, chemical resistance
- Urethanated alkyd resins (urethane alkyds)
  → Surface drying, water, chemical and weather resistance, abrasion resistance
- Silicone-modified alkyd resins
  → Heat and weather resistance
- Polyamide-modified (thixotropic) alkyd resins
  → Stability of the wet film
- Aluminium-reinforced alkyd resins
  → Full curing

We should remember that whilst a modification will improve the properties mentioned above, it may well cause other properties to deteriorate. In the case of styrenation, for example, accelerated drying is achieved at the expense of a reduction in crosslink density and hence in the solvent resistance of the cured coating film.

Water-thinnable alkyd resins

Short and medium oil carboxylic acid resins were first offered as true solutions with a high amine content. More prevalent nowadays, though, are colloidal secondary dispersions which have a lower amine content and contain alkyd resins that have been modified to different extents. Advantages over the strongly ionic, true solutions include better hydrolytic (storage) stability, a more favourable rheological behaviour and lower levels of amine emissions. Delivery forms are essentially the same as for water-thinnable, saturated polyesters (see section 2.1.4.1).

Conventional alkyd resins can be stabilised in water by means of external emulsifiers, acting either alone or in a supporting capacity. Such alkyd resin emulsions are used primarily to protect wood, but are also employed in industrial and house paints.

2.1.4.5 Acrylic resins (AC, AY)

General introduction – chemistry and properties

Acrylic resins (polyacrylates, acrylate resins, polyacrylate resins) are generally polymer solutions or solid resins obtained by radical *solution polymerisation* but also by *bulk* or *bead polymerisation* of acrylic monomers (see Table 2.6) for more detailed explanation).

The acrylic monomers are both simple, not especially functionalised esters of acrylic or methacrylic acid and derivatives of (meth)acrylic acid with specific reactivity. A significant expansion of the product range for acrylic resins can be achieved by copolymerisation with non-acrylic monomers such as styrene or maleic anhydride. Table 2.6 provides an overview of some of the main acrylic monomers.

"*Pure acrylates*", consisting solely of acrylic monomers, are characterised by good chemical and photochemical resistance, from which their exceptional weather resistance is derived. Important physical application-related properties are determined by the average molecular weight, the molecular weight distribution and the glass transition temperature (T_g).

The average molecular weight and the molecular weight distribution can be determined from the reaction course during polymerisation. Lack of space prevents us from discussing this in more detail. The number average of the molecular weight can range from around 1000 to 3000 g/mol in the case of high solid resins to several tens of thousands of g/mol in the case of the generally higher

Table 2.6: Monomers for acrylic resins (selection) and glass transition temperatures of homopolymers

Monomer (R)	**T_g [K] of**	
	acrylate $CH_2=CH-C(=O)-O-R$	**methacrylate** $CH_2=C(CH_3)-C(=O)-O-R$
non-functional:		
methyl	282	378
ethyl	249	338
n-butyl	219	295
2-ethylhexyl	223	263
dodecyl (lauryl)	n. k.	208
functional:		
2-hydroxyethyl	258	328
H (acrylic/methacrylic acid)	379	501
glycidyl	n. k.	319
non-acrylic:		
styrene	373	
α-methyl styrene	441	

n. k.: not known

molecular solid resins. A narrow molecular weight distribution is normally desirable in the interests of obtaining the lowest possible solution viscosity (and hence a higher solids content).

Provided that the molecular weights are not too low, the glass transition temperatures for the copolymers can be roughly calculated using *Fox's* formula (→ 2.1.1.12) from the glass transition temperatures for the homopolymers of the relevant monomers; in the case of physically drying (thermoplastic) acrylic resins they are at approximately room temperature or slightly higher.

In general the following is true:

- The higher the T_g, the harder (but also the less flexible) is the film.
- The glass transition temperature rises as a result of chemical crosslinking.
- The glass transition temperature of polymethacrylates is higher than that of polyacrylates.

Acrylates are very frequently copolymerised with styrene to produce "*styrene-acrylates*". Styrene is considerably cheaper than the acrylate monomers and promotes greater hardness, hydrophobia (water resistance) and alkali resistance; on the other hand it leads to increased susceptibility to yellowing and chalking and to greater embrittlement.

Non-crosslinking ("thermoplastic") *acrylic resins* - often supplied in flakes or beads - are characterised by elevated molecular weights as a means of achieving adequate film strength. They have (virtually) no functional groups; in the interests of improved adhesion, however, a certain content of carboxyl groups is often provided (by the incorporation of free (meth)acrylic acid).

Externally crosslinking acrylic resins are usually more or less highly hydroxyfunctional ("polyacrylic-polyols"); in addition, a certain quantity of acid is incorporated by polymerisation to catalyse the crosslinking reaction, to improve adhesion and possibly to obtain water solubility.

Self-crosslinking acrylic resins contain, in addition to hydroxyl and carboxyl groups, N,N-bis-butoxy-methylamide groups which react with them:

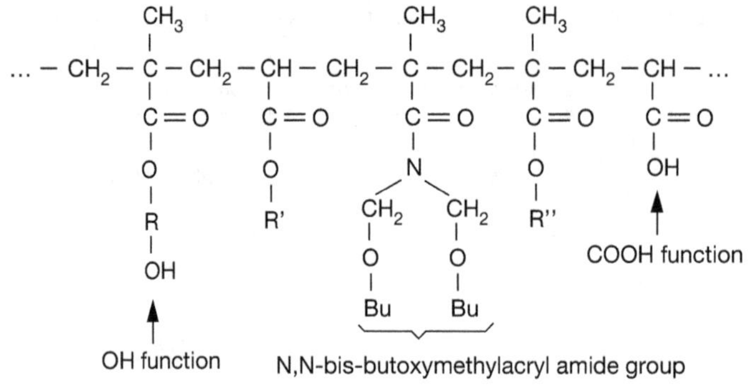

self-crosslinking acrylic resin (schematic)

The course of a simple solution polymerisation reaction can be described as follows:

1. Place an inert organic solvent in a stirred tank reactor and heat to reaction temperature.
2. Allow the monomer blend to feed in in parallel with the initiator solution (peroxide) over a number of hours. (Polymerisation proceeds with development of heat.)
3. Following inflow, keep at reaction temperature for several hours (re-initiate if necessary).
4. Finally adjust to final solvent form if necessary.

The reaction mixture remains homogeneous (clear), i.e. single-phase, throughout the entire synthesis.

The polymerisation of monomers in undiluted state is known as *bulk polymerisation.* A modern variant of this process is the continuous bulk polymerisation of acrylic monomers at high temperature without the addition of initiators.

In *bead polymerisation*, drops of coarsely emulsified monomer/initiator mixture are polymerised directly into resin beads in an aqueous solution containing emulsifiers and other additives. The beads are washed and dried.

(The course of the emulsion polymerisation reaction is covered in the grey box in section 2.1.4.6).

Coating applications

Non-crosslinking acrylic resins are sold commercially as 40 to 60 % viscous solutions or as solid bulk or bead polymers. They can be found as sole or additional film formers in physically drying coatings for a range of applications on virtually any substrate. Typical properties include: water resistance, resistance to alkalis, flexibility, good adhesion and - in the absence of styrene - excellent resistance to yellowing and gloss retention in exterior applications. Other areas of application for non-crosslinking acrylic resins include flexographic and intaglio printing inks as well as pressure-sensitive adhesives.

Externally crosslinking acrylic resins are used either in the form of one-component stoving coatings or as air drying or forced drying two-component systems. For environmental reasons there is a general trend towards medium solid or high solid systems.

Externally crosslinking stoving coatings are usually based on a combination of polyacrylic-polyol/ melamine resin or blocked polyisocyanate. Other crosslinking chemisms are also used, especially the epoxy-carboxy reaction needed for curing acrylic powder coatings. These generally very strong and durable coatings are used in industrial coating systems (including automotive primers) and coil and can coatings.

Coatings obtained from the acrylic resins which self-crosslink via N-butoxymethyl and hydroxyl groups have very good chemical resistance and find application both as stoving primers and topcoats for white goods (washing machines, refrigerators, dishwashers and the like). However, they have ceded a great deal of ground to powder coatings on account of their high solvent content.

Two-component systems cure with free polyisocyanates. Aliphatic polyisocyanates should be used to meet demands for maximum yellowing and weather resistance. Two-component systems are used wherever outstanding coating properties are required and stoving is not possible (vehicle repair, heavy vehicles, aircraft, high-quality kitchen furniture, etc.).

Water-thinnable acrylic resins

When acrylic copolymers have an adequate content of free carboxyl groups e.g. of free (meth)acrylic acid incorporated by polymerisation, they are water-soluble after neutralisation with ammonia or amine. The solvent arising from solution polymerisation is either additionally incorporated into the "water-borne coating" as co-solvent or is removed by distillation after polymerisation. The resistance to hydrolysis of the acrylic resins means that there are no real problems with the storage stability of the aqueous products.

Aqueous one-component acrylic resin coatings, like the conventional products, are generally stoved with water-thinnable curing agent resins (melamine resins of the HMMM type, blocked polyisocyanates, etc.). The carboxyl groups act to accelerate the reaction. However, two-component water-borne coatings with free polyisocyanate incorporated by emulsification have also become state of the art now. As a consequence of the anionic polyelectrolyte structure of water-soluble acrylic resins, they can also be used as film formers in anodically precipitable electrodeposition coatings, although these are admittedly of only limited importance nowadays.

2.1.4.6 Polymer dispersions (emulsion polymer)

General introduction – structure, chemistry and properties

Polymer dispersions, often simply known as "dispersions", are always emulsion polymers. (See below for a more detailed discussion of the principle of emulsion polymerisation).

A polymer dispersion is a translucent (semi-transparent), bluish or pure-white liquid that usually has a relatively low viscosity. The average particle size increases in the order given above, from 0.03 to over 1 µm. In microscopic terms it is a micro-heterogeneous system essentially consisting of roughly spherical polymer particles ("latex particles") which are evenly distributed in a usually aqueous[1] phase. (In the language of colloid chemistry these particles are described as the disperse or internal phase, the liquid as the cohesive or external phase or the dispersant. The term "dispersing agent" should be avoided because of the danger of confusion with the additive used to promote pigment dispersion, which is frequently referred to by the same term.)

In dispersions the polymer chemical characteristics of average molecular weight (generally above 100,000 g/mol) and molecular weight distribution are joined by *average particle size* (usually between 0.05 and 0.8 µm) and *particle size distribution*. There is no direct correlation between molecular weight and particle size. Since the molecules are fully contained within the latex particles, the molecular weight of a dispersion has no direct influence on the viscosity of the system.

As a minimum, the aqueous phase contains *stabilisers* such as surfactants[2] (emulsifiers) and/or protective colloids together with decomposition products of the polymerisation initiator. It may also contain buffer substances (to regulate the pH), defoaming agents and preservatives to guard against microbial attack.

In the case of dispersions used in the paints and coatings sector, the latex particles are always copolymers. The main types are:

- *Styrene-butadiene* copolymers
- *Vinyl acetate* copolymers (e.g. with vinyl propanoate, *"Vinyl versatate" ("VeoVa")*, ethylene)
- *Pure acrylate* copolymers ("pure acrylates")
- *Styrene-acrylate* copolymers ("styrene acrylates").

We will discuss the characteristic properties of these basic types in more detail below.

The applicational properties of dispersions are characterised by a series of different values, such as non-volatile matter (solids content), viscosity/flow properties, average particle size, pH/stability, freeze-thaw, electrolyte (salt) and shear resistance as well as residual monomer content. There are also a number of properties relating to the dried dispersion films.

Developments in recent years aimed at controlling the internal structure. The principal objective behind current developments is to control the internal structure, morphology and size distribution of the latex particles and thus to obtain improved or even completely new application-oriented dispersion properties (keywords: core-shell dispersions, microgels, bimodal dispersions, etc.). It was hoped that core-shell technology in which a soft shell is polymerised onto a hard core particle would close the gap between low MFT[3] on one hand and high blocking and soiling resistance on the other. This goal was achieved, albeit not always wholly satisfactorily. Improvements were also made with regard to the other properties mentioned earlier.

---

1) *NADs* (non aqueous dispersions) contain highly volatile, aliphatic hydrocarbons as dispersant.
2) Low-molecular substances with hydrophilic and hydrophobic molecule sections, which accumulate at the interfaces between aqueous and non-aqueous phases and thus stabilise them.
3) Minimum film-forming temperature (→ 3.10.2.2)

The principal objective behind current developments is to control the internal structure, morphology and size distribution of the latex particles and thus to obtain improved or even completely new application-oriented dispersion properties (keywords: core-shell dispersions, microgels, bimodal dispersions, etc.).

Polymer dispersions are formed by a reaction known as *emulsion polymerisation*, which can be described in simplified form as shown below:

The only slightly water-soluble monomer blend is added to an aqueous solution of emulsifier (surfactant) and/or protective colloid (polyvinyl alcohol, water-soluble cellulose derivatives, *inter alia*), water-soluble initiator (usually peroxide disulphate) and optionally other additives. The monomer blend disperses in the water phase, mostly undissolved as "monomer droplets" but a small part completely dissolved. When the blend is heated to reaction temperature (usually 70 to 90 °C), the initiator forms radicals which start polymerisation with the dissolved monomer molecules. Further chain growth takes place in the emulsifier micelles[1)], with monomer molecules continually moving from the monomer droplets through the water phase and new radicals moving from the water phase into the micelles. The original micelles (primary particles) grow into the latex particles of the dispersion until the monomer droplets are used up.

Emulsion polymerisation can be performed using a number of different procedures to obtain a range of product properties from the same monomer composition. Generally speaking, emulsion polymerisation is a complex synthesis procedure involving expensive plant and elaborate measuring, control and regulating techniques.

Film-forming and how to influence it

Polymer dispersions for use in water-borne coatings can often be cured at elevated temperature and/ or as two-pack formulations; UV-curable and oxidatively curing dispersions are also available. Most dispersions, however, undergo purely physical film formation by *coalescence*, which is described in section 3.10. Where necessary, emulsion paints can be crosslinked to a degree (this is called interparticulate crosslinking) by adding crosslinking agents, such as *polycarbodiimides, polyaziridines, epoxysilanes* and metal complexes. The outcome is a marked reduction in dirt pick-up by, e.g., facades, and greater film strength.

Polymer dispersions usually undergo rapid physical setting - this either can be intended or may lead to problems with production and application in the form of encrustation or stringing.

Areas of application and properties of the main types of dispersions

Styrene-butadiene (SB) dispersions, generally containing 35 to 40 % butadiene and chemically modified to include a small quantity of free carboxyl groups, display outstanding alkali resistance and water repellency. The isolated double bonds of butadiene units remaining in the polymer create a capability for oxidative post-curing, which can be accelerated using driers. Typical applications for SB dispersions in the coatings sector include primers, anti-corrosion paints and coatings to protect against stone chipping. Their strong tendency to become chalky and yellow when exposed to light (due to the presence of reactive double bonds and styrene) means that SB dispersions are unsuitable for use in colourfast exterior top coats.

Vinyl acetate homopolymer dispersions[2)] are no longer used in the paints sector because of their relatively high brittleness and low resistance to hydrolysis. Copolymers with vinyl esters of higher

1) Micelles are loose associations (congregations) of many surfactant molecules. They take the form of submicroscopic hollow bodies with a hydrophilic exterior (facing the aqueous phase) and a hydrophobic interior.

2) Vinyl acetate has the chemical formula $CH_3-C(=O)-O-CH=CH_2$

carboxylic acids, acrylates, maleinates or ethylene display superior elasticity, weather and water resistance and are widely found in emulsion paints for a variety of substrates and in synthetic resin plasters.

Pure acrylate dispersions, which tend to be amongst the most expensive, also display the best all-round performance in the dispersions category. Their outstanding gloss retention, yellowing resistance, flexibility and good saponification resistance mean that they can be used across the entire range of water-thinnable coating materials - from glossy automotive paints to high-grade masonry paints. Chemical crosslinking capability can be achieved by the incorporation of reactive groups (carboxyl, hydroxyl, amide, glycidyl, etc.).

Styrene-acrylate dispersions exhibit even greater hydrolysis and alkali resistance than do pure acrylate polymers, although their lightfastness and flexibility are inferior. Other than in topcoat applications, styrene-acrylate copolymers often represent a more cost-effective alternative to pure acrylate copolymers.

Finally we should add that blends of dissolved film-formers and dispersions, usually primary dispersions, are often used for water-borne coatings. The resin solution contributes gloss, pigment wetting and, if desired, substrate penetration while the dispersion essentially accelerates physical setting and lowers the undesirable amine content in the coating. Another modern blend of film-forming agents consists of a primary, acrylic dispersion and a secondary, polyurethane dispersion (see section 2.1.4.9).

2.1.4.7 Formaldehyde condensates

Formaldehyde (and to a lesser degree also other aldehydes) has a tendency to bond its electron-deficient (electrophilic) carbon atom to electron-rich (nucleophilic) atoms. The main products of this process are reactive *methylol groups* or monomers bearing methylol groups ("*methylols*"), which can then be reacted further by means of condensation reactions to produce polymers and optionally also three-dimensional crosslinked structures.

Depending on the parent substance with which the formaldehyde is reacted, we speak of *phenolic, melamine, benzoguanamine, urea* and *carbamic ester resins*. The last four types are collectively referred to as "*amino resins*" (also "amine resins", "amido resins" or "aminoplastics"), although chemically speaking this is not strictly accurate.

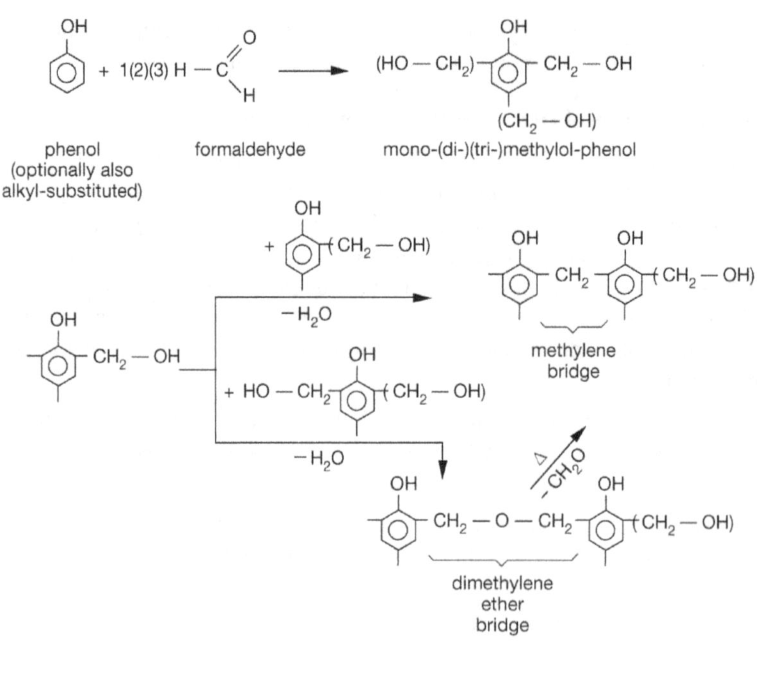

Phenolic resins (phenol-formaldehyde resin, PF)

Phenol and alkyl-substituted phenols (monomethylphenols (cresols), dimethylphenols (xylenols), p-tert.-butylphenol, etc.) form methylol phenols with formaldehyde; depending on the temperature and pH of the reaction solution these condense more or less rapidly, forming methylene and/or dimethylene ether bridges, to produce the oligomeric substances, phenolic resins (phenolic/formaldehyde resins, PF).

A molar excess of phenol in respect of formaldehyde and an acid reaction environment produces *"novolaks"* consisting of linear molecules having no methylol groups.

The use of novolaks (non-curing thermoplastics) in the coatings sector is restricted to polishes, insulating coatings, barrier primers and flexographic printing inks. Crosslinkable *"epoxy novolaks"* containing glycidyl groups are formed by reaction with epichlorohydrin.

An excess of formaldehyde in an alkaline reaction mixture results in branched, reactive *"resols"* containing still unreacted methylol groups.

Simple ("unplasticised") resols are reactive (self-crosslinking and externally crosslinking), more or less water-soluble, scarcely soluble in coating solvents and very brittle in the self-crosslinked state known as *"resite"*. They can be modified in various ways to make them more suitable for use in coating applications; the most important modifications are listed below:

- Partial etherification of methylol groups with n- or isobutanol
  → improved solubility and compatibility, plasticisation
- Use of alkyl-substituted phenols (e.g. p-tert.-butylphenol)
  → improved solubility and compatibility, plasticisation, development of (tung) oil reactivity
- Boiling with plasticising partners (fatty oils, fatty acids, polybutadiene oils and many others)
- Combination with flexible film formers (epoxy resins, polyesters, alkyd resins, polyvinyl butyral, acrylic polymers), by boiling if necessary
- Reaction with colophony, followed by esterification
  → readily soluble and compatible hard resins

By far the most important use of resols in the coatings sector is in the crosslinking of hydroxy-functional film formers. This involves two parallel processes: external crosslinking between resol and hydroxy resin (polyol) and self-crosslinking of the resol:

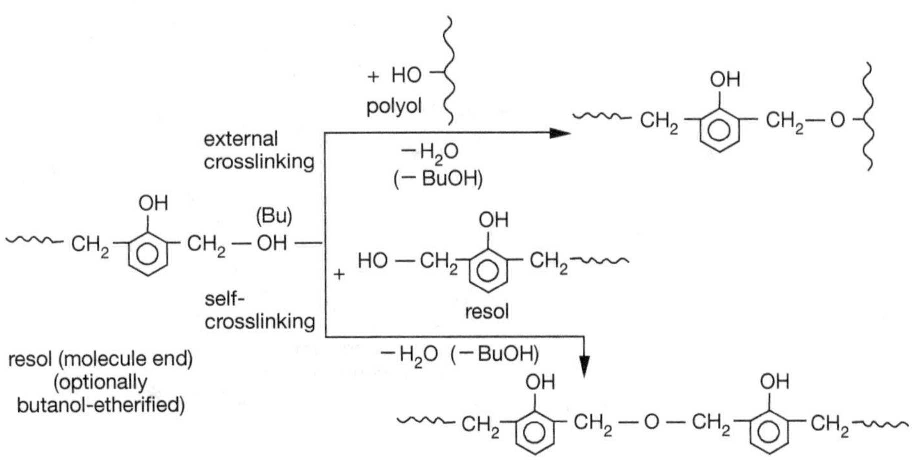

Crosslinking is performed either by baking at around 180 °C or in the presence of acid (hydrochloric, toluene-p-sulphonic or phosphoric acid) as a catalyst at reduced temperature down to room temperature. The main application of *acid curing* is in timber coating systems, where a two-component process is involved, the acid solution acting as hardener.

The strong inherent coloration of phenolic resins (yellow-brown) prevents them from being used in decorative paints. Their main application is in chemically resistant protective coatings and primers. Examples include:

- Blends with epoxy resins ("gold coatings") as flexible can linings
- Blends with polyvinyl butyral in "wash primers" (primer coats for metals)
- Blends with polyvinyl formal in electrical insulating or wire coatings
- Blends with flexible polyhydroxy resins (polyesters, short oil alkyds, acrylic resins, etc.) in chemically resistant primers (acid curing in the case of wood primers).

Plasticised resols, produced by boiling primarily with tung oil but also with other oils and free fatty acids, can also be used as sole binders. Oxidatively drying coatings with improved chemical, water and heat resistance can be formulated using long oil alkyd resins.

The glycerol or pentaerythritol esters of colophony-modified resols ("synthetic copals") are hard resins which improve the hardness, gloss and mechanical resistance of physically and oxidatively drying coatings. They are mainly used in printing inks, however.

Specially formulated water-soluble resols with free neutralisable carboxyl groups are available for use in water-borne coatings such as anodic electrodeposition coatings. Water-thinnable products can also be obtained by the pre-condensation of resols with other water-soluble film formers.

Melamine resins (melamine-formaldehyde resins, MF)

Melamine resin chemistry is technically much the same as phenolic resin chemistry. Melamine, a very high-melting, thermally stable solid that is scarcely soluble in water, can be reacted with formaldehyde - in the same way as phenol - via the methylol stage to form the soluble oligomers, melamine resins (melamine/formaldehyde resins, MF), finally being cured to produce thermosets.

Only those resins that have been etherified with low alcohols, usually butanols or methanol, are of relevance to coating chemistry. Product properties such as reactivity (when stoved and/or acid-cured), viscosity and solubility are determined by the average molecular weight, the degree of branching, methylolisation and etherification and by the type of etherification alcohol. An example of the synthesis of a melamine resin is illustrated by the following reaction scheme:

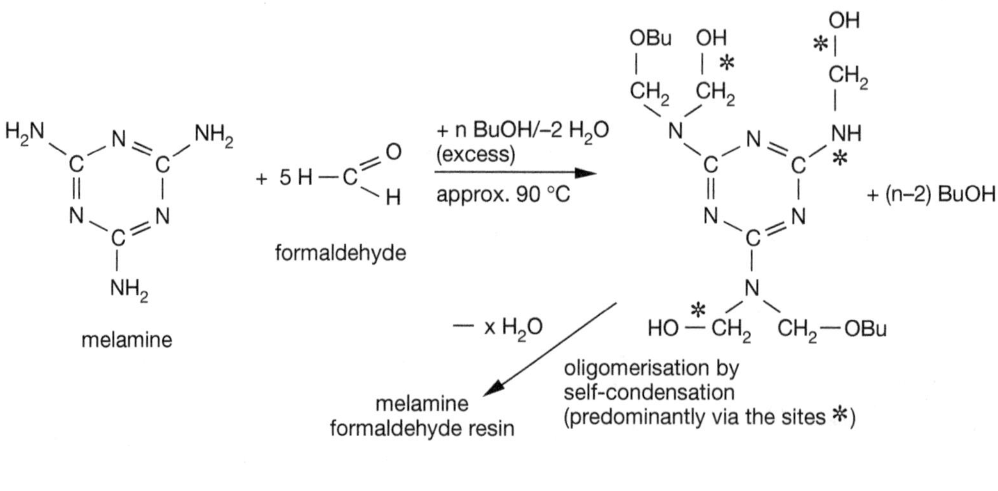

Butanol, which is present in considerable excess, functions as reactant, entrainer (to remove the reaction water) and solvent. The resins are generally sold as butanol solutions with solids contents of around 60 % (or higher). Unlike phenolic resins, melamine resins are virtually colourless and undergo almost no yellowing either during crosslinking or thereafter.

Melamine resins, like resols and urea resins, are capable of self-crosslinking, but the films produced would be too brittle. Curing is therefore conducted in combination with flexible polyhydroxy resins by means of stoving or (more rarely) acid curing. Here, self-crosslinking of the melamine resin is accompanied by external crosslinking with the excess polyol (polyester, alkyd, acrylic, epoxy resin); the reactions correspond in principle to those involved in resol curing (see reaction equations on page 72). In the case of highly reactive melamine resins, curing temperatures may be as low as 80 °C; more typical, however, are stoving conditions of for example 30 minutes at 140 °C. The presence of acid or acid groups in the resin accelerates the reaction or reduces the curing temperature.

High-grade stoved coatings crosslinked with melamine resin are characterised by hardness, good bond strength, high gloss, outstanding water and weather resistance and by resistance to oils, fuels and other chemicals. They are principally used in automotive top coats and one-coat paints and other higher-grade industrial products, especially domestic appliances, as well as in coil coating.

Acid curing with melamine resins is relatively uncommon and found only in wood coatings. The coatings do not display the excellent properties of the stoved systems and, in particular, have no outdoor weather resistance.

Resins etherified with methanol represent a special class of melamine resins. The lowest-molecular member of this class of resins is *hexamethoxymethyl melamine*, or *HMMM* for short:

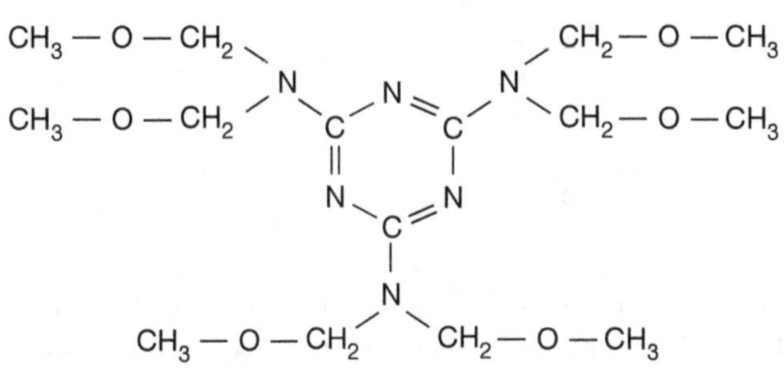

Industrial HMMM resins contain oligomers in addition to monomeric HMMM; they are liquid and are usually sold in practically 100 % form. Because of their solubility in water and their low average molecular weight, HMMM resins can be used as curing agent resins - both in water-thinnable and in high solid coatings. The high degree of etherification makes HMMM resins chemically inert, however. Stoving temperatures of over 180 °C are required if no catalyst is used. The addition of blocked acid catalysts can reduce the stoving temperature to 120 °C. (More reactive, methanol-etherified melamine resins (with free NH bonds) are currently under development.)

It is worth mentioning at this point that the use of catalysts and similar reaction-accelerating or reaction-initiating substances should be kept to a minimum, since these low-molecular and often chemically aggressive or at least reactive foreign components weaken the stability of the film. Greater yellowing during stoving, more rapid film degradation under weathering or reduced water resistance with loss of adhesion may result.

Benzoguanamine resins

In benzoguanamine one of the NH_2 groups in melamine is replaced by a phenyl group (benzene ring); unlike melamine it therefore contains only two NH_2 groups and its maximum functionality is four. Coating films cured with benzoguanamine resins are therefore more flexible and have better adhesion but are also somewhat less weather resistant than films cured with melamine resins. For this reason they are primarily used in primers and fillers.

Urea resins (urea-formaldehyde resins, UF)

The structural formula for urea is: 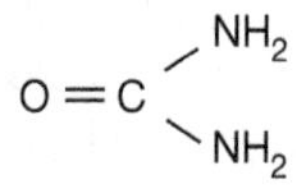

The synthesis of butanol-etherified urea resins (urea/formaldehyde resins, UF) proceeds in a similar way to the synthesis of melamine resins.

Since the films produced by the self-crosslinking of urea resins are too brittle, polyhydroxy resins are added to the urea resins as plasticising and reactive partners. Unlike melamine, urea is readily soluble in water and only tetra-functional. These two properties mean that coating films cured with urea resins are less water and weather resistant than those crosslinked with melamine resin.

The use of urea resins is limited to low-cost coating materials for interior applications, e.g. acid-curing wood varnishes. "*Plasticised urea resins*" produced by precondensation with polyesters can be used as sole binders.

Glycoluril resins

Glycoluril, a condensation product of glyoxal (OHC-CHO) and urea, can be etherified with methanol to yield *tetramethoxymethyl glycoluril (TMMGU)*, a solid:

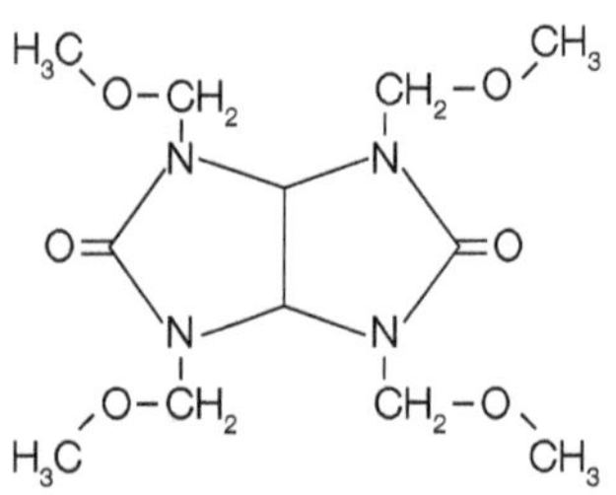

This serves as a crosslinker for hydroxy-functional polyester powder coatings. The blended, liquid methyl/ethyl and butyl ethers are used in liquid coatings (including water-borne types) as crosslinkers for producing strongly adhering, flexible metal coatings employed primarily in can and coil coating.

Carbamic ester resins

Esters of carbamic acid (urethanes) with unsubstituted NH_2 group:

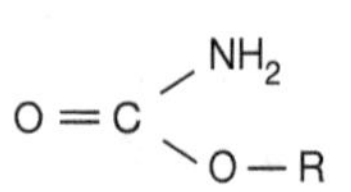

R = alkyl

can thus be condensation polymerised with formaldehyde to produce thermoplastic soft resins known as "carbamic ester resins" ("carbamide resins", "urethane resins").

They are used as unsaponifiable polymer plasticisers in cellulose nitrate, urea resin and other coatings.

2.1.4.8 Epoxy systems

The epoxy group

The central structural feature of epoxy film formers is the three-membered ring formed from two carbon atoms and one oxygen atom, which is known as the epoxy group, epoxide group or oxiran. The pronounced tendency of epoxy groups to react both with nucleophiles such as amines and with electrophiles such as protons (hydrogen ions from acid groups), with ring scission, is attributable to their bond angle tension and polarity:

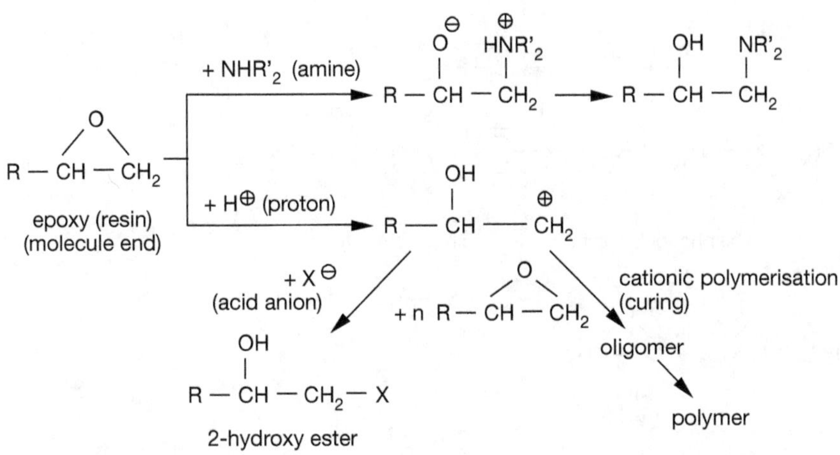

In the second case a multiple layering of epoxy groups, i.e. a cationic ring-scission polymerisation, may ensue, largely due to the ionic UV curing of epoxides (see page 81).

Epoxy resins (EP) and other epoxy compounds

Historically speaking, epoxy resins are one of the newest classes of resin; they only came onto the market in 1947. The broad range of possible applications for epoxy resins, as the basis for casting and moulding compounds, circuit boards, adhesives, coatings and much more, has led to a rapid growth in production; world production in 1995 was estimated at 850,000 t.

Epoxy resins are compounds having at least two epoxy groups per molecule. Some display a resinous solidity, whilst other very low-molecular examples are to a greater or lesser extent liquid. Certain highly pure "liquid resins" exhibit a tendency to crystallise.

By far the most important class of epoxy resins are the "bisphenol A-epichlorohydrin resins" ("bis-A resins" for short) condensed from "bisphenol A" [2,2-bis(4-hydroxyphenyl)propane] and epichlorohydrin (1-chloro-2,3-epoxypropane) in the presence of caustic soda. These consist primarily of molecules having the following structure:

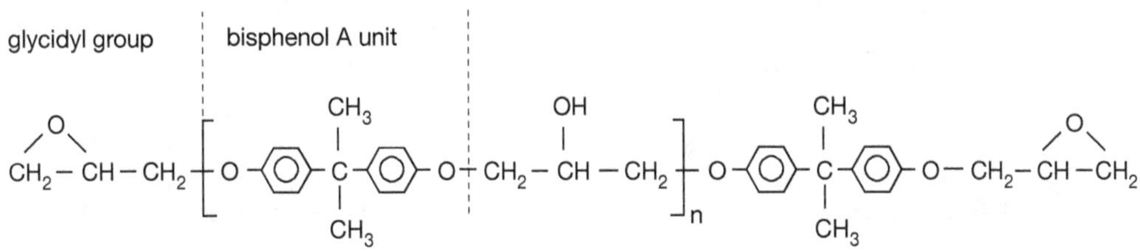

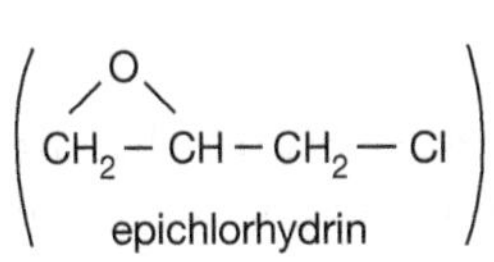

Depending on the size of n, these may also be referred to as "diepoxide 0", "diepoxide 1", etc. Since industrial epoxy resins are polydisperse, however, the mean values for n are fractional numbers. For example, the lowest-molecular industrial product has a value for n of approx. 0.1; the highest n is around 26, corresponding to an average molecular weight of about 8000 g/mol. The threshold between liquid and solid resins lies at around n = 1.1, corresponding to an average molecular weight of approx. 650 g/mol.

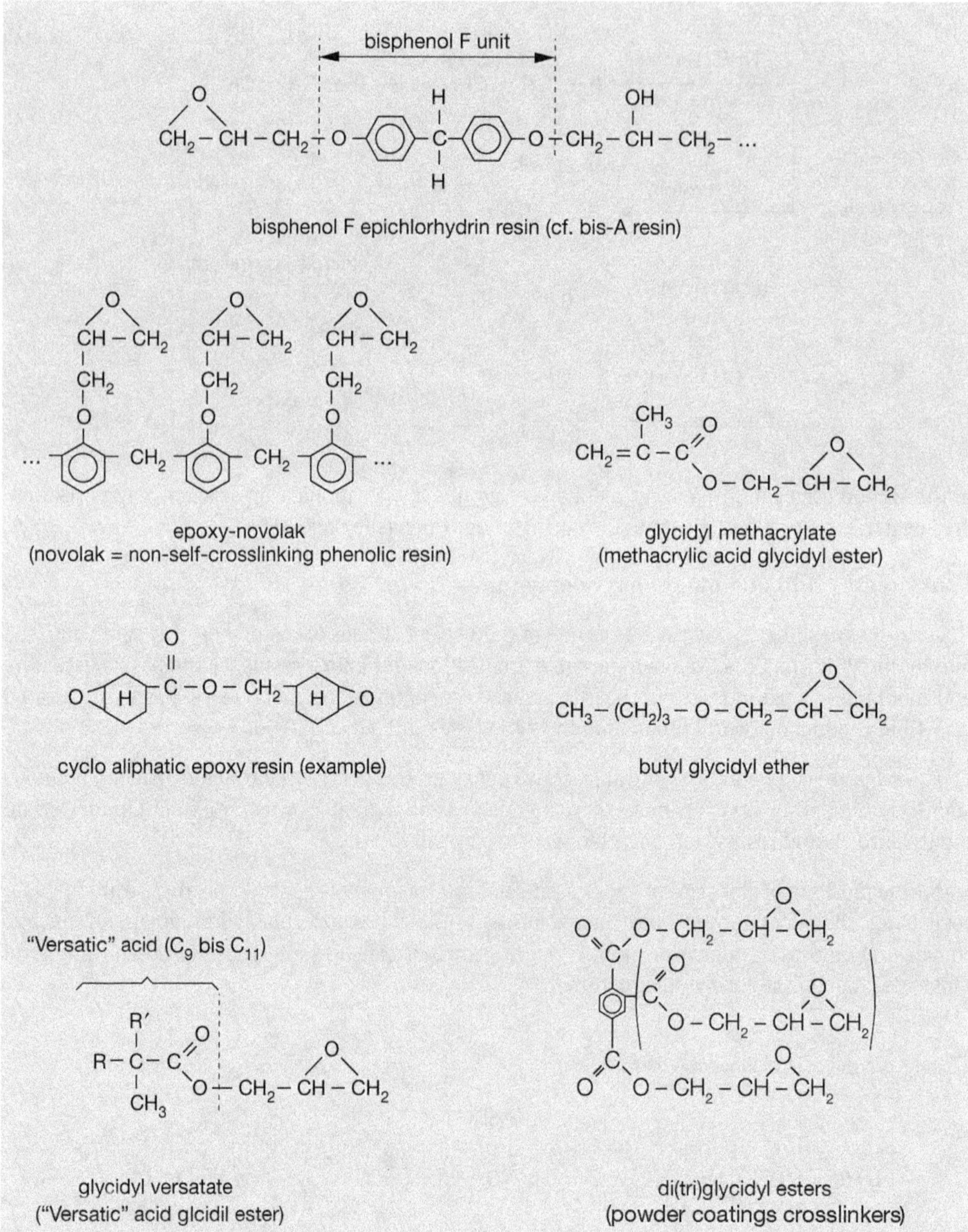

Figure 2.20: The main epoxy raw materials (selection)

Examples of epoxy resins and other epoxy compounds of relevance to coating chemistry which do not belong to the bisphenol A-epichlorohydrin class are listed below. Their formulae can be found in Figure 2.20.

- *Bisphenol F-epichlorohydrin* resins: liquid grades have lower viscosity and do not crystallise as readily as bisphenol A types.
- *Epoxy-novolaks*: produced from novolaks and epichlorohydrin; higher than bifunctional (→ films with higher crosslink density).
- *Cycloaliphatic* epoxy resins: outdoor weather resistance superior to that of aromatic-containing grades (bis-A-/F-) and undergo cationic polymerisation more readily (e.g. UV curing).

- *Triglycidyl isocyanurate* (TGIC)or epoxy substitutes[1]: crosslinkers for outdoor weather resistant powder coatings.
- *Glycidyl "versatate"*: used to incorporate the "Versatic" acid group in other synthetic resins.
- *Glycidyl methacrylate*: used to incorporate glycidyl or epoxy groups in polymerisation resins.
- Various liquid (low-molecular), largely monofunctional epoxy compounds such as butyl, isooctyl or o-cresol glycidyl ether are used as reactive thinners.

Technical epoxy raw materials fall into two chemical classes. Products bearing the glycidyl group (see formula above) are derived from epichlorohydrin. The other class is formed by epoxidising olefinic C-C double bonds with peroxy acids and are therefore also known as olefins or cyclo-olefins.

With a few exceptions, epoxy resins are sold undissolved as liquid, semisolid or solid substances. The liquid resins are mainly used to produce solvent-free coatings, adhesives and related products. The solid grades are either dissolved in organic solvents in the coating factory or are processed in their undissolved state directly to form powder coatings.

Water-thinnable epoxy resins are available as directly emulsifiable liquid resins or as aqueous solid-resin dispersions. Epoxy resins modified with amine are an important binder for electrodeposition coatings (see page 81).

Curing of epoxy resins

With a very few exceptions, epoxy resins are chemically cured. (The exceptions relate to instances where epoxy resins are used as additives, e.g. as stabilisers in PVC plastisols).

Depending on the reaction partner, curing originates either exclusively or predominantly from the epoxy groups or from the hydroxyl groups positioned along the molecule. In the first case ring scission is caused by the NH bonds of amines or the carboxyl groups of carboxylic acids, producing a new hydroxyl group which may itself undergo further reaction (see reaction scheme on page 74).

Longer-chain bisphenol A or bisphenol F-epichlorohydrin resins contain, in addition to the terminal epoxy groups, secondary hydroxyl groups located equidistantly along the molecule chain; these can be reacted with resols, melamine, urea or guanamine resins, polyisocyanates and polyanhydrides.

Amine-type hardeners

First of all let us take a closer look at curing with amine compounds. In principle an epoxy group reacts with an NH bond of an amine hardener, i.e. the number of NH bonds in an amine molecule is identical to its functionality. Crosslinked structures can only be produced, of course, if the hardener is more than bifunctional. The stated stoichiometry of amine curing must be taken into consideration in the processing of two-component coatings. The "epoxy equivalent weight" is the central characteristic here:

> The epoxy equivalent weight (EEW) is the weight of an epoxy resin containing exactly one mole of epoxy groups. (Unit: g/mol).

Another measure for epoxy content is the "*epoxy value*". This indicates how many moles of epoxy groups are contained in 100 g of epoxy resin.

The equivalent weight of the hardener can be calculated as the quotient of the molecular weight of the amine and the number of NH bonds per molecule[2].

1) TGIC has recently come under suspicion of being mutagenic and since mid-1998 has had to be labelled with a skull. Replacement products are in the process of being developed (see also section 3.7.1).

2) A further value, known as the *amine value*, is often quoted for amines or aminic hardeners. It has no direct correlation to the NH equivalent weight and is defined as follows: the amine value is the quantity of potassium hydroxide in mg that can bond as many protons of a very strong acid as 1 g of substance. The amine value thus provides an indication of the total content of protonatable (basic) nitrogen atoms in a substance.

The following simple rule applies:

For normal curing (1:1 curing), the resin and hardener must be mixed according to the ratio of their equivalent weights. In order to modify the film properties, the 1:1 ratio can be shifted by about 10% in either direction (*undercuring* or *overcuring*).

All hardeners containing free primary (-NH_2) or secondary amino groups (-NH-) are processed with epoxy resin as a two-component system, where the pot life (processing time) may be anything from minutes to days, depending on hardener structure, temperature (at least 15 °C) and, where applicable, catalysis.

Low-molecular acyclic polyamines are strongly basic and relatively volatile liquids; they exhibit the highest reactivity with epoxides. Cycloaliphatic, araliphatic and especially aromatic amines are chemically more inert and require either elevated temperatures and/or special catalysts for curing. *Ketimines* (*polyketimines*) produced by *blocking* primary amines with low ketones, cure only after *deblocking* by the action of atmospheric moisture on the wet film, which significantly increases the pot life. The formulae of a number of amine hardeners are shown in Figure 2.21 (page 82).

Simple amines do produce densely crosslinked coating films with good chemical resistance, especially to alkalis, but they also present a number of serious defects: high volatility, unpleasantly fishy odour and - in the case of the aromatic amines - toxicity. The high volatility of the lower amines leads not only to severe odour development during processing of the coatings, but also to an incomplete cure of the coating surface. This problem can be alleviated by for example first allowing the amine to pre-react with a stoichiometric deficit of epoxy resin components to form an "*in-situ adduct*", which is then mixed with the remaining residue of epoxy resin immediately before processing. During formation of the adduct, the amine molecules bond to the epoxy resin via some of their NH bonds, which means that they can no longer evaporate. This pre-reaction also prevents another cause of poor surface curing: *carbamate formation* from primary amino groups and atmospheric CO_2. During formation of the adduct almost all of the primary amino groups are converted to secondary groups and are therefore no longer available for carbamate formation.

Adducts of epoxy resin and excess amine are also sold as isolated "*in-situ adduct hardeners*".

Another important group of amine hardeners are the *polyaminoamides*. These are polyamides which have been formed by the reaction between dicarboxylic or polycarboxylic acids - mostly fatty acid dimers (dimerised, unsaturated fatty acids) - and excess diamines or polyamines and which have terminal, basic NH_2 and perhaps also incorporated NH groups (see Figure 2.21 for typical formula). Polyaminoamides are non-volatile and provide good flexibility, in contrast to the amine-cured films, albeit with reduced alkali resistance.

A hardener commonly used for epoxy resin powder coatings is *dicyandiamide* (DICY or DCD) in free or "modified" form (see Figure 2.21, page 82 for formulae). During stoving, complex reactions are initiated in which 4 to 5 epoxy groups are reacted per molecule of DICY. The required stoving temperature of around 180 °C can be reduced by the use of accelerators (catalysts).

Polycarboxylic anhydrides and carboxylic acid resins

Aside from the class of amine, i.e. basic, hardeners for epoxy resins, there are also acid crosslinking agents, which in turn may be either polyanhydrides or polycarboxylic acids in the form of acid resins. Polyanhydrides react at elevated temperature (undergoing "anhydride ring" scission) first with the hydroxyl groups and then to some extent with the epoxy groups in the resin, according to the following reaction scheme:

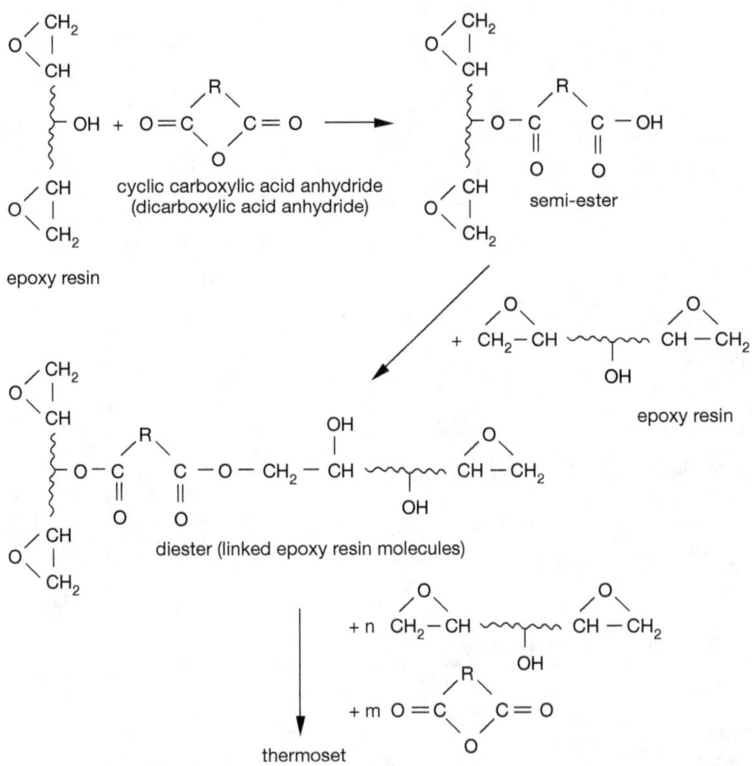

Polyanhydrides are used in solid form for powder coatings and as solutions for liquid coating materials. The stoved films are resistant to chemicals, more resistant to acids than alkalis, compared with amine-crosslinked films, flexible, free from odour and flavour; this explains the use of anhydride-EP systems for can linings.

Allied to anhydride curing is the stoving of epoxy resins with (carboxy) *acid polyesters* for powder coating applications; the corresponding products are also known as *"hybrid powders"*. Other acid polymers such as polyacrylates are also used occasionally for epoxy resin curing.

Other curing agents for epoxy resins

Phenolic resins (etherified resols), melamine, urea and benzoguanamine resins are used as curing agent resins in stoving epoxy resin coatings. As it cures, the (higher-molecular) epoxy resin reacts principally via its hydroxyl groups with the etherified methylol groups to produce highly flexible films. The required stoving temperature of up to 200 °C and above can be reduced considerably by the addition of acid catalysts. The epoxy/phenolic resin blends known as "gold coatings" are of particular importance; because of their excellent chemical and food contact resistance, flexibility and adhesion, they are predominantly used for can linings. Epoxy/amino resin blends are much less strongly coloured than the gold coatings, but are also less media-resistant.

Higher-molecular (solid) epoxy resins can be cured in two-component systems with free polyisocyanates, mainly the more reactive aromatic grades, at temperatures down to around 0 °C to form densely crosslinked, resistant and yet highly flexible films. In this case the reaction proceeds exclusively via the free hydroxyl groups.

One-component stoving coatings can be formulated with blocked polyisocyanates. In broader terms, this class of coatings also includes the *cathodic electrodeposition coatings* based on amine-modified epoxy resins, which after neutralisation with acetic acid or other lower monocarboxylic acids are capable of dilution with water.

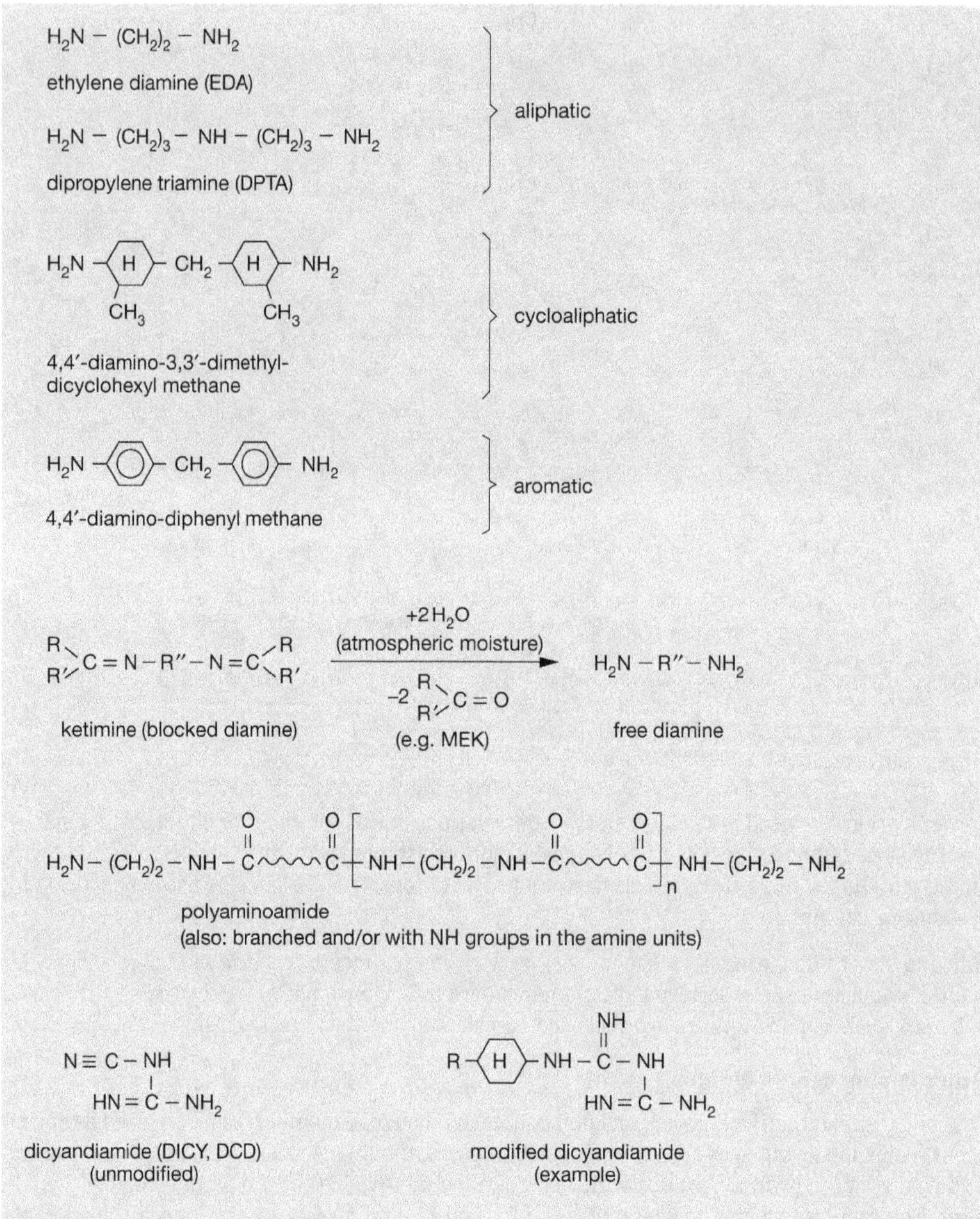

Figure 2.21: Amine epoxy resin hardeners (selection)

Simple example:

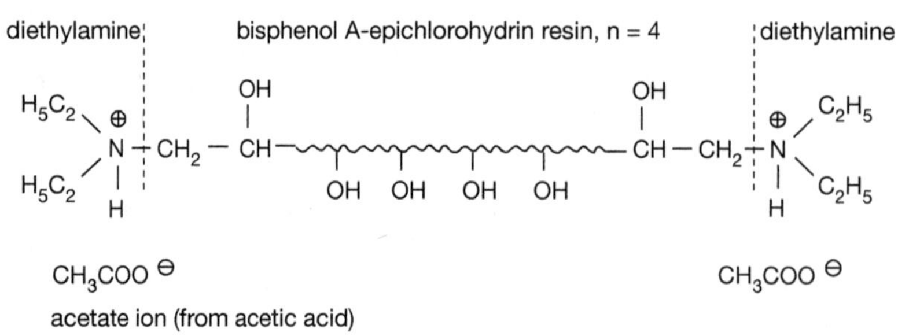

Radiation curing

Let us conclude by mentioning the *cationic UV radiation curing* of epoxy resins, preferably the cycloaliphatic grades. Substances forming very strong acids by photolysis are used as photoinitiators here. These trigger a cationic ring-scission polymerisation, as shown in the reaction scheme on page 81).

Ionically curing UV coatings offer a number of advantages over the radically curing products described in section 2.1.4.3, notably freedom from air sensitivity during cure, low shrink and thus improved adhesion to metals, as well as post-curing in the dark. Disadvantages include their lower curing speed and a limited range of raw materials.

Areas of application of epoxy systems

The characteristic basic properties of epoxy resin coatings - already described in detail in the previous section - such as good to excellent chemical resistance, high flexibility, good adhesion to metals and many other substrates, yellowing and degradation (chalking) in sunlight, suggest the following principal applications:

- Corrosion protection, building protection, physical protection (two-component, high-build; also in combination with tar pitch or other water-repellent additives)
- Primer for metals (one- or two-component, depending on the size of the object)
- Primer for absorbent, generally mineral substrates (two-component)
- Liquid and powder one-coat or top-coat paints with good protective effect for interior use (one-component, liquid coatings, in some cases also two-component; non-aromatic epoxy resin systems also suitable for exterior use)
- Ionic UV curing.

When working with epoxy resin products it should be noted that epoxy resins having molecular weights below 700 g/mol (liquid resins) are a skin irritant and may have a sensitising effect. Amine and acid hardeners are likewise irritant and in some cases even corrosive.

Epoxy resin esters

By epoxy resin esters (epoxy esters) we mean epoxy resins, generally based on bisphenol A, that are partially esterified with fatty acids. Each epoxy group can bond up to two acid molecules, and each hydroxyl group naturally one acid molecule. As with alkyd resins, we differentiate between short oil, medium oil and long oil grades of epoxy resin esters. Parallels with alkyd resins also exist in terms of coating applications: long oil grades are used in air-drying coatings because they cure by oxidation, whereas the short oil products are hot cured with urea resins (for interior use) or melamine resins (for exterior use).

Films based on oxidatively drying epoxy resin esters display greater hardness, bond strength, elasticity and water resistance than alkyd resin coatings, although epoxy resin esters are more expensive. They are mainly used as anti-corrosive coatings or primers and heavy-duty exterior paints and coatings. For exterior applications the chalking tendency typical of bisphenol A-based products must be taken into consideration.

Given their outstanding adhesion and flexibility, heat-curing epoxy resin esters are predominantly used in protective coatings for metals, e.g. in tube and stamping coatings.

Epoxy resin esters can in some cases be chemically modified to further improve their properties. Water-thinnable carboxylic acid grades capable of amine neutralisation can be obtained in this way.

2.1.4.9 Polyurethane systems

Isocyanate and urethane groups, polyurethanes (PUR)

If an *isocyanate* - a compound with the (linear) atomic grouping –N=C=O– is reacted with an alcohol, a *urethane*, i.e. an ester of carbamic acid, is obtained. Unlike free carbamic acid, urethanes are both thermally and chemically very stable.

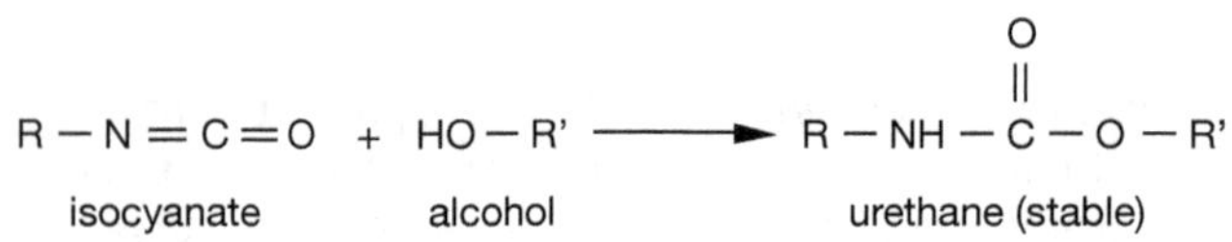

Now if a *diisocyanate* is reacted with a bifunctional alcohol (diol), a "*linear polyurethane*" is obtained. If higher functional isocyanates and/or polyols are also used, a branched or even crosslinked polymer is formed.

Polyurethanes - like epoxy systems - are extraordinarily versatile and correspondingly widely used. The annual world production of polyurethane raw materials for the coatings sector alone was estimated at 400,000 t in 1992. The overall tonnage for all polyurethane products must be many times higher than this.

Reactions of the isocyanate group

The most important reactions of the isocyanate group in regard to coating chemistry are shown in Figure 2.22. In addition to poly(urethane) formation, poly(urea) formation initiated by the addition of water is of particular interest (moisture-curing systems).

Classification of polyurethane coating systems

Polyurethane coating systems can be conveniently separated first of all into *one-component* (*1K*) and *two-component* (*2K*) systems. It is useful to subdivide one-component systems again by curing temperature and finally by curing chemism.

- One-component systems:
  - Oxidatively drying at room temperature or under forced conditions (80 °C): *urethane alkyds, urethane oils*
  - Polyurea formation at room temperature: *moisture-curing urethane prepolymers* containing free isocyanate groups

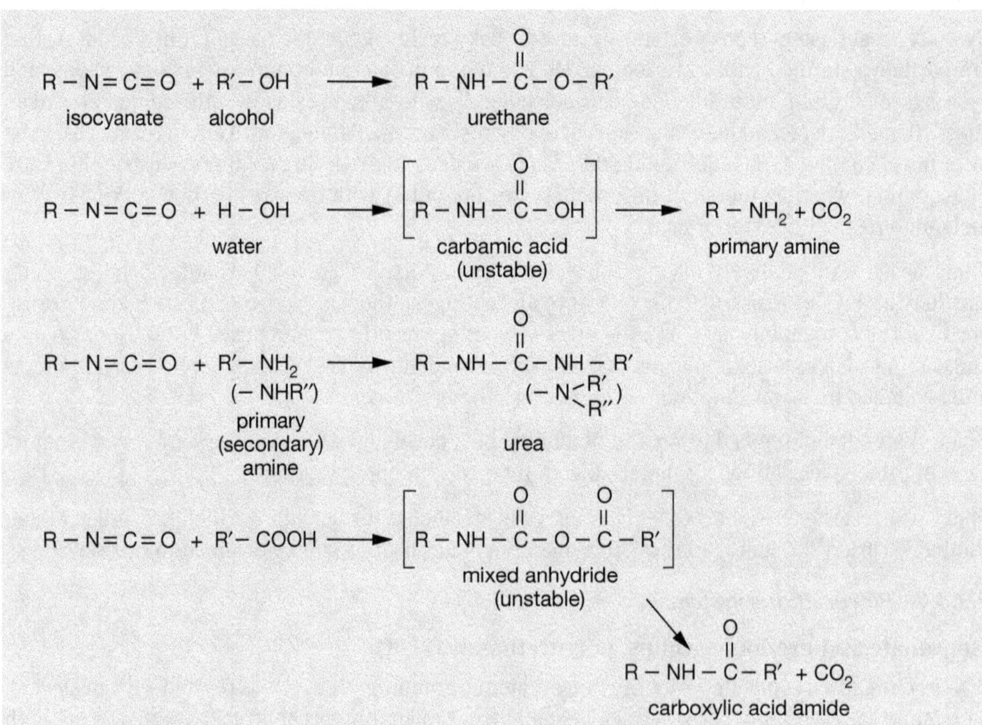

Figure 2.22: The most important reactions of the isocyanate group for coating chemistry

- Polyurea/polyurethane formation at room temperature:
moisture-curing blends of polyisocyanates and *bis-oxazolidines*
- Physically drying at room temperature or under forced conditions, solvent-containing:
thermoplastic, unreactive *polyurethanes*
- Physically drying at room temperature, aqueous:
polyurethane dispersions (PUR dispersions or PU dispersions)
(also in functionalised form, which can be cured chemically, oxidatively or by radiation)
- Polyurethane formation at 110 to 200 °C:
polyhydroxy resins plus *blocked polyisocyanates*
- Polyurethane formation at 100 to 160 °C:
polyhydroxy resins plus *micro-encapsulated polyisocyanates*

- Two-component systems:
  - Polyurethane formation at room temperature to 130 °C, solvent-containing:
  polyhydroxy resin plus polyisocyanate
  - Polyurethane/polyurea formation at room temperature to 130 °C, aqueous:
  polyhydroxy resin (water-thinnable) plus emulsified polyisocyanate

Key isocyanates used in coating chemistry

The formulae for the key diisocyanates and polyisocyanates used in coating chemistry are shown in Figure 2.23. The volatile diisocyanates *toluylene diisocyanate* (TDI), *hexamethylene diisocyanate* (HDI) and *isophorone diisocyanate* (IPDI) are used in the curing of coatings not in monomer form but as oligomers produced by various means. Common oligomerisations include addition with trimethylolpropane, *biuret formation* and trimerisation to *isocyanurate* (see Figure 2.23).

The vapour pressure is reduced by molecular enlargement to such an extent that practically no harmful polyisocyanate vapours are released at room temperature. Nevertheless:

Skin contact with polyisocyanates and especially inhalation of spray mists containing isocyanates must be avoided under all circumstances!

Blocked (*capped*) polyisocyanates are used for one-component stoving coatings or powder coatings. These are polyisocyanates in which H acid compounds (substances with dissociable $H^{\oplus}$ ions) such as ε-caprolactam (see formula below), butanoneoxime, various acyclic and heterocyclic secondary amines, low alcohols and H acid compounds, such as malonic ester or acetoacetic ester, sometimes in combination with each other. The polyisocyanates blocked with the various blocking agents differ in their deblocking and curing temperature (between 100 and 180 °C), thermal yellowing, retention of blocking agent in the film, storage stability (in the case of 1-pack coatings) and solution or melt viscosity. Deblocking causes the bulk of the blocking agent to be released into the oven air, as the following example for ε-caprolactam shows:

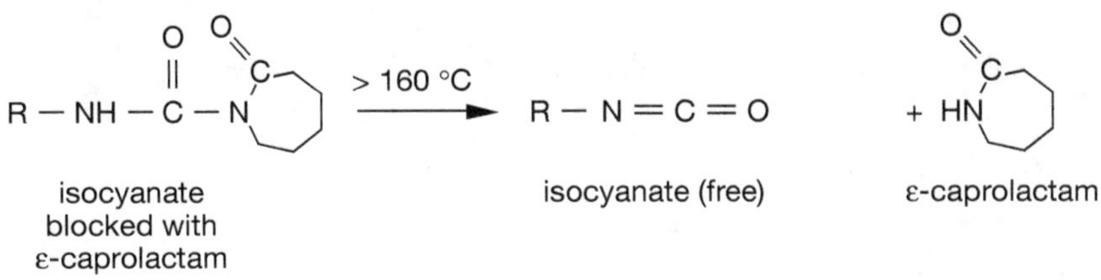

The *polyuret diones* of diisocyanates, in which two isocyanate groups effectively block each other by the formation of four-membered rings, are free from blocking agents. The unfavourably high scission temperature of the polyuret diones can be reduced from over 170 °C to around 140 °C through the use of appropriate catalysts, the process proceeding as follows:

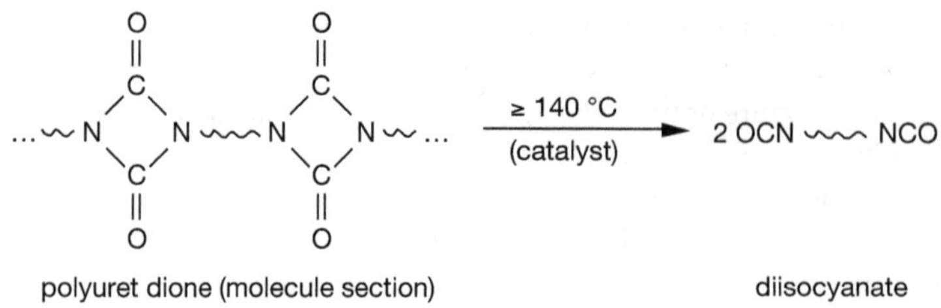

Cyclopentanone 2-carboxylic acid esters (CPME or CPEE) also function as blocking agents, with no cleavage involved. Isocyanates blocked with them also exhibit little yellowing and have a low viscosity.

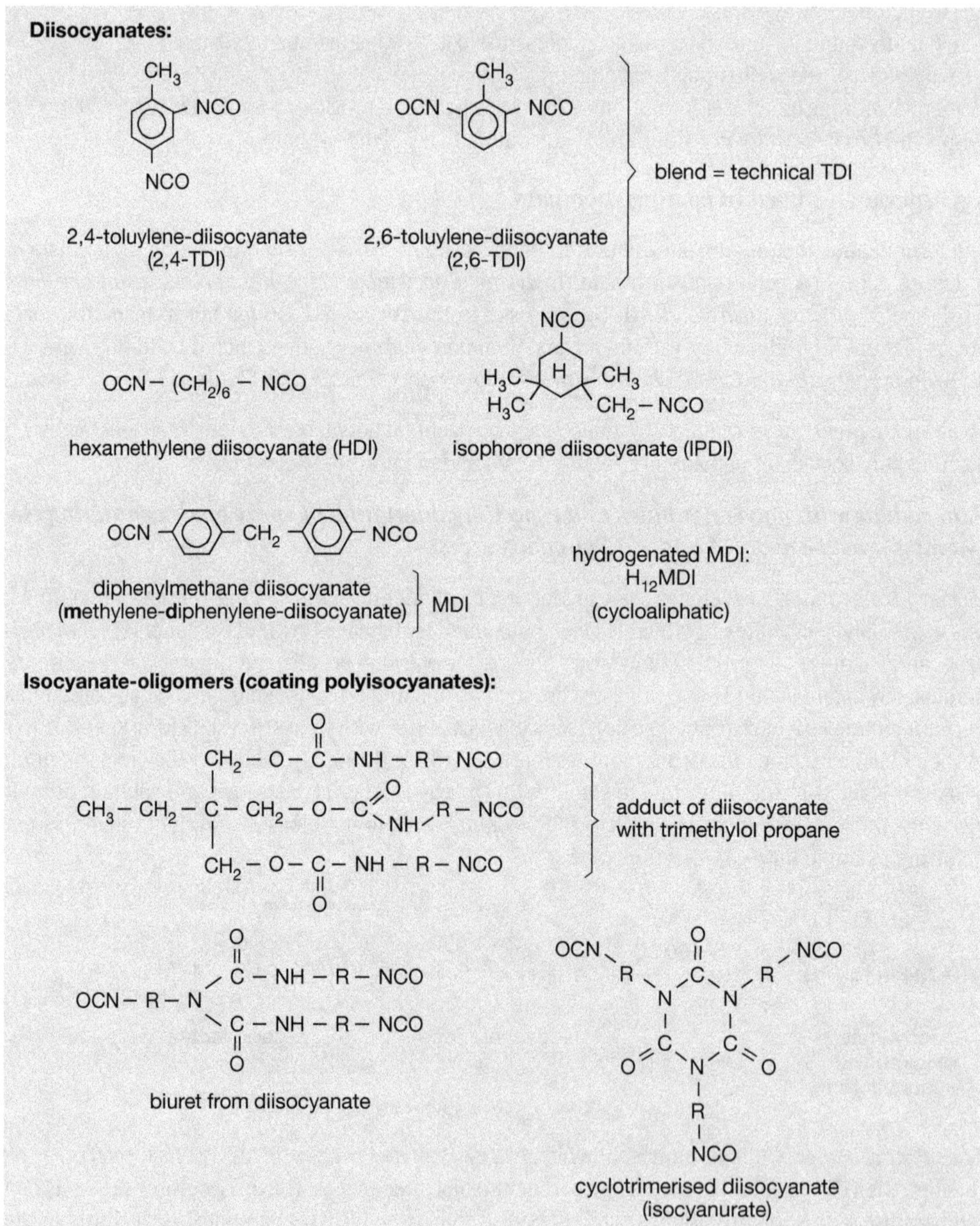

Figure 2.23: Key di- and polyisocyanates used in coating chemistry

Further information about the most important one-component polyurethane systems

Moisture-curing urethane prepolymers are 100 % (liquid) or dissolved resins prepared from polyols and an excess of di- or polyisocyanates. Curing takes place by the chemism described above, with formation of urea groups. These are converted to solvent-free or solvent-containing, non-pigmented or pigmented, highly resistant coating and impregnating materials for mineral substrates, e.g. concrete floors, for timber (products) and for flexible substrates. (To prevent the thickening or even gelling of the final products in drums under the influence of moisture, they contain water-binding additives such as *triethyl ortho-formiate, molecular sieves* or *p-toluene sulphonyl isocyanate* (TSI)).

Polyurethane dispersions are a relatively recent class of binder, but are widely used as water-thinnable coating materials and related products. The basic chemical structure of the film former of an anionic PU dispersion is shown in simplified form below:

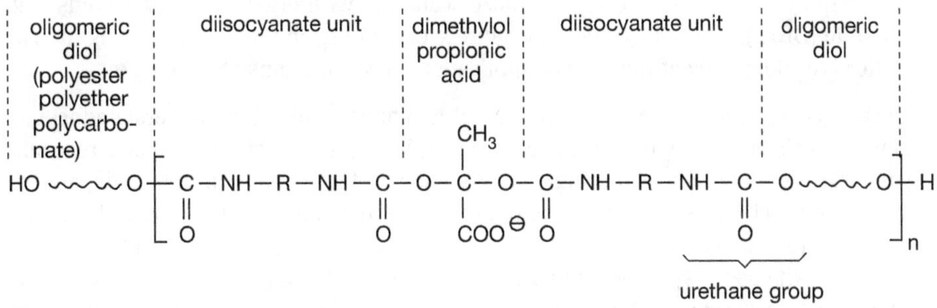

Mostly, the chain contains additional diamine units which act as chain extenders and are incorporated via urea groups. There are also (cationic) PUR dispersions which have been stabilised by sulphonate -($-SO_3^-$)- or by ammonium groups.

The particular strengths of PU dispersions are

- good environmental compatibility, being aqueous
  (in some instances they contain small quantities of solvent such as N-methylpyrrolidone)
- can be processed as one-component systems, being physically drying
  (in some cases chemical crosslinking may also be possible or necessary)
- high mechanical resistance (abrasion resistance) and flexibility
  (chemical and yellowing resistance is dependent on dispersion raw materials)
- absence of external emulsifiers
  (therefore films are particularly water-resistant on evaporation of neutralising agent).

These benefits do not come cheaply, for which reason PU dispersions are commonly used in combination with polyacrylate dispersions and other water-thinnable film formers. Directly synthesised *PU-acrylate dispersions* are also commercially available.

Traditional applications of PU dispersions include the coating of flexible substrates such as textiles, leather and paper. High-grade wood (parquet floor) varnishes and masonry paints may also contain PU dispersions. The use of PU dispersions in undercoats for automotive paint systems is probably the most important application naturally, physical drying here is complemented by chemical curing, e.g. with blocked polyisocyanate. For other application areas, too, where physically dried films would not be resistant enough on their own, there are PU dispersions which can be cured in one-pack or two-pack form or by UV radiation.

The one-component polyurethane systems include stoving systems with blocked isocyanates as curing agents. Cathodic electrodeposition coatings and PU powder coatings are important members of this product group. In the first example, the blocked polyisocyanate, which has first been emulsified in water and then deposited on the object together with the amine-modified epoxy resin, reacts during stoving with the hydroxyl groups in the epoxy resin.

Polyurethane powder coatings are based on a blend of solid polyester or polyacrylic polyol and blocked polyisocyanate or polyuret dione (see above). Stoving temperatures range from 140 to 200 °C. PU powder coatings demonstrate good mechanical and chemical resistance; those based on aliphatic polyisocyanate also display good weather resistance.

Two-component polyurethane systems

Two-component PU coatings[1)] represent the oldest and at the same time most significant class of coatings within PU coating technology. Hydroxy-functional polyesters or acrylic resins in particular, but also short oil alkyd resins, higher epoxy resins and polyethers, are used in the form of approx. 50% solution up to 100 % liquid resin as stock resin components, which cure chemically at room temperature or under forced conditions with coating polyisocyanates as hardeners. Whilst aromatic polyisocyanates (NCO group directly on the benzene ring) react rapidly at room temperature, aliphatic polyisocyanates require elevated temperatures and/or catalysis with organic tin compounds, e.g. dibutyl tin dilaurate (DBTL), zinc salts or tertiary amines, depending on the desired curing time or pot life. The latter are added only at the time of application - e.g. by means of injected air.

In addition to their characteristic general properties of toughness, flexibility, universal adhesion, chemical and water resistance, polyurethanes based on aliphatic polyisocyanates are also resistant to yellowing under the influence of weathering. The prominence of the cited properties is dependent not only on the chemical character of the two binder components but also on the crosslink density in the film, which in turn is a function of the hydroxyl or isocyanate content of the components. As with epoxy resins, the best way to determine this is to consider the equivalent weights of stock coating and hardener. These relate to the hydroxyl or isocyanate content (proportion by weight) as shown below:

$$\text{OH equivalent weight [g/mol]} = \frac{17}{\text{OH-content [\%]}} \cdot 100$$

$$\text{NCO equivalent weight [g/mol]} = \frac{42}{\text{NCO-content [\%]}} \cdot 100$$

The hydroxyl content of polyols is generally given in the form of the hydroxyl value, which is defined as follows:

The **hydroxyl value** (OH value) is the weight of potassium hydroxide in mg which contains the same quantity of hydroxyl groups as does one gram of substance. (Unit: mg/g).

The hydroxyl content can be calculated from the hydroxyl value using the following formula:[2)]

$$\text{OH content [\%]} = \text{OH value [mg/g]} \cdot 0{,}0303$$

1:1 crosslinking occurs when the stock resin or coating and hardener are combined in the ratio of their equivalent weights. Decreasing the proportion of (expensive) polyisocyanate, i.e. undercuring, increases the flexibility of the film but at the same time reduces its mechanical and chemical resistance. Overcuring increases the crosslink density through the formation of urea groups with atmospheric moisture, but also makes the coating more expensive.

Two-component PU coatings are mainly used in applications which place high demands on the performance of the coating but are unsuitable for stoving. Typical applications include: heavy vehicle and aircraft paints, automotive refinishing paints, some primary automotive paints, industrial coatings (machines, plastic components, high-quality furniture). Two-component PU materials are also to be found in decorator's, masonry and DIY coatings in areas where excellent resistance or

1) Traditionally these coatings are sometimes known as "DD coatings", which strictly speaking is a tradename.

2) Readers are strongly recommended to derive these and similar conversion formulae for themselves to aid understanding.

durability is required of the coating. It is worth mentioning that industrial two-component coating systems require appropriate two-component plants with separate feed systems for stock coating and hardener.

Increasingly stringent environmental protection conditions have led to the development of more environmentally compatible products in two-component PU technology too. *High solid coatings* (HS coatings) represent one such development. The low molecular weights and narrow molecular weight distributions of polyol resins allow the production of coatings with lower solution viscosities and consequently higher solid contents. In spraying consistency, the solids contents (non-volatile components) in clear coatings is around 60 to 65 wt.%, in pigmented coatings - depending on pigment density and pigmentation level - around 65 to 80 wt.%. When working with high solid coatings, the absence of physical surface drying, the slower solvent evaporation and the greater dry film thickness with the same application must be taken into account in particular (risk of runs and overcoating).

The late 1980s saw another - in chemical terms rather surprising - development in the area of environmentally friendly products, in the form of *water-thinnable two-component PU systems*. The underlying principle of these coatings can be briefly outlined as follows: A conventional (hydrophobic) or self-emulsifying (hydrophilic) polyisocyanate is incorporated by emulsification into the clear or pigmented basecoat component of a water-thinnable two-pack coating based on an anionic polyol (solution, secondary or primary dispersion, mostly polyacrylate) immediately before processing. The mixture is then applied within the time recommended by the manufacturer (the pot life). In this case, there is no clear rise in the viscosity of the mixture to signal the end of the pot life. In microscopic terms, this is a complex curing mechanism which largely prevents the polyisocyanate from reacting with the water (and the neutralisation amine) instead of the polyol to form polyurea. While water-borne two-pack PU coatings have not displaced conventional or high solid, two-pack coatings, they have at least replaced or complemented them in some areas.

Non-isocyanate systems

As we have already mentioned, isocyanate hardeners can cause irritation, asthma and dangerous sensitisation in contact with the skin or by inhalation of spray mist. Around the mid-1980s, triggered by this fact and by the looming statutory regulation of the use of coating polyisocyanates, the research area known as "*non-isocyanate chemistry*" (NISO chemistry) developed.

By non-isocyanate systems we mean coatings which are similar to two-component polyurethane coatings in terms of processing methods, properties and applications, but which do not contain isocyanate. Despite many partial successes, no adequate replacement for two-component PU coating systems has as yet been made commercially available. This may be due in part to the fact that the isocyanate problem, which can be controlled by technical protective measures, has faded into the background in comparison with the much more urgent issue of environmental protection and solvents.

2.1.4.10 Film formers based on silicon

This class of film formers essentially includes the *polysilicates* (water glasses), the *polysiloxanes* (silicones) and the products of reactions between *silicic acid esters* and *silanes* (*sol-gel binders*). There are also countless hybrids formed from among these materials and with purely organic components. The semi-metal silicon is essentially tetravalent, like carbon, but can temporarily bond five or six further atoms (coordination number: 6) and forms Si-O-Si (siloxane units) in preference to Si-Si bonds as they are more stable. These facts can be used to summarise the chemistry behind the use of silicon in coatings as follows:

- **Molecular structures**

Chains and networks of –Si–O–Si– linkages

with unreactive and all kinds of reactive terminal groups.

- **Formation of the structures** (syntheses, crosslinking/curing):

$$\text{-Si-O-R} + H_2O \rightarrow \text{-Si-O-H} + \text{ROH} \qquad (1)$$

$$\text{-Si-O-H} + \text{H-O-Si-} \xrightarrow{\text{Room temp.}} \text{-Si-O-Si-} + H_2O \qquad (2)$$

$$\text{-Si-O-R} + \text{H-O-Si-} \xrightarrow{\text{Elevated temp.}} \text{-Si-O-Si-} + \text{ROH} \qquad (3)$$

All three of these reactions are accelerated by acids and bases (and temperature increase), with reaction (1) boosted more by acid and reaction (2) more by base. The reaction rates for (1) and (3) decrease with increase in the alkyl chain length from $R = CH_3$ on.

Water glass

Water glass is an alkali silicate which is sold in the form of a viscous, highly alkaline, aqueous solution. In the coatings sector "potassium water glass" is used almost exclusively. Film forming mainly occurs by absorption of atmospheric carbon dioxide with precipitation of polysilicic acids, so-called *silicification*, which ultimately leads to vitreous silica:

$$\underset{\text{Potassium water glass (typical formula)}}{K_2Si_2O_5} + CO_2 \longrightarrow 2\,SiO_2 + K_2CO_3$$

The film is characterised by high alkalinity because of the alkali carbonate (potassium carbonate) released. Water glass-based *"silicate paints"* produce extremely hard coatings that are resistant to weathering, light and chemicals and permeable to gas and water vapour. The main area of application for these products is in the protection and restoration of buildings, monuments, etc., and in corrosion protection. Silicate paints with added polymer dispersion (e.g. styrene acrylate) are known as *"silicate emulsion paints"*.

Alkyl silicates, ethyl silicate

Alkyl silicates (silicic acid esters) differ from water glass in that the ionically (salt-like) bound alkali metal ions (K^+) of the waterglass have been replaced by simple alkyl groups covalently bonded to terminal oxygen atoms. As a result of reactions (1) and (2) above, the alkyl silicates are ultimately transformed in the presence of water or atmospheric humidity into SiO_2 and alcohol (which evaporates). Alkaline substances are not formed.

Ethyl silicate (tetraethyl orthosilicate (TEOS), $Si(OC_2H_5)_4$, or corresponding di- and tri-silicates), which is used almost exclusively, is generally pre-hydrolysed and pre-condensed with some water and acid and used as a colloidal solution in ethanol. The acid acts as a stabiliser due to positive charging of the colloidal condensate particles. The pigment, such as zinc dust, is admixed just prior to application. In the film, the stabilisation is impaired and curing with atmospheric humidity occurs. The primary applications of ethyl silicate are to be found in masonry protection and heavy-duty corrosion protection. The coatings are also extremely heat resistant due to their inorganic structure. Other than as a binder for coating materials, ethyl silicate makes a good reactive impregnating agent, e.g. in natural stone conservation.

Silicone resins

Silicone resins belong to the class of *silicones* (SI), scientific name *polysiloxanes*, with alkyl, aryl and perhaps functional groups attached direct to the Si atom. According to their molecular structure and average molecular weight, silicones can mainly be found in the form of oils, pastes ("silicone greases")[1], elastomers or resins. Figure 2.24 provides a schematic illustration of the basic molecular structure of silicone resins.

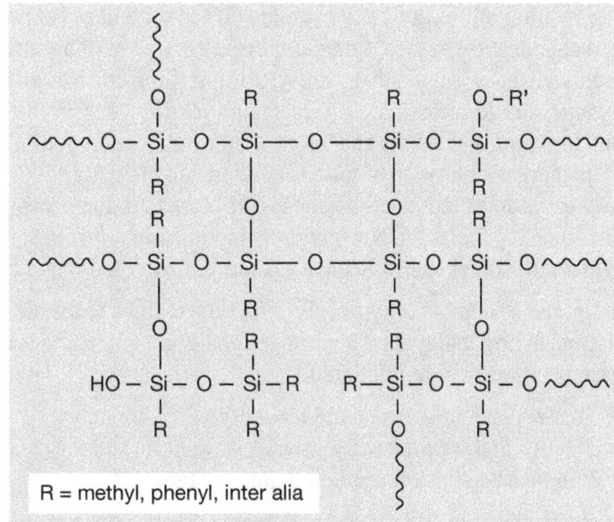

Figure 2.24: Silicone resin (schematic formula)

The variability of the overall structure of the molecules, their average molecular weight and the type and quantity of organic groups (methyl, phenyl, etc.) and of functional groups where applicable (hydroxy, alkoxy, vinyl), means that a wide range of silicone products are available. The silicone resins normally used as additional film formers in coatings are partially crosslinked substances with a certain heat reactivity, in a wide range of high molecular weights; they occur in the form of solutions, emulsions and solid resins. Film forming generally takes place by stoving at temperatures above 200 °C, usually in combination with other film formers.

The typical incompatibility of many silicones with other polymers can frequently be overcome by boiling "*silicone intermediates*" (reactive silicones) with basic binders such as polyesters, alkyd resins, acrylic resins, epoxy resins, etc., to form silicone blend resins ("*silicone polyesters*", "*silicone alkyds*", etc.). The use of silicone resins in paints and coatings improves

- heat resistance
- weather resistance
- pigment wetting, flow, surface smoothness and gloss retention.

A further important function is the hydrophobising and dirt-repellent effect that silicones possess by virtue of their low surface tension. The minimum proportion of silicone resin which the complete binder needs for transferring the positive silicone properties to the coating is around 30 per cent by weight. The relatively high cost of silicones must be borne in mind, however; this, together with their limited compatibility, rules out their use as film formers in many cases.

In marginal areas of coating technology, in building protection and electrical insulation for example, silicone resins are used as binders, saturants, water repellents and impregnating agents. (Modified silicones are important coating additives.)

Sol-gel binders, silylated film formers

As described above, ethyl silicate is mostly used in a partially hydrolysed and pre-condensed form as an aqueous-alcoholic, colloidal solution (sol). After application of this pure solution, the hitherto inhibited molecular enlargement accelerates and a porous network (gel) is formed. Stoving ultimately gives rise to a brittle, cracked film. Sol-gel technology was developed in the late 1990s when *trialkoxy silanes*, which have one organic group that may or may not be functional and which were already being used as adhesion promoters, were integrated into sol formation. Typical alkoxy silanes

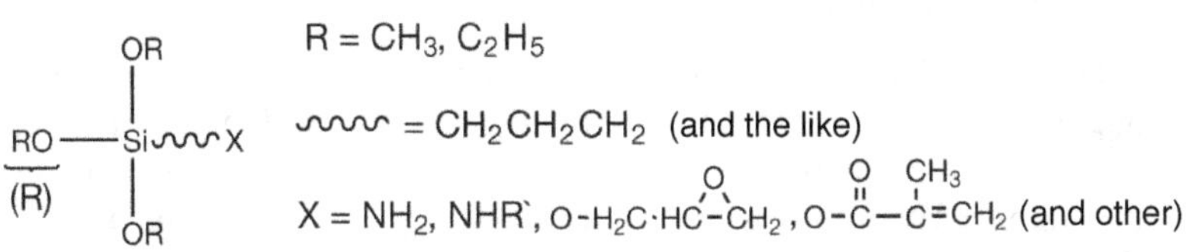

[1] "Silicone greases" are mixtures of silicone oils and thickening substances.

The resultant films are more flexible and "coating-like" yet retain the bulk of their high mechanical and chemical resistance. Compatibility with any auxiliary binders present is also improved. These highly vitreous, sol-gel materials yield thin films and can be organically modified by incorporating organic film formers, possible reactions taking place at the SiOH and SiOR groups or at organic functions introduced via the silane. After application and gel formation, final cure takes the form of further condensation of the SiOH groups to yield Si-O-Si linkages. Reactions may also take place between other functional groups incorporated via the organic binder and added hardener (curing agent) or UV radiation. For inorganic modification during sol manufacture, alkoxides (alcoholates) of elements other than silicon, e.g. titanium, zirconium and aluminium, can be used.

An immense number of synthesis routes and recipes have now grown up around sol-gel and inorganic-organic hybrid materials. The primary goals and applications of these developments, some of which have been realized in full, include

- *Easy-to-clean* properties and ("super-hydrophobicity")
- Plastics and coatings with greater scratch resistance
- Enhanced corrosion protection for metals

Further effects include heat resistance, optical effects (anti-reflection layers on spectacle lenses) and anti-microbial functions.

While sol-gel materials are already finding widespread use for high-quality specialty applications requiring coatings just a few micrometers thick, they are making little headway into paint technology in general. The primary reasons are the sols' relatively low solids content and associated high solvent content (alcohol), unfavourable rheological and application properties, a very limited shelf life and a relatively high price. Moreover, the expected positive properties have not always lasted any great length of time and in some cases never materialised to the extent expected.

The conversion of conventional film formers with functional alkoxy silanes via the respective organo-functional group on the silane or via direct polymerisation of methacrylol-functional silane leads to *silylated* products, which crosslink with atmospheric humidity after application via reactions (1) and (2) above, possibly in addition to undergoing conventional curing. Examples are silylated isocyanate oligomers and trialkylsilyl-functional acrylic resins. Trialkylsilyl polymers can even be produced as aqueous dispersions.

2.1.4.11 Other film formers

Numerous other polymers and oligomers not yet described in this book may be present in coating materials and related products where they serve as base raw materials or as additives for improving chemical, water, weathering and/or heat resistance, adhesive strength, flexibility and filling power ("build").

Hydrocarbon resins

Unsaturated hydrocarbons can be polymerised in the presence of strong acids at elevated temperature and under pressure to form hydrocarbon resins. Depending on the origin of the starting materials, we distinguish between

- Petrol(eum) resins (raw materials obtained from petroleum refining)
- Coumarone-indene resins (raw materials obtained from coal-tar oil)
- Terpene resins (raw materials obtained from oil of turpentine).

Hydrocarbon resins are thermoplastic, highly resistant to chemicals and, in particular, unsaponifiable, non-polar oligomers or polymers with average molecular weights of around 3000 g/mol; consistency and inherent coloration are very variable, depending on the raw material basis and manufacturing conditions. Hydrocarbon resins are used *inter alia* as binder components in coatings to increase their chemical and water resistance. They are also used in adhesives and printing inks.

Xylene/formaldehyde resins (XF resins), which are mainly used as polymer plasticisers ("soft resins") are also classed as hydrocarbon resins in the broadest sense.

Ketone resins

The alkali-catalysed condensation polymerisation of lower ketones such as methyl ethyl ketone, cyclohexanone or acetophenone with formaldehyde produces *ketone/formaldehyde resins* (ketone resins). When added to coating systems these light, unsaponifiable hard resins improve gloss, hardness and body ("build").

Aldehyde resins

The traditional aldehyde resins produced from auto-condensed aldehydes are very rarely used now in coatings. The same is not true of urea-modified aldehyde resins (*isobutyraldehyde/formaldehyde/urea resins*) which, as well as being used as coating additives to improve gloss, body and UV resistance, are known as ideal "*grinding resins*" for pigment pastes (paste resins) because of their high pigment binding power and universal compatibility.

Phenoxy resins

These relatively high-molecular aromatic polyethers formed from diphenols - like bisphenol A - and an excess of epichlorohydrin are distinguished from the corresponding epoxy resins by the absence of epoxy groups. They can be thermally cured by means of the hydroxyl groups distributed through the resin molecule, with amino resins, phenolic resins (resols) or blocked isocyanates for example. Physically drying coatings can also be formulated with higher-molecular phenoxy resins. The coatings display properties similar to those of epoxy resin systems.

Polyvinyl acetals

The incomplete hydrolysis ("saponification") of polyvinyl acetate produces polyvinyl alcohol with residual acetate groups. The hydroxyl groups formed in this way are then reacted with formaldehyde - again incompletely - to give *polyvinyl formal* (PVFM) or with n-butyraldehyde to give *polyvinyl butyral* (PVB). A section from a PVB molecule has the following formula:

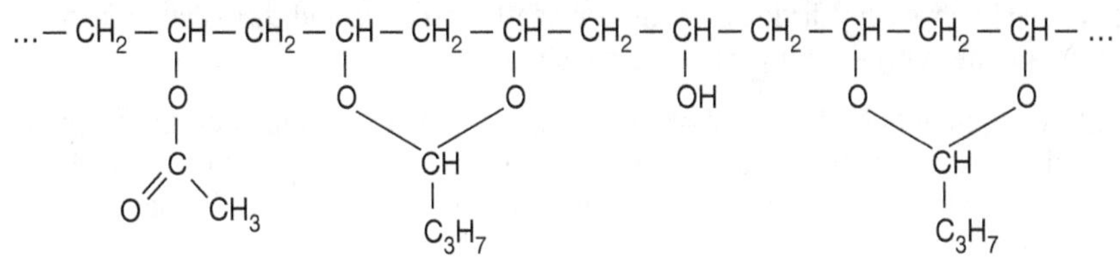

polyvinyl butyral (molecule section)

The average molecular weight, acetyl content and content of free hydroxyl groups determine the solubility and compatibility, the thermoplastic properties and the reactivity of the products.

Polyvinyl formal is primarily used - in combination with phenolic resins (resols) and free phenols - in stoving coatings for wire insulation. Polyvinyl butyral is softer and more flexible but less resistant than polyvinyl formal. A particular feature is its good adhesion to metals, glass and wood. Low-molecular polyvinyl butyral is used *inter alia* as a plasticising component in stoving coatings for the coating of metals. Polyvinyl butyral of average molecular weight - possibly together with phenolic or epoxy resin - is used as a film former in *wash primers*[1], where the phosphoric acid added as hardener initiates acetal cleavage, which is responsible for curing and for bonding with the substrate.

As a thermoplastic, flexible and highly adhesive film former, polyvinyl butyral is used in many other applications in the coatings and adhesives sector.

1) A wash primer is an anti-corrosive primer coat to be applied in a very thin layer to cleaned metal substrates.

Polyvinyl chloride, vinyl chloride copolymers

Because of its insolubility in typical coating solvents, polyvinyl chloride (PVC) in homopolymer form is used in the coatings sector only as a PVC plastisol and organosol and as a fluidised bed powder coating.

A *PVC plastisol* is essentially a paste-like blend of PVC powder and plasticiser, which gels at approx. 200 °C to form a tough and resilient, homogeneous film. An organosol additionally contains organic solvent.

The solubility of copolymers of vinyl chloride with vinyl esters, ethers and/or other comonomers is much better than that of PVC. These are used as flexible, thermoplastic film formers - either dissolved in organic solvent or in the form of dispersions - for physically drying paints and coatings.

Polyvinylidene fluoride

Organosols based on polyvinylidene fluoride (PVDF, monomer formula: $CH_2=CF_2$) produce extremely weather-resistant metal coatings. Unlike PVC, PVDF is not subject to degradation under natural UV radiation.

Chlorinated polymers

Whereas natural rubber used to be converted to "chlorine rubber" by means of chlorination, nowadays it is really only synthetic polymers that are still chlorinated, with the aim of improving chemical resistance and solubility. Base polymers are polyisoprene (synthetic rubber), polyethylene, polypropylene and ethylene-vinyl acetate copolymer. The chlorination process involves substitutions (of hydrogen atoms by chlorine atoms), possibly with additions (of chlorine molecules to double bonds, if present) and, to a varying degree, cyclisations (formation of ring-shaped molecule sections). *Chlorinated polyisoprene* contains 65 to 67 wt.% of chlorine; polyolefins used in coating applications are chlorinated to roughly the same level; polyethylene vinyl acetate absorbs significantly less chlorine.

Chlorinated polymers are highly resistant to water, chemicals and weathering, compatible with many other film formers and soluble in coating solvents such as aromatics and esters.

They are principally used in physically drying coatings for weatherproofing and corrosion protection, where they represent an alternative to vinyl chloride copolymers.

In the presence of basic pigments such as zinc oxide or lithopone (barium sulphate/zinc sulphide blend) and/or elevated temperature, chlorinated rubber in particular tends to decompose with release of hydrogen chloride. For that reason the long-term service temperature of chlorine rubber coatings should not exceed 60 °C.

Polyvinyl acetate, vinyl acetate copolymers

Polyvinyl acetate (PVAC) and vinyl acetate copolymers are used principally in the form of dispersions but also in solid form (as "PVAC resin") in paints, coatings and adhesives.

At 28 °C the glass transition temperature of the VAC homopolymer is unfavourably high for many applications and it is relatively readily saponified. These weaknesses can be alleviated by copolymerisation with suitable monomers; this greatly extends the compatibility range too. Incorporation of free acids, generally crotonic acid, improves adhesion to metals.

PVAC and VAC copolymers can be used in physically drying paints and coatings as principal film formers or as additives to improve adhesion, lightfastness, gloss and body.

Polyamides, polyaminoamides

High-molecular partially crystalline polyamides, which as ductile plastics are extremely important, have only a limited significance in marginal areas of coating technology (wire coatings, fluidised bed coatings, flame spraying) because of their poor solubility.

The polyamides (PA) of interest to us are condensation polymerisation products of dicarboxylic acids and diamines; the monomer units are bonded to one another by means of the non-basic amide group (–C(O)NH–).

The solubility of low-molecular polyamides condensed from dimeric fatty acids[1)] and diamines is significantly better. Boiling with alkyd resins produces highly thixotropic coating resins. These flexible polyamides, which adhere well to non-absorbent substrates, are also found in flexographic and gravure printing inks, overprinting inks and heat-sealing coatings.

Polyaminoamides are low-molecular polyamides with excess (i.e. free) amino groups. These basic substances are used as hardeners for flexible epoxy resin coatings (→ 2.1.4.8).

Polyvinyl ethers

Polyvinyl methyl, ethyl and isobutyl ethers in combination with cellulose nitrate or hard resins act as unsaponifiable polymer plasticisers ("soft resins"); they are also used as the raw material for pressure-sensitive adhesives. A special feature of the methyl ether is its solubility in water.

Polybutadienes, modified polybutadienes

Of the various homopolymers of butadiene, only the relatively low-molecular *polybutadiene oils* are of any interest for the coatings sector. Characteristic parameters include the moderate molecular weight, the ratio of 1,2 to 1,4 bonds and the cis/trans ratio. Since polybutadiene oils cure oxidatively, they are used as a more saponification-resistant alternative to fatty, drying oils, in anti-corrosion primers, for example.

As with drying oils, the reactivity of polybutadiene oils is also utilised for chemical modifications such as maleination, epoxidation, hydration (to produce hydroxyl groups), styrenation and acrylation. Maleination produces carboxylic acid resins, which are water-thinnable after neutralisation with amine and are used as film formers for anodic electrodeposition primers, for example.

2.2 Solvents

2.2.1 Classification and definitions

The vast majority of coating materials contain solvents, i.e. volatile compounds which serve primarily to adjust the viscosity of the coating material to the optimum level for a given application, usually within the range from around 0.05 to 1 Pa·s, and to control this level during application and film forming. The nature and quantity of solvent used largely depend on the film former system and the manner of application. Table 2.7 (page 96) provides a summary of typical solvent contents in various coating materials.

The composition of a solvent blend has a much greater influence on the properties of the final coating than would first be supposed. As well as its fairly obvious influence on the occurrence of film defects such as running or poor flow, it also influences adhesion to the substrate and anti-corrosive effect, for example. A number of factors must therefore be taken into account in choosing a solvent, including volatility, rheological behaviour, surface tension, combustibility, toxicity, odour, environmental compatibility and, not least, cost.

According to ISO 4618,
a **solvent** is a single liquid or blend of liquids, which is volatile under specified drying conditions and in which the binder is soluble.

1) Dimeric fatty acids are unsaturated fatty acids dimerised to long-chain, chemico-structurally non-uniform dicarboxylic acids.

Table 2.7: Typical solvent content of paints and coating materials

Coating material		Solvent content in %
Coating powders, silicate paints, silicate plasters		0
Emulsion paints for interior use		0 to 2
Emulsion paints for exterior use, synthetic resin bonded plasters, primers		3 to 5
Electrodeposition coatings		1 to 4
Latex coatings (environment-friendly)		7 to 9
Water-borne coatings (industrial)		3 to 18
High-solids coatings		20 to 30
Alkyd resin coatings, polyester resin coatings, epoxy resin coatings, polyurethane coatings, polystyrene resin coatings, phenolic/urea/melamine resin coatings		40 to 50 (exceptionally > 55)
Dilutions	for conventional coatings	100
	for water-borne coatings	0 to 50

Dissolution therefore does not chemically change the binder. Whilst this standard differentiates between solvent, thinner, extender for solvents, reactive solvent and reactive thinner, in practice all liquids used to make a coating material processable are known as solvents, and the only further distinction made is between extenders and reactive solvents.

According to ISO 4618,
a **thinner** is a single liquid or blend of liquids, volatile under specified conditions of use, added to a coating material to reduce viscosity or influence other properties

According to this standard,
a **diluent** is volatile liquid, single or blended, which, whilst not a solvent, may be used in conjunction with the solvent without causing any deleterious effects.

Common to all three definitions is the fact that the substances are liquid and are volatile under film forming conditions. The differentiation under DIN 55 945 roughly corresponds to the terms "active solvent", "solvent blend" and "latent solvent" that are used in industry, although the term "latent solvent" is taken rather further. We describe as *latent* a solvent which cannot dissolve a binder by itself but can dissolve it when blended with a second, active or latent solvent.

According to DIN 55 945,
a **reactive solvent** is a solvent which undergoes a chemical reaction during film forming to become part of the binder and thus loses its properties as a solvent.

A **reactive thinner** behaves with a reactive solvent in the same way as a thinner behaves with a solvent.

In very general terms we can say that solutions are obtained by mixing liquid, solid or gaseous components in liquids. In these cases the liquid is always characterised as a solvent. If a solution comprises two or more liquids, the liquid present in excess is normally described as the solvent. This is in distinction to the aforementioned standard, according to which the volatility is what distinguishes the solvent from the binders (which, by definition, are not volatile under the specified drying conditions). A solvent is generally required to be light and colourless, volatile with no residue, neutral, unreactive or only slightly reactive, to have a slight or at least not unpleasant odour, and generally to be anhydrous, scarcely toxic, biodegradable and moderately priced.

During dissolution, the molecules or ions of the substance to be dissolved are ideally completely separated from each other and solvated by solvent molecules. The viscosity of such solutions is virtually independent of the concentration of the dissolved substance. By contrast, polymer and usually oligomer solutions do not exhibit ideal behaviour. Their viscosity is heavily dependent on the concentration of the dissolved substance and there is often no clear saturation limit. They constitute a special type of colloidal solution (see section 2.1.1.5). In the case of large molecular aggregates (50 to 100 nm), a two-phase system is formed, which is a dispersion, not a solution. Accordingly, the continuous phase cannot be termed a solvent (nor, according to the pertinent standards, can be it called a thinner). The literature describes a whole series of models for establishing the parameters which determine either quantitatively and/or qualitatively the solubility or dissolution process of one substance in another. The solubility parameter model, as discussed in section 2.1.1.6, is one of the most common.

2.2.2 *Characterisation and classification of solvents*

2.2.2.1 *Solvating power*

A number of organic compounds are used as solvents in coating materials, either alone or blended with other liquids. These are generally low-molecular and low-viscosity liquids. Key criteria for formulators choosing a solvent are the solvating power and influence on viscosity and volatility. A fairly rough way of estimating solvating power is to classify solvents by the way they participate in hydrogen bonds:

- Solvents, which do not participate in hydrogen bonding.
- Solvents which react in hydrogen bridge bonds only as hydrogen acceptor (solvents with moderately strong hydrogen bridge linkage) and
- those which can react in hydrogen bridge bonds as both hydrogen acceptor and hydrogen donor (solvents with strong hydrogen bridge linkage).

The *hydrogen bridge linkage parameter* γ offers an alternative means of characterising solvents participating in hydrogen bridge bonds. It is obtained by IR spectographic measurements and describes the varying strength of hydrogen bridge bonds in each solvent. It should not be confused with the solubility parameter δ_H.

2.2.2.2 *Solvents, which do not participate in hydrogen bonding*

The group of solvents having weak hydrogen bridge linkage includes

- hydrocarbons,
- chlorinated hydrocarbons,
- nitro compounds and
- nitriles.

Hydrocarbons can themselves be subdivided into two large groups: aliphatic and aromatic hydrocarbons.

As aliphatic compounds, aliphatic benzines or paraffin hydrocarbons are chemically very resistant and display good dissolving power for relatively non-polar film formers such as mineral oils, many fatty oils, waxes and paraffin. They also dissolve such polar polymers as polyacrylic acid, polybutyl methacrylate or polyvinyl ether. Most other coating film formers are insoluble in them, however. See Table 2.8 for typical properties.

- **Special boiling point spirits**, according to DIN 51 693, are

dearomatised refined petroleum fractions having a flash point (according to DIN 51 755) below 21 °C and a fixed boiling range. A distinction is primarily made between grades 1, 2 and 3 of special boiling-point spirits.

Special boiling point spirits, (not quite standard), known also in the coatings sector as *specialist spirits*, are used in quick-drying coatings because of their rapid evaporation. In view of their flash point below 21 °C, they must be processed in explosion-proof areas only. Special boiling point spirits also include petroleum ether (see DIN 51 630).

Table 2.8: Physical properties of some hydrocarbons

Solvent	Boiling point in °C	Density d_4^{20} in g/cm³	Flashpoint in °C	Evaporation index
hexane (tech.)	65 to 70	0,644	–22	1,4
special boiling-point spirit grade 1	60 to 95	0,680 to 0,700	–30	2,0
special boiling-point spirit grade 3	100 to 140	0,740 to 0,750	–5	8
white spirit, grade 1	130 to 185	0,766	approx. 22	25 to 30
white spirit, grade 3	150 to 190	0,785	approx. 40	85 to 90
balsam terpentine	150 to 180	0,861	32	38
toluene	110 to 111	0,873	6	6,1
xylene (mixture of isomers)	137 to 142	0,874	25	17
solvent naphtha (II)	150 to 195	0,870	40 to 45	40 to 45

• **Test spirits**, according to DIN 51 632, are more specifically refined petroleum fractions having a flash point according to Abel of at least 21 °C and a fixed boiling range.

Test spirits are also known as white spirits. The DIN standard referred to above differentiates between grades 1 to 5 of white spirits, although this does not entirely correspond to the classification conventionally used in the coatings sector, which distinguishes between grades 21, 30, 40 and 60. White spirits are mainly used as solvents or extenders for oil-based coatings, alkyd resin coatings and chlorinated rubber coatings. They may contain a few per cent of aromatic hydrocarbons. Mineral "spirit of turpentine" is likewise an aliphatic hydrocarbon blend, although it is not in fact made from terpenoid hydrocarbons as its name suggests. For this reason it is also known as *turpentine substitute*. **V**arnish-**m**akers' and **p**ainters' naphtha (VMP naphtha), consisting of highly volatile aliphatic compounds with a boiling range between 100 and 150 °C, is also used. White spirits and VMP naphtha are used principally as extenders.

• **Cycloaliphatic hydrocarbons** are ranked between aliphatic and aromatic hydrocarbons in terms of their dissolving power; they are miscible with most organic solvents but are insoluble in water. As well as those resins capable of being dissolved by aliphatic compounds, they can also dissolve alkyd resins, bitumen and other substances.

• **Spirit of turpentine, wood turpentine** and **terpenoids** belong to the terpenoid hydrocarbons obtained by distillation of various conifers. They have a higher dissolving power than white spirits and are used in oil-based and alkyd resin coatings as high-boiling solvents. The higher-boiling constituents of oil of turpentine are also called **pine oils**, and serve as high-boiling solvents in stoving enamels.

• **Aromatic hydrocarbons** are generally more expensive than aliphatic solvents but have a higher dissolving power. They can dissolve alkyd resins, saturated polyester resins, polyacrylates, a range of vinyl (co)polymers and many less polar resins. For toxicological reasons benzene is not used as a coating solvent. Toluene and in particular xylene, one of the most important coating solvents, are commonly used. In this form xylene is a blend of the various dimethylbenzene isomers with a small proportion of toluene and larger amounts of ethylbenzene. Toluene and xylene are principally used in alkyd resins and in heat-curing coatings based on the various phenolic or aminoformaldehyde resins. When blended with esters they can also dissolve chlorinated polymers such as (post-chlorinated) PVC. Solvent naphtha is a blend of alkyl and dialkyl benzenes with three to five C atoms in the side chains. Nowadays, the use of aromatic hydrocarbons as solvents is often avoided, where possible.

Table 2.9: Physical properties of some ketones and esters

Solvent	Boiling point in °C	Density d_4^{20} in g/cm³	Flashpoint in °C	Evaporation index
acetone	56.2	0.792	−19	2
methyl ethyl ketone	79.6	0.805	−14	2.6
methyl isobutyl ketone	115.9	0.800	14	7
methyl amyl ketone	147.0 to 153.5	0.818	46	55
isophorone	215 to 220	0.922	96	230
ethyl acetate	76 to 78	0.900	−4	2.9
butyl acetate	123 to 127	0.880	25	11
1-methoxypropyl acetate	143 to 150	0.953	60	34
butyl glycol acetate (2-butoxy ethyl acetate)	190 to 198	0.945	75	250
ethyl diglycol acetate	210 to 220	1.011	107	> 1,200
N-methyl pyrrolidone (NMP)	202	1.027	95	

• Although their polarity means that they have a very good dissolving power for most resins, **chlorinated hydrocarbons** are no longer used as solvents in the coatings industry. Dichloromethane (methylene dichloride) is still used in stripping pastes or baths for the removal of coatings, but here too it is increasingly being replaced by aqueous systems.

2.2.2.3 *Solvents with moderately strong hydrogen bridge linkage*

Both esters and ketones are widely used as solvents for coating materials. Their carbonyl group makes them hydrogen acceptors and they generally exhibit excellent dissolving power. Under comparable vapour pressure, ketones are usually cheaper than esters. Unlike glycol ethers, ethers are little used in coating technology because of their high volatility. Table 2.9 shows typical properties for this class of solvents.

Ketones used as solvents are water-clear, readily mobile liquids having a characteristic odour. Lower ketones such as acetone or methyl ethyl ketone (MEK) dissolve polar resins and a range of less polar substances. In higher ketones the hydrocarbon character is more pronounced, making them good solvents for non-polar resins and (co)polymers. Acetone in fast-drying cellulosic coatings, methyl ethyl ketone (MEK), a low boiler often used in place of ethyl acetate, methyl isobutyl ketone (MIBK), a multi-use medium boiler, and methyl amyl ketone (MAK), a high boiler with excellent dissolving power, are all frequently used in coating materials. Isophorone and trimethyl cyclohexanone are used as a high-boiling solvent for physically and oven-drying coatings, where it improves substrate and pigment wetting (where applicable), flow, gloss and bond strength.

Like ketones, esters are also neutral and highly resistant clear liquids. Their inherently good dissolving power for polar substances diminishes as the length of the hydrocarbon chains grows, but increases similarly for less polar resins. Some of the lower esters are soluble in water. The esters used in the coatings industry are generally acetates. Esters are often preferred to ketones with roughly the same solubility and evaporating characteristics because of their generally more pleasant, often fruity, odour. Ethyl acetate is the most important solvent for quick-drying coatings such as those based on cellulose nitrate, and is also used in PU systems.

Butyl acetate is the most important medium boiler in the coatings industry. It is just volatile enough to be able to leave a coating film quickly without giving rise to the cratering, blushing (see page 101) or flow problems typical of excessive volatility. Solvent blends of butyl acetate with aromatic hydrocarbons and with butanol, which significantly increases the dissolving power, are important. Its low inherent viscosity also makes it extremely suitable for use as an auxiliary solvent in high solid coatings.

Table 2.10: Physical properties of some alcohols and glycol ethers

Solvent	Boiling point in °C	Density d_4^{20} in g/cm³	Flashpoint in °C	Evaporation index
methanol	64.7	0.791	6.5	6.3
ethanol	78.3	0.789	16	8.3
2-propanol (isopropanol)	82.3	0.786	15	11
n-butanol	117.7	0.811	34	33
2-methyl-1-propanol (isobutanol)	107.7	0.802	28	25
ethyl glycol (2-ethoxyethanol)	132 to 137	0.930	42	43
propyl glycol (2-propoxyethanol)	150.5	0.911	51	75
butyl glycol (2-butoxyethanol)	167 to 173	0.901	67	163
ethyl diglycol (2-(2-ethoxyethoxy)ethanol)	194.2	1.021	90	1,200
butyl diglykol (2-(2-butoxyethoxy)ethanol)	196 to 205	0.990	98	>1,200

1-methoxypropyl acetate is a pleasant-smelling solvent for many natural and synthetic resins, is miscible with organic solvents and to a limited extent with water and is less toxic than methyl or ethyl glycol acetate. Butyl glycol acetate is miscible with organic solvents but not with water. As a high boiler it improves flow in stoving coatings. Ethyl diglycol acetate is used in industrial, automotive and coil-coating paints to improve flow and to prevent blistering.

Esters should not be used as solvents for resins having primary or secondary amino groups because of the risk of amide formation by means of aminolysis. In water-borne coatings there is the additional risk of saponification or aminolysis due to the amines used as neutralising agent.

Nitroparaffins such as 2-nitropropane increase the electrical conductivity of a coating because of their high polarity. They are mainly used as co-solvents to improve pigment wetting, rheological behaviour, flow and electrostatic processability. 2-nitropropane is carcinogenic, however.

N-methylpyrrolidone (NMP), as well as being found in stripping agents, is used especially as an additive in polyurethane dispersions and as a true solvent for high-molecular polyurethane coatings.

2.2.2.4 Solvents with strong hydrogen bridge linkage

Alcohols are used when solvents are required which react both as hydrogen donor and hydrogen acceptor in hydrogen bridge bonds. Their dissolving power depends on the length of the non-polar hydrocarbon chain and the position of the hydroxyl group(s). Table 2.10 provides a summary of the typical properties of these solvents.

Lower alcohols, such as methanol or ethanol, have a pronounced dissolving power for strongly polar film formers, such as phenolic and amidoformaldehyde resins, cellulose derivatives or polyvinyl acetate. Non-polar or weakly polar substances are not dissolved.

In the case of higher alcohols, the dissolving power for polar resins diminishes as the hydrocarbon chain grows. They are therefore mainly used as extenders with the corresponding acetates. These alcohols are also used in top coats applied to a primer, where their gentle dissolving power means that the primer is neither softened nor lifted. On plastics too they cause only a slight surface swelling of the substrate, thus ensuring good adhesion of the coating without softening the plastic.

In the coatings industry it is mainly the relatively volatile alcohols up to the butanols that are used. Although n-butanol is not a good solvent for alkyd resins, the addition of even small quantities to alkyd resin or oil-based coatings produces a marked reduction in viscosity as well as improved flow and spreadability. It shows this same viscosity-reducing effect when used as an auxiliary solvent in water-borne coatings.

Ethylene and **propylene glycol ethers** are very good solvents for a number of polar resins since they carry both ether and alcohol groups ("ether alcohols"). They are miscible with a large number of both non-polar and polar solvents and for that reason represent excellent *coupling agents* in blends of only poorly miscible resins or solvents. Their alcohol group also makes them miscible with water and so they are widely used as a co-solvent and coalescing agent in water-borne coatings and water-thinnable systems. Their main disadvantage is their relatively low volatility, which somewhat restricts their use. They improve the wetting, flow and surface quality of a coating. For toxicological reasons monoethylene glycol ethers have been replaced in the paints and coatings industry by the corresponding monopropylene glycol ethers and also by butyl glycol.

When alcohols and glycol ethers are used as solvents, it must be remembered that they may react with isocyanates. Melamine resins undergo ether interchange with solvent alcohols, which may alter the speed of the curing reaction. However, problems of this nature can be minimised by the use of less reactive secondary or tertiary alcohols. The presence of monofunctional alcohols stabilises such coatings by reducing the probability of the melamine resins reacting with the hydroxyl groups of the polyester.

2.2.3 Properties

2.2.3.1 Volatility

During application and film forming, the volatile components of a coating material are intended to evaporate. The speed with which this occurs influences not only the drying time but also the appearance and the physical properties of the final coating.

Solvents are conventionally classified by boiling range. We differentiate between

- Low boilers with boiling ranges below 100 °C,
- Medium boilers with boiling ranges between 100 and 150 °C and
- High boilers with boiling ranges over 150 °C.

In the paints and coatings industry, however, it is at least as important to know the volatility of a solvent below boiling point as well as its boiling range.

According to *Clausius-Clapeyron* the vapour pressure of a liquid is principally dependent on the molar enthalpy of vaporisation, as shown by the following relation:

$$\ln p_L = A - \frac{\Delta_v H_m}{R \cdot T}$$

where $\Delta_v H_m$ is the molar enthalpy of vaporisation, p_L is the vapour pressure of a liquid and A is a constant.

This shows that vapour pressure rises as the temperature T increases. According to the *Pictet-Trouton* rule, the boiling temperature of a solvent T_v

$$\frac{\Delta H_v}{T_v} \approx \text{const.}$$

is proportional to the enthalpy of vaporisation. Although this suggests that both vapour pressure and boiling point of a solvent are dependent on the enthalpy of vaporisation, there is no general correlation between these two quantities. For example, solvents forming hydrogen bridge bonds – like water, alcohols or amines – are less volatile than other solvents with the same boiling point. Within a class of chemically very similar solvents, however, the rate of evaporation generally falls as the boiling point rises. The rate of evaporation of a solvent is also dependent on

- vapour pressure,
- specific heat,
- enthalpy of vaporisation and
- surface tension,
- molecular weight,
- humidity.

Since most of the factors that determine the rate of evaporation are in turn dependent on one another, it is almost impossible to make theoretical predictions about the precise extent of solvent release from the coating film. The empirical Knudsen equation

$$w\left[\frac{g}{cm^2 \cdot s}\right] = K \cdot P_V \cdot \sqrt{\frac{2\pi RT}{M}}$$

where the transmission coefficient K ≈ 1 and the vapour pressure is Pv, states that the rate of evaporation w is proportional to the square root of the temperature and inversely proportional to the square root of the molecular weight.

Characterisation of evaporation

In practice we compare the rate of evaporation of a solvent with that of diethyl ether.

According to DIN 153 170
the **evaporation index** (EI) is the ratio of the evaporation time measured for the test liquid to the evaporation time for *diethyl ether* ($C_2H_5OC_2H_5$) as reference liquid.

$$EI = \frac{t_{sv}}{t_{ether}}$$

where t_{sv} is the evaporation time for the solvent and t_{ether} is the evaporation time for diethyl ether. The evaporation indices for a number of solvents are set out in Table 2.11.

In comparing evaporation indices it is important to check the reference liquid. In the USA a similar classification uses *butyl acetate* as the reference quantity, according to the following relation:

$$E\ (\text{evaporation rate}) = \frac{t_{90}\ (\text{test solvent})}{t_{90}\ (\text{n–butyl acetate})}$$

where t_{90} is the time taken to evaporate 90% of a test sample.

Table 2.11: Examples of the evaporation index of solvents (for diethyl ether = 1)

Solvent	Boiling point in °C	Evaporation index
white spirit K 30	145 to 200	35
xylene (mixture of isomers)	106.2	17
1-butanol	82.3	33
butyl glycol	167 to 173	63
butyl diglycol	224 to 234	> 1,200
methoxypropanol	122.8	38
butyl acetate	123 to 127	11
2-butoxyethyl acetate	190 to 198	250
methyl isobutyl ketone	115.9	7
N-methylpyrrolidone	202	95

Table 2.12: Classification of solvents according to their volatility

In Germany		In the USA	
classification	**evaporation index (EI)**	**classification**	**evaporation rate (E)**
highly volatile	< 10	rapid evaporation	< 0.8
moderately volatile	10 to 35	moderate evaporation	0.8 to 3.0
difficultly volatile	35 to 50	slow evaporation	> 3.0
scarcely volatile	> 50		

Solvents can be classified according to their evaporation index as shown in Table 2.12.

The evaporation indices apply to pure solvents. However, if higher-molecular substances are dissolved in a solvent, the rate of evaporation falls and is also dependent on the nature of the dissolved substance. We then need to consider the solubility parameter model (→ 2.1.1.6).

In the solubility parameter zone, if the solubility parameter triplet for the solvent is close to the centre of the polymer's solubility volume then it is a very good solvent for the polymer molecules. The solvent therefore leaves the drying polymer film only slowly and in extreme cases there will be solvent retention. The further the solubility parameter triplet of the solvent is shifted towards the margin of the solubility volume, or the further outside this margin it appears, the weaker these interactions become.

Evaporation of solvent blends

For the optimum drying of a coating film, a blend of different solvents is generally required. *Raoult's law* approximately describes the vapour pressure of the individual components in this case.

$$p_i = x_i \, p_i^0$$

where p_i is the partial pressure of component i, x_i is the mole fraction of component i and p_i^o is the vapour pressure of the pure components.

This law strictly applies only to ideal blends (i.e. blends in which the intermolecular forces between all of the components are of the same size). In real blends, deviations from the ideal behaviour occur; these can be attributed to the interactions between the molecules of the blended components and are different for each solvent blend.

Particularly in the case of blends containing water as one of the components, the strong interactions with this component can often lead to marked deviations from Raoult's law. At the same time the rate of evaporation of the water is strongly influenced by the relative humidity (r.h.). Whilst at a relative humidity of below 5 % and a temperature of 25 °C water evaporates at roughly the same rate as methoxypropyl acetate with an evaporation index of 34, its evaporation index at 100 % r.h. is infinite. For example, if a blend of butyl glycol and water is evaporated, butyl glycol accumulates in the remaining mixture at low r.h., whereas water accumulates in the mixture at high r.h. At a given r.h. (around 80 % in this case), both components evaporate at the same rate and the composition of the liquid remains constant. This is then known as *critical relative humidity*.

Many water-soluble solvents form *azeotropes* with the water. In such a case the evaporation of the solvent-water mixture may be greatly accelerated in comparison to the evaporation of the individual components. Here too, of course, the relative humidity influences the composition of the evaporating mixture.

Cooling effects

As a consequence of the high heat of evaporation of water, the rate of heating of a hydrous coating film appears lower than that of an anhydrous solvent-containing film. This may cause a delay in the start of the crosslinking reaction during stoving.

As the solvent evaporates from a coating film, it causes the film to cool. If the temperature of the coating is below the *dew point* of the ambient air, water condenses from the air onto the surface of the coating. The risk of this occurring is naturally greater in conditions of high relative humidity than in low relative humidity. If the coating contains components that are capable of absorbing water - at least in small quantities - then the condensed water will also be absorbed by the coating film and will be more or less homogeneously distributed within it. This is not usually apparent in the coating. If on the other hand the condensate remains in the surface region, it is visible as a haze known as "blushing". Blushing can be countered by the use of solvents forming an azeotrope with water, such as many aromatic hydrocarbons or butanol.

All in all, the rate of evaporation of a solvent depends not only on the vapour pressure but also on the temperature, the surface to volume ratio and the movement of air over the surface.

The decisive temperature for the rate of evaporation of a solvent is that at the surface of the given medium, e.g. of a freshly applied coating film. This initially corresponds to the ambient temperature but falls during evaporation. The temperature gradient formed between the surface of the medium, the gas phase above it (ambient air or similar) and the underlying layers is balanced out again by heat exchange. If this takes place rapidly, the surface layer more or less retains the temperature of its surroundings; otherwise a sharp jump in temperature occurs at its surface. This cooling is most marked in solvents which evaporate quickly or which have a high enthalpy of vaporisation.

The influence of the surface to volume ratio is based on the fact that all evaporating molecules must pass through the interface between coating film and gas phase. Thus a particular quantity of liquid evaporates in a narrow vessel much more slowly than the same quantity spread out over a large surface area. If we apply this principle to conditions during coating, this means that in the case of spray application, where atomisation into many small droplets creates a very large surface area per volume, a large part of the solvent has already evaporated during the short flight time from the spray head to the surface to be coated, before the coating settles on the surface.

For a given substrate surface, a 100 μm thick coating film initially dries as quickly as one only half as thick. However, since it contains twice the amount of solvent, after a given time its solvent content is higher than that of the thinner film. The viscosity in the thicker film increases correspondingly more slowly, resulting in different flow and dripping behaviour.

Movements of air over the surface of the coating influence the drying speed because the rate of evaporation is governed by the partial pressure of the solvent above the coating film. In still air the solvent accumulates in the gas phase, its partial pressure is therefore greater and it evaporates more slowly. Air movements transport already evaporated solvent away from the surface of the coating, preventing the solvent from accumulating, as a result of which its partial pressure drops. Spraying with compressed air produces much greater air movements than airless spraying, as a consequence of which solvent losses during spraying are significantly higher for the same solvent composition. For this reason the solvent composition chosen for airless spraying must be different from that used for spraying with compressed air.

2.2.3.2 Polarity

Although no unambiguous physical definition exists for the term polarity and various parameters, such as dipole moment, polarisability or tendency to participate in hydrogen bridge bonds, are also included, in practice solvents are frequently categorised according to their polarity into *polar* and *non-polar* solvents as a means of describing their dissolving power. Highly symmetrical molecules such as tetrachloromethane have no dipole moment whatsoever; others - such as aliphatic or aromatic hydrocarbons - have only a small dipole moment for the same reason.[1] However, even good solvents such as dioxan, whose dissolving power is comparable to that of the highly polar DMSO (dimethyl sulphoxide), often display only a small permanent dipole moment.

It is therefore impossible to predict dissolving power solely from the size of the dipole moment. A more suitable mechanism is the *solubility parameter model*, which can also be used to determine the interactions between the dipoles. Together with the other two values in the three-dimensional solubility parameter, the parameter δ_p is thus a useful means of considering the dissolving power of a solvent attributable to its *polarity*. (See section 2.1.1.6 for a more detailed explanation).

2.2.3.3 Surface tension

The influence of the surface tension of a solvent or coating material, firstly on the wetting of the substrate (and of the pigments and fillers) and secondly on the formation of the surface of a coating system, is discussed in connection with surfactants. These highly effective additives are added in only small quantities to the coating material, however. By contrast, coating materials usually contain large amounts of solvent. For this reason, it is natural that the surface tension of the solvent also influences that of the entire coating material.

Surface tension and rate of evaporation are ultimately both determined by cohesive forces. There should therefore be a connection between the surface tension of a solvent and its simple solubility parameter, which after all is defined as the square root of the cohesive energy density. The following empirical relation has been found for polar and non-polar solvents, which very neatly describes this connection:

$$\delta = 2.1 \cdot K \cdot \left(\frac{\gamma_0}{V^{1/3}} \right)^a$$

where V = molar volume of the solvent [cm^3/mol], K = substance-dependent constant (approx. 3.6), a = substance-dependent constant (approx. 0.56), δ = solubility parameter [$(J/cm^3)^{1/2}$] and γ_0 = surface tension [mN/m].

2.2.3.4 Density

The density of a solvent is normally determined at 20 °C. Since it is influenced by impurities, it can be used in the same way as the boiling course and the refractive index to check the degree of purity of a solvent. With the exception of chlorinated solvents, the densities of organic solvents are generally lower than that of water. The difference in density between two solvents may be of economic interest if solvents are bought by *weight* but the manufacturing costs for a coating material are calculated by *volume* or if the final coating material is sold by *unit volume*.

Most environmental and health and safety regulations state limits in mass per unit volume, e.g. mg/m^3. Such regulations encourage the use of lower-density solvents because in the paints and coatings industry the use of solvents is largely based on effects that depend upon the volume content of the solvent in the coating material.

2.2.3.5 Viscosity

The viscosity η of solvents normally rises in an homologous series with increasing molecular weight, e.g. methanol (0.61 mPa·s), ethanol (1.19 mPa·s), n-propanol (2.26 mPa·s), n-butanol (3.00 mPa·s) and hexanol (4.3 mPa·s). It decreases with rising temperature.

1) A molecule displays permanent dipole moment if the centre of negative charge distribution no longer coincides with that of positive charge distribution. Just as a magnet has a north and a south pole, such molecules then have a negative and a positive pole.

Coating formulators can adjust the viscosity of a coating material by their choice of film former, the use of rheological additives and quite fundamentally by the type and quantity of added solvents. The influence of solvents on the viscosity of a coating material can essentially be attributed to two factors:

- the viscosity of the solvent itself and
- the interactions between film former and solvent and also between various solvents in a blend.

If we compare the viscosity of polymer solutions with that of pure solvent, the relatively small differences in the viscosities of the various solvents at first appear to be negligible. For the viscosity range between 0.1 and 10 Pa · s the dependence of the viscosity of a solution on that of the pure solvent can be described by the following relation:

$$\eta_{sol} = K' \cdot \eta_{sv}$$

where K' is a constant dependent on the type and quantity of dissolved polymer, η_{sol} is the viscosity of the solution and η_{sv} is the viscosity of the solvent.

However, in the case of a 50 % polymer solution in two solvents, even small differences in the viscosities of the solvents, of around 0.1 mPa · s, can produce a viscosity difference in the solutions some 3 to 4 orders of magnitude greater.

The flow properties of polymer solutions are generally dependent on the volume content of the participating substances. For this reason it is best to assume equal volume contents of solvents when comparing the viscosity of solutions containing various solvents. However, since we know only the volumes of the pure components and polymer solutions only rarely behave ideally, the data obtained in this way is clouded by a certain degree of uncertainty. Thus, whilst we can recognise a connection between the viscosity of the polymer solution and that of the pure solvent in the esters listed in Table 2.13, a comparison with xylene clearly shows that this principle cannot be applied to all solvents, particularly those which undergo very different interactions with the polymer.

Interrelations

The *interactions between molecules of different solvents* are generally small, except in the case of blends containing water or alcohols. These are characterised by a relatively high viscosity in comparison with other solvents of comparable molecular weight. The development of intermolecular hydrogen bridge bonds leads to the formation of high-molecular complexes.

When solvents of this type are added to other solvents in relatively small proportions (below 40 %), these bridge bonds are loosened. For this reason the viscosity of the blend does not increase by an amount equivalent to the relatively high viscosity of the alcohols. This effect is even more pronounced in water-solvent blends.

Table 2.13: Viscosities of solutions of a high solid acrylate resin in various esters at 25 °C (c_{sv} = 400 g/l solution)

Solvent	η_{sv} [1] **(mPa · s)**	ρ_{sv} [2] **(g/cm³)**	η_{sol} [3] **(mPa · s)**
ethyl acetate	0.46	0.894	121
n-butyl acetate	0.71	0.883	202
isobutyl isobutyrate	0.83	0.851	367
xylene	0.66	0.877	367

1: viscosity of solvent; 2: density of solvent; 3: viscosity of solution

In contrast to the interactions between molecules of different solvents, *interactions between solvent and polymer molecules* generally have a major influence on viscosity. Although these effects are not yet fully understood, they are essentially attributed to two factors that we shall discuss here.

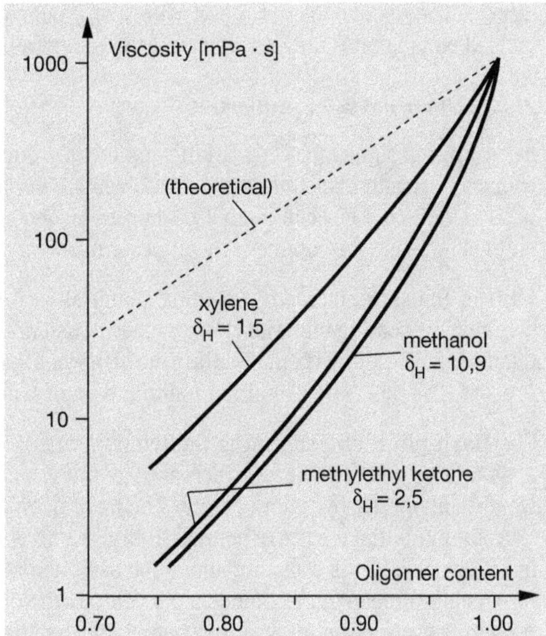

Figure 2.25: Comparison of the reduction in viscosity on dilution of a solution of an OH-functional UV-curing oligomer with xylene, MEK and methanol, according to [38].

Hydrogen bridge bonds

Most of the film formers used in paints and coatings contain *polar substituents*, some of which can participate in hydrogen bridge bonds. Typical of these are carboxyl, hydroxyl or ester groups. The resin molecules form intermolecular complexes via these groups, causing the viscosity of such solutions to rise significantly. To reduce it, the interactions between these groups must be weakened. Polar solvents that can only react as H acceptors in hydrogen bridge bonds, such as ketones or esters, are most capable of doing this. Being so abundant, the solvent molecules occupy all of the groups in the resins that are capable of forming complexes. Since they themselves are still only participating in a single hydrogen bridge bond, they can no longer form complexes between the resin molecules and so the viscosity falls correspondingly.

Figure 2.25 demonstrates this by means of the viscosity of solutions of a hydroxy-functional epoxy oligomer in three different solvents with differing solubility parameters δ_H. Both methyl ethyl ketone (MEK) and methanol reduce viscosity much more markedly than xylene, which has only a low value for δ_H. Methanol, whose δ_H value is much greater than that of MEK, reduces viscosity only slightly more effectively than MEK.

Volume effects

The viscosity of a polymer solution depends to a great extent on the *hydrodynamic*[1] *volume* of the dissolved polymer molecules. The greater this hydrodynamic volume, the higher the viscosity for the same concentration. Solvents having strong interactions with the dissolved polymer extend the polymer coils, whereas only weak interactions cause the coils to contract accordingly. Solvents for which the latter is the case should therefore reduce the viscosity of a solution more effectively than good solvents. This is the reason for the marked viscosity-reducing effect of benzines as extenders, for example. If the interactions between solvent and polymer molecules are very weak, however, the interactions also existing between the resin molecules cannot be neutralised; polymer complexes form and the viscosity rises again. Since the addition of a diluent shifts the solubility parameters of the solvent blend towards the edge of the solubility range of the binder, there is a risk that the solubility parameters of the solvent blend will move out of the solubility range of the binder and that haze will occur.

Most coatings contain a blend of solvents, making the situation even more complicated. In this case the viscosity is essentially determined by whichever solvent displays the strongest interactions with

1) This is taken to be the volume of a molecule in respect of its flow-mechanical resistance to movements in the solution. It is for example high when the molecule coil is extended and at its lowest when such a coil has undergone the maximum possible contraction.

the resin. The dependence of the viscosity of a polymer solution on the molecular weight of the dissolved polymer, its concentration and the temperature were covered in section 2.1.1.8.

2.2.3.6 Other physical properties

- In coatings designed for electrostatic application, **conductivity** can be an important factor in the choice of solvent. The conductivity of solvents correlates with their dielectric constant hydrocarbons exhibit very low conductivity, whereas alcohols, nitroparaffins and even small quantities of amines increase the conductivity of a coating.

- Most of the solvents used in coating materials are readily combustible. Whilst fires involving combustible coating materials have caused considerable damage and great care should always be taken when handling them, the **flammability** of a solvent plays only a minor role in the selection process. The flash point, ignition point and explosive limits are key safety parameters, however.

- The **flash point** of a solvent is the temperature at which a solvent vapour/air mixture can be set alight by a naked flame. It increases as the vapour pressure drops, i.e. with rising molecular weight and increasing boiling point. The flash point is the basis on which solvents are classified - according to the degree of combustibility - into *danger classes*. What is important here is that in blends of solvents it cannot simply be assumed that the flash point is set by the most readily flammable component. In blends of solvents with very diverging solubility parameters δ_H it may be significantly lower; on the other hand it is often higher in the case of blends including chlorine-containing solvents.

- Above a certain temperature, solvent vapour/air mixtures can ignite even without the presence of a naked flame. This **ignition temperature** (or ignition point) is used in accordance with the German VDE regulation 0171 to group solvents into *temperature classes* or *ignition groups*. A solvent vapour/air mixture can only explode within a certain range of blends characterised by the lower or upper **explosive limit**. Below this range the solvent vapour concentration is too low for explosive combustion, above it the oxygen concentration is too low. The explosive limits are given in vol.% or g/m^3 at 1013 hPa.

2.2.3.7 Physiological properties

Although solvents are inert in respect of the dissolved substances, they can enter the human body in the form of solvent vapour via the lungs and in liquid form through the skin or mouth and can of course also be absorbed by animal or plant organisms. The severity of their effect varies, in part according to the amount of solvent absorbed and the exposure time. In all cases care should be taken when handling solvents to avoid physiological damage, especially in the following situations:

- Chronic effects are dangerous and familiarisation can often prevent them from being recognised in time.

- Acute solvent poisoning can be recognised by confusion, headache and dizziness, even extending to loss of consciousness.

Chronic poisoning by a given solvent can cause damage to specific organs after extended periods. Repeated contact with solvents degreases the skin, leading to dryness and cracking through which microorganisms and dirt can easily penetrate. Some solvents also cause direct skin irritation in the form of reddening and swelling. A number of solvents, some of them toxic, are absorbed very easily through the skin and thus into the body.

A number of solvents have *carcinogenic* (cancer-causing), *teratogenic* (damaging to the embryo), or *mutagenic* (gene-altering) properties; they can also trigger allergic reactions in the case of sufficient sensitisation. The appropriate safety instructions and technical directions must always be followed when handling solvents.

Almost all solvents have a more or less pronounced characteristic odour. This should be regarded as a form of advance warning when handling solvents. The odour threshold at which a vapour can be detected depends on the solvent itself and differs greatly from one individual to the next according to their perceptive faculty. Many odours which in small concentrations seem quite pleasant are damaging in high doses or under constant exposure. Others which at first seem unpleasant are grown accustomed to after extended exposure. The fact that most solvents with an unpleasant or strong odour are harmful to health does not mean that we can assume that those characterised by a pleasant or slight odour are safe.

2.2.4 Solvents in coating materials

2.2.4.1 Influences of solvents on the properties of coatings and coating systems

The main purpose of solvents in coating materials is to adjust their viscosity for application and to control it during the drying process. At the same time, however, solvents also influence a whole series of other properties and processes involved in the manufacture, application and drying of coatings, as the following examples show.

- The influence of coatings on the *wetting of pigments* is often underestimated. They influence values such as the interfacial tension between the film former solution and pigment surface and the viscosity of the solution, both of which determine the rate of pigment wetting, and they compete with the film former and dispersing additives for adsorption space on the surface of the pigment. The nature and strength of the interactions between the binder or solvent on the one hand and the pigment surface on the other are dependent here on the type of pigment and the solubility and hydrogen bonding parameters of the binder or solvent.
- When true solvents are the least volatile component of a solvent blend, the gloss of a coating is generally increased because it is critically dependent on the surface condition of the coating. Such high-boiling true solvents allow the individual coating particles to merge well with one another during drying and also prevent the formation of eddies in the still liquid film due to evaporation of the solvent for example, thus producing a smooth, even film.
- Finally, solvents can also have a pronounced effect on the mechanical properties of a coating film. This is attributable to several factors. Depending on its dissolving power, the solvent may be able to align the film former molecules (see Figure 2.4) or to prevent their alignment. In reactive coatings certain solvents can partially react with a film former component to reduce the crosslink density, which corresponds to internal flexibilisation. If solvent is additionally withheld from the coating film, it acts as an external plasticiser.

2.2.4.2 Solvents in low solid and medium solid coatings

The demands made on the rheological behaviour of a coating by the application process, drying temperature and drying time for the coating are critical for the composition of the solvent. With low boilers, drying after application can be forced, whereas medium and high boilers remain in the coating for longer periods. High boilers in particular should instigate the formation of a perfect coating film surface. The following solvent composition is typical for a coating that is physically drying at room temperature:

- approx. 45 % low boilers
- approx. 45 % medium boilers and
- approx. 10 % high boilers.

It has proved to be beneficial for film-forming if the composition of the solvent blend - containing true solvents and extenders - changes such that during evaporation the position of its solubility parameter triplet is shifted from the solubility margin towards the solubility centre. This leads to rapid surface drying of the coating film on the one hand and on the other hand ensures that the binder does not "precipitate" during the drying process, i.e. the coating is able to dry without becoming hazy. On the other hand, excessive compatibility between residual solvent and binder harbours the risk of solvent retention.

To prevent the coating from "boiling" during stoving, coatings that are dried at elevated temperatures contain only small amounts of low boilers, if any, and a correspondingly greater proportion of high boilers.

2.2.4.3 Solvents in high solid coatings

According to ISO 4618-1, a high solid coating material is a term applied to coating materials in which the content of volatile matter is kept to a minimum, consistent with the maintenance of satisfactory application properties. They generally contain relatively little solvent (roughly half as much as in conventional coatings). As a consequence of this, the solvents used here must be capable of greatly reducing the viscosity of the resin solution even when added in the smallest possible quantities. Their low average molecular weight and relatively high content of polar functional groups make high solid film formers soluble in virtually all solvents with the exception of benzines. For this reason it is impossible to determine precisely the limits of the film former's solubility range in the solubility parameter zone. For the same reason it is also almost impossible to estimate the influence of the solvent composition on the film former/solvent interactions.

It is generally the case that solvents having a lower inherent viscosity demonstrate a strong viscosity-reducing effect. In order to obtain good flow properties it is essential to include high boilers with good dissolving power. For this reason butyl acetate and butanol are mainly used in high solid coatings, usually in combination with glycol ethers and glycol ether acetates.

2.2.4.4 Solvents in water-borne coatings

Modern water-borne coatings contain between around 0.1 and 15 % auxiliary solvent (known as co-solvent) which acts as solubiliser. In the presence of the film former, at least, this must generally be unlimitedly miscible with water. The viscosity anomaly known as the water mountain (discussed in 2.1.1.9) that occurs in water-borne coatings is reduced or even completely eliminated by the use of auxiliary solvents, and the flow improved. In many cases they also have a favourable influence on the drying behaviour of the coating, reduce foaming - at the time of application - and they largely prevent the risk of boiling during stoving.

In order to achieve the best possible dissolving power throughout the entire film-forming phase, the auxiliary solvent(s) should leave the coating film roughly at the same time as the water. To prevent incompatibility, it is vital for the drying coating film always to contain enough co-solvent until all the water has escaped from the film.

The most important solvents of this type used in water-borne coatings are glycol ethers such as butyl glycol or methoxy propanol, alcohols such as the various butanols or propanols and, increasingly, N-methylpyrrolidone. Often, the amines used to neutralise the carboxyl groups of the binder have solvent properties as well. Esters cannot be used in water-borne coatings because they are saponified by the neutralising agent. That said, in certain cases it may be advantageous to use solvents that do not form an azeotrope with water. Tert.-butanol or isopropanol, for example, leave the coating film before the water, thus increasing viscosity and preventing the liquid paint from dripping. Higher boiling solvents such as dialkylene glycol ether remain in the film longer than the water, keeping it highly liquid and thus improving flow. At the same time they keep the film open for longer, reducing the risk of boiling.

In aqueous dispersions too, solvents are often used in small quantities as coalescing and flow control agents. To ensure good film formation, coalescing agents need to remain in the film for longer than the water. On the other hand a film remains relatively soft whilst it still contains solvent. Since the rate of evaporation of water is heavily influenced by relative humidity, in conditions of high relative humidity every high boiling solvent may theoretically leave the coating film before the water, thus preventing a complete film formation. For this reason, the composition of the solvent blend that would produce the best range of properties under the anticipated film-forming conditions must often by determined by experiment.

2.3 Pigments and fillers

2.3.1 Definitions and classification of pigments

Under DIN 55 943 (General terms) colorant is the generic term for all colour-imparting substances. Colorants can be categorised from a technological standpoint into dyes, pigments, sole pigments, pigment blends, pigment preparations, etc., or from a coloristic and chemical standpoint (DIN 55944).

According to DIN 55 943 a **pigment** is
a substance consisting of particles which is practically insoluble in the application medium and is used as a colorant or by virtue of its corrosion-inhibiting or magnetic properties.

In contrast, according to the same standard, a **dye** is
a colorant that is soluble in the application medium.

This standard also makes a distinction between the terms filler and pigment.

A **filler** is
a substance consisting of particles which is practically insoluble in the application medium and is used to increase volume or to improve technical properties and/or to influence optical properties.

Note: The term "extender" should be avoided in this context. The terms "extender pigment" and "pigment extender" are incorrect.

On this basis, whether a substance should be regarded as a filler or a pigment is determined by its application.

From a coloristic standpoint, pigments - separated into organic and inorganic types - are subdivided into:

- white pigments (organic species currently of no practical importance)
- coloured pigments
- black pigments
- special effect pigments
- fluorescent pigments.

For coating applications it has proved helpful to add a further level of classification for coloured pigments, by hue. In order to understand the properties of pigments, however, classification from a chemical standpoint is more useful.

In addition to these methods of classification, there is one further classification system known as the Colour Index (C.I.). Here a distinction is made between

- the generic term (C.I. I) or
- the chemical constitution (C.I. II).

In the C.I. I, the *first* letter indicates the *type of colorant*, where P stands for pigment and S (solvent) for water soluble dyestuffs. The *second* capital letter (or *second* group of letters) indicates the *hue*, as shown in Table 2.14. In the case of some pigments, additional

Table 2.14: Hue codes in the C.I. I

Hue	Code
blue	B
black	Bk
brown	Br
green	G
metal	M
orange	O
red	R
violet	V
white	W
yellow	Y

distinctions are made between various modifications, e.g. P.B. 15:2 for a copper phthalocyanine pigment in the a-modification, which has additionally been stabilised against flocculation. To this combination of letters is added a *1 to 3 digit* number which precisely identifies the *pigment*. For example, iron oxide red pigments are designated by P.R. 101.

The C.I. II is composed of a five-digit number indicating the chemical compound class. The numbers 10,000 to 10,299 are reserved for nitroso compounds, for example, 11,000 to 19,999 for monoazo compounds, 74,000 to 74,999 for phthalocyanine compounds. Oxide pigments have a C.I. II of between 76,000 and 76,999; for other inorganic pigments it is between 77,000 and 77,999.

Not every organic pigment has a C.I. II since the chemical constitution has not been published for all of them, but they all have a C.I. I. In practice this is also more commonly used, since it groups all pigments of the same type under the same designation. Such pigments then have the same basic properties but may differ considerably on detailed inspection. Thus, for example, all titanium dioxide pigments, irrespective of whether they are anatase or rutile types, are classed under the Colour Index P.W. 6.

2.3.2 Physical principles

The colour-imparting properties of pigments, such as colour intensity, hiding power and transparency, as well as their influence on gloss, are dependent not only on their chemical composition but also on physical properties such as crystalline structure and the shape and size of the crystals. In the disperse state the state of distribution in the application medium and interactions with the film former and other components of the coating also play an important part in determining the application-oriented behaviour of a pigment.

2.3.2.1 Pigment morphology

Pigments are produced in the form of more or less finely divided powders. The powder particles consist of single crystals, coalescent crystals, aggregates and agglomerates.

The terms used in the morphological characterisation of pigments are defined in DIN 53 206 (Figure 2.26).

On that basis,

- a **particle**

is a separable unit of a pigment. It may be of any form and any structure.

- **primary particles** or **individual particles**

are particles that can be recognised as individual entities by suitable physical processes (e.g. using a light microscope or electron microscope).

- an **aggregate**

is an intergrown association of primary particles aligned side by side, where the total surface area is smaller than the sum of the surface areas of the primary particles.

- an **agglomerate**

is a non-intergrown association of primary particles aligned, for example, along edges and corners and/or of aggregates, where the total surface area is little different from the sum of the individual surface areas.

- a **flocculate**

is a (liquid-filled) agglomerate occurring in suspensions (e.g. in pigment-binder systems), which can be broken up by means of low shear forces.

Since aggregates cannot be further divided into individual particles under the shear forces occurring during dispersion, individual particles and aggregates are occasionally also referred to as primary particles, contrary to this standard. Since in aggregates the individual particles are aligned side by side, there are no cavities between them, which means that aggregates - unlike agglomerates and flocculates - have no inner surface either.

The smallest entity occurring in pigments is the *crystal*. Typical ionic crystals such as NaCl and many other inorganic compounds are held together overwhelmingly by ionic bonds. The pigments belonging to this group are titanium dioxide, iron oxides and lead chromates. Molecular crystals are largely held together by hydrogen bridge bonds and/or by dipole and dispersion forces. As well as the pure organic pigments, this group also includes a whole range of inorganic compounds together with organo-metallic compounds such as the Cu-phthalocyanines (Figure 2.27).

The distinction made in DIN 55 944 between inorganic and organic pigments is roughly in line with this classification, with pigment blacks and metallic pigments occupying a special position.

Both inorganic and organic compounds can occur in various different crystalline structures (modifications). Titanium dioxide, for example, is used in both the rutile and anatase modification, iron oxide and aluminium oxide occur in an α and a γ form. Many organic pigments likewise occur in various crystalline modifications, such as the Cu-phthalocyanine and quinacridone pigments. The various crystalline modifications usually have quite different optical and application-oriented properties. For example, anatase is distinguished from rutile by a lower refractive index (→ hiding power), its photochemical activity is significantly higher (→ photochemical degradation) and its light absorption in the range from around 400 to 430 nm is somewhat smaller (→ tint).

The diameters of pigment particles cover a broad size range. Figure 2.28 compares this range with the size range for various other particles (down to molecules), the wavelengths for various types of electromagnetic radiation and the analytical methods suitable for specific particle size ranges. The smallest particles are found in pigment blacks, whose particle diameters lie in the range from 10^{-9} to 10^{-7} m. They are followed by organic pigments

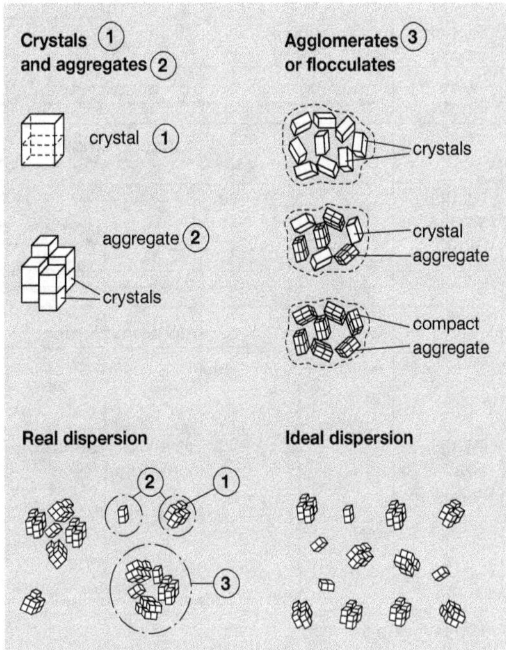

Figure 2.26: Pigment model according to DIN 53 206 for crystals, aggregates and agglomerates, and distinction between real and ideal dispersions

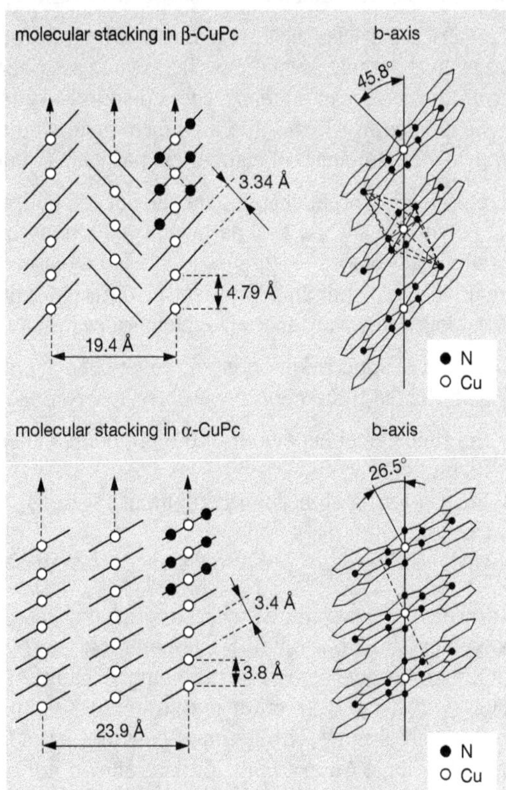

Figure 2.27: Schematic view of the molecular crystals of α- and β-Cu-phthalocyanine (β-CuPc), according to [1]

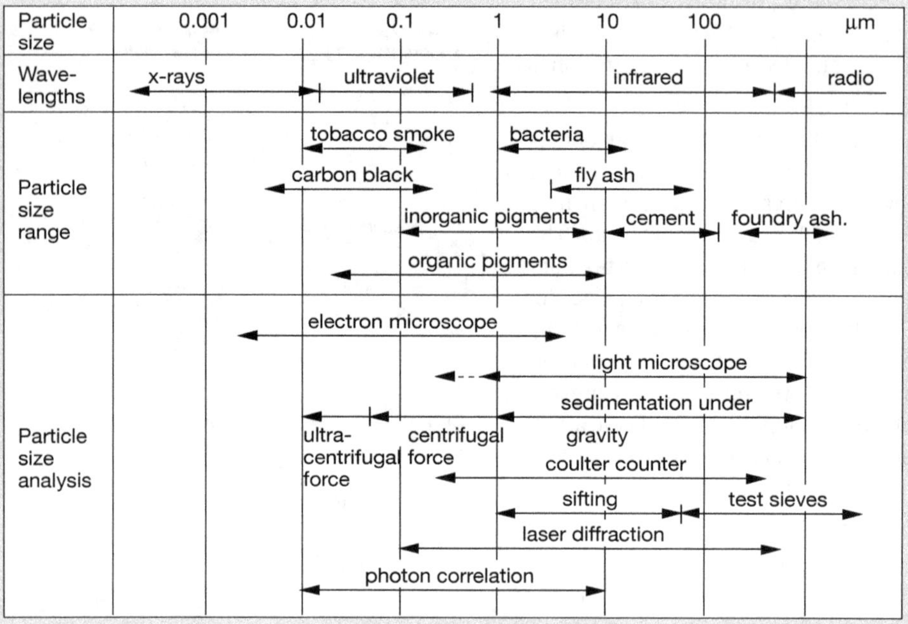

Figure 2.28: Particle sizes of pigments in comparison to sizes of other particles and analytical methods, according to [40, Bayer AG]

with diameters normally between 10^{-8} and 10^{-6} m. Inorganic pigments are generally considerably more coarse, with diameters of between 10^{-7} and 10^{-5} m.

In spherical particles the term diameter is unambiguous. However, pigments occur in the most diverse forms, from largely *isometric* particles such as spheres or cubes through to extremely *non-isometric* forms like needles or platelets. For non-spherical but regularly shaped particles the *effective diameter* D_{eff} is defined in DIN 55 206 as the characteristic measure of magnitude. Frequently, though, the equivalent diameter D of a spherical particle is quoted.

As engineering products pigments do not crystallise monodispersely, i.e. with uniform particle size, but polydispersely, i.e. in a particle size distribution. Measurement of the particle size distribution usually takes place in suspension. Since the methods used to measure the particle size distribution are unable to distinguish between the different types of particle, the best possible dispersion must be achieved to ensure that all agglomerates are divided into individual particles and aggregates.

A number of application-oriented properties of a pigment in a coating depend on the size and hence also on the distribution of its particles, for example:

- the rheological behaviour of the pigmented coating,
- its hue,
- the scattering ability of the pigment,
- the colour intensity and also
- various different types of resistances (fastnesses).

In principle, crystalline structures can only be determined using sufficiently well developed single crystals in X-ray diffractometers. It is often difficult to grow these single crystals, especially in the case of organic pigments, because of their desirable insolubility in many solvents. However, X-ray diffraction measurements using the Debye-Scherrer process, which can be conducted directly with pigment powders, are sufficient to distinguish between the crystalline modifications. An example of this is shown in Figure 2.29.

Diagrams of this type can also be used with mixtures of various modifications of a pigment to calculate the individual proportions of the modifications.

Particle size (distribution) can be determined by means of

- Various *sedimentation processes,*
e.g. centrifugal and ultra-centrifugal sedimentation analysis.

- *Optical methods,*
using the interaction of light with particles whose diameter is in the same order of magnitude as the wavelength of the light. Two such processes are photon correlation spectroscopy (QELS = quasi-elastic light scattering) and Fraunhofer light diffraction.

Both of these methods can provide only approximate calculations of the particle size distribution, since such calculations are normally based on an assumption of spherical, non-absorbent particles, which in reality do not exist. Since these methods are also based on different physical effects, it is not surprising that there may be discrepancies in particle size distributions measured on a single sample with different instruments.

- *Electron microscope*
Individual particles can be identified directly only with transmission electron micrographs (TEM), a process which unlike most other methods also frequently enables a distinction to be made between aggregates and agglomerates. The majority of the particle sizes occurring in pigments can be covered by a TEM examination, but it still detects only a limited number of particles, leading to a statistically inevitable uncertainty in the particle size distribution. Furthermore, only two dimensions are normally detected, which means that the third is likewise subject to uncertainty.

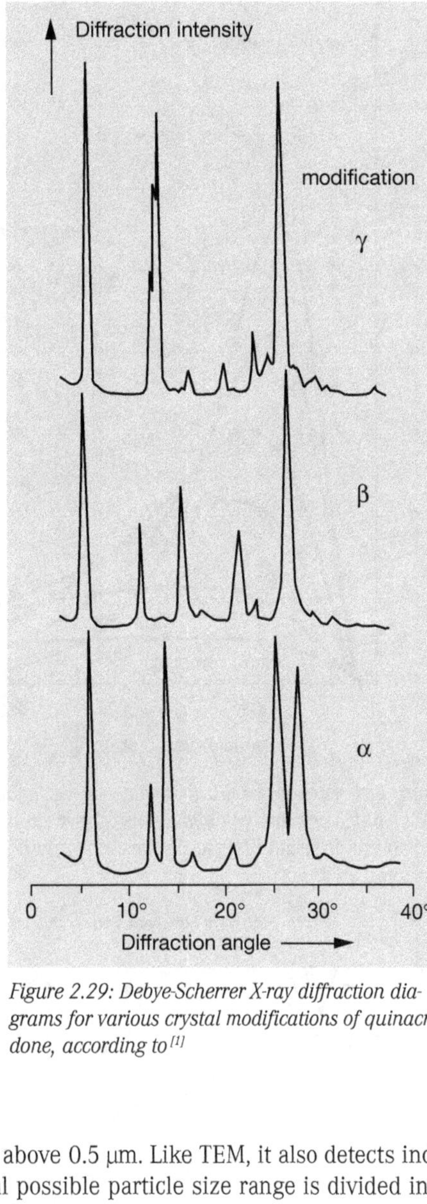

Figure 2.29: Debye-Scherrer X-ray diffraction diagrams for various crystal modifications of quinacridone, according to [1]

- *Counting processes*
The Coulter Counter is also suitable for particle sizes above 0.5 µm. Like TEM, it also detects individual particles whereas with other methods the total possible particle size range is divided into classes whose relative proportions are then determined.

The measured particle size distribution does not necessarily correspond to the actual distribution of primary particles and aggregates, reflecting only the current distribution state in the suspension. The actual distribution of primary particles and aggregates is ultimately a threshold that is approached by the best possible dispersion but can never be exceeded. In Figure 2.30 such an ideal particle size distribution is shown as a dotted curve. The curves drawn through it show actual particle size distributions. The shape of these particle size distributions is moreover typical for products of comminution processes and can be described mathematically as a standard logarithmic distribution. The statistical terms and evaluation processes for particle size distributions of this type are standardised in DIN 66 143 to 66 145.

Since the processes discussed so far for measuring particle sizes can only be used in strongly diluted and hence optically largely transparent suspensions, the pigment concentrations in these suspensions are far removed from those found in coatings. For that reason it must not be assumed that

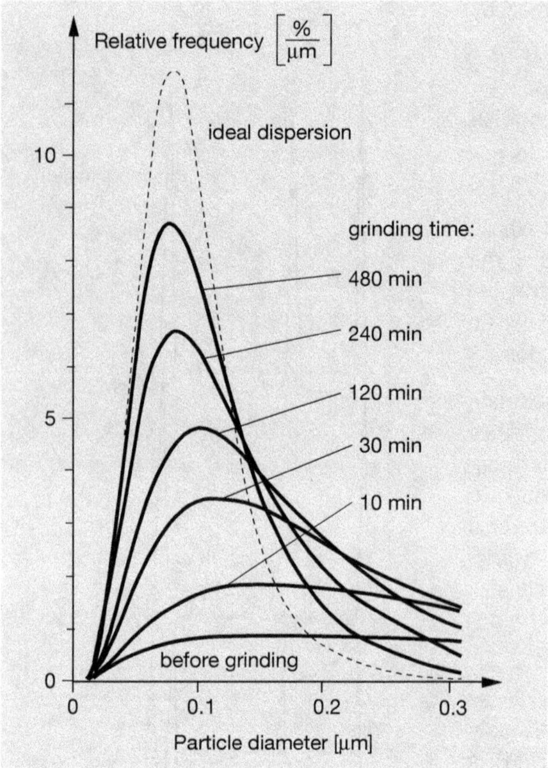

Figure 2.30: Particle size distribution of a wet-grinding series of β-Cu-phthalocyanine (β-CuPc) in water after various dispersion times with relative frequency density and particle diameter, according to [1]

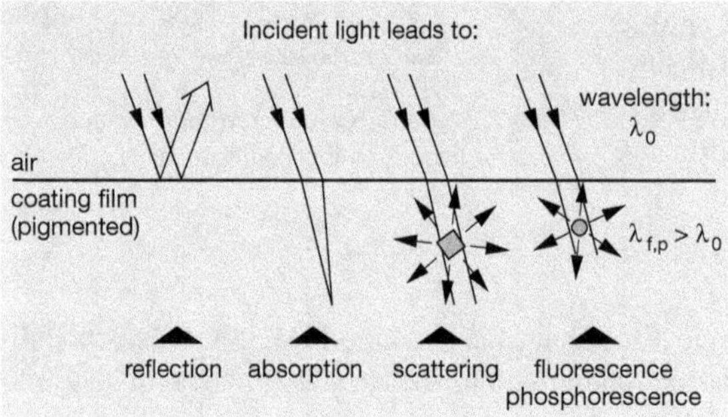

Figure 2.31: Interactions between light radiation and matter (non-primary light sources)

the particle size distributions measured in this way correspond to those in the final coating. They can only reflect the best achievable state. Only TEMs can be used to determine the true particle size distribution in the dried coating.

2.3.2.2 Appearance of pigments

In colloquial language we use the term "colour" to convey various meanings, e.g. a colour-imparting sub-stance, a type of radiation or a sensory perception. Under DIN 5033 (Colour measurement) colour is a sensory perception conveyed through the eye, i.e. a visual impression.

The sensory perception *colour* is - more precisely - a subjective impression influenced by all sorts of individual and situation-related factors, e.g. spectral transmission capacity of the eye, angle of view or surroundings. There is no clear connection between colour impression and the chromatic stimulus which provokes an impression of colour by retinal stimulus (see Colorimetry for more detailed information). Sources of chromatic stimuli may be primary light sources or non-primary light sources. The latter require an external light source to function as a source of colour stimulus whereas primary light sources provoke colour stimuli by themselves.

Interactions

Non-primary light sources can interact with visible light in various ways (Figure 2.31).

- They can reflect light directed at the interface between two media having different refractive indices, e.g. air/coating film. This is a reflection which may cause a colour to be produced. If the light reflection is not the same for all wavelengths, we call it selective reflection. (This effect occurs with coloured metals such as Cu, for example).
- Absorption occurs when light suffers a loss of intensity on passing through a medium. Absorption of incident light may be either dependent on or independent of wavelength. In the first case, this is called selective absorption and, in the second, non-selective absorption.
- Incident light may be deflected diffusely at so-called scattering centres, e.g. pigments. The individual scattering centres then act as sources of non-directional light radiation or scattered radiation. The scattering may also be wavelength-dependent, i.e. selective. Scattering is also said to occur when light is reflected in different directions, i.e. diffusely, at a phase boundary such as the coating surface. A matt surface is then perceived.

- Finally, fluorescence and phosphorescence are phenomena whereby light is first absorbed and then emitted again after a brief interval often most with a longer wavelength.

At the transition between two media having different refractive indices, light is deflected from its original direction, i.e. refracted, even if no other interactions occur.

In terms of the optical effects of pigments it is primarily their *optical constants*, i.e.

- their refractive index - and hence their light-scattering power - and
- their absorption coefficient (and hence their absorption capacity) that are important.

Light scattering occurs whenever a medium includes even the smallest areas having a different refractive index. The extent of light scattering is determined by the difference in the refractive indices of the intercalated medium and the surrounding medium. The aggregate conditions in the participating media are mostly of little importance here. In coatings pigments are generally the intercalated phase (with the exception of paints with supercritical PVC in which the pigment and filler particles are in contact with one another and the cavities between them are mainly no longer filled with binder but with air). For that reason their refractive index - or preferably their refractive index function, i.e. their refractive index as a function of the wavelength of the light - is decisive for their light scattering power.

The imparting of colour by coloured pigments is based on their selective absorption processes. The cause of light absorption ultimately lies in a change in the electronic state of the light-absorbing substance. The main causes of colour in inorganic pigments are shown in Table 2.15.

In organic substances colour is produced by an extended system of conjugated double bonds which causes the energy difference between the HOMO (**h**ighest **o**ccupied **m**olecular **o**rbital) and the LUMO (**l**owest **u**noccupied **m**olecular **o**rbital) to be shifted into the region of visible light.

The imparting of colour is greatly influenced on the one hand by the extension of this type of conjugated π electron system (chromophoric system) and on the other by what are known as chromophores and auxochromes. These are atoms or groups of atoms in the molecule which function as electron acceptors or donors.

An expert can read the key colour potentials and qualities of a pigment from the dependence of the absorption coefficient on the wavelength of the light. For example, there is a correlation between the steepness of the absorption edge of a pigment and its level of colour. Other important factors include the position of the absorption maxima, their width or the residual level in the non-absorbing range (see Figure 2.32, page 118).

To simplify greatly, we can say that the refractive index n_λ of a pigment has a decisive bearing on light scattering and hence hiding power, whilst the absorption coefficient k_λ determines absorption and colour intensity. The link between these material constants and the coloristic behaviour of coatings is covered in section 8.6.

Table 2.15: Various causes of colour in inorganic solids

Cause	Typical examples	Technical application
S_x radical anions	Ultramarine	ultramarine pigments
transitions between energy bands	ZnS, CdS, CdSe, HgS, Fe_2O_3	cadmium sulphide pigments, iron oxide pigments
crystalline field bands (d – d · or f – f transitions)	Oxide	iron oxide, chromium oxide or mixed-phase oxide pigments
charge transfer bands	CrO_4^{2-} MoO_4^{2-} MnO_4^-	chromate pigments molybdate red pigments manganese blue
change in oxidation state	Fe^{2+}/Fe^{3+}	iron oxide black, iron blue

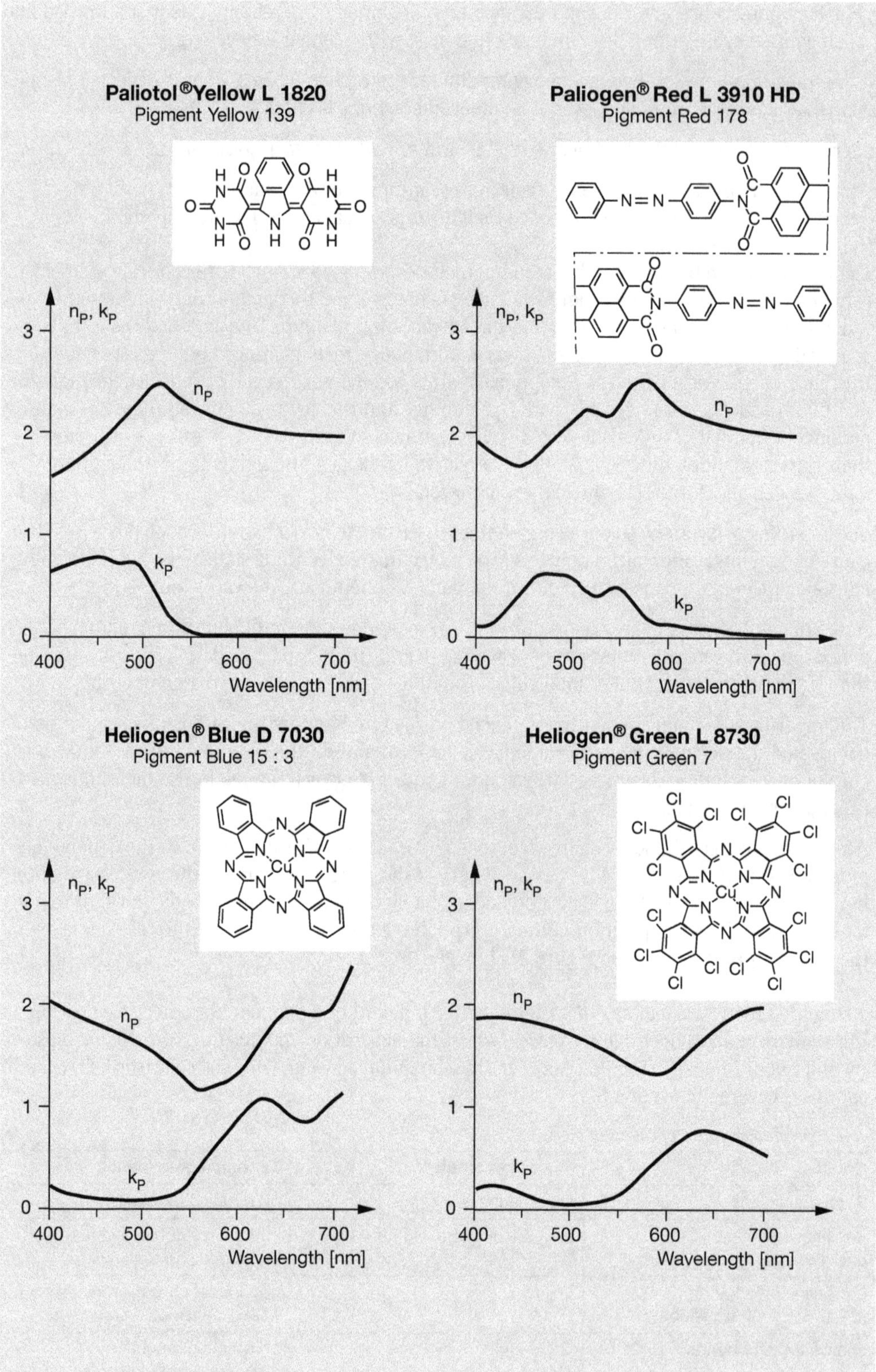

Figure 2.32: Dependence of refractive index n_p and absorption coefficient k_p on wavelength for various organic pigments, according to [1]

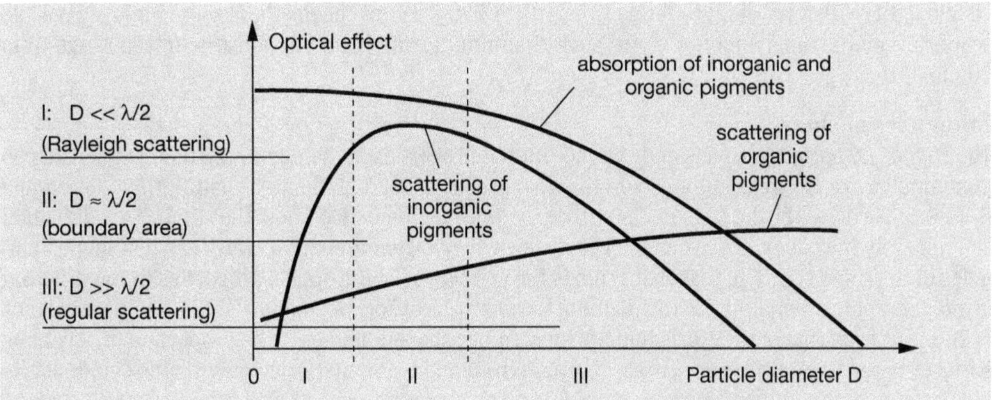

Figure 2.33: Schematic view of the relations between scattering power and/or absorption and particle diameter

Particle size influence

The relation between optical effect and pigment particle size is sketched in Figure 2.33. Below a certain particle size, very small particles are optically active through their entire volume. The absorption does not increase further with further size reduction. Above this limiting particle size, only the outer areas of a pigment particle are absorbent. As the particle size increases, the relative proportion of this outer layer in relation to the total volume of a pigment particle falls. Absorption also falls when the quantity of pigment is constant. The location of the limiting pigment size depends on the absorption coefficient k. It is lower with a high value for k than with a low value. In general it can be assumed for coloured pigments that with particle sizes below 1 µm, as the particle size decreases the shade is shifted towards the medium wavelengths, i.e. from blue to green and from red to yellow. These influences are particularly noticeable in the brightening of whites.

The hiding power of a pigmented system is primarily determined by the scattering power, which depends on the instantaneous particle size (distribution) of the pigment in the coating, in addition to the difference in the refractive indices of the pigment and the binder.

Under DIN 53 164 the **scattering power** is
the ability of a pigmented medium to reflect a part of incident light diffusely. When identifying the pigment as the cause of the light scattering of the pigmented medium, we refer to the scattering power of the pigment.

Note: Scattering power is generally dependent on the pigment concentration.

Single-crystal colourless substances are clear and transparent. As the particle size decreases, the scattering power at first increases in inverse proportion, passes through a maximum and then decreases again below this maximum with a reciprocal of the particle size to the power of 3. The position of this maximum is dependent on the wavelength of the scattered light and on the difference in refractive index between pigment and binder.

The theoretical principles for calculating light scattering in particles were developed by G. Mie. According to these principles, for spherical isotropic particles the particle diameter at which maximum light scattering occurs (d_{max}) is approximately

$$d_{max} \approx \frac{2\lambda}{(n_{pigment} - n_{binder}) \cdot \pi}$$

For pigments, the theory of light scattering must be extended to deal with a number of particles and a standard logarithmic distribution of diameters. Then, using the example of titanium dioxide pigments (with refractive index $n_P = 2.75$) in a binder (with refractive index $n_B = 1.55$), a maximum for scattering power is obtained at a median d_M in the range $\lambda/3$ to $\lambda/1.5$. For the scattering of visible light (with a wavelength range from 400 to 750 nm) this means that for the highest possible scattering power the median of d_M for such titanium dioxide pigments must lie in the range from 200 to 400 nm.

Influences of shape

The *shape of a particle* also has an influence on the properties of the pigment. Thus, needle-shaped pigments have a stronger influence on the flow properties of a pigmented coating than do roughly isometric particles. In the case of Cu-phthalocyanine pigments we know that their β modification contains roughly cuboid (i.e. isometric) particles which appear greener in tone than bar-shaped particles of the same size. The reason for this is the anisotropy of the optical properties in the different crystallographic orientations of the pigment particle. The absorption behaviour of their basal planes is different from that of their prismatic planes. In bar-shaped pigments the proportion of prismatic surfaces in the total surface area is greater than in cuboid pigments. Consequently the colour of this type of pigment is influenced more strongly by the absorption spectrum of the prismatic surface. This optical behaviour is known as *dichroism*.

Since the scattering power also differs along the various axes of the crystal, particle shape influences brightness as well as shade. In the same way as scattering power, other physical properties, such as adsorption behaviour or electrical charging ability of the various crystal surfaces, also differ from one another. When such non-isotropic pigments are aligned during application of a coating, e.g. when applied with a brush, the resulting paint film displays stripes of different depths of colour according to the angle from which it is viewed. This effect is known as “changing”, “bronzing” or “silking effect”.

Gloss effects

Gloss is a visual impression gained when light is reflected at a surface in a preferential direction. The gloss of a coating system is determined by a number of factors. Pigments can influence the gloss of a coating through their concentration, their particle size distribution and their degree of dispersion in the binder system. Only those pigment particles lying directly below the surface of the coating are involved in this process. They appear as small bumps in the even film surface, with the result that at these points incident light is reflected diffusely rather than directly. The higher the concentration of pigment, the greater the number of such gloss-reducing particles and the lower the gloss. The size of these bumps depends on the size of the particles. Whether these are primary particles, aggregates, agglomerates or flocculates is of no importance here. Gloss is influenced only by the current particle size distribution in the coating, itself naturally depending on the particle size distribution of primary particles and aggregates, and by the degree of dispersion and the extent of flocculation.

2.3.2.3 Interactions between pigment and surrounding medium

Pigments are never used alone but always in an application medium, e.g. a coating or plastic. At their shared interface interactions occur between pigment and medium. The surface of the pigment is therefore also involved at this interface, its properties - e.g. size or polarity - influencing these interactions accordingly; they in turn govern many properties of the entire system.

Specific surface area

The size of the surface area of a given amount of pigment depends on its density and particle size distribution. The lower the density and the smaller the particles, the larger the surface area. Assuming that the particles are spherical in shape, 1 cm^3 of pigment with a particle diameter of 0.4 μm has a surface area of 15 m^2/cm^3, or 60 m^2/cm^3 with a particle diameter of 0.1 μm. The specific surface area “S” is the surface area of one gram of pigment and is therefore expressed in m^2/g. It can range from 2 to 20 m^2/g for inorganic pigments, over 10 to 80 m^2/g for organic pigments and up to 700 m^2/g for carbon blacks. The specific surface area can also be used to determine to an approximation of the average particle size according to the following relation:

$$d = \frac{f}{\rho \cdot S}$$

where S = specific surface area [m^2/g], d = diameter [mm], ρ = density of the pigment [g/cm^3] and f = form factor.

For isometric particles such as cubes or spheres, the form factor is 6000, for long cylinders or needles it is 4000, where d corresponds to the cylinder diameter, and for extended platelets such as are found in metallic pigments it is 2000, with d representing the thickness of the platelet.

Since the particles in pigment powders are not of a uniform size, the specific surface area can be calculated only approximately from the particle size distribution. It is therefore determined using indirect methods.

- **BET method**

The most important method is that developed by *Brunauer, Emmett* and *Teller* (*BET method*). The amount of nitrogen (N_2) required to adsorb a monomolecular layer of N_2 molecules on the surface of the pigment is determined. The specific surface area of the pigment can then be calculated from this amount and from the surface area requirement of a N_2 molecule.

The BET method for determining the specific surface area uses N_2. This is particularly suitable in this instance because the nitrogen molecule is bound to the surface of the pigment only by non-specific van der Waals forces. If instead we use substances having special interactions with the surface of the pigment, e.g. chlorinated hydrocarbons, alcohols or water, we can learn about other properties of the pigment surface such as polarity and hydrophilicity.

Immersion calorimetry can be used to determine the *heat of wetting* of pigments with liquids. This can then be used as a measure for the affinity between pigment and liquid, the heat of wetting with water serving as a measure for polarity and hydrophilicity. It must be remembered, however, that water is normally already adsorbed at the surface of the pigment and this must first be expelled before wetting with another liquid. This expulsion can take a long time in non-polar liquids and is often incomplete.

- **Carman method**

This method is more commonly used with fillers. Here the resistance to fluid flow of a powder packing is determined. This is dependent on the size of the cavities remaining between the particles, the size of the former being smaller, the smaller and more fissured the individual particles. Thus the resistance to fluid flow increases with decreasing particle size and increasing pigment surface area.

- **Oil absorption value**

Another quantity frequently used to characterise the size and properties of a pigment surface is the *oil absorption value* (DIN EN ISO 787 T5). This indicates the amount of refined linseed oil absorbed under fixed conditions by a sample of pigment or filler and is given in g linseed oil/100 g pigment. The larger and more oleophilic the surface of the pigment, the higher the oil absorption value. It can range from below 5 g/100 g for inorganic fillers up to well over 100 g/100 g for certain pigment blacks.

Dispersion quality

A pigment-binder dispersion is a suspension before drying and a solid suspension after drying. Consequently the terms and regularities of colloidal and interfacial chemistry are used to describe pigment-binder systems. The dispersion process for pigments is extremely complex and is described in more detail in section 2.4.6. The quality of the dispersion, i.e. the current particle size distribution present in the pigment suspension, generally has a decisive influence on the properties of the system. The viscosity of the dispersion as well as a series of optical properties, principally shade, gloss, fog and, particularly, colour intensity, are all highly dependent on the dispersion state. They are all to a greater or lesser extent suitable (according to DIN 53238 and DIN EN ISO 8781) for determining the quality of the dispersion.

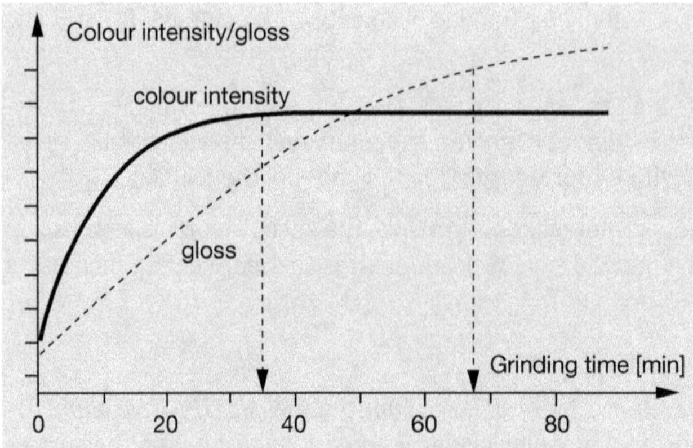

Figure 2.34: Schematic view of colour intensity and gloss development as a function of dispersion time

The quality of dispersion in a given film former system is dependent on the one hand on the interactions between binder and pigment surface and on the other on the type of dispersing unit used (→ 4.9). The interactions between pigment surface and binder system may in turn be significantly influenced on the pigment side by modifying the surface of the pigment particles (post-treatment → 2.3.3.2 and 2.3.4.2) and on the binder side by the use of dispersing and wetting agents (→ 2.4.6).

In addition to the achievable quality of dispersion, the speed with which a given dispersion quality can be achieved is also critical for the coating manufacturer. Whether a pigment is difficult or easy to disperse, i.e. whether a large or small amount of energy is required to achieve the dispersion quality needed to develop its coloristic properties to the full, can be roughly estimated from the dispersion hardness.

Dispersion hardness

This is determined by measuring the colour intensity development (→ 8.6.2) for a mixture of the test pigment with white over a progressive dispersion time (see Figure 2.34). The dispersion hardness DH or under DIN EN ISO 8781 the increase in colour intensity IS is calculated from the ratio of the colour intensity after a short dispersion time to that after a longer dispersion time:

$$IS = \left(\frac{(K/S)_2}{(K/S)_1} - 1 \right) \cdot 100$$

where K = absorption coefficient and S = scattering coefficient.

The two dispersion phases are fixed such that phase 2 lies near the final colour intensity and phase 1 generally after around 10 % of the time for phase 2. When comparing the dispersion hardness of various pigments, the same phase 1 and phase 2 should be used. If, in the case of fully saturated pigmentations, for example, the purity of the hue, a high gloss or low fog is of greater interest, the development over time of the relevant variable should be observed in the same way as when determining the dispersion hardness. These variables react much more sensitively with small amounts of agglomerates that have not yet been broken down; the latter have practically no influence on colour intensity but often require additional dispersion time to break down.

Table 2.16: Refractive indices for various white pigments and fillers

Substance	Refractive index n_D
calcium carbonate	1.48 to 1.65
mica	1.58 to 1.61
barium sulphate	1.64
lithopones	1.84 to 2.08
zinc oxide	2.01 to 2.09
zinc sulphide	2.37
titanium dioxide, anatase	2.55
titanium dioxide, rutile	2.75
organic binders	approx. 1.40 to 1.70

2.3.3 *White pigments*

The optical effect of white pigments is based on their very low light absorption and high unselective light scattering. In chemical terms they can be categorised into oxides (TiO_2, ZnO (zinc white), ZrO_2, SnO_2), carbonates ($2PbCO_3 \cdot Pb(OH)_2$ (lead white), $2ZnCO_3 \cdot 3Zn(OH)_2$), sulphates ($2PbSO_4 \cdot Pb(OH)_2$) and sulphides (ZnS, ZnS

+ $BaSO_4$ (lithopones)), of which titanium dioxide pigments are by far the most important in coating materials because of their high refractive index. Other white pigments are now used only in special circumstances. The decisive criterion for the suitability of a substance as a pigment is its refractive index; Table 2.16 shows a number of examples.

2.3.3.1 *Titanium dioxide pigments*

Table 2.17: Key physical properties of TiO_2 (see text for explanation)

Melting point		> 1,800 °C
refractive index n_D	see Table 2.16	
density	rutile:	4.2 g/cm³
	anatase:	3.9 g/cm³
TiO_2 content		> 88 %
light absorption	rutile:	$\lambda \leq 415$ nm ($\Delta E = 3.05$ eV)
	anatase:	$\lambda \leq 385$ nm ($\Delta E = 3.29$ eV)
Mohs' hardness	rutile:	6.5 to 7
	anatase:	5.5 (compared with steel: 4)

Physical and chemical properties

The world annual production of titanium dioxide pigments is around $4 \cdot 10^6$ t, more than half of which is used in coatings. Slightly less than half of current production is based on the traditional sulphate method, slightly more than half on the more modern chloride process, the two having different raw material requirements. The sulphate process can be used to manufacture both rutile and anatase pigments, whilst the chloride process produces only rutile pigments, albeit with a somewhat higher scattering power.

Of the three natural titanium dioxide modifications, two are found in titanium dioxide pigments:

- *rutile,* which is of greater importance for coatings
  and
- *anatase,* which - as mentioned - can only be obtained via the sulphate process.

Under normal conditions the rutile modification is thermodynamically the most stable. Some of the key physical properties of TiO_2 pigments are summarised in Table 2.17.

Titanium dioxide absorbs electromagnetic radiation with $\lambda \leq 415$ nm (rutile) or 385 nm (anatase). This corresponds to an energy difference between the valence band and the conductivity band of ΔE = 3.05 eV or 3.29 eV. Shorter-wave UV radiation is absorbed (see photocatalytic behaviour of TiO_2), longer-wave radiation is scattered. The reflectance spectra are illustrated in Figure 2.35.

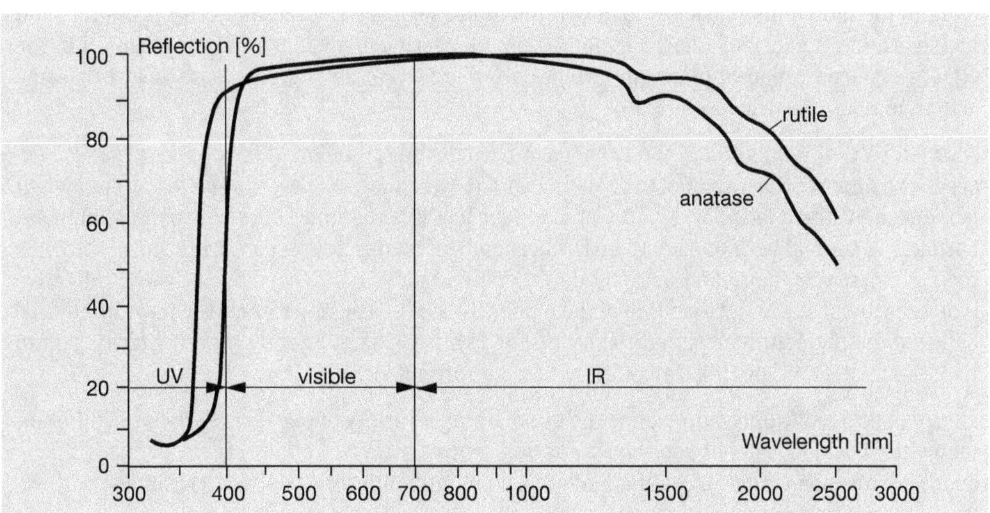

Figure 2.35: Reflection spectra for rutile and anatase, according to [1]

Foreign substances are added to the titanium dioxide pigments during → post-treatment processes, through which the TiO_2 content can fall to as low as 88 %. The most important foreign atoms are Al and Si.

Under the conditions to which coating materials are usually exposed, titanium dioxide is chemically very resistant and is insoluble in water and organic solvents. Although TiO_2 itself is non-toxic, its fine dust content arising from its method of production means that when working with titanium dioxide pigments the limiting values set for all workplaces exposed to inert fine dusts must be complied with.

Optical properties

The optical effect of white pigments is based principally on their high unselective scattering power. The conditions for the scattering of light were covered in section 2.3.2.3. The two key factors are the *difference in refractive index* ($n_{pigment} - n_{binder}$) and the *particle size.* Of all the white pigments titanium dioxide has the highest refractive index, contributing to its superiority as a white pigment. As mentioned earlier, for optimum light scattering, the median of the particle size distribution should lie in the range between 200 and 400 nm, with as narrow a particle size distribution as possible.

Both the scattering power and the scattering coefficient of a pigmented coating depend on the pigment concentration.

According to DIN 53 164 the **scattering coefficient** S is
the percentage of the radiation flux, **Φ**, which relative to the coating thickness is back-scattered at very small coating thicknesses h, i.e.:

$$S = \frac{1}{\Phi} \cdot \frac{d\Phi}{dh}$$

This is the characteristic physical quantity for the scattering power of a pigmented medium and is expressed in mm^{-1}.

The pigment concentration is conventionally cited as pigment volume concentration (PVC) in % (→ 3.2 and 3.3). Figure 2.36 shows the relation between scattering coefficient and PVC.

To understand the maximum S for a PVC of around 30 %, it is important to realise that the average distance between particles decreases rapidly as the PVC increases. In the case of spherical particles, at a PVC of 10 % this distance is already less than the average particle diameter and by around 25 % it is less than half the particle diameter. Below a distance of half the wavelength of light, the individual particles no longer function as separate scattering centres. As the PVC increases, more and more particles fall below this "critical" distance, which means that the effective concentration of pigment particles above a PVC of 25 % grows more slowly than the actual number of particles.

Above a PVC of around 30 % the number of particles to fall below this "critical distance" exceeds the number that are effectively being added, which means that the scattering coefficient is reduced. At the *critical* PVC (CPVC) the binder just fills the interstices between the pigment particles in contact with one another. These islands of binder now act as scattering centres. As the PVC increases, more and more binder in the interstices is replaced by air whose refractive index of around 1 is lower than that of the binder. This increases the refractive index differential between the pigment matrix and the scattering centres - causing the scattering coefficient to rise - until at a PVC of 100 % all of the interstices are filled with air.

These models only apply, however, to the use of a single variety of pigment. Particles of different pigments continue to act as separate scattering centres even when they are in contact with one another. This means that in coating materials containing fillers as well as TiO_2 pigments, these considerations do not apply.

As long as the pigment particles act as scattering centres, i.e. roughly up to the CPVC, the scattering coefficient for a more finely divided pigment is larger than that for a more "coarsely" divided pigment. Both d_M values (see section 2.3.2.2) must be larger than the optimum particle size, however. As soon as the intercalated binder - or in the case of even higher PVCs - the air islands begin to act as scattering centres, their size determines the scattering power. This means that the ratios in the supercritical PVC range are reversed.

According to DIN 44 943

the **hiding power** of a pigmented medium is its ability to cover the colour or the difference in colour of the substrate.

The *hiding power index* is derived from the hiding power. According to the DIN 55 984 standard, it describes the surface area of a contrasting substrate that can be covered opaquely by 1 l or 1 kg of a pigmented medium. It is expressed in m^2/l or m^2/kg. The criterion for opacity is an agreed colour difference between the two contrasting fields of the coated contrasting substrate. The reciprocal of the hiding power index indicates the thickness of the coating that is classed as opaque on the basis of the underlying opacity criterion.

In both opaque and non-opaque coatings, the wavelength-dependence of the scattering can lead to colour changes. Generally speaking the scattering increases as the wavelength decreases.

- For opaque coatings containing *only ideally scattering* pigments, this is of no consequence. The colour of the reflected light corresponds to that of the incident light.
- Opaque coatings containing *only ideally absorbing* pigments appear black, since as the depth of penetration increases, the light is more and more readily absorbed until its intensity approaches zero; the same is true of coloured pigments.
- In opaque coatings containing *both scattering and absorbing* pigments, the distance travelled by the light between entering and leaving the coating decreases with decreasing wavelength. This means that with the same ab-sorption coefficient, short-wave violet light is absorbed less readi-ly than longer-wave light. This gives a grey coating a blue tint.

In completely opaque coatings the colour of the substrate is irrelevant. In this case the index ∞ is added to the symbol for reflectance, e.g. R_∞.

In non-opaque coatings both the colour of the substrate and the coating thickness influence the colour of the reflected light. In the case of white coatings, as the coating thickness increases only the longer wavelength light ultimately reaches the substrate, where it is reflected or absorbed. The hiding power of a coating is therefore wavelength-dependent. If the coating contains pigments which absorb in the wavelength ranges capable of reaching the substrate, this reflected light is also weakened. The hiding power of a coating containing only scattering pigments can thus be improved by the addition of absorbing pigments.

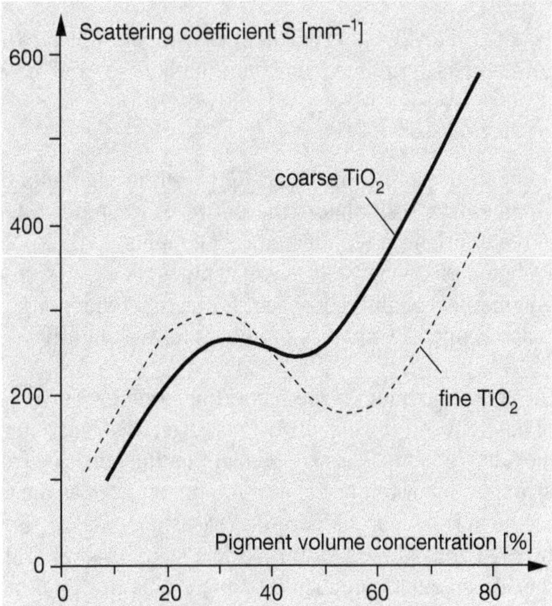

Figure 2.36: Schematic view of the relation between scattering coefficient and PVC for two different TiO_2 pigments, according to [1]

According to DIN 55 943,
the **whitening power** is the capacity of a pigment to increase the lightness of a coloured, grey or black medium.

The whitening power of a white pigment is determined by adding a quantity q_p of this pigment to the pigment paste to be lightened and measuring the standard colour value Y of an opaque coating containing this pigment paste. A calibration curve is then used to calculate the weighed portion of a reference pigment q_{ref} with which the same value is obtained. The *whitening power index* can thus be calculated as

$$\text{Whitening power index} = \frac{q_{ref}}{q_p} \cdot 100$$

The whitening power index is thus not an absolute quantity but must always be considered in relation to an agreed reference pigment in a particular system. No generalisation can be made regarding the behaviour of this pigment in other pigmented systems.

The whitening power is attributable to the scattering power of the pigment. As the concentration of the scattering pigment increases, the distance travelled by the light between entering and leaving the coating reduces because it is scattered more strongly. According to *Lambert-Beer's law*, absorption decreases as the distance travelled by the light through the coating decreases, which means that more light is reflected. The dependencies of the scattering power of the pigment are the same as those for the hiding capacity.

According to DIN 55 980,
the **tint** of an almost white or colourless sample is the small amount of chromatic colour by which the colour of a sample deviates from ideally white or (ideally) achromatic. It is described as saturation[1)] and desaturation[1)]. The colour location of the illuminating type of light is taken to be the tint-free (achromatic) reference point.

An ideally white sample according to this standard is a sample which completely reflects incident light in the wavelength range from 380 to 780 nm. Although an achromatic sample does not reflect the light completely, the proportion of light reflected in this wavelength range is likewise independent of the wavelength.

In the CIELAB system of colour co-ordinates the size of the tint is expressed by the distance C^*_{ab} of the measured hue from achromaticity ($a^* = 0$, $b^* = 0$), which is calculated as follows:

$$C^*_{ab} = \sqrt{a^{*2} + b^{*2}}$$

As can be seen from Figure 2.35, rutile pigments display a higher reflectance than anatase pigments across almost the entire wavelength range for visible light. They therefore have a greater brightness than anatase pigments. At the short-wave violet end of the visible wavelength range both display a lower reflectance than in the rest of the range. This gives titanium dioxide pigments a slight yellow tint. Since this reduction in reflectance is slightly less marked in the case of anatase, anatase pigments have a slightly lesser yellow tint than rutile pigments.

Since the maximum of the scattering power for a given particle size is dependent on the wavelength of the light and consequently for a given wavelength on the particle size, the yellow tint of titanium dioxide pigments is also dependent on the particle size distribution. Close to the optimum particle size this yellow tint becomes more pronounced as the average particle size increases, and weakens as the average particle size decreases. The yellow tint of TiO_2 pigments cannot be entirely offset by regulating the optimum particle size, however. This means that more finely divided titanium dioxide pigments produce a stronger blue tint than more coarsely divided pigments when mixed with grey, for example.

1) See DIN 5033 part 1 for definitions.

Post-treatment

Full advantage of the optical properties of titanium dioxide pigments can only be taken if the pigment is properly distributed in the binder, i.e. it is dispersed as thoroughly as possible. The dispersion of pigments is discussed in section 2.4.6. The surface of the pigment has a considerable influence on the main chemical and physical processes and parameters here.

The dispersion of titanium dioxide pigments can be greatly facilitated by means of suitable post-treatment, although titanium dioxide pigments tend to be among the more readily dispersible pigments anyway.

Surface treatment has no real influence on the first phase of dispersion, the breaking down of the agglomerates. In order to achieve good wetting, the forces of adhesion between pigment surface and dispersion medium must be greater than the forces of cohesion within the dispersion medium. These forces of adhesion can be influenced by modifying the pigment surface. After distribution of the wetted pigment particles, the suspension still needs to be stabilised. Post-treatment can alter both the type and size of the charge at the surface of the pigment particles (*zeta potential*) and the adsorption behaviour in respect of stabilising polymers. In other words, both types of stabilisation can be improved - be it electrostatic or steric.

A better dispersibility can be achieved nowadays by post-treating the titanium dioxide pigments with inorganic substances such as aluminium oxide hydrate and silicon oxide hydrate, and/or organic substances such as polyols or alkanol amines. In aqueous systems interface-active substances can often improve the wetting of the already inherently hydrophilic surface of titanium dioxide pigments. Their use in organic media is generally not beneficial, however, and is more likely to lead to undesirable effects such as flocculation.

Weathering influences

Like most organic materials, when coatings pigmented with titanium dioxide are used in exterior applications they undergo ageing under the influence of weathering and sunlight; this leads to changes in appearance and in physical and chemical properties and can also impair the effectiveness of the coating. Typical damage includes reduction in gloss, fading, embrittlement and chalking.

Titanium dioxide pigments intervene in these degradation processes in two different ways.

- They protect the coating against the direct influence of UV light, since both anatase and rutile absorb UV radiation (see Figure 2.35). This means that UV radiation can penetrate a coating containing TiO_2 pigment by only a few μm before being absorbed. The binder regions located below the pigment particles are protected against photo-oxidative degradation. Only the thin, uppermost layer is ever attacked and over time destroyed; this can be observed as chalking, for example.
- The photoactivity of TiO_2 pigments can also cause the degradation of organic binders, however. The photoactivity of anatase pigments is significantly higher than that of rutile pigments. On absorbing UV radiation, both set in train a photochemical reaction chain (photocatalytic cycle) that can lead to degradation of the binder in the coating.

Although the processes involved in the weathering of polymer substances have not yet been fully explained, the theory of the photocatalytic cycle at the surface of titanium dioxide offers a relatively simple and plausible explanation of all the weathering phenomena observed in materials containing titanium dioxide pigments. The following reaction stages are involved:

$$h\nu_{abs} \rightarrow e^- + p^+$$

$$p^+ + TiO_2 \cdot OH^- \rightarrow TiO_2 + {}^{\bullet}OH$$

$$e^- + Ti^{4+} \rightarrow Ti^{3+}$$

$$Ti^{3+} + O_2 \rightarrow Ti^{4+} \cdot (O_2^-)_{ads}$$
$$(O_2^-) + H_2O \rightarrow OH^- + HO_2^\bullet$$
$$TiO_2 + OH^- \rightarrow TiO_2 \cdot OH^-$$

They enable a detailed interpretation to be made. By absorption of a photon with an energy above 3.05 eV (415 nm) in the case of rutile or above 3.20 eV (385 nm) in the case of anatase, an electron-hole pair is produced when an electron is lifted into the conductivity band, leaving a gap in the valence band. Normally the electron drops back into the valence band very quikkly and recombines with a hole. Alternatively the hole p^+ in the valence band may migrate to the surface of the titanium dioxide particle where it oxidises an absorbed hydroxyl group to a hydroxyl radical. The electron e^-, which is likewise moving in the conductivity band, reduces a Ti^{4+} to Ti^{3+}, which in turn reduces the oxygen adsorbed on the surface of the crystal to an oxygen radical anion. This is chemisorbed at the surface of the crystal and can be reacted with water to a perhydroxyl radical and a hydroxide ion.

As the total reaction, therefore, we obtain hydroxyl radicals and perhydroxyl radicals from water and oxygen; these radicals can diffuse into the surrounding organic medium and provoke degradation reactions there.

$$H_2O + O_2 \xrightarrow{h\nu / TiO_2} OH^\bullet + HO_2^\bullet$$

This photocatalytic cycle can be interrupted by reducing the mobility of electrons and holes in the crystal and/or by suppressing the reactions of holes and electrons at the surface of the crystal by means of an "isolating" coating (Figure 2.37).

An interruption of this photochemical cycle can be achieved by means of lattice stabilisation (incorporation of foreign ions, such as Al^{3+} or Zn^{2+} in the crystal lattice) or by applying a thick layer of colourless, inert oxides of Si, Zr, V, Al or Ce during post-treatment. The best weathering results are obtained with lattice-stabilised and post-treated titanium dioxide pigment types.

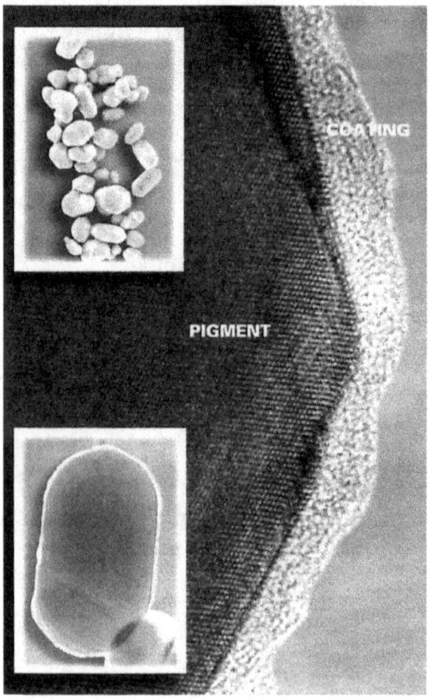

Figure 2.37: Cross-section through a TiO_2 pigment particle coated with SiO_2, according to [40, DuPont]

Titanium dioxide pigment types

The TiO_2 pigments manufactured by the sulphate or chloride process are sold in both untreated and post-treated forms. The type of post-treatment is governed by the intended use of the pigment. Various particle size distributions extend the range even further. Around 400 different types of titanium dioxide pigments are available worldwide today, no two of which are likely to be identical. DIN 55 912 distinguishes between five groups of titanium dioxide pigments (Table 2.18).

Powdered pigments are sold in 25 kg sacks or in 0.5 to 1 t sacks known as *big bags*. Aqueous suspensions or *slurries* with a solids content of 68 to 75 % are also available, however; these are particularly advantageous when used in water-compatible systems.

"Micronised" titanium dioxide pigments

Aside from the titanium dioxide pigments sold so far, several manufacturers have been selling finely divided titanium dioxide grades having average particle sizes of between 15 and 50 nm. Their small particle size makes these clearly yellow-tinted pigments transparent or at

Table 2.18: Classification of titanium dioxide pigments (according to DIN 55 912)

Property	Requirements				
	Type A (anatase)		Type R (rutile)		
	A1	A2	R1	R2	R3
TiO_2 content	≥ 98 %	≥ 92 %	≥ 97 %	≥ 90 %	≥ 80 %
content of H_2O-soluble matter	≤ 0.6 %	≤ 0.5 %	n. v.	≤ 0.5 %	≤ 0.7 %
volatile components	≤ 0.5 %	≤ 0.5 %	≤ 0.5 %	n. v.	n. v.

least semi-transparent. At the same time they are strong UV absorbers, which means that they are being considered as an alternative to organic UV stabilisers in clear coatings for wood substrates. When combined with aluminium pigments in metallic coating systems they produce a colour-change effect from yellow to blue, known as frost effect. This is a consequence of the strong wavelength-dependence of light scattering at very small particles.

The intensity here is roughly inversely proportional to the wavelength by the power of 4 (Rayleigh scattering), which in this case means that blue light is around 5 to 6 times more strongly scattered than red light. The proportion of blue light is therefore increased in scattered light in comparison with white light and reduced by an equivalent amount in light reflected from aluminium pigments. Consequently a coating of this type appears yellowish close to the glancing angle and bluish when viewed from remote angles.

2.3.3.2 Other white pigments

The other white pigments are less important than titanium dioxide pigments because their optical performance is far inferior.

Zinc oxide pigments are used as active base pigments in primers and anti-corrosion coatings. Under DIN 55 943, **zinc white** is a zinc oxide pigment with a purity of over 99 %, which is produced via the indirect or French process.

Zinc sulphide pigments are characterised by good brightness, pure hue and low binder requirement. Because of their low abrasiveness and limited UV absorption, these non-acid-resistant pigments are used alone or together with TiO_2 in, for example, UV-curing coatings, printing inks and special applications, e.g. with brighteners or daylight fluorescent colours.

Lithopones produced by precipitating $BaSO_4$ and ZnS together carry greater importance. The incorporation of 0.02 to 0.5 % cobalt provides excellent light stability. These pigments have low binder requirement and good wettability. Lithopones are therefore used in primers, fillers and stoppers and for wall and floor coatings, as well as in combination with TiO_2.

2.3.4 Black pigments

2.3.4.1 Classification

In black pigments the absorption coefficient determines optical quality just as the refractive index does in white pigments. The particle size and particle size distribution also play a part, particularly in relation to colour intensity and hiding power.

The most important black pigments are

- pigment blacks,
- iron oxide blacks and
- various mixed phase oxide pigments such as spinel black.

For systematic reasons we will discuss the last two groups together with inorganic coloured pigments.

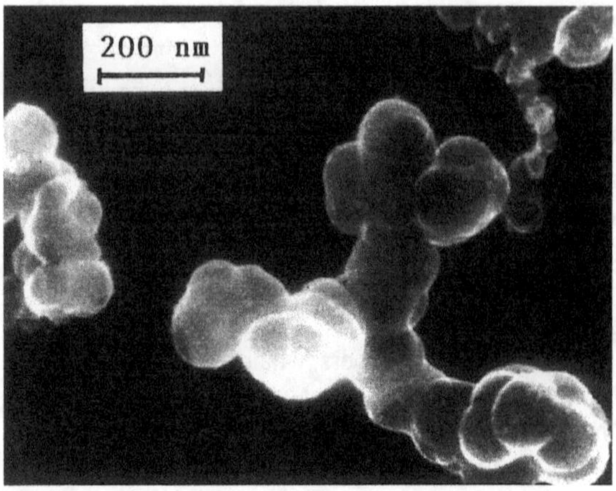

Figure 2.38: Electron micrograph of a pigment black aggregate, according to [7]

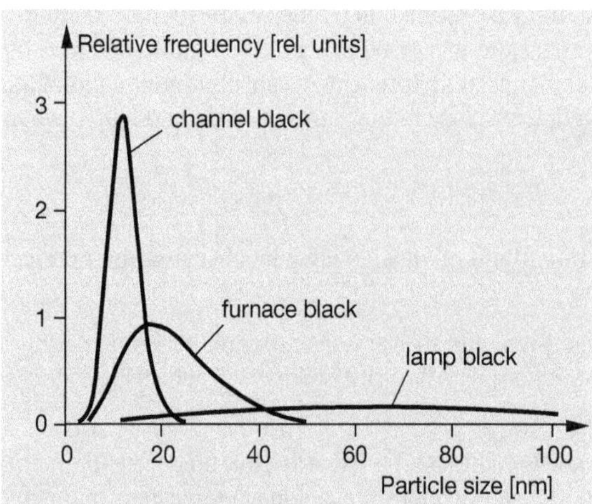

Figure 2.39: Typical particle size distributions of various pigment blacks, according to [4]

2.3.4.2 *Pigment blacks*

Their outstanding properties make pigment blacks the most commonly used black pigments in the paints and coatings industry. The annual world production of carbon black is over 4 million tonnes, of which in fact only about 5 % is used as pigment. The leading manufacturing countries of carbon black in Europe are Germany, France, Italy and Britain.

Physical and chemical properties

Carbon black is a finely divided, partially microcrystalline carbon, whose almost spherical primary particles have grown into larger aggregates which form chains or clusters (Figure 2.38). The extent of this growth is known in the industry as *structure*. The most important variable for pigment blacks is the particle size or particle size distribution. Figure 2.39 shows typical particle size distributions for pigment blacks produced by a number of different processes. The primary particle size is between 5 and 500 nm, that for pigment blacks lying in the range from 10 to 100 nm (titanium dioxide: approx. 300 nm). For this reason and because of its lower density, even a relatively coarse carbon black such as lampblack contains 60 times more particles per gram than a titanium dioxide pigment, whilst the finely divided channel blacks contain 28,000 times as many.

Nowadays carbon blacks are manufactured by the pyrolysis of hydrocarbon-containing substances. Depending on their manner of manufacture, we distinguish between *lampblacks*, *furnace blacks* and *channel blacks*. The pigments produced by these three methods differ in their particle size and structure as well as in the properties of their particle surfaces (see Table 2.19).

Table 2.19: Manufacturing processes for pigment blacks

Process	**Raw material**	**Atmosphere during manufacture**	**Quantities that can be influenced**		
			Primary particle size	Structure	Surface chemistry
furnace black process	petrochemical residues	reducing	within broad limits 17 to 70 nm	within broad limits	pH 8 to 10, cannot be influenced
lampblack process	oils rich in aromatic compounds	oxidising	50 to 210 nm	cannot be influenced	pH 7, cannot be influenced
channel black process	high-boiling oils rich in aromatic compounds	oxidising	10 to 30 nm	cannot be influenced	pH 2 to 5, can scarcely be influenced

The application-oriented properties of pigment blacks can be influenced by oxygen-containing functional groups on the surface of the carbon black particle.

- Their method of manufacture means that channel blacks usually carry a certain excess of acid groups on their surface, e.g. carboxyl, lactone or carbonyl groups.
- Furnace blacks, by contrast, exhibit oxygen in the form of basic groups such as basic oxides or pyrone-type structures on their surface.
- The surface of lampblacks reacts more or less neutrally.

Modifications

In order to optimise their application-oriented properties, a number of carbon blacks can be post-treated by oxidation, either by heating with a certain quantity of oxygen or by treating with strong oxidants, e.g. nitric acid, nitrous gases, ozone or NaOCl. This increases the oxygen content by up to 20 %, with a correspondingly low C content. These carbon blacks with polar surfaces are distinctly hydrophilic and are more readily wetted and dispersed in polar coating systems.

A measure of the degree of oxidation and hence of the surface polarity of carbon blacks is the loss on ignition at 950 °C under exclusion of air, according to DIN 53 552 (see Table 2.20). Like carbon, carbon black is exceptionally resistant to chemicals and is attacked by only a few chemicals or solvents, e.g. by O_2 only above 300 °C.

The particle size of a carbon black determines its application.

- The *finest* carbon blacks, with particle sizes from around 10 to 30 nm, are used as pigment blacks in coating materials or as reinforcing fillers in rubber. This range is covered by the channel blacks on the one hand and furnace blacks on the other.
- *Coarser* carbon blacks, up to around 100 nm, such as are largely found in lampblacks and to a certain extent also in furnace blacks, are used as shading blacks and as partially active fillers in rubber.
- *Even coarser* carbon blacks are used as non-active fillers in the rubber industry.

The size of the *specific surface area* depends on both the particle size and the porosity of the carbon black, those having a specific surface area of over 150 m^2/g being porous with pore diameters of less than 1 nm.

Structure-related characteristics

Information about the structure of a carbon black is also provided by the values for *dibutyl phthalate (DBP) adsorption* and *oil requirement*. These values are dependent on both the specific surface area and the wetting behaviour of the surface of the carbon black. With a *density* of around 1.86 g/cm^3 (1.8 to 2.1 g/cm^3), carbon black is less dense than graphite (2.27 g/cm^3), due to its lower crystalline order.

Table 2.20: Key physical and chemical properties of pigment blacks

Structure	Graphite (para-crystalline)	chemical nature of surface	neutral, basic or acid, depending on manufacturing process
primary particle size	10 to 100 nm	solubility	insoluble in H_2O, acids, bases and coating solvents
spezific surface area	10 to > 1,000 m^2/g	C content	80 to 99.5 %
density	approx. 1.86 g/cm^3	O content	0.5 to 15 %
light absorption	max. 99.8 % (also in UV and IR)	resistance	very good, sensitive to O_2 at elevated temperatures
hardness	very soft	content of volatile components	channel blacks: 4 to 8 % furnace blacks: 1 to 3 % post-oxidised blacks: < 25 %
electrical conductivity	< 0.125 $m/(\Omega \cdot mm^2)$ (graphite)		

The *electrical conductivity* of pigment blacks is below that of graphite at 0.125 m/Ω mm^2 (copper: 60 m/Ω mm^2) and is governed by the method of manufacture, the specific surface area and the structure of the carbon black.

The widespread use of pigment blacks is explained by their excellent *absorption* of visible light, up to 99.8 %. Here black colours may also appear blue- or brown-tinted, depending on the refraction of light, the wavelength of the incident light and the medium surrounding the pigment. Since carbon blacks also absorb UV and IR radiation, they are also used as UV stabilisers or in IR camouflage paints.

Pigment blacks in coating materials

Pigment blacks owe their widespread use as a black colouring for plastics, printing inks and coating materials and as a shading black in combination with coloured and white pigments to their outstanding pigment properties.

- Finely divided channel blacks[1)] are generally used for the *black colouring* of paints and coatings since in most coating systems they display a number of advantages over furnace blacks, such as readier dispersibility, lower flocculation tendency, higher gloss and improved weather resistance. These grades of carbon black are suitable for both fully saturated and transparent colours. Rheological behaviour and dispersibility can be improved even further by means of an oxidative post-treatment. This is accompanied by a greater depth of black and a shift in hue towards blue.
- By contrast, lampblacks are mainly used in *tinting pastes* and *bright colours* because their lesser depth of colour makes them easier to add.
- Carbon blacks with a high structure (see page 131), known as conductive carbon blacks, are used in *electrically conductive coatings.*

The use of furnace black is determined by its depth of colour, which is dependent on the particle size. The finer grades can often compete with channel blacks, whereas the coarser grades cover much of the same range as the lampblacks.

The concentration of carbon black in coating materials does not normally exceed 3 to 5 %, although a significantly higher concentration is often required in electrically conductive coatings.

Colouring and blackening effects

In addition to their chemical resistance, pigment blacks are also characterised by good light and weathering stability as well as excellent depth and intensity of colour.

According to DIN 53 235
the **depth of colour** is a measure of the intensity of a colour impression, which rises with increasing saturation and generally falls with increasing brightness.

Note: colours produced with the same depth of colour appear to the observer as if they were produced with equal concentrations of equally intensely coloured colorants.

In this connection we should also mention the term *standard depth of colour* (SD), which in the same standard is defined as the conventionally established level of depth of colour. Commonly used standard depths of colour in the coatings sector are 1/3 SD and 1/9 SD.

In evaluating black colours, depth of colour is taken to mean the sensory impression described as intensity of *blackening*. In coatings coloured with carbon blacks, the lower the level of light reflection, the greater the intensity, whilst in achromatic tones (grey tones) depth of colour means the opposite of brightness. The intensity of blackening is characterised coloristically by the **black index M_Y**. DIN 55 979 describes the method for determining the black index.

1) Pigment blacks produced by incomplete combustion of methane and separation in water-cooled channels.

According to DIN EN ISO 787, part 24,
the colour intensity is the ability of a pigment to absorb incident light and in this way to impart colour to or to darken e.g. a white paint into which it has been incorporated.

The colour intensity is always determined relative to a reference pigment and in terms of content corresponds to the whitening power in titanium dioxide pigments. Colour intensity and depth of colour do not necessarily progress in parallel, since the intensity of colour depends on the structure and dispersion state of the carbon black as well as on the particle size or particle size distribution. Above all it has a practical relevance as an indicator of the yield of shading blacks.

The key parameters influencing the application-oriented behaviour of pigment blacks are

- the average particle size or particle size distribution
- the structure of the carbon black and
- the chemical properties of the surfaces of the carbon black particles.

Pigment blacks are usually selected from a coloristic standpoint. Although pigment blacks do not occur in the form of primary particles, their coloristic properties are still principally determined by the average particle size. As the average particle size decreases, the depth of colour in coatings coloured with carbon blacks increases. The tint is also influenced by the primary particle size; see page 125 for its definition according to DIN 55 980.

Application-oriented properties

In fully saturated colours, more finely divided carbon blacks have more of a blue tint than more coarsely divided carbon blacks, which tend to have a brown tint. In mixtures with white, in transparent colours and in metallics, this behaviour is reversed. Here more coarsely divided carbon blacks tend to have a blue tint, whilst more finely divided grades tend to have a brown tint. The colour intensity behaviour remains unaffected by this. With finely divided carbon blacks we obtain brown to brown-red colours with a high transparency; as the particle size increases this transparency shifts towards bluish-opaque settings. The particle size of the carbon blacks also influences other application-oriented properties such as dispersion behaviour, rheology and weather resistance of paints.

It is evident that more finely divided carbon blacks require a greater dispersion effort than more coarsely divided grades. At the same time the viscosity of the millbase of the latter is unsurprisingly lower for the same quantity. In paints coloured by more finely divided carbon blacks, the reduction in gloss under weathering is less pronounced than in the case of paints coloured with more coarsely divided grades.

A number of application-oriented properties of pigment blacks can be improved by oxidative post-treatment (see above).

- Dispersion is often made considerably easier, since the more polar surface can be wetted more readily by the binder or special dispersing agent.
- This also improves flocculation stability.
- The viscosity of the millbase and of the final coating material is lower in the case of surface-treated carbon blacks than with corresponding carbon blacks that have not been post-treated.
- In the final coating post-treated carbon blacks lead to higher gloss which is also retained for longer because of its improved weathering stability.

2.3.5 Inorganic coloured pigments

2.3.5.1 General properties

Coloured pigments differ from black and white pigments in that

- their absorption and scattering coefficients are dependent on the wavelength of light and
- their absolute values can fluctuate dramatically.

Table 2.21: World production capacities for inorganic coloured pigments (as at 2005)

Pigment	Production in 10^3 t/a
iron oxide pigments	700
chromium oxide green pigments	84
chromate yellow pigments	60
iron blue pigments	13
molybdate red pigments	20
ultramarine pigments	40
metal pigments	35
cadmium pigments	2

The dependence of the two coefficients on the wavelength of light, on particle size, particle size distribution and particle form determines the colour, colour intensity and hiding power of coloured pigments. This is true both of organic and inorganic coloured pigments, of course. The causes of the colour of inorganic compounds have already been set out in Table 2.15.

Inorganic coloured pigments can be classified according to their colour, e.g. into red, yellow, green and blue pigments, or - as here - into chemical groups. We can differentiate between oxides and oxide hydroxides (iron oxide pigments, chromium oxide pigments) including complex inorganic coloured pigments (CICPs), bismuth, cadmium, cerium sulphide, chromate, ultramarine and iron blue pigments.

After titanium dioxide pigments, inorganic coloured pigments are now the largest group of pigment products. World production capacities are listed in Table 2.21. Only around a third of the total production volume of inorganic coloured pigments is used in paints and coatings, however.

As with white and black pigments, particle size and particle size distribution (together with average value and breadth of distribution) are among the most important parameters affecting the desired application-oriented properties of coloured pigments too. They influence gloss, film strength, hiding power, whitening power or colouring power and purity of hue.

2.3.5.2 Oxide and oxide hydroxide pigments

General introduction

Inorganic coloured pigments belonging to the group of oxides and hydroxides are either coloured inorganic single compounds or mixed phases whose colour is based on the incorporation of colour-giving cations in often inherently colourless host lattices.

This group includes iron oxide pigments, chromium oxide pigments and mixed phase pigments with rutile or spinel structure.

Iron oxide pigments

The production quantity of iron oxide pigments is greater than that of all other inorganic coloured pigments put together. They are manufactured in yellow, orange, red, brown and black colours. There are still natural iron oxide pigments too, of course, such as ochre, umber or burnt sienna, although these are of no relevance in the coatings industry.

Apart from their good value for money, synthetic iron oxide pigments are characterised by a number of outstanding application-oriented properties. They are insoluble in water, organic solvents and organic binders, and also in inorganic binders (such as cement or water glass); they are also fast to light and water and resistant to alkalis. The fastness to water and light of inorganic coloured pigments, unlike that of organic coloured pigments, is virtually independent of particle size. Iron oxide pigments exhibit a relatively high colour intensity for inorganic pigments. Their hiding power is very much dependent on particle size. Likewise the coloristic and all other properties are heavily influenced by particle size. The following regularities can be seen in iron oxide red pigments, for example (see Table 2.22):

- Very fine iron oxide red pigments, with a $d_M < 0.05$ µm, are transparent.
- Coarser pigments with a d_M of between 0.1 and 1 µm, on the other hand, have outstanding hiding power because of their good scattering power.

Table 2.22: Relation between the properties of iron oxide red pigments and particle size[1]

Properties	Particle diameter in µm					
	0.001	0.01	0.1	1.0	10.0	100.0
pigment type	transparent iron oxide red		opaque iron oxide red		micaceous iron ore	
hue	yellowish red		yellowish red	violet	metallic	
hiding power	transparent		very good		low	
spec. surface area	←					
tendency to settle out						→
tendency to flocculate	←					
dispersibility						→
oil absorption value	←					

- By contrast, micaceous iron ores, with a d_M over 10 µm, display relatively poor hiding power.
- The hue changes with increasing particle size from a yellow-tinted to a blue-tinted red.
- In the same way the specific surface area and with it of course too the oil absorption value decrease.
- The larger mass of coarser particles means that the tendency to settle out increases as the particle size increases; dispersibility improves.
- The gloss of pigmented coatings is lower with coarser pigments than with finer grades.

Iron oxide pigments manufactured by industrial production methods do not have a uniform particle size. Their fineness is therefore defined by a particle size distribution. As we have just seen, however, pigment particles of different sizes result in different hues, which means that commercial pigments are a mix of particles of differing hues. The narrower the particle size distribution in such pigments, the purer the shade. Synthesis can be regulated to produce pigments having a narrower particle size distribution than would be achievable by grinding batches of coarser particles. For this reason, natural iron oxide red pigments generally demonstrate a lower saturation than synthetically manufactured versions.

Iron oxide pigments can be used to colour coating systems based on organic film formers, such as primers, emulsion paints, topcoat paints or wood protection systems, as well as paints based on inorganic film formers, such as silicate paints, both easily and cost-effectively.

Iron oxide yellow pigments (P.Y. 42)

Iron oxide yellow pigments (α-FeO(OH), lepidocrocite), with a pronounced needle-like structure, are more a green-tinted yellow. As the particle cross-section increases, the shade shifts towards orange and becomes dirtier. Under high shear forces the needles can break, causing the hue to shift towards a dirty yellow. This risk should be alleviated by more recent spherulitic iron oxide yellow pigments, whose spherical particles are less vulnerable to grinding.

At temperatures above 180 °C all iron oxide yellow pigments expel bonded water and convert to iron oxide red with a corresponding shift in hue. The same is naturally true of iron oxide yellow pigments contained in iron oxide brown pigments. The stated stability limit of 180 °C should be regarded as a guide value for the pure dry pigment. In the case of pigments incorporated in coatings, this temperature may be substantially exceeded for short periods without initiating any change in hue.

Iron oxide red pigments (P.R. 101)

Of the various modifications of Fe_2O_3 available, it is principally haematite, α-Fe_2O_3 (corundum structure), that is of interest as a coloured pigment. Iron oxide red pigments, whether produced by the aniline process or by calcining or hydrolysis, have cuboid or spherical particles. Unlike iron oxide yellow pigments, they display excellent resistance to high and low temperatures in addition to their other outstanding fastnesses and resistances. Unfortunately they cannot be used to produce brilliant hues. The relation between coloristic and application-oriented properties and particle size is shown in Table 2.22.

The needle-shaped iron oxide red pigments obtained by calcination of needle-shaped iron oxide yellow pigments generally display a higher intensity of colour than the approximately cuboid or spherical types of a similar shade. Their hue depends on the calcination temperature and also, as with iron oxide yellow pigments, on the cross-section of the needles.

Iron oxide black pigments (P.Bl. 11)

The empirical formula for iron oxide black pigments roughly corresponds to that for magnetite, Fe_3O_4. As the particle size increases, the colour intensity falls and the hue shifts from a brown-tinted black in pigments having a d_M of around 0.1 µm to a blue-tinted black at a d_M of approx. 0.6 µm. The depth of colour in these black pigments is nowhere near that of pigment blacks, but they are much more readily dispersed. Micronised grades should be used for glossy coatings. Above 180 °C iron oxide black pigments are converted in the presence of oxygen first into brown γ-Fe_2O_3, which above 350 °C is transformed into red α-Fe_2O_3. This does not apply to some grades containing foreign metals, which are stable up to around 1000 °C.

Iron oxide brown pigments (P.Br. 6)

Iron oxide brown pigments are usually mixtures of red, yellow and black iron oxide pigments. The mixing ratio determines the hue and also the application-oriented properties. Brighter shades contain a higher proportion of yellow, whereas darker shades predominantly contain iron oxide black. The resistance to high and low temperatures is restricted by the stability limits for the iron oxide yellow and black components.

Transparent iron oxide pigments

Iron oxide pigments with particles of around 0.01 µm in size cease to have practically any hiding power in organic binders. They produce transparent, coloured films. Commercial transparent iron oxide pigments display particle sizes in the range from 0.001 to 0.05 µm and can therefore only be dispersed with some difficulty. Not only do they display the outstanding fastness properties common to most iron oxide pigments, but as red and yellow pigments they also provide metallic coating systems with a pronounced brightness flop. Aside from metallic coatings their good absorption capacity for UV radiation means that they are also used for pigmenting wood glazes.

Chromium oxide pigments

There is only one important chromium oxide pigment, *chromium oxide green*, Cr2O3; *chromium oxide hydrate green*, CrO(OH), is of little significance nowadays. Like iron oxide red, chromium oxide green (P.G. 17) has outstanding fastness properties matched by no other green pigment. It consists of approximately isometric, very hard particles (Mohs' hardness 8 to 9), which results in high abrasiveness during dispersion and spraying. Those grades whose d_M is around 0.03 µm display the best properties.

This highly opaque pigment is used in chemically resistant coatings, in topcoat paints for steel structures, often in conjunction with micaceous iron ore, in exterior finishes and, because of its high reflectivity in the near IR range, in infra-red reflective coatings.

Complex inorganic coloured pigments (CICPs)

The complex inorganic coloured pigments or mixed metal oxide inorganic coloured pigments (set out in Table 2.23) are derived from other white or coloured oxide pigments such as TiO_2, Fe_2O_3 or spinel, $MgAl_2O_4$, by virtue of the fact that some of the host lattice metal ions are replaced by other metal ions without changing the basic structure of the host lattice. The type and number of ions incorporated determine the colour of these pigments. Host lattices with good fastness properties and stabilities should be chosen in order to obtain highly stable pigments. The most important mixed-phase oxide pigments are the *rutile* and *spinel mixed-phase pigments.*

Complex inorganic coloured pigments (CICPs) are consequently characterised by outstanding fastness properties. They are resistant to light, weathering, acid and alkalis and also to most chemicals. Their method of manufacture means that they necessarily display excellent temperature stability. However, they do all demonstrate a relatively high Mohs' hardness (rutile pigments: 6 to 7, spinel pigments: 7 to 8).

The lemon-yellow *nickel titanium yellow* and the much more red-tinted *chromium titanium yellow* are the most important Complex inorganic coloured pigments (CICPs) used in the coatings industry. In comparison to the iron oxide yellow pigments, both have a much purer hue. They differ markedly in terms of hiding power. Whilst nickel titanium yellow has only 10 to 30 % of the hiding power of a standard rutile pigment, that of chromium titanium yellow pigments can be 1.5 to 2 times higher.

Cobalt blue and *cobalt green pigments* are the most important spinel-structured Complex inorganic coloured pigments (CICPs). The red-tinted hue of cobalt blue pigments can be altered to a green-tinted blue by the partial replacement of Al with Cr. Cobalt blue and cobalt green pigments cover approximately the same range of hues as the admittedly much more intensely coloured phthalocyanine pigments. So far, however, despite their outstanding resistances and good purity of hue, they have not been able to establish themselves in the coatings industry. Their excellent weathering stability, even when heavily whitened with titanium dioxide, is worth mentioning and is superior to that of all organic blue and green pigments. Very finely ground grades with particle sizes below 0.1 µm are suitable for transparent coatings with outstanding weather resistance in a range of hues where organic pigments are often unable to meet the necessary requirements.

The colouring power of *spinel black pigments* is roughly comparable to that of iron oxide black pigments. They are readily dispersible and have only a very slight tendency towards flocculation, which means that even grey coatings pigmented with spinel black, unlike those containing carbon black pigments, often display little or no rub-out effect (see page 372). Complex inorganic coloured

Table 2.23: Complex inorganic coloured pigments (CICPs)

Pigment	Colour	Formula	Host lattice/crystalline structure	C.I.
nickel titanium yellow	lemon yellow	$(Ti,Ni,Sb)O_2$	rutile	P.Y. 53
chromium titanium yellow	yellow ockre	$(Ti,Cr,Sb)O_2$	rutile	P.Y. 24
cobalt green	emerald green	$(Co,Ni,Zn)_2(Ti,Al)O_4$	inverse spinel	P.G. 19
cobalt blue	green-hinted blue	$Co(Al,Cr)_2O_4$	spinel	P.B. 36
cobalt blue	red-hinted blue	$CoAlO_4$	spinel	P.B.28
zinc iron brown	brown	$(Zn,Fe)FeO_4$	spinel	P.Y. 119
chromium iron brown	brown	$(Fe,Cr)_2O_4$	spinel with lattice defects	P.Br. 29
iron manganese black	black	$(Fe,Mn)(Fe,Mn)_2O_4$	spinel	P.Bk 26
spinel black	black	$Cu(Cr,Fe)_2O_4$ $Co(Cr,Fe)_2O_4$ $Cu(Cr,Fe,Mn)_2O_4$	between normal and inverse spinel	P.Bk. 22 P.Bk. 27 P.Bk. 28

Table 2.24: Properties of cadmium pigments

Name	Colour	Formula	C.I.	Density	Particle size
cadmium yellow	green-tinted yellow	(Cd,Zn)S	P.Y.35	4.3 to 5.3 [g/cm^3]	0.1 to 0.3 [μm]
cadmium yellow	red-tinted yellow	CdS	P.Y. 37		
cadmium orange	orange	Cd(S,Se), Se < 10 %	P.O. 20		
cadmium red	red	Cd(S,Se), Se > 10 %	P.R. 108		

pigments (CICPs) are suitable for pigmenting all types of coatings which have to meet high demands in terms of light, weathering or chemical resistance. Key areas of application include exterior finishes or coil coating paints. Their good temperature stability and ready dispersibility means that they have also proven themselves in powder coatings.

2.3.5.3 Cadmium pigments

Cadmium pigments are also known as sulphide/selenide pigments because of their chemical composition. All are based on *cadmium sulphide* and crystallise in hexagonal wurtzite lattices. They are characterised by particularly brilliant shades, high hiding power, good intensities of colour, good temperature stability and absolute migration resistance. Further properties are listed in Table 2.24.

Cadmium pigments have long been the subject of critical debate because of their Cd content. Despite these reservations and also despite their relatively high price, they held on for quite some time in certain sectors of the coatings industry, for example in high-temperature resistant, overspray-resistant stoving coatings, powder coatings or coil-coating systems, where their replacement by high-grade combinations of organic and inorganic pigments can even now only be achieved at the cost of losing certain coloristic properties. Nowadays cadmium pigments can only be used if no adequate substitution is possible.

2.3.5.4 Chromate pigments

Pigments in this group are also known as *lead chromates* and exist in the colour range yellow, orange and red. The composition and hues of the invidiual pigments are shown in Table 2.25.

Blends of chromium yellow and phthalocyanine blue are known as *true chromium green*, those with iron-blue pigments as *chromium green*. Both are rarely used today.

Lead chromate and lead molybdate pigments are classed as hazardous to health. Appropriate safety precautions must be followed when manufacturing and handling these pigments. In accordance with the recommendations of the German workplace monitoring commission, they are Class 3B carcinogens. For regulations and measures, see the GESTIS materials database (www.dquv.de/bgia/stoffdatenbank) .

Chromate pigments have proved themselves over many years in automotive paints, industrial coatings and masonry paints because of their hiding power and good gloss retention. Now they are being replaced in many countries for toxicological and ecological reasons by other, generally more expensive pigments and pigment blends. They may not be used in plastics and coatings for toys.

2.3.5.5 Bismuth vanadate pigments

In contrast to the Complex inorganic coloured pigments (CICPs), bismuth vanadate pigments are a mixture of two different substances. Their composition corresponds to a range in composition, extending from $BiVO_4$ through to mixed pigment 4 $BiVO_4 \cdot 3Bi_2MoO_6$, where $BiVO_4$ is the pigmenting component. They demonstrate good colour intensity, produce brilliant colours and have a high coating power. With their greenish-yellow hue, out of all the inorganic pigments they come closest to cadmium yellow and chromium yellow pigments in their coloristic behaviour. Although significantly more expensive than nickel titanium yellow pigments, their use can still be justified

Table 2.25: Name and properties of lead chromate pigments

Name	Colour	Formula	Crystal system	C.I.	Density [g/cm³]	Particle size [µm]
chromium yellow	green-tinted yellow	$Pb(Cr,S)O_4$	orthorhombic	P.Y. 34	5.5 to 5.7	0.2 to 0.6
chromium yellow	pale yellow to golden yellow	$Pb(Cr,S)O_4$-$PbCrO_4$	monoclinic, acicular	P.Y. 34	5.4 to 6.0	0.1 to 0.8 l:d < 3:1
molybdate red	red	$Pb(Cr,Mo,S)O_4$	tetragonal	P.R. 104	5.5 to 6.2	0.15 to 0.25
molybdate orange	orange					0.4 to 1.0
chromium orange	orange	$Pb_2[CrO_4(OH)_2]$		P.O. 21	6.6 to 7.1	0.1 to 1.0
chromium red	red					1 to 12
chromium green	green	chromium yellow + iron blue		P.G. 15		
true chromium green	green	chromium yellow + phthalocyanine blue		P.G. 48		

because of their purer hue and four to five times higher colouring power. They are used in lead-free, weather resistant, brilliant yellow automotive finishes and industrial coatings. They are suitable for pigmenting many other coating systems and, when blended with other pigments, also for hues in the orange and red range.

2.3.5.6 *Iron-blue pigments*

Iron-blue pigment products (C.I. Pigment Blue 27, 77150) are also known under names such as "Milori blue", "Vossen blue", "Berlin blue", "Prussian or Turnbull's blue". They have the composition $M[Fe^{II}Fe^{III}(CN)_6] \cdot x\ H_2O$, where M = Na, K or NH_4. Green to violet-tinted shades as well as bronzing or non-bronzing types can be produced by varying the cation and the manufacturing conditions.

Being finely divided they have good intensity of colour but are difficult to disperse. They can be heated up to 180 °C for short periods. They have outstanding light and weathering stability, but are sensitive to concentrated acids and oxidants and especially to alkalis. They were formerly used in automotive finishes but today are primarily used in the printing inks sector, although their use is in decline.

2.3.5.7 *Ultramarine pigments*

Ultramarine pigments are aluminosilicates containing coarse-meshed skeletons of AlO_4 and SiO_4 tetrahedra. Sulphur compounds as chromophoric groups are intercalated in the cavities of these skeletons. Their composition approximately satisfies the formula $Na_{6+x}Al_{6-y}Si_{6+y}O_{24} \cdot S_n$ where n = 2 to 4 and y is between approx. 0.2 and 1.2. The charge of the anionic sulphur is offset by the x-excess Na ions.

Depending on the structure of the S_n species, blue (C.I. Pigment Blue 29, 77007), green, red (C. I. Pigment Red 15, 77007) or violet (C. I. Pigment Violet 15, 77007) ultramarines are obtained, of which those of a red to green-tinted blue are the most brilliant. The pigments are very light-fast and are resistant to temperatures of up to around 400 °C. Their alkali resistance is adequate for lime but not for cement. On the other hand, they react with all acids. They are used *inter alia* in printing inks, emulsion paints, stoving coatings, industrial coatings and powder coatings. They are not suitable for air-drying coloured coatings for exterior use.

2.3.6 Organic coloured pigments

2.3.6.1 General properties

Organic pigments have grown in importance in the coatings industry since the restriction in the use of some inorganic pigments due to their heavy-metals content. In 1993 the world production of organic pigments was around 170 kt, about half of which was used in printing inks and a quarter in paints and coatings. As Table 2.26 shows, organic pigments differ from inorganic pigments in a number of their properties.

Organic pigments are up to ten times more finely divided than inorganic pigments. Although their scattering power is lower, they are generally more intensely coloured and more brilliant than inorganic pigments. Their chemical structure is such that they exhibit mostly hydrophobic surfaces. Organic pigments are often relatively sensitive to heat, and even the most heat-resistant types, like almost all organic compounds, are stable only up to a maximum of 300 °C in the presence of air. Many organic pigments have only a limited lightfastness and weather resistance, and their resistance to chemicals is very much dependent on their chemical nature.

As with inorganic pigments, crystalline structure and particle size again influence a number of key pigment properties and resistances. In the three modifications of unsubstituted copper phthalocyanine (P.B. 15) to have achieved industrial prominence (referred to as α-, β- and ε-CuPc), the hue shifts from the red-tinted blue of the ε-modification through the relatively neutral blue of the α-modification to the rather green-tinted blue of the β-modification, for example. The particle size influences the hiding power, colour intensity, hue, various resistances, dispersibility, flocculation tendency and flow of the pigmented medium.

Again as with inorganic pigments, more coarsely divided organic pigments are more readily dispersed; the stability of such dispersions is then greater, their rheological behaviour less problematic and their resistances better than is the case with more finely divided types. At the same time their hue shifts and their colour intensity is reduced. Since the average particle size of organic pigments is usually below the optimum value for scattering power (see Figure 2.33), more coarsely divided organic pigments exhibit a greater hiding power.

Organic pigments are manufactured by synthesis and are dried to form powdered products. The individual syntheses vary enormously and in some cases are extremely complex, accounting for the often high price of coloured organic pigments.

Table 2.26: Comparison of inorganic and organic pigments

Property	inorganic pigment	organic pigment
primary particle size	tend to be large	tend to be small
scattering power	usually higher	usually lower
colour intensity	usually lower	usually higher
surface character	hydrophilic	hydrophobic
heat resistance	generally higher	generally lower
generalised range of properties	lightfast*) opaque weather resistant*) readily dispersible	intensely coloured transparent brilliant
*) Lightfastness is evaluated according to the 8-level wool scale (1 = lowest, 8 = highest), weather resistance according to the 5-level grey scale (1 = lowest, 5 = highest, no colour change).		

2.3.6.2 Classification of organic pigments

General introduction

Organic pigments can be roughly divided into three groups, as shown in Table 2.27.

Many organic pigments are derived from organic dyestuffs, the main concern here being to reduce or to eliminate entirely by various means their solubility in water or organic solvents. By "pigmenting" acid groups we obtain scarcely soluble metal salts, for example. Additionally, polar groups, notably acid amide and Cl groups, can be incorporated into the pigment molecules in order to suppress solubility. Finally, low solubility can also be achieved by expanding the molecules.

The most practical method of classification is from a chemical standpoint, as laid down in DIN 55 944.

Azo pigments

Azo pigments with the general formula R^1-N=N-R^2 are so called because they contain at least one azo group -N=N-. In terms of quantity they are the most important group of organic pigments (approx. 60 % of the total production volume), yet are relatively inexpensive. A distinction is made between *monoazo pigments* having only one azo group per molecule and *disazo pigments* with two such groups (R^1-N=N-R^2-N=N-R^1). These two sub-groups can be subdivided further according to the chemical nature of the organic radicals R^n, which contain more or less extended π electron systems. Azo pigments cover the red to yellow colour range. Table 2.28 (page 142) provides some typical examples. The performance of simple azo pigments in respect of current requirements relating to lightfastness and overcoatability is generally only poor or at best moderate, whereas good resistances can also be achieved with high-grade types.

Table 2.27: Classification of organic pigments

Pigment group	Examples	Colour range	C.I. I
azo pigments			
monoazo pigments		green-tinted yellow – yellow – orange – red – violet/brown	
	acetoacetarylide	green-tinted mid-yellow	P.Y. 1
	naphtol AS	yellow-red, burgundy/brown	P.O. 36
	benzimidazolone	yellow-tinted red burgundy/carmine/violet/brown	P.R. 112
	pigmented β-naphthol dyes	yellow-tinted red	P.R. 53:1
disazo pigments		green-tinted yellow – yellow – orange – red/brown	
	azo condensation pigments	yellow/orange/scarlet/red/brown	P.R. 144
	dipyrazolone	orange/yellow-tinted red	P.O. 13
polycyclic pigments		yellow – orange – red – violet – blue – green/chestnut/brown	
	quinacridone	blue-tinted red, violet	P.V. 19
	dioxazine	violet	P.V. 23
	perylene	orange/red	P.R. 149
	diketopyrrolo-pyrrole	yellow-tinted red	P.R. 254
	isoindoline	yellow – orange – red	P.Y. 139
metal complex pigments			
	Cu-phthalocyanines	red-tinted blue – yellow-tinted green	P.Bl.15

Table 2.28: Properties of various azo pigments

Pigment group	Light fastness	Overcoatability, solvent resistance	Hue
monoazo yellow and monoazo orange	moderate – good	usually low	green-tinted to strongly red-tinted yellow
β-naphthol	moderate	low	yellow-tinted orange – blue-tinted red
naphthol AS	low – good	low – good	yellow-tinted – very blue-tinted red
β-oxy naphthoic acid, pigmented	moderate	good	yellow-tinted – blue-tinted red
benzimidazolone	good	very good	green-tinted yellow – very blue-tinted red, brown
diaryl yellow	usually low	moderate – good	very green-tinted – very red-tinted yellow
disazo condensation	usually good	good	green-tinted yellow – red – violet

Polycyclic pigments

Polycyclic pigments encompass a chemically diverse range of compounds whose common feature is that they all contain carbocyclic and/or heterocyclic structures but no azo groups. Although copper phthalocyanine pigments really belong to this group too, they are sometimes grouped separately along with other metal complex pigments.

It is often difficult to make generalising statements about polycyclic pigments because of their great diversity. Their suitability for special areas such as coil coatings or powder coatings should be chekked where necessary. They are without exception solvent-resistant and overcoatable. Furthermore, they usually exhibit good resistance to high and low temperatures, good chemical resistance, light stability and weather resistance. Quinacridone and diketopyrrolopyrrole pigments display the best resistances. The lightfastness, weather resistance and solvent resistance of isoindoline pigments are rather poorer. The often high-priced polycyclic pigments are frequently used in automotive finishes and high-grade industrial coatings.

Metal complex pigments

Copper phthalocyanine pigments (CuPc pigments) are the most important of the metal complex pigments. Other metal complexes now play only a minor role. CuPc pigments cover the range from a yellow-tinted green to a red-tinted blue. Their colour can be controlled by the degree and the type of halogenation. The non-substituted or just low-chlorinated types are blue. Blue-tinted green pigments (P.G. 7) are obtained by substituting 14 to 16 H atoms per molecule with Cl atoms. The more of these Cl atoms that are substituted with Br atoms (8 to 13), the more pronounced the yellow tint of the pigments becomes (P.G. 36). Their good resistances and relatively low price has made them the second largest group of organic pigments after the azo pigments. They are used in all types of coating materials.

2.3.6.3 Optical properties of organic pigments

As we have already mentioned, the hiding power of coloured pigments is determined by the absorption and scattering of light. Given their mostly, high refractive index and associated high scattering power, inorganic coloured pigments principally cover by light scattering. Organic coloured pigments, by contrast, exhibit scattering noticeably only at certain wavelengths (see Figure 2.32). Their hiding power is usually principally determined by absorption. Organic blue and violet pigments absorb large parts of the incident light, which means that correspondingly opaque coatings appear dark.

The refractive indices of organic pigments often rise sharply along the short-wave absorption edge and then fall again as the wavelengths increase, approaching the refractive index of the binder. The larger the range in which the refractive index of the pigment differs little from that of the binder, the lower the contribution of light scattering to the hiding power of an organic pigment. The influence of particle size on scattering power has already been covered in section 2.3.2.3.

As we illustrated there, the appearance of a coated object is determined by reflection, scattering and absorption in the coating film and at the substrate. In opaque systems the substrate by definition cannot be seen. Plain hues are generally obtained here by means of highly opaque inorganic coloured and white pigments. Any organic coloured pigments that may be present have a primarily colouring effect.

In transparent coating systems, on the other hand, the colour and scattering power of the substrate also have an important part to play. Thus, for example, a transparent coating may appear as a red-tinted yellow on a highly scattering white substrate and as a gold hue on a selectively reflecting aluminium metal. A coating that appears blue when applied to a white substrate is a violet-tinted black on a black substrate. Light that reaches the substrate is completely absorbed. Since blue pigments are strongly absorbent in the red and yellow range, the blue and violet components dominate in the small proportion of reflected light resulting from the very poor scattering.

2.3.6.4 Fields of application for organic pigments

The field of application of a pigment is not determined by its chemical identity alone, since chemically identical pigments can differ substantially in many of their physical properties, such as crystalline structure and particle size, and in the criteria that depend upon these, including hiding power or colour intensity, for example. Any generalising information about areas of application for the individual classes of pigment should therefore be taken as a guideline only.

▪ Very high resistances are often required for use in *high-grade industrial coatings*. The stringent requirements for *automotive production-line* and *refinishing paints* are well known. Since they are frequently applied by automated equipment the coatings must flow well and have consistent rheological properties. In particular, high solid coatings require that the influence of pigments on the viscosity of the coating - which is greater in the case of organic pigments because of their smaller particle size than in the case of inorganic pigments - must be reduced by optimising the particle size and, if necessary, by providing a surface treatment. Good stability of the pigment dispersion is a prerequisite for high gloss, in automotive coatings for example. With the exception of isoindoline pigments, almost all polycyclic and CuPc pigments are suitable. Apart from a few disazo condensation and benzimidazolone pigments, azo pigments are rarely used.

▪ The requirements for pigments used in *general industrial coatings* are not usually so high. Generally good light stability, weathering and solvent resistance are expected, however. In some cases, e.g. air drying coatings, the requirements pertaining to solvent resistance are less stringent. In inexpensive products compromises are often made in respect of the other resistances too. In the case of coil coatings, adequate resistance to high temperatures on the part of the pigments is vital, of course, such as is achieved in only a few high-grade organic pigments. This very broad range of requirements means that almost all pigments are used in the area of general industrial coatings. Polycyclic pigments are used only when requirements for high resistances make it absolutely necessary, however.

▪ Dispersion of organic pigments in the *aqueous phase* is often difficult because of their hydrophobic surface. An appropriate surface treatment of the pigments can facilitate wetting, but surface-active dispersing agents generally have to be added too. In addition to steric stabilisation, electrostatic stabilisation is also an option in water. In such cases, given its straight-forward use, the coating is often formulated using pigment preparations. Expensive polycyclic pigments are used only for tinting or in application areas such as outdoor paints and automotive OEM finishes where good weather resistance is required. Otherwise many azo pigments are used here.

▪ Powder coatings contain neither water nor organic solvents. Here the pigments are incorporated into the reactive binder systems by means of an extruder. When processing shear-sensitive pigments, care should be taken since the particles can easily be buckled or broken in the extruder. The very high stoving temperatures to which powder coatings are exposed require correspondingly heat-resistant pigments.

2.3.7 Lustre pigments

DIN 55 944 divides inorganic lustre pigments into *metallic pigments, pearlescent pigments* and *iridescent pigments*. The optical effect of these predominantly flat-structured pigments is based on the following phenomena:

- In metallic pigments a largely selective reflectance occurs at oriented pigment particles,
- in iridescent pigments the colour-imparting effect is based entirely or principally on the phenomenon of interference of light at thin, highly refractive layers.
- Pearlescent pigments, which belong to the iridescent pigments, produce their characteristic pearly lustre by means of the multiple reflection of light at oriented, highly light-refractive particles.

Since special optical effects are obtained with all of these lustre pigments, they are classed as so-called special effect pigments.

2.3.7.1 Metallic pigments

Introduction

Metallic pigments are generally platelet-shaped particles of mostly non-ferrous metals. In addition to the metallic platelets (flakes), other particle shapes are also known, however, including that of spherical zinc dust or dendritic copper powder. The platelets can have diameters ranging from a few millimetres (spangles) to several micrometres (bronze powders). They can be used either alone or in combination with more or less transparent colorants as colouring and effect-imparting pigments in coatings, printing inks and plastics. They cover a broad range of applications in the coatings industry, being used in metallic coatings, heat-resistant paints, primers, hammer finish paints and anti-corrosive coatings, for example. They can also take on a functional role in such applications.

For historical reasons metallic pigments are divided into silver and gold bronzes. Whereas earlier they were actually Cu/Sn or Cu/Al bronzes, they now consist principally of Al, Cu or Cu/Zn alloys and, more rarely, of silver or nickel alloys.

Most metallic pigments are manufactured by the Hall process, whereby aluminium (with a purity of around 99.5 %) is atomised into a fine aluminium grit. This intermediate product is sifted and then comminuted in ball mills, with the addition of white spirit and grinding agents, and formed into platelet-shaped particles. The purpose of the grinding agents, which act in the same way as lubricants, is to prevent the Al platelets from undergoing cold fusion during grinding. The length of grinding and the type and quantity of grinding agent can be adjusted to tailor the Al pigments to their ultimate area of application.

• Leafing pigments

If stearic acid, for example, is used as grinding agent, the carboxyl groups bond to the surface oxides of the metallic pigment. The long hydrocarbon radical protrudes from the surface. Further stearic acid molecules intercalate in a head-to-tail arrangement into this hydrocarbon layer, which is additionally stabilised by the formation of hydrogen bridges. In this way a double layer of stearic acid molecules is formed on the surface of the pigment, the polar ends of which are directed outwards (Figure 2.40).

The pigment surface is then hydrophilic or oleophobic. Compatibility with the binder is reduced and the pigment floats on the surface (leafing effect) due to the high interfacial tension in the coating.

These leafing grades form a dense mirror of metallic platelets on the surface oriented parallel to the surface of the coating. *Leafing pigments* are principally used in applications requiring a good physical diffusion barrier as well as reflection of heat and UV radiation. Such pigments are not suitable for use as primers.

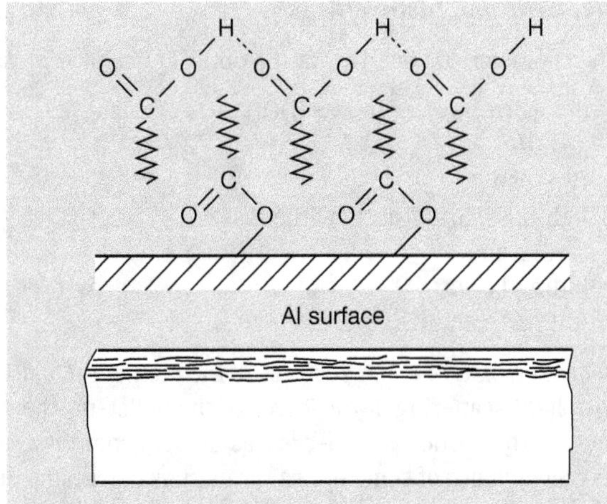

Figure 2.40: Production of a hydrophilic surface by coating with stearic acid

- **Non-leafing pigments**

If branched or unsaturated fatty acids, e.g. oleic acid, or polar substances such as fatty amines are used as lubricant in place of stearic acid, the carboxyl groups again bond to the surface oxides of the metallic pigments. In this case, however, the hydrocarbon radicals do not protrude from the surface but instead lie along the surface, thus preventing the formation of a double layer. The surface acquires a more non-polar, oleophilic character (Figure 2.41). Such non-leafing pigments are wetted well by the binder, do not float on the surface and are distributed evenly across the entire film layer. The coarse fraction is decocted and then with the exception of special cases the pigment suspension is adjusted to the conventional commercial composition of 65 % solids and 35 % solvents.

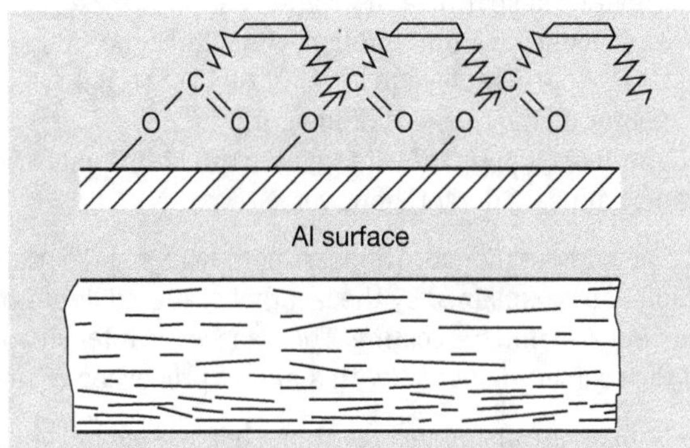

Figure 2.41: Production of an oleophilic surface by coating with oleic acid

The following pigment types are also available for use in water-borne coatings: phosphated, chromated and/or pigments encapsulated with a protective layer (inorganic, e.g. SiO_2, or corrosion inhibitors, such as phosphoric acid esters, or polymeric).

Properties of metallic pigments

The *chemical properties* of metallic pigments naturally depend on their chemical composition. Al pigments react with acids and alkalis with formation of hydrogen, causing them to lose their metallic lustre. Cu and Cu/Zn pigments are considerably less reactive. They are often sensitive to acids but rarely to alkalis. Complex formation is possible in an alkaline environment, whilst thickening or, more rarely, a green coloration by Cu ions may occur in the presence of acid.

Particle shape and particle size have a critical influence on the *optical properties* of metallic pigments. The ratio of thickness to diameter of the platelets is termed the form factor and is generally between 1:50 and 1:500. The shape of the platelets varies according to the manufacturing process between almost circular and irregular. Aluminium pigments produced by conventional manufacturing processes display a more or less random particle size distribution, with a d_{50} value ranging from 5 to 50 µm. The thickness of the platelets influences not only the hiding power but also above all the mechanical stability of the pigments. Too great a shear stress during dispersion or in closed cycles can cause the platelets to be deformed or even to be reground, leading to loss of optical quality, speck formation and possibly to chemical reactions.

Coloristic and lustre effects

The characteristic features of the optical effect are

- distinctiveness of image (DOI)
- sparkle
- two tone
- lightness, brightness, whiteness
- saturation
- hiding power
- tinting strength.

Reflection of light at the surfaces of the pigment platelets causes the metallic effect, which overlaps with light scattering at the edges of the platelets. The ratio of edge length to pigment surface area rises as the particle size decreases and is generally greater in irregularly shaped particles. Hence the dependence of some optical properties of metallic pigments on particle shape, particle size and particle size distribution.

- The larger a pigment particle, the smaller the fine fraction; the more regular its shape, the greater the sparkle and brightness.
- The smaller the fine fraction, the higher the saturation or chromaticity in pigmentations.
- Coarse particles lead to lower gloss in coatings, which manifests itself in reduced if not actually poorer distinctiveness of image.
- The finer a pigment, the greater the hiding power of the coating system; at the same time it also appears darker, however.

A fine pigment therefore appears both whiter and darker than a coarse pigment. This apparent contradiction is explained by the fact that scattered light is radiated not only outwards from the coating film but also into the coating. This causes the intensity of the reflected light to be reduced in respect of the coarser pigments (with a higher proportion of reflection directed out from the coating).

Al bronzes are distinguished from other achromatic pigments by their metallic lustre. Cu-containing bronzes are classed by hue into the following “natural hues”: copper, bleached gold (Cu/Zn = 90/10), rich bleached gold (Cu/Zn = 85/15) and rich gold (Cu/Zn = 75/25). In the case of fire-coloured bronzes, oxidation of the pigment surface can produce a range of other hues such as sea-green, mint gold or flame red.

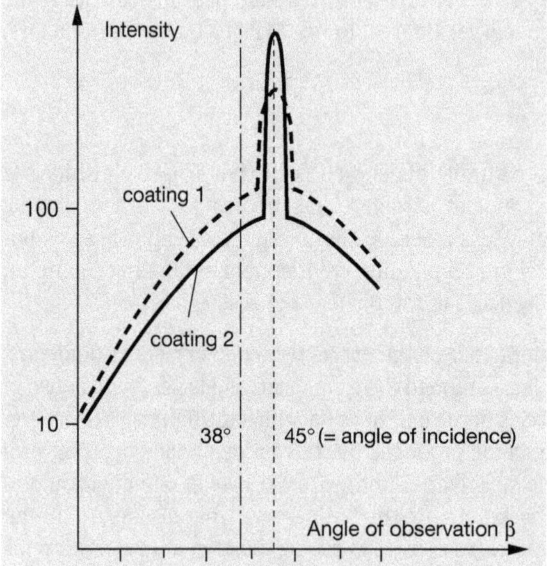

Figure 2.42: Goniophotometer curves for two metallic coatings

Two-tone effects

The characteristic optical feature of metallic coatings is the dependence of the brightness on the angles formed in respect of one another by the light source, coating surface and observer. If the angle of observation is roughly the same as the angle of incidence of the light (glancing angle), the coating appears significantly brighter than it does from angles of observation differing greatly from the angle of incidence. In order to quantify this effect (known as two-tone effect), a test piece coated in metallic paint is illuminated from a constant angle of 45° and the reflectance is measured according to the angle of observation (Figure 2.42 and Figure 2.43).

The two-tone value (ME value) is calculated from the following formula:

$$\text{ME value} = \frac{R(45°/38°)}{R(45°/0°)} \cdot 100$$

The sharper the drop from the glancing angle, the more pronounced is the two-tone effect, and the higher the characteristic shoulder in the goniophotometer curve at around 38°, the more sparkling the appearance of the pigment. Since the two-tone effect is determined by the ratio of directly reflected light to diffusely scattered light, it is more pronounced in coarser pigments than in finer examples. For this reason, two-tone effect and distinctiveness of image also run counter to one another.

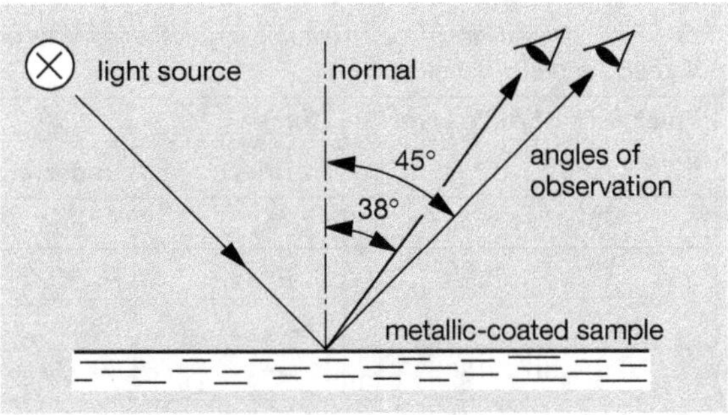

Figure 2.43: Measuring geometry for determining two-tone effects

2.3.7.2 Pearlescent and iridescent pigments

Pearlescent pigments too owe their optical effect to their platelet structure. Incident light is reflected - selectively in parts - at the binder/pigment boundaries - and where applicable at phase boundaries inside the particle - according to the difference in the refractive index of the participating phases. At the same time, unlike metallic pigments, pearlescent pigments are translucent. A typical feature of their effect is that a large part of the incident light penetrates the pigment particle and is then partially reflected again at the next particle below it. This repeated partial reflection creates the impression of a *deep-down* lustre.

In pearls this impression is due to their laminar construction of thin aragonite lamellae separated by thin layers of proteins with a lower refractive index. Natural pearlescent pigments have been known for centuries in the form of "pearl essence". Only synthetic products, principally coated mica, are used in the coatings industry, however.

Figure 2.44 shows the structure of such pigments in schematic form. The mica particles are coated with a transparent metal oxide (e.g. TiO_2) layer of a precisely defined thickness. Not only does this multilayer structure cause the light to be reflected repeatedly, but the reflected rays of light can also interfere with one another if the layer is of the correct thickness. In thin layers (< 50 nm approx.) this type of interference only occurs at oblique angles of observation, however, and the path difference between the two participating rays of light must lie in the range of the wavelength of visible light.

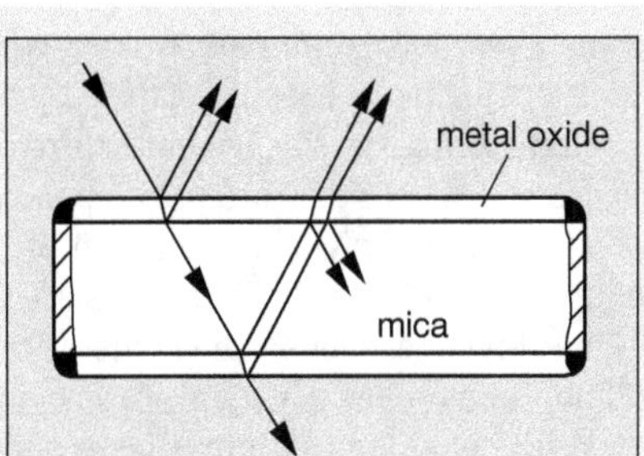

Figure 2.44: Schematic structure of a pearlescent pigment

Depending on the angle of observation and the thickness of the layer, varying colours also occur due to mutual quenching and intensification, as shown in Table 2.29 (page 148). These pigments are only coloured when coated with relatively thick layers of TiO_2, however, and the colour occurs not only in the top view but also in the undertone, where the complementary colour appears.

The following relation according to thin-layer physics applies to the interference colour:

$$\lambda = 4(2z - 1) \cdot d \cdot \sqrt{n^2 - \sin^2\alpha}$$

where z = order of interference (1, 2, 3, ...), α = angle of incidence, d = thickness of layer, n = refractive index and λ = wavelength.

Table 2.29: Dependence of the colour of pearlescent pigments on the thickeners of the TiO_2 coating

Thickness of rutile layer [nm]	Colour	
	top view	undertone
40 to 50	white	white
72 to 80	gold	blue
92 to 100	red	green
110 to 130	blue	yellow (gold)
140 to 150	green	red

The effect can be improved by using synthetic aluminium or silicon oxide flakes whose surfaces are smoother than those of the mica particles. If mica platelets are coated with a coloured oxide layer, e.g. Fe_2O_3, in addition to or in place of the colourless TiO_2 layer, more intensely coloured pigments are obtained. The colour is derived from interference colour and absorption of the coloured metal oxide layer. With a few exceptions these mica-based special effect pigments are not opaque. However, if a thin Fe_2O_3 layer of a thickness > 10 nm is applied to Al platelets, opaque, metallic lustre, iridescent pigments are obtained in hues ranging from yellow and orange-red through to violet.

Pigments intended for exterior use are frequently also coated with an additional layer which optimises the interactions between pigment and binder and, in the case of pigments with a TiO_2 surface, reduces the latter's photoactivity (→ 2.3.3.2). Furthermore, the metal oxide surface of pearlescent or iridescent pigments is additionally given a transparent, colourless coating with an organometallic compound.

A range of other mono- and multi-phase iridescent pigments is now available, e.g. aluminium pigments coated with Fe_2O_3, special iron oxide platelets with haematite structure, platelet-shaped copper phthalocyanines, graphites, 1,4-diketopyrrolopyrroles and also liquid crystal pigments, which can sometimes produce incredible effects.

2.3.7.3 Incorporating special effect pigments into coatings

Sensitivities

Strong shear forces must be avoided when incorporating both metallic and also pearlescent and iridescent pigments because of their fragility. This does not represent any great problem in the case of liquid coating systems, given their particle size and relative ease of dispersibility. Depending on the coating system they are pre-dispersed in alcohols, esters, ether esters or in aromatic hydrocarbons in a ratio of pigment to solvent of 1:1 to 1:2; wetting can be facilitated by the use of special additives. Chlorinated hydrocarbons are unsuitable for making metallic pigment pastes (because of Grignard-type reactions). The binder is then added whilst stirring. If some of the agglomerates have still not been broken down, this can be done using a high-speed mixer at speeds of below 1000 rpm for periods of around 10 min. Higher speeds or longer dispersion periods damage the pigments.

Sedimentation tendency

Since some sedimentation of the incorporated lustre pigments is scarcely to be avoided given the size of the particles, the formation of a stubborn (difficult to agitate) deposit must be prevented. Increasing the viscosity of the coating system slows down sedimentation, although the processing settings for the viscosity must be taken into account. In the case of pigments stabililsed electrostatically with carboxyl- or amine-containing additives, the like-charged pigment particles repel one another. If the surface of the pigment is covered with long-chain organic molecules, this makes the formation of a stubborn deposit more difficult. The success of these measures (either alone or in combination) is very much dependent on the binder system as well as on the incorporated pigment and must be tested for each individual case.

In aqueous systems, specially stabilised metallic pigment types must be chosen which do not grey or gas in contact with water (→ 2.3.7.1).

Powder coating pigmentation

Special effect pigments are almost always incorporated into powder coatings using the "dry blend" process. Given their particle size and shape, they would not survive the extrusion process undamaged. The differing density and different particle forms of special effect pigments and powder coating particles (platelet-shaped and more or less spherical, respectively) means that separation occurs to a varying extent in such powder coatings, however - particularly if applied electrostatically - leading to fluctuations in colour and lustre in the final coating.

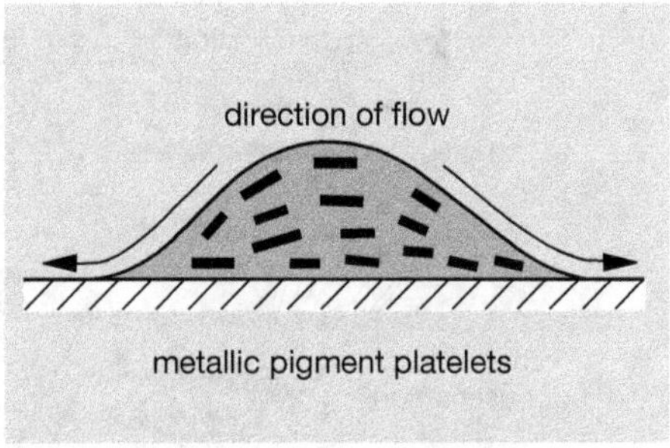

Figure 2.45: Orientation of platelet-shaped particles during flow of coating droplets

For this reason we try to fix the special effect pigment particles completely to the powder coating particles. The "elaborate bonding" process is a thermo-mechanical operation, which makes the manufacture of "bonded" powder coatings cost intensive. Additional advantages over the dry-blend method are superior dispersion of the pigment particles in the powder coating which enables the pigment volume concentration to be lowered by up to 30%, and the recyclability of the metallic powder coatings made in this way.

2.3.7.4 Formation of the special effect

In order for lustre pigments to form their special effect in a coating, the pigment platelets must be oriented parallel to the surface. Only then is the incident light reflected in a preferential direction rather than diffusely to all sides. The first prerequisite for this parallel alignment is of course the best possible breaking down of the pigment agglomerates as they are incorporated into the coating. Depending on the type of application, the pigment particles are then already more or less strongly aligned on application (Figure 2.45). The solvent content of the coating has a key influence on the formation of the effect. Volume shrinkage during evaporation of the solvent encourages parallel alignment of the pigment particles.

A good orientation is obtained in this way in high solvent low solid and also in medium solid systems, whereas the alignment in high solid and powder coating systems is less satisfactory. In the latter systems the best possible alignment must be achieved during the application process itself, e.g. by the flow of the coating droplets as they meet on the object being coated.

If a coating contains transparent coloured pigments in addition to metallic pigments, the impression of colour is created by a combination of selective light reflection at the surface of the coating (1a) (see Figure 2.46, page 150) and at the metallic pigments (1b), light scattering (2) at the metallic pigments and light absorption (3) at the coloured pigments; it also depends on the angle of view of the observer.

Close to the glancing angle (A), the selective reflection of the metallic pigments predominates; at all other angles of observation (B) diffuse scattering and light absorption predominate. In the first instance the coating appears glossy and bright, whereas in the second it appears dark and similar to the saturated shade of the coloured pigment.

Depending on the angle of illumination, the hue of the saturated shade in such coatings can also change. This is explained by the longer distance travelled by the light in the paint at a flat angle, which also means that more light is absorbed (beam paths a and b in Figure 2.47, page 150). In a blue-tinted red pigment this makes the hue, for example, more yellow and hazy.

The other lustre pigments too are often combined with transparent organic or inorganic coloured pigments. In this way and also by combining various pearlescent pigments, a vast range of options of colours and effects is opened up to the designer. In the case of pearlescent pigments, not only is the low hiding power a factor here, but it must also be borne in mind that in most applications an additive colour blend is produced in mixtures of different pearlescent pigments.

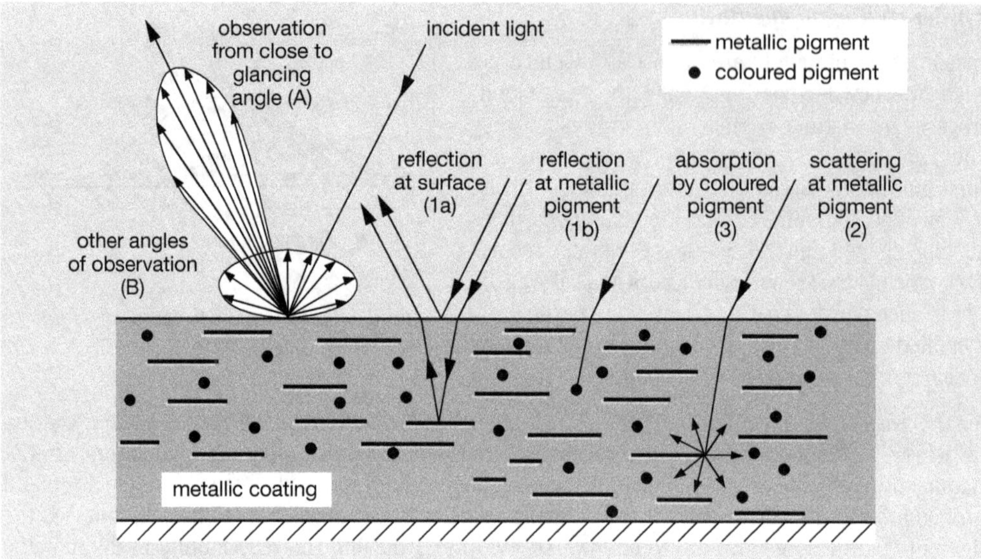

Figure 2.46: Formation of the impression of colour in metallic coatings

2.3.8 Functional pigments

The terms "functional pigments" or "special pigments" are usually used to encompass groups of pigments that are used in coating materials not for their colouring properties but for other, e.g. corrosion-inhibiting, magnetic or electrical properties. For this reason, in addition to these anti-corrosive pigments, magnetic pigments and conductive pigments, daylight fluorescent pigments too, *inter alia*, are classed with these pigments having special properties.

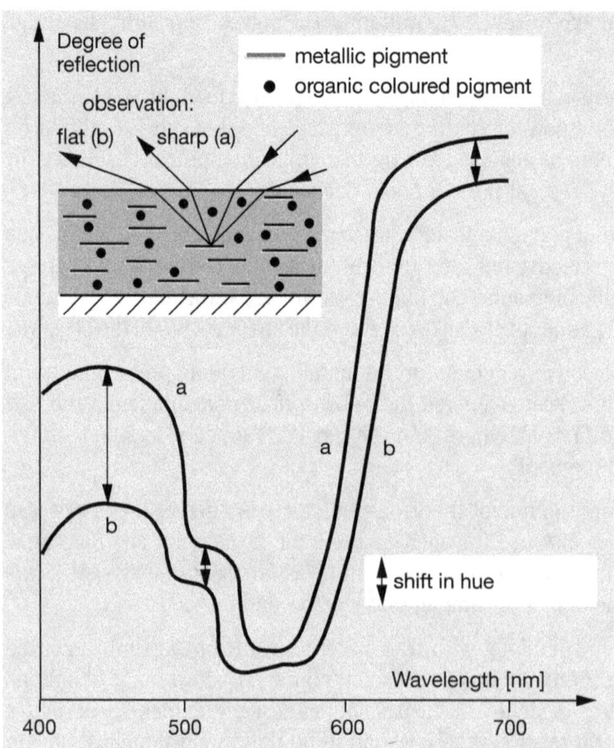

Figure 2.47: Formation of an angle-dependent shift in hue in metallic coatings

The following discussion is restricted to *anti-corrosive pigments* and *conductive pigments.*

2.3.8.1 Anti-corrosive pigments

According to DIN 50 900, **corrosion** is the reaction of a material with its surroundings that causes a measurable change in the material and can lead to an impairment in the function of the component or in an installation.

According to this definition corrosion is not restricted to metallic materials but can also occur in mineral materials, e.g. concrete, or in plastics. Anti-corrosive pigments are used in coating materials designed to protect metal substrates, mainly steel, zinc or aluminium, against corrosion. According to DIN 55 943, an **anti-corrosive pigment** is understood to be a pigment that when used in primer coats on metals inhibits or prevents the corrosion of the metal surface generally due to a chemical or physico-chemical action.

Corrosion in metals is essentially attributable to

- chemical corrosion reactions, e.g. scaling of iron (simplified: $3\ Fe + 2\ O_2 \rightarrow Fe_3O_4$) and to
- electrochemical corrosion reactions.

The presence of an electrolyte on the surface of the metal and the existence of a potential difference between different points on the surface of the metal are prerequisites for electrochemical corrosion reactions. Both, that is the existence of an electrolyte and of a potential difference, then lead in the presence of a suitable oxidant, usually oxygen and/or protons, to the formation of a corrosion cell.

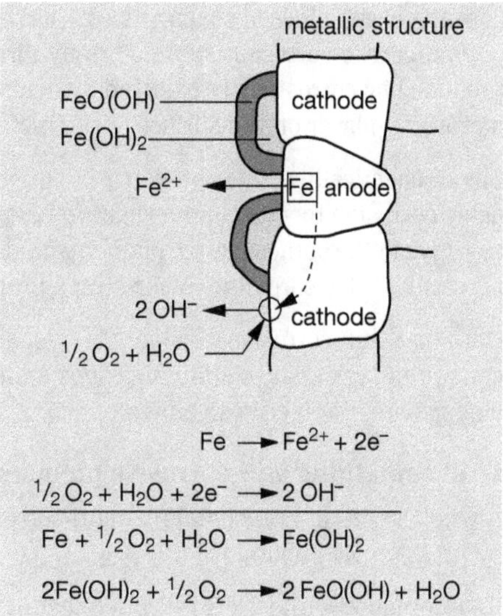

Figure 2.48: Principle behind oxygen corrosion

The reason for a potential difference may be, *inter alia*, contact between different metals, heterogeneity in the metallic structure, the presence of various substances on the surface of the metal, differing mechanical stress or deformation ratios, local differences in concentration in the attacking medium or local temperature differences.

The processes involved in the electrochemical corrosion of iron by oxygen or acids are shown schematically in Figure 2.48. The corrosion may additionally be accelerated by corrosion stimulators, e.g. chloride or sulphate ions. In the processes illustrated here anti-corrosive pigments intervene.

The mode of action of anti-corrosive pigments may be of a

- physical,
- chemical and/or
- electrochemical nature.

As shown in Figure 2.49, all anti-corrosive pigments have a physical protective action simply by virtue of the fact that they extend the diffusion distance for water and aggressive substances, e.g. oxygen or corrosive ions, from the surface of the coating to the surface of the metal. In addition to this *barrier effect*, some of these pigment types improve the *adhesion* of the coating to the substrate. Some pigments protect the binder against photochemical breakdown by *reflection* or *absorption of UV radiation*.

In order to exert a chemical or electrochemical effect, appropriately reactive anti-corrosive pigments must have a *specific solubility*. Many anti-corrosive pigments set an alkaline pH on the metal surface. At the same time this neutralises acid substances, e.g. fatty acids formed by autoxidation of the binder or inorganic acids diffusing into the coating from the attacking medium. The protective action

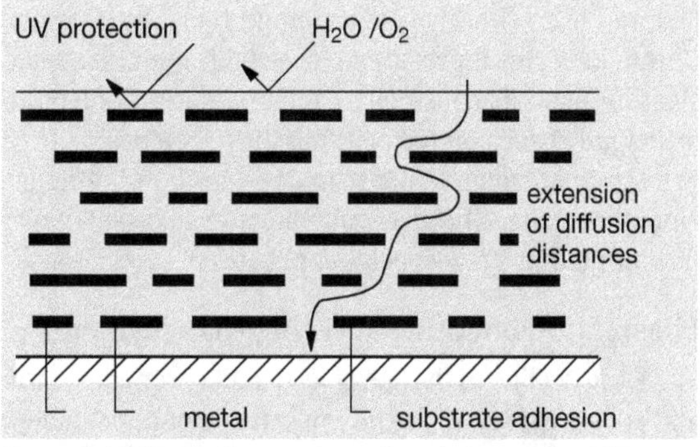

Figure 2.49: Physical protective action of pigments and fillers (schematic view), according to [22]

of scarcely soluble lead soaps arising from the reaction of PbO with fatty acids is well known. Many pigments trap corrosion stimulators by forming scarcely soluble compounds. Corrosive Fe^{2+} ions produced by corrosion are oxidised by oxidants and then as scarcely soluble $F_2O_3 \cdot x\ H_2O$ they form a protective layer on the surface of the metal at high pH values.

Electrochemically active anti-corrosive pigments passivate the surface of the metal. Pigments that delay corrosion by their high oxidation potential, e.g. chromate-containing pigments, are described as active in the cathode zone. Pigments that are active in the anode zone, e.g. phosphate-containing pigments, delay corrosion by forming a protective layer on the surface of the metal.

Nowadays anti-corrosive pigments are usually divided according to their chemical nature into lead-containing, chromate-containing, phosphate-containing and metallic pigments as well as miscellaneous other anti-corrosive pigments.

Lead-containing anti-corrosive pigments

A whole series of lead-containing anti-corrosive pigments has been developed, of which red lead is by far the most important.

In accordance with DIN 55 916, red lead (Pb_3O_4) contains at least 93.2 % Pb_3O_4, the sum of Pb_3O_4 and free PbO being at least 99 %. This standard differentiates between three grades: red lead non-setting, red lead highly disperse and red lead highly disperse ultra, which differ in their fineness. In the coatings industry the highly disperse grades are normally used. Depending on the grade, the oil absorption value is between 5 and 8 g/100 g. The electrochemical activity of red lead is largely based on its Pb(IV) content, which has a passivating effect. With fatty acids, red lead forms lead soaps ($PbO + 2\ HOOC\text{-}R \rightarrow Pb(OOC\text{-}R)_2 + H_2O$), which over time pass through the coating film in the form of lamellae and improve its mechanical strength, its water resistance and its adhesion to the substrate. With corrosion stimulators such as sulphate or chloride ions, red lead forms scarcely soluble lead sulphate or lead chloride (e.g. $PbO + SO_4^{2-} + H_2O \rightarrow PbSO_4 + 2\ OH^-$).

Other lead-containing anti-corrosive pigments worthy of mention here include lead silicochromate, calcium plumbate, lead cyanamide and dibasic lead phosphite.

Anti-corrosive pigments containing lead are nowadays being superseded by other, less environmentally harmful pigments.

Chromate-containing anti-corrosive pigments

Various zinc chromates and strontium chromate, $SrCrO_4$, are used as chromate-containing pigments. The technical conditions for the supply of zinc chromate pigments are laid down in DIN 55 902.

- Zinc chromate type 1a, $K_2CrO_4 \cdot 3ZnCrO_4 \cdot Zn(OH)_2 \cdot 2H_2O$, is usually known as basic **zinc potassium chromate**. Type 1b is differentiated from type 1a by its lower chromate content.
- Zinc chromate type 2, $ZnCrO_4 \cdot 4Zn(OH)_2$, is known as **zinc tetrahydroxychromate**.

Like red lead, chromate-containing pigments have both a chemical and an electrochemical action. Chromate forms a protective layer with metal oxides on the surface of the material and has a passivating effect because of its high oxidation potential. The chemical activity of zinc chromates is principally based on the fact that they set a basic pH. Strontium chromate forms strontium sulphate with sulphate ions. In order to develop its full efficiency, it has to be combined with zinc oxide or zinc phosphate. Chromate pigments are classed as carcinogenic, as a result of which their use is now limited to a few special applications.

Phosphate-containing anti-corrosive pigments

Since the decline in the use of lead- and chromate-containing anti-corrosive pigments on toxicological and ecological grounds, the importance of phosphate-containing pigments has grown dramatically. As they still cannot replace the "traditional" anti-corrosive pigments in every respect, efforts are being made to improve both their efficiency, by combining various phosphates or by adding other substances such as zinc oxide or zinc borate, and their reactivity - by micronisation, for example. Some of the most important members of this group include zinc phosphate ($Zn_3(PO_4)_2 \cdot 2$ to $4H_2O$), chromium phosphate ($CrPO_4 \cdot 3H_2O$), aluminium triphosphates, calcium magnesium phosphates, barium phosphate and aluminium zinc phosphate.

In addition to their chemical activity, the electrochemical activity of phosphate-containing pigments in the anode zone is under investigation. Small quantities of pigment hydrolyse under the influence of moisture, producing for example zinc hydroxide and *sec.* phosphate. A protective layer on the surface of the iron forms from these, consisting of scarcely soluble complex iron phosphates. It is assumed that reactions of the pigment with inorganic ions or with carboxyl groups from the binder lead to the formation of basic complexes which in turn react further with metal ions to form what are known as inhibitor complexes.

Since the hydrolysis reaction is a prerequisite for the activity of phosphate-containing pigments, their action does not begin until after the start-up phase. Phosphate-containing pigments must therefore be combined with other corrosion-inhibiting compounds such as zinc oxide or organic inhibitors, which can take over corrosion protection during this start-up phase.

Metallic anti-corrosive pigments

Zinc and lead are used in metallic form as anti-corrosive pigments. Of the two, zinc in the form of zinc dust is by far the more important.

According to DIN EN ISO 3549, the content of metallic zinc in zinc dust pigments must be at least 94 % and the total zinc content at least 98 %. The free-flowing blue-grey powder consists of almost spherical particles whose average particle size, depending on the grade, is between 1 and 10 µm.

The action of zinc dust in primers is based on both chemical and electrochemical processes. High-volume corrosion products, principally zinc hydroxide, are formed on the surface of the zinc dust particles under the influence of moisture and atmospheric oxygen. This increase in volume compacts the coating and increases the barrier effect. At the same time acid substances, such as sulphuric acid from atmospheric SO_2, are neutralised by the basic zinc hydroxide.

The electrochemical protective action of zinc dust consists in the cathodic protection of the iron by the baser zinc, which acts as sacrificial anode. A prerequisite for this, however, is metallic contact between the zinc dust particles and the surface of the iron and also between one another. For this reason the zinc dust content in primers of this type must be at least 94 to 96 %.

Zinc dust primers are widely used in the protection of steel structures as well as under water and in shipbuilding. In addition to organic binders, inorganic systems such as alkali silicates or alkyl silicates are also used as binders. However, primers of this type - in the same way as those containing water-thinnable organic binders - must be designed as multi-component systems.

Lead dust pigments are principally used in protection against aggressive chemical influences.

Other anti-corrosive pigments

Since the phosphate-containing anti-corrosive pigments cannot offer a protective action in the cathode zone, attempts are being made to remedy this disadvantage by the use of *molybdates*, *wolframates*, *zirconates*, and also *vanadates*, which are chemically related to the chromates. Their electrochemical protective action is heavily pH-dependent. However, none of these substances demonstrates the high oxidising capacity of the chromates, which is critical to their electrochemical action in the cathode zone. In terms of price too they do not represent a real alternative. They are also found in combination with phosphate-containing pigments.

As a non-toxic alternative to chromates, *ion-exchanger pigments* have also been developed. Their action consists in the ionic exchange of calcium ions, which are bonded to zeolites or amorphous siliceous carrier material, for hydrogen ions. In this way the pH in the coating film and on the surface of the metal is kept in the alkaline range.

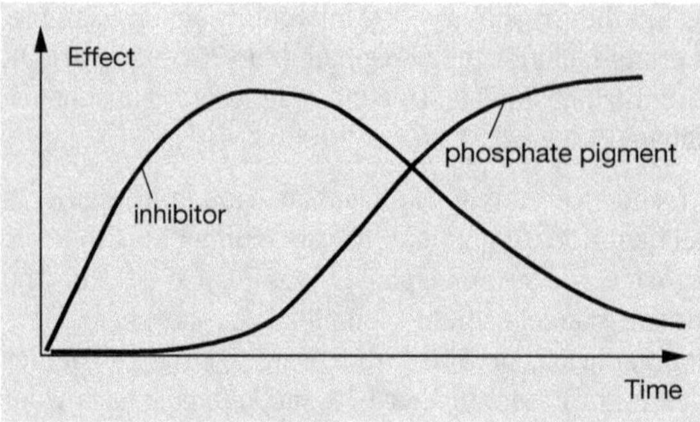

Figure 2.50: Synergistic effect arising from the combination of phosphate-containing anti-corrosive pigments with organic inhibitors

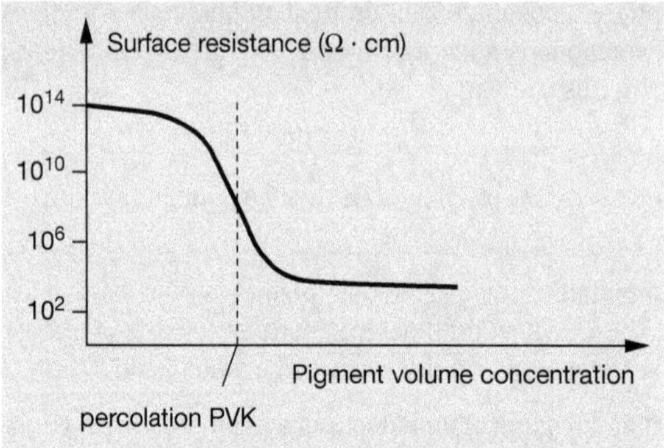

Figure 2.51: Dependence of the electrical conductivity of a coating on the PVC of a conductive pigment

Micaceous iron ore represents an exception to the other anti-corrosive pigments discussed here in that its action is purely physical.

According to DIN EN ISO 10 601, a **micaceous iron ore pigment** is understood to be a purified mineral (known as mirror haematite) or a synthetic product consisting predominantly of iron(III) oxide (Fe_2O_3). It is grey in colour, has a metallic gloss and is more or less lamellar in shape.

Because of its platelet structure micaceous iron ore provides coatings with a high resistance to weathering and atmospheric influences, even if the film formers used are not otherwise resistant. For the same reason, however, micaceous iron ore is also sensitive to shear. During grinding there is a risk that it will transform into conventional non-lamellar iron oxide red. Its main area of application is in heavy-duty corrosion protection.

Organic corrosion-inhibitors

In addition to the inorganic anti-corrosive pigments discussed so far, a series of organic compounds having an inhibiting action are also used in anti-corrosive coatings. They differ from the soluble inhibitors that are added to the aggressive medium in their poor solubility. Unlike the inorganic anti-corrosive pigments, they are often also known as inhibitors.

The most important member of this group is the zinc salt of 5-nitrophthalic acid. When added in small quantities (< 2 %) the inhibitors are supposed to eliminate the delay in protective action that occurs when phosphate-containing anti-corrosive pigments are used (synergistic effect), as shown in Figure 2.50.

2.3.8.2 Conductive pigments

In certain cases specific electrical properties are expected of surfaces. Antistatic surfaces do not attract dust from the air so strongly, for example, which is useful in clean rooms. In other cases, in the area of explosives or when working with microelectronic components, for example, electrostatic charging must be avoided, since this can lead to sparks on discharging. Organic antistatics can only ever provide a temporary antistatic effect, whereas coatings having an electrical derivation capability can provide a surface with the long-term conductivity (conductive primers) necessary for antistatic behaviour or charge derivation.

Coatings can be made electrically conductive using special conductive pigments, generally carbon blacks with a high structure, metal pigments or, if a light coating is required, using mica pigments coated with Sb-doped SnO_2, whose structure is similar to that of pearlescent pigments.

A coating pigmented in this way is only electrically conductive if the electrically conductive pigments form a kind of network of pigment particles in contact with one another throughout the entire coating. The minimum concentration of electrically conductive pigments needed to achieve this is called the *percolation PVC*. In the area of the percolation PVC the conductivity of a coating increases by many powers of 10 (see Figure 2.51). For carbon blacks, depending on the grade, it lies between 7.5 and 25 %; for the considerably more expensive mica pigments it is around 10 %.

2.3.9 Fillers

2.3.9.1 Definition and classification of fillers

The term *filler* was defined in section 2.3.1 according to DIN 55 943. The similarity to the definition of the term "pigment" (→ 2.3.1) is indicative. The notes to this standard expressly state that the designation of a substance as a "pigment" or a "filler" is dependent only on its application.

The fillers principally used in paints and coatings are predominantly light-coloured, mainly inorganic, powder-form substances that are inert in respect of the given binder and differ from white pigments primarily in their lower refractive index.

Fillers differ from pigments in their particle size too. Whilst white pigments are optimised in order to obtain the best possible scattering power, the objective with fillers is to ensure that they fill hollow spaces as effectively as possible, sometimes in a blend, in order to build a compact, stable skeleton structure with the pigments in the coating material. At mostly around 1 to 10 µm in coating materials, their average particle size is generally larger than that of pigments. As a result of these space-filling properties, expressed as PVC, fillers influence quite a number of the characteristic features of a coating (→ 3.2 and 3.3). They are used *inter alia* for

- reinforcing,
- improving flexural, adhesive and tensile strength,
- controlling the degree of gloss of coatings (flatting) and/or,
- since they are generally less expensive than pigments, reducing the price of a coating material.

The risk of cracking in the coating can be greatly reduced by the use of fibrous or platelet-shaped fillers. The shape and size of filler particles naturally also influence the rheological behaviour of a coating material. Like pigments, fillers with strongly non-isometric particles can also form skeleton structures by means of electrostatic interactions; these structures then have a thixotropic effect. Furthermore, the economic aspects of the use of fillers should not be ignored. Table 2.30 provides a summary of the most commonly used range of fillers.

Fillers can be categorised from a number of different standpoints. In terms of the possible applications of a filler, particle size and particle size distribution are a key factor. A distinction is made in this respect between

- coarse grinds with particle sizes above 250 µm,
- medium grinds with particle sizes between 50 and 250 µm,
- fine grinds with particle sizes between 10 and 50 µm and
- ultra-fine grinds with particle sizes below 10 µm.

At the same time, however, particle shape is very important for a number of the properties which are said to be improved by the use of fillers (see above). In addition to more or less isometric fillers such as chalk or barytes, there are also non-isotropic fillers, e.g. fibrous fillers such as mineral fibres, flakes, like mica, or those with more of a platelet shape, such as talc.

Fillers are generally now categorised from a chemical-mineralogical standpoint, according to which they are usually classed by anionic component.

In calcium carbonates, barium sulphates and silicon dioxide and silicas, a further distinction - between natural and synthetic products - must be made.

The principal areas of use for fillers in the coatings sector are in primers, stoppers, filling compounds, emulsion paints and synthetic resin bonded plasters. They have only a limited application in topcoat paints.

Table 2.30: Properties of the fillers most widely used in paints and coatings

Chemical class	**Filler**			**Properties**			
		synthetic	natural	density [g/cm³]	Mohs' hardness	pH-value [1)]	refractive index
carbonates	chalk		+			8 to 10	1.5 to 1.6
	calcite (ccn)		+	2.7	3	8 to 10	1.59
	calcite (ccp)	+		2.7	3	9 to 10	1.59
	dolomite		+	2.9		8 to 10.5	1.6
silicon dioxide	silicate flour cristobalite		+ +	2.2 to 2.7 2.35	6.5 6.5	7 8.5	 1.48
silicic acids	kieselgur		+	2.3 to 2.65	5 to 7	7 to 9	1.55
	diatomaceous earth		+	1.9 to 2.3	6	6.5 to 9.5	
	precipitated	+		1.9 to 2.1	6	3.5 to 8	
	pyrogenic	+		2.2	6	2.2	
silicates	talc		+	2.7 to 3.5	1	8.5 to 9.9	1.54 to 1.59
	kaolin, china clay		+	2.1 to 2.6	2.5	4.5 to 5.6	1.56
	mica, muscovite		+	~2.8	2.5	8.4	1.58 to 1.61
	precipitated aluminium silicate	+		2.1		10	1.46
sulphates	barytes		+	4.0 to 4.5	3 to 4	6 to 10	1.64
	blanc fixe	+		4.1 to 4.5	3 to 4	3.8 to 10	1.64

1) of a suspension

2.3.9.2 Manufacture of fillers

The overwhelming majority of fillers used in coating materials today are natural in origin. The following stages are involved in their preparation:

- mining
- sorting of the raw stones
- comminution in crushers
- wet or dry grinding
- sizing.

The products from one deposit usually have varying fineness of grain and also varying breadth of particle size distribution. The colour and purity of the powders obtained naturally depends on the deposit. The colour of talc, for example, can vary - depending on the deposit - from pure white through light grey, grey to greyish brown; even greenish brown types are found.

Table 2.31: Paint properties influenced by fillers

Paint property	Filler	Critical filler property
volume (build) (CPVC)	calcium carbonate dolomite (barium sulphate/lamellar fillers)	density, packing density (particle size distribution)
flow behaviour (thickening, intrinsic viscosity, thixotropy, antisettling effect)	pyrogenic silicas kaolin wollastonite fibre fillers	tendency to form framework structures, particle shape
Reinforcement (reduction in susceptibility to cracking	fibre fillers mica wollastonite	particle shape
corrosion protection barrier effect	mica micaceous iron oxide talcum	particle shape
neutralisation	calcium carbonate	pH
matting	all fillers, especially: kieselguhr precipitated silicas precipitated Al silicate calcined kaolin talcum	particles shape, particle size, framework structures
adhesive strength	talcum mica kaolin	particle shape, surface chemistry
surface hardness, abrasion resistance and non-slip characteristics	quartz flour cristobalite mica (surface hardness only)	hardness particle shape
sandability	talcum	hardness
abrasion resistance (of emulsion paints)	talcum mica	particle shape
pigment distribution ("spacing" for TiO_2)	kaolin talcum mica precipitated Al silicate barium sulphate precipitated calcium carbonate	particle shape, particle size
UV protection	micaceous iron oxide mica	UV absorption reflection
density	barium sulphate hollow micro spheres	density

The market for synthetic fillers lies in applications requiring particular purity (and hence lightness) or especially fine grades. They are generally obtained by means of precipitation reactions from solutions of appropriate raw materials (the exception being pyrogenic silicic acids).

Again, as with pigments, the dispersion and wettability of both natural and synthetic fillers can be improved by post-treatment.

2.3.9.3 Some commonly used fillers

Carbonates

In terms of volume, *calcium carbonates*, chemical formula $CaCO_3$, are the most important group of fillers used in coating materials. Calcium carbonate occurs naturally in the form of chalk, calcite and aragonite, both chalk and calcite finding use as fillers. Synthetic forms are obtained by precipitation from milk of lime with carbon dioxide as calcite or aragonite. Common to all calcium carbonates are their low hardness and poor acid resistance as well as their alkaline pH, which can be used for buffering in some applications, however. Calcium carbonates are non-toxic and very light (lightness L: ≥ 80 for chalk, ≥ 85 for natural calcite and ≥ 95 for synthetic calcite).

According to DIN EN ISO 3262-4,
chalk is a natural, calcium carbonate composed of weakly cemented sediments from the chalk group. Chalk is characterised by microcrystalline calcite crystals up to 1 μm in diameter and consists mostly of the shells and skeletons of small maritime organisms, e.g. foraminifera and coccoliths. Fragmented shell remnants are therefore a key feature of chalk.

The term chalk may not be used to described other forms of naturally occurring or precipitated calcium carbonate.

Depending on purity, the product is classed as chalk, calcium carbonate type KA (min. 96 % $CaCO_3$) or type B (min. 95 % $CaCO_3$).

The largely amorphous, lightly compacted and readily degradable sedimentary rocks used for chalk production can be both wet and dry ground. The especially fine "prepared chalk" is obtained by elutriation followed by slicing the sedimentary product into layers. Its loose structure means that chalk develops a pronounced absorbent effect which, like its acid sensitivity, makes it suitable for use only in interior paints.

According to DIN EN ISO 3262-5,
calcite (calcium carbonate type C) is a natural calcium carbonate derived from limestone or marble. The trigonal, rhombic crystals are generally larger than those of chalk.

The widespread use of calcite is explained by its good all-round properties, such as low binder requirement, good weather resistance and good compatibility with pigments. Its most important areas of application are in primers, anti-corrosive paints, emulsion paints, stoppers, silk-finish paints and fillers, for which calcite grades having average particles sizes of around 2 to 3 μm are predominantly used.

According to DIN EN ISO 3262-6,
precipitated calcium carbonate (calcium carbonicum praecipitatum, ccp) is a synthetic calcium carbonate consisting of trigonal crystals (corresponding to those of calcite) or rhombic bipyramidal crystals (corresponding to those of aragonite).

Since it is whiter and generally also more finely divided than the natural product, it is mainly used in applications requiring lightness and fine-particle character. It is used as a means of economising on white pigments and in very fine grades (approx. 0.06 μm) to prevent dripping.

According to DIN EN ISO 3262-7,
dolomites are naturally occurring calcium magnesium carbonates that contain between 1.18 and 1.23 parts by weight CaCO3 for every 1.0 part by weight $MgCO_3$.

Silicon dioxide and silicic acids

Silicon dioxide (silicic acids), chemical formula SiO_2, can be divided into natural crystalline silicon dioxide such as silica flour, natural amorphous silicon dioxide, such as kieselguhr (diatomaceous earth), and synthetic precipitated or pyrogenic silicon dioxide.

Under the terms of DIN EN ISO 3262-13,
natural quartz is a substance composed of the pulverised, low-temperature modification of quartz of theoretical density 2.65 g/cm^3.

Grades having a SiO_2 content of over 99 % are often used in coating compounds. Striking features include chemical inertia and great hardness. Silica flour is an inexpensive filler which is used to improve abrasiveness in stoppers, synthetic floor and wall coatings (plasters) and road-marking paints. Post-treated, silanised quartz types are more easily wetted by organic binders and because of their improved barrier effect they also give better results in anti-corrosive paints.

Cristobalite fillers are made from quartz by thermal modification. According to DIN EN ISO 3262-14, they must contain at least 60 % cristobalite and 98 % silica. Particles of cristobalite are rounder than those of quartz. Otherwise, they are comparable to the quartz fillers. However, their higher price restricts their use to applications requiring greater brightness, smaller yellow value and less abrasiveness.

Amorphous natural silicic acids such as *kieselguhr*, are generally used as a low-cost flatting agent in emulsion and road-marking paints, etc. They also improve the drying, grindability and interlayer adhesion of coating films.

Pyrogenic silicic acids are used in the coatings sector less as fillers than as thixotropic or antisettling agents. They consist of coagulated spherical silicon dioxide particles with diameters of between 5 and 50 nm. This extremely fine-particle character, together with their low refractive index, means that the flatting effect of pyrogenic silicic acids is much lower than that of the significantly coarser **precipitated silicic acids** used as flatting agents.

Silicates

The properties of silicates vary much more widely than those of the classes of fillers discussed so far. Naturally occurring silicates are characterised by a variety of different crystal lattices, from nesosilicates through inosilicates, cyclosilicates and phyllosilicates, to three-dimensional network structures. These crystal structures have a substantial influence on both the appearance and a number of chemical properties of such minerals. The three main silicate fillers used in the coatings sector, talc, kaolin and mica, are phyllosilicates characterised by their lamellar, platelet-shaped appearance. Bentonites are used as rheological additives and are covered in the appropriate section (→ 2.4.3).

Talc, idealised chemical formula $Mg_3[Si_4O_{10}(OH)_2]$, is a naturally occurring magnesium silicate hydrate. The Mg^{2+} ions are layered in every second interlayer between the silicate layers, thus forming a kind of sandwich structure. At the interlayers not filled with Mg^{2+} ions, these "sandwiches" are held together only by van der Waals forces, explaining the platelet-like structure and ready cleavability of talc. This lamellar structure, with its organophilic surface, is the reason for the relatively high oil absorption value of talc and enables the flow properties of a coating system to be selectively influenced. This structure, together with its chemical inertia, is the basis for the use of talc in anti-corrosive paints. Talc improves adhesion in primers through its OH groups. It is very soft, with a Mohs' hardness of 1, as a consequence of which it improves grindability in stoppers and fillers.

Kaolin is an aluminium silicate hydrate with the general formula $Al_2O_3 \cdot 2SiO_2 \cdot 2H_2O$. It is formed by the weathering of feldspat-rich rocks in the form of microcrystalline hexagonal platelets and is also known as china clay or ASP. The water of crystallisation is eliminated by calcination, through which somewhat whiter and harder grades are obtained. Kaolins in the finer particle size ranges are supplied as talc, they are characterised by good wettability but have a high binder requirement. Thanks to its acid and weather resistance, this filler is mainly used in coatings for extending white pigments. Other areas of use include emulsion paints, electrodeposition coatings, primers and fillers. Due to its slightly acid pH, the use of non-calcinated kaolin in anti-corrosive primers is limited.

Natural **mica** occurs principally in the form of muscovite, with the idealised composition $K_2O \cdot 2Al_2O_3 \cdot 6SiO_2 \cdot 2H_2O$. The brownish, pearlescent platelets are transparent. As a highly chemically and weather resistant filler, mica is used in coatings for which excellent chemical resistance, heat resistance, UV protection and electrical insulating properties are required. It is also used in primers, fillers, emulsion paints and anti-corrosion systems. The crack-bridging action of mica when added to coatings is well known. Unfavourable properties include its often very dark colour, relatively poor dispersibility and comparatively high price.

Synthetic aluminium silicate fillers, especially precipitated sodium aluminium silicate, have a whiteness of over 95 %, and very high brightness combined with a neutral yellow value. They are used in emulsion paints to improve dry hiding power and brightness, as well as to lower the titanium dioxide content

Sulphates

Of the large group of sulphates, only barium sulphate - chemical formula $BaSO_4$ - is used as a filler in the paints and coatings industry.

According to DIN EN ISO 3262-2,
barytes is naturally occurring barium sulphate, $BaSO_4$.

It is available in various particle sizes and lightnesses. Discoloration by traces of Fe, Cu or Mn compounds can be largely eliminated by means of chemical treatment (bleaching). Its chemical inertia, low oil absorption value and high density are characteristic of barytes. It is predominately used in knifing fillers.

As with calcium carbonates, the general trend towards purer and therefore lighter fillers with defined narrow particle size distribution is increasing the demand for precipitated barium sulphate, **blanc fixe**, which according to DIN EN ISO 3262-3 must be free from additives, e.g. from natural barytes. It is produced by precipitation of barium sulphate from a barium sulphide solution with Na_2SO_4 in the form of compact crystals with average particle diameters of between 0.5 and 4 µm. In combination with platelet-shaped fillers such as talc or kaolin, blanc fixe can be used to achieve an optimum packing density in coatings. This is the basis for its use in fillers, primers and stoppers. Advantage is taken of its high chemical resistance in industrial coatings. For applications requiring high gloss values, micronised grades of blanc fixe are also available with narrow particle size distributions in the range from 0.03 to 0.06 µm. These relatively expensive fillers are used *inter alia* in high-grade topcoat paints.

2.3.9.4 Nanoparticles

In addition to "traditional" fillers, it is becoming increasingly common to incorporate solid, inorganic nanoparticles to obtain special effects. However, there is no sharp transition between them and very finely divided variants of, e.g. blanc fixe, calcite or silica. Table 2.31 shows several examples and the effects which can be obtained with them.

The surfaces of nanoparticles are often treated or functionalised with organofunctional silanes to render the particles easier to incorporate into the polymer matrix. Uses of such nanoparticles in the coatings material sector are self-cleaning facade paints ("catalytic clean" effect), UV protection in clearcoats for wood substrates and non-scratch, automotive clearcoats.

Table 2.32: Commercial nanoparticles and their application

Nanoparticles	Characterisation	Effect	Typical application
TiO_2	rutile	UV absorption, hard	improved scratch resistance, anti-reflection coatings
TiO_2	anatase	photocatalysis, hard	self-cleaning surfaces, bactericidal treatment, non-fogging coatings
Al_2O_3	corundum	hard	improvement in scratch resistance
ITO (indium tin oxide)	blue	electrically conducting, UV and IR absorption	conductive and anti-static coatings, IR anti-reflection coatings
colloidal silver	round	bactericidal	bactericidal coatings

2.3.10 Dyes

Table 2.33: Comparison between application-oriented properties of pigments and fillers

Property	Dye	Pigment
processability	+	–
transparency (↔hiding power)	+ (–)	– (+)
sparkle	+	–
colouring power	+	–
light fastness	–	+
weather resistance	–	+
overcoatability	–	+

The distinction between dyes and pigments was discussed in section 2.3.1. Since dyes by definition are soluble in the application medium, their use does not involve any potentially complex dispersion processes. As dissolved substances they have no scattering power, which means that they cannot be used by themselves to manufacture coloured opaque coatings, but because of their molecular distribution in the application medium they generally have a stronger colouring power than pigments. At the same time, however, their main resistances are inferior to those of most pigments for this reason. Table 2.33 compares a number of application-oriented properties of pigments and dyes.

Classification

Natural dyes such as indigo or purple have been known for many centuries. In general, however, they cannot meet the quality criteria set for dyes used in modern coatings. The synthetic dyes used today are divided according to their chemical structure into

- metal complex dyes,
- anionic dyes and
- azo dyes.

Metal complex dyes consist of a metal atom, generally Cr, Cu, Co or Ni, which is coordinated with chelating ligands, mainly azo dyes, via O and N atoms. We differentiate between 1:1 and 1:2 complexes depending on the number of colour-imparting ligands. The free coordination sites in the central atom are occupied by water or hydroxide ions.

Metal complex dyes display relatively good resistances in comparison to other dyes. They are soluble in polar solvents such as alcohols, glycol ethers and esters, and ketones.

Anionic dyes are azo dyes containing one or more sulphonic acid groups. They used to be known as "acid dyes". Their resistances do not reach the level of those of the metal complex dyes. They are soluble in water and many blends of water with water-miscible organic solvents, e.g. glycol ethers or alcohols.

Azo dyes are of no real importance in the coatings sector. They display poor lightfastness and tend to sublimate at temperatures above 100 °C.

Solubility

After colouring power, *solubility* is the most important property of dyes. In the case of chemically pure substances we assume that under defined conditions they dissolve without residue to a specified concentration known as the saturation concentration. The dyes discussed here are industrial products, however. They often contain insoluble components which form a slight deposit even below the saturation concentration; this can lead to inaccurate results when assessing their solubility. The process for determining the solubility of a dye is described in DIN EN ISO 7579. A saturated solution of the dye is prepared and the solids content in g/l is determined by evaporation. Solubility data with no added substances indicate that for the corresponding weighed portion at least 90 % of the dye dissolves.

The solubility behaviour of dyes varies, as shown in Figure 2.52 (page 162).

- In *systems with a clear saturation limit*, the dissolved quantity of dye remains constant above the saturation limit (curve 1).

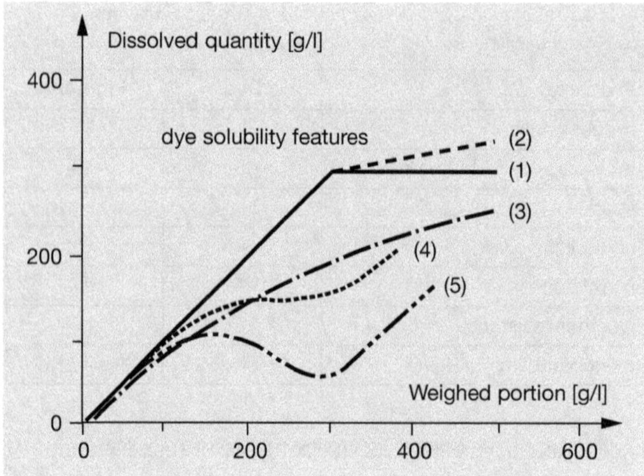

Figure 2.52: Schematic view of the solubility behaviour of dyes. (Symbols explained in text)

- In *systems with no clear saturation concentration*, there is a break in the solubility curve at the saturation concentration. Above this point only a part of the excess dye is dissolved (curve 2).
- *Systems with no saturation concentration* cannot be evaluated by the process described above. The quantity of dissolved dye generally increases as the weighed portion increases (curve 3).
- In *systems with a solubility plateau*, the dissolved quantity of dye appears to reach saturation concentration at a given weighed portion. As the weighed portion increases further, the quantity of dissolved dye remains constant over a certain range and then starts to rise again as the weighed portion continues to increase (curve 4).
- In *systems with extreme solubility values*, the solubility reaches a maximum at a certain concentration and then falls again (curve 5).

In the last three cases the solubility can only be described graphically.

Solubility data of this type always relate to individual solvents. They are rarely transferable to solvent blends such as are conventionally used in the coatings industry. Selecting the composition of suitable solvent blends or optimising them in respect of the solubility behaviour of the dye can be a very complex process.

Dyes are principally used in furniture mordants today, although with the growing interest in special effect coatings their use in other areas is also increasing. Attractive, highly pure and brilliant hues can be obtained when dyes are used in conjunction with special effect pigments. Unfortunately their generally poor water resistance prevents their wider application - in automotive production line coatings, for example.

2.4 Additives

2.4.1 Classification and definition

In addition to film formers, solvents and pigments and fillers, coating materials generally also contain one or more additives. DIN 55 945 A1 defines

an **additive** as a substance added to a coating material in small quantities in order to impart specific properties to the coating material itself or to the coating produced from it.
Note: the term "auxiliary" is also used with the same meaning.

These additives are generally subdivided into various groups, the most important of these being:

- Defoaming (and deaerating) agents
- Wetting and dispersing agents
- Surface-active additives
- Flatting agents
- Rheological additives

- Corrosion inhibitors
- Light stabilisers
- Driers (+ anti-skinning agents) and catalysts or accelerators
- Biocides.

Additives can be classified according to other systems too, particularly as some additives can influence several properties at the same time; for that reason the above classification is also sometimes rather arbitrary. Moreover, the boundary between additives and other coating components is not always very clear, since some film formers, pigments, solvents or even fillers are occasionally used like additives in terms both of function and of quantity added.

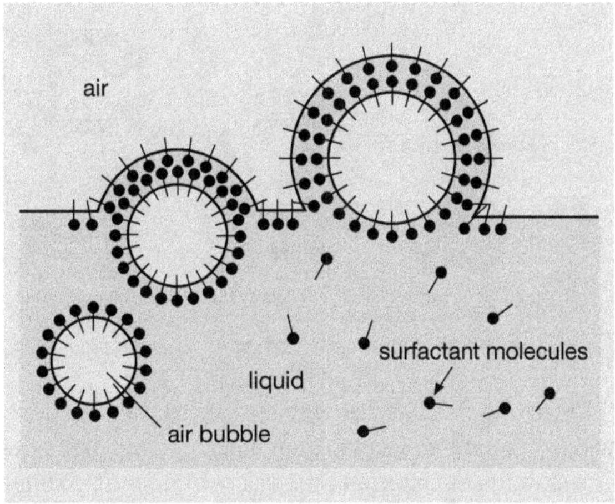

Figure 2.53: Rise and stabilisation of air bubbles in a surfactant-containing liquid, according to [40, Tego Chemie Service GmbH]

The discussion of wetting and dispersing agents is bound to overlap with the section on pigments and fillers, in the same way as there are bound to be overlaps between the coverage of driers and catalysts and the section on film formers.

Since the function of corrosion inhibitors is comparable to that of anti-corrosive pigments, they are covered with the latter in the section on special pigments.

2.4.2 Interface-active additives

2.4.2.1 Defoaming and deaerating agents

Defoaming agents

Very many coating materials contain interface-active substances that are used to obtain particular effects. Examples include emulsifiers used in aqueous coatings to emulsify water-insoluble film formers, or additives to improve the wetting of pigments and fillers or of the substrate. Such interface-active substances either act like or are surfactants and also reduce the interfacial tension at the phase boundaries between coating and air (surface tension). As a consequence of this, however, an undesirable side effect of these additives is that any air introduced into the coating material during manufacture, packing or application does not completely leave the coating again quickly enough and instead is stabilised in the form of foam. This not only causes optical defects in the coating surface but can also mean that containers are not properly filled at the packaging stage, and it can ultimately also diminish the protective function of the coating.

Foam formation

Foam (strictly speaking liquid foam) is the term given to the dispersion of a gas in a liquid medium. Characteristic of such foams is their extremely large - in comparison to other physical states - interface between the gaseous and liquid phases, which separates the gas bubbles from one another in the form of a lamella (Figure 2.53).

In the interests of energy conservation any liquid normally endeavours to keep its surface area as small as possible. An air bubble in a liquid thus assumes a spherical shape and rises to the top. At the surface of a pure liquid these air bubbles burst and the trapped air is released. For that reason pure liquids do not foam. In surfactant-containing liquids, surfactant molecules accumulate at the liquid/gas bubble phase interface and at the surface of the liquid, thus reducing the interfacial tension that normally prevents the formation of foam. Gas bubbles enclosed by surfactant molecules likewise rise to the top.

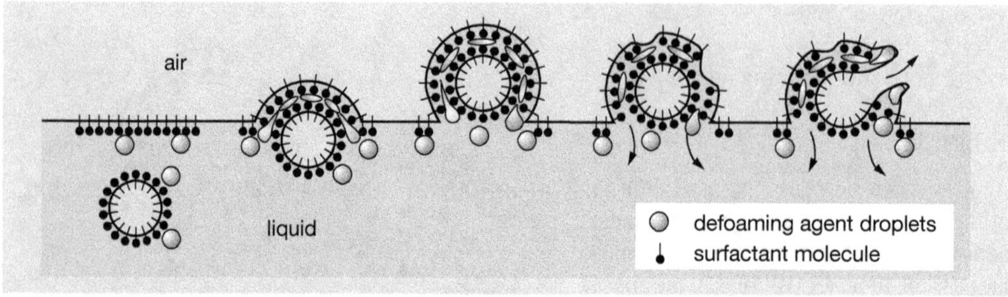

Figure 2.54: Model of the mode of action of defoaming agents, according to [40, Tego Chemie Service GmbH]

On reaching the surface (which is likewise covered with a layer of surfactant), a lamella stabilised with surfactant molecules then develops, which in aqueous systems has a thickness of several µm. The air bubbles collect at the surface and form a foam which is stabilised by various effects, such as a slow drain-off of liquid due to the narrowness of the lamella layer or electrostatic forces between ionic surfactant molecules.

Foam alters its structure over time. At first it still contains a relatively high proportion of liquid surrounding the spherical, scarcely deformed gas bubbles. Such foams are thus also known as "wet foams" or "spherical foams". The liquid drains off over time, causing the foam lamellae to become thinner, move closer together and gradually mutually deform to polyhedra. This foam is then known as "dry foam" or "polyhedral foam".

How defoamers work

For this, it must be virtually insoluble in the liquid to be defoamed. Total incompatibility, however, would lead to film defects, which are to be avoided. In order to be effective, a defoaming agent must be able to enter the foam lamellae. In the foam lamella, it has to spread out the layer of surfactant stabilising the lamellae, producing a film that is much less elastic than before, whereupon the foam collapses. At the same time, as a result of the new defoamer/liquid interface in the lamella, surfactant molecules that stabilise the lamella become bound and are no longer available to stabilise the lamella (see Figure 2.54).

The action of defoaming liquids can be further improved by the addition of ultra-finely divided solids particles that are incompatible with the liquid to be defoamed. The defoaming liquid then also serves as a transport medium, carrying the solids particles into the foam lamellae. These particles act as foreign bodies in the liquid, disrupting the cohesive forces in the lamella and at the same time absorbing surfactant molecules. As occurs with the defoamer liquid itself, the lamella becomes destabilised until finally it tears. It should be emphasised that no generally applicable model of the mode of action of the defoaming process has yet been derived, however.

In physico-chemical terms, the defoaming agent - as shown in Figure 2.55 - must demonstrate

- a positive penetration coefficient ($P > 0$) and also
- a positive spreading coefficient ($S > 0$)

in the medium to be defoamed.

The following relations apply here:

$$P = \sigma_L - \sigma_D + \gamma_{L/D}$$

$$S = \sigma_L - (\sigma_D + \gamma_{L/D})$$

where σ_L = surface tension of the liquid, σ_D = surface tension of the defoaming agent and $\gamma_{L/D}$ = interfacial tension between liquid and defoaming agent.

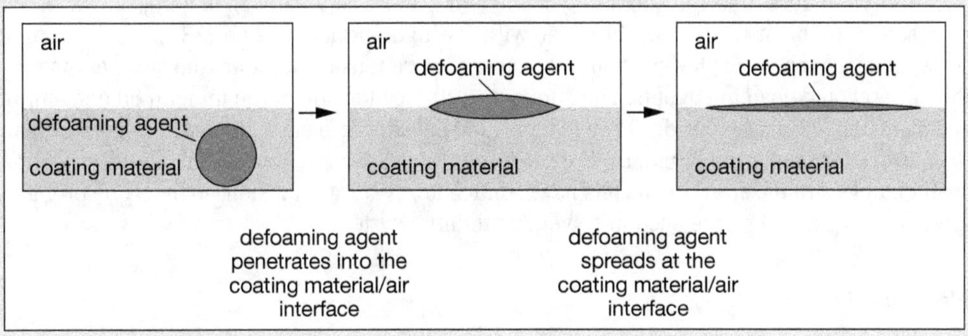

Figure 2.55: Basis for the definitions of penetration coefficient and spreading coefficient

A distinction is frequently made between defoaming agents and deaerating agents. A defoaming agent leads to the destruction of the macrofoam at the surface of a liquid. Deaerating agents on the other hand are supposed to remove the air that has been mixed into the coating film during application as quickly as possible. They are discussed in more detail below.

Types of defoaming agents

Aqueous coating systems

Two main product groups are used as defoaming agents in modern aqueous coating systems:

- Mineral oil defoaming agents
  These consist largely of mineral oil (carrier oil). Aromatic mineral oils were mainly used at one time, but for physiological and ecological reasons these are now increasingly being replaced by aliphatic mineral oils. A hydrophobic solid such as hydrophobised silicic acid or a polyurea compound together with small amounts of emulsifiers and other additives may also be added to the carrier oil if required.
- Silicone defoaming agents
  Such products are generally considerably more expensive than the mineral oil defoaming agents. They are usually dimethyl polysiloxanes, often modified with hydrophobic polyethers. As with the mineral oil defoaming agents, they may also contain hydrophobic solids in order to improve their efficiency. They are characterised by superior compatibility, as a result of which the susceptibility of the coating to cratering is also reduced, and what is more they cause no gloss reduction in high-gloss systems.

Solvent-containing coating systems

Mineral oil defoaming agents cannot be used in solvent-containing coating systems. Defoaming agents with an extremely low surface tension, now usually silicone-based products, have to be used here.

A number of different test procedures can be used to select an appropriate defoaming agent. Above all it must display a reasonable compatibility with the coating system. Defoaming agents that are too compatible lose their efficiency, whilst those that are too incompatible lead to cratering. Suitable examples are, firstly,

- polysiloxanes with a viscosity of between 5,000 and 50,000 mPa·s and, secondly,
- specially modified siloxanes.

A pre-selection can be made by incorporating the defoaming agent into the pure binder solution and observing how quickly the resulting bubbles burst or how quickly the volume of foam formed reduces.

According to Figure 2.55 the spreading coefficient S must be greater than zero; this does not represent a problem in the case of water with its high surface tension (73 mN/m). Organic solvents, by contrast, display substantially lower surface tensions that are too close to those of the mineral oil defoaming agents. The sum of the surface tension of the mineral oil defoaming agent and the interfacial tension between liquid and defoaming agent is then usually greater than the surface tension of the liquid, i.e. the solvent in this case, which would make the spreading coefficient less than zero. Silicone oils have a much lower surface tension, so the same problem rarely arises when they are used in solvent-containing systems.

Selection criteria

A finished coating contains numerous components that may themselves influence foam forming or foam stability. For that reason it is vital to test the defoaming agents identified by the pre-selection process in the coating system itself. In highly filled systems the most practical method may be to determine the foam content of a sample by measuring its density. To this end air must be incorporated into the coating by reproducible means, causing foam to form. This can be done by shaking it in a separating funnel, stirring in air in a high-speed mixer or rolling it up with an open-cell foam roller, for example. The reduction in foam volume is then observed and any defects caused by the foam in the applied coating film examined. As we have mentioned, however, since defoaming agents must be incompatible to some degree with the coating system, a side effect of their use is the risk of

- gloss reduction
- haze in a clear coating
- a tendency towards cratering and
- an influence on interlayer adhesion.

Since many defoaming agents lose their efficiency after being stored for some time, this must be checked again where necessary following extended storage periods.

Deaerating agents

Whereas *macrofoam* occurs principally in aqueous coating systems, *microfoam* can occur in any coating system. This consists of small spherical bubbles surrounded by a relatively large quantity of liquid. In the finished coating microfoam can lead to the introduction of air and the formation of *pinholes*. These are the channels through which small air bubbles rise to the surface, which can no longer be closed because the coating has become too viscous. Both air bubbles and pinholes impair the function of a coating, since the coating film is very thin at these defect points and the pinhole may extend as far as the substrate, allowing the penetration of corrosive substances, for example. This is particularly dangerous if the pinholes are too small to be seen by the naked eye, as a consequence of which no remedial action can be taken.

Pore formation

A number of possibilities for the introduction of air into a coating system were described in the section on macrofoam. In microfoam, in addition to these, small bubbles of gas can also evolve during chemical reactions in which volatile products are formed. Such reactions include the reaction of isocyanates with atmospheric moisture, forming CO_2, and also the elimination of blocking reagents in corresponding coating systems.

In pure binder solutions, small gas bubbles of this type - which only have a limited buoyancy, of course - are prevented from being released only by the viscosity of the coating system, which continues to rise during curing. This would be disadvantageous during evaporation of a freshly applied clear coating, for example. In formulated coating systems with their large number of components, some of which are also interfacially active, the gas bubbles are often stabilised as microfoam. The speed of cure of the coating is then usually greater than the speed at which the bubbles rise in the coating film, and entire air bubbles - or the channels through which they rose, in the form of pinholes - remain.

To prevent the formation of pinholes, the rising velocity of the gas bubbles has to be increased. From Stokes' law we can deduce that the rising velocity v of gas bubbles is proportional to the square of the bubble radius r.

$v \sim r^2/\eta$

where η is the dynamic viscosity.

For that reason, substances that contribute to or accelerate the coalescence of individual bubbles to a larger bubble act as deaerating agents. This process is achieved using non-polar substances that are just soluble in the coating system and thus accumulate at the gas/coating interface. The gas bubbles are then enveloped by a layer of non-polar deaerating agent. This layer displaces the surfactant stabilising the microfoam from the surface of the air bubbles, which in itself accelerates the microfoam's rise. Moreover, if two gas bubbles approach one another their layers of deaerating agent can readily fuse since they display only slight interaction with the binder or binder solution. This results in coalescence, and the air bubble that is forming rises more quickly to the surface.

Deaerating agents are thus designed primarily to accelerate the rise of gas bubbles to the surface of the coating. Some of the requirements of deaerating agents are similar to those of defoaming agents. Once again, no one substance is suitable for all coating systems.

The following substances are currently used as deaerating agents:

- special organic polymers, such as polyethers or polyacrylates,
- dimethyl polysiloxanes,
- special organically modified polysiloxanes and
- fluorosilicones.

Selection criteria

The procedure for selecting a suitable deaerating agent for a particular coating material can be extremely complex. Since the differences in density between samples with and without microfoam are very slight, density measurements cannot normally be used. A visual examination must therefore be carried out on the coating film itself - using coatings applied to transparent films, for example.

The samples are examined under transmitted light for bubbles and pinholes and under incident light for any craters or orange peel structure. The finest microfoam cannot be detected by eye, however. In such cases it is recommended that the coating material to be examined be applied in a realistic coating thickness to an irradiated iron sheet. Once fully cured, the coated side is exposed to a copper sulphate solution for 24 hours. After rinsing with water the presence of fine pores is indicated by red dots; this is elementary copper, formed from copper ions reduced by iron (see also → pore testing).

In addition to the efficiency of the deaerating agent, its compatibility with the coating material must of course also be tested. As with defoaming agents, incompatibility can cause surface defects. Sometimes two deaerating agents have to be combined in order to achieve the optimum efficiency and compatibility. If the gas bubbles then rise to the surface but do not burst, it may be helpful to combine the deaerating agent with a silicic acid-containing product, which then acts in the same way as in defoaming agents.

2.4.2.2 Surface-active additives

The term "surface-active additives" is applied to additives that are used to eliminate any surface defects that may form in the coating. Such defects include:

- poor flow
- inadequate substrate wetting
- cratering due to fall of spray mist or dust
- sensitivity to draughts
- Bénard's cells
- ghosting.

These defects are caused by differing surface tensions within the coating film, which in the liquid coating can generate flows from areas of low surface tension to those of higher surface tension.

Formation of Bénard's cells

An important phenomenon in this context is the occurrence of *Bénard's cells* (Figure 2.56). As the solvents evaporate, relatively solvent-rich (i.e. low-viscosity) and lower density coating flows up from the lower layers of the coating to the surface, where it spreads out and the solvents evaporate. This means that the density of the uppermost layer of coating is greater than that of the lower areas and so it sinks back down again. These flows create eddies which may entrain pigments of different densities to a varying degree. Since the viscosity of the coating increases as the solvents evaporate, these flows come to a stop as the coating dries out. Ideally a honeycomb structure of more or less regular hexagons, known as Bénard's cells, is formed on horizontal surfaces. On vertical surfaces these cells flow into stripes (silking).

The occurrence of Bénard's cells is particularly disruptive in mixed-pigment coatings. The varying density of the pigments leads to different particle mobilities. This means that the pigments are not all entrained to the same extent by the flows. The lighter pigments accumulate at the edges of the eddy, the heavier ones in the centre, where the rate of flow is lower, as a result of which the pigments in the coating become separated. In flatted clear coatings the flatting agent can play the part of the less mobile pigment, once again leading to honeycomb structures or to silking on vertical surfaces. In unpig-mented systems an irregular surface will be formed.

Surface-active substances

In addition to these differences in surface tension occurring in the coating itself, defects can also be caused by differences in surface tension between substrate and coating and between coating and grinding dust or coating spray mist. All of these phenomena can be counteracted with surface-active agents, which keep the surface tension at a constantly low level throughout the entire film-forming and curing process. Surface-active substances are used to this end. Substances that are classed as *surface-active* can reduce the surface tension of a liquid by accumulating at its surface.

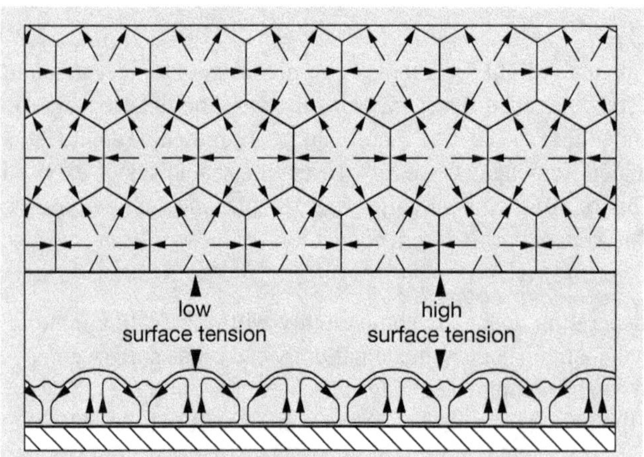

Figure 2.56: Formation of Bénard's cells, according to [40, Tego Chemie Service GmbH]

In *aqueous systems* with their relatively high surface tensions (σ_{H2O} = 73 mN/m), countless substances are surface-active. The most well-known of these are the surfactants.

The situation is somewhat different in *solvent-containing systems*, which already demonstrate a relatively low surface tension. Fluorinated compounds, silicone oils and modified polysiloxanes in particular are surface-active in these cases. Modification of polysiloxanes (with polyether or polyester chains and aromatic groups) makes them more compatible with the coating system. Alkyl groups have a very strong influence on surface tension, with methyl groups having a more pronounced surface tension-reducing effect than longer alkyl chains.

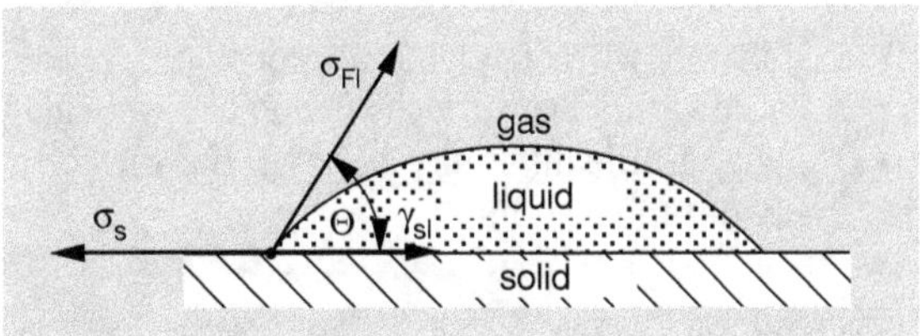

Figure 2.57: Distribution of forces at the solid/liquid phase boundary, according to [10]

Flow

The formation of a uniform, smooth surface, from which all application-related irregularities (brush strokes, etc.) have been evened out, is called *flow*. The prerequisites for good flow are perfect wetting of the substrate, good spreading of the liquid coating and uniform evaporation of the solvents across the entire surface of the coating. A *uniform level of surface tension* is more important for flow than a high surface tension, which would only be effective on a smooth coating surface with the smallest possible surface area. This makes the formation of Bénard's cells as described above more difficult. Flow is also extremely dependent on the flow behaviour of the coating, which can be controlled by suitable blends of solvents, for example. Conversely, however, the action of a flow-control agent can also influence the the flow behaviour of the coating.

To summarise, therefore, we can say that the addition of surface-active substances sets a uniformly low surface tension across the entire surface of the coating, which changes little during evaporation. This prevents the development of surface tension gradients at the surface of the coating, and solvents can evaporate uniformly over the entire surface of the coating, producing the desired even surface.

Wetting and flow-control agents

In addition to the classes of substances mentioned above, acrylate copolymers are also used as flow-control agents. They are only slightly soluble in the coating system and accumulate as expected at the surface, although their surface activity is generally not as great as that of the silicones. Moreover, not only the surface tension effects need to be taken into account when evaluating flow, since different rates of evaporation can also occur due to draughts (→ sensitivity to draughts).

The low surface tension of a coating to which a flow-control agent has been added generally also leads to improved *wetting* of the substrate, another precondition for good flow. Complete wetting is understood to be the spreading of a drop of liquid on the surface of a solid. Good or bad wetting manifests itself here in the size of the contact angle Θ formed by the drop on the substrate (Figure 2.57). A closed coating film is formed if Θ is not too great (<< 90°). The following relation is valid for the angle Θ:

$$\cos \Theta = \frac{(\sigma_S - \gamma_{S/L})}{\sigma_L}$$

where σ_S is the surface tension of the substrate, σ_L is the surface tension of the liquid and $\gamma_{S/L}$ is the interfacial tension between liquid and solid.

A liquid thus spreads on a substrate if the spreading coefficient

$$S = \sigma_S - (\sigma_L + \gamma_{S/L}) = (\sigma_S - \gamma_{S/L}) - \sigma_L$$

is positive. Then $(\sigma_S - \gamma_{S/L}) > \sigma_L$ and hence cos Θ < 90°. Since the interfacial tension is usually low in comparison to the surface tensions, it can often be ignored. The following empirical rules are then obtained from the above equations:

- A substrate with a high surface tension σ_S is easily wetted.
- A liquid with a low surface tension can wet easily.
- The greater the difference between the surface tensions of the substrate and the liquid, the better the wetting.

Table 2.34 sets out the surface tensions of a number of solvents, substrates and surfactant solutions in water (0.1 %, 25 °C).

If we assume that the substrate has received the ideal pretreatment (cleaned, degreased, phosphated if applicable, corona-treated, etc.,), wetting can be improved by reducing the surface tension of the coating. This can be done by means of *substrate wetting additives*, whose efficiency is expressed by their capacity to reduce surface tension in even the smallest possible concentrations. It must be ensured, however, that this does not give rise to any undesirable side effects, such as uncontrollable foaming, reduced water resistance or poor interlayer adhesion (overcoatability).

This last effect can be explained as follows. As surface-active substances, wetting agents accumulate at the surface of a coating, causing the tension at the surface to be lower than that inside the coating. If the surface-active substances are then permanently bonded there during stoving, e.g. through incorporation into the binder network via reactive groups, they can no longer migrate into the next layer during overcoating. This subsequent coating layer thus forms a substrate which, given its low surface tension, is only sparingly wettable, leading to poor interlayer adhesion.

Table 2.34: Surface tension σ of various solvents, film formers and substrates

Liquid	**σ (mN/m)**	**Surfactant solution in H_2O (0,1 %ig, 25 °C)**	**σ (mN/m)**
water	73	nonyl phenol ethoxylate	35
butyl glycol	30	silicone surfactant	31
toluene	29	polyether siloxane	21
2-propanol	22	non-ionic fluoro surfactant	17
n-octane	21		
hexamethyl disiloxane	16		
isopentane	14		
Coating resin	**σ_1 (mN/m)**	**Substrate**	**σ_s (mN/m)**
epoxy resins	45 to 60	glass	73*)
melamine resins	42 to 58	untreated steel	29*)
chlorinated rubber	57	phosphatised steel	43 to 46*)
alkyd resins	33 to 60	untreated aluminium	33 to 35*)
poly(meth)acrylates	32 to 41	polyester	43
polyvinyl alcohol	37	polyethylene	36
polyvinyl acetate	36	polypropylene	30
		paraffin wax	26
		PTFE	20

*) critical surface tension σ_{cr}:
If we measure cos Θ for an homologous series of liquids i on the same surface and plot cos Θ against σ_i, we obtain a straight line which by extrapolation to cos Θ = 1 (Θ = 0°) gives the value for σ_{cr}.

The suitability of a wetting agent cannot be determined directly from the surface tension values, but first indications can frequently be obtained by means of a simple spreading test. A drop of the test liquid is placed on the substrate, once with the additive and once without, and the surfaces are compared.

Orange peel effect

A flow problem that is specific to application by spraying and also occurs with powder coatings is the orange peel effect. This occurs, *inter alia*, in the presence of too highly volatile solvents, which evaporate from the coating droplets on the way from the spray gun to the substrate. As a consequence, the arriving droplet of coating no longer has a sufficiently low viscosity to be able to achieve optimum flow. Differing electrostatic charges during transfer from the gun to the object can also cause this effect. Surface-active additives such as silicones can often resolve this problem, although they must be added with caution to avoid the risk of cratering.

Cratering

The term crater is applied to small, trough-shaped depressions in coating films that are usually attributable to defective wetting properties in the liquid coating. There may be a number of reasons for this. If the substrate to be coated has been subject to localised contamination by substances having a lower surface tension, e.g. fats or oils on metals, residues of lubricants or mould release agents on plastics, or dust, the applied coating wets the substrate less well at these points or may even retract. If these impurities are soluble in the coating, as are, for example, many fats or oils in organic solvents, there are two possibilities. If the dissolved impurities are able to disperse reasonably evenly through the liquid coating, the substrate is wetted as expected and no surface defect remains. If, on the other hand, the coating exhibits a lower surface tension at the contaminated point, this generates flows in the coating away from the contaminated point into the areas of higher surface tension, as a result of which the layer of coating is thinner at the contaminated point and a hole (crater) may even form in the coating film.

Craters may also develop if the applied coating is contaminated with air-borne particles of lower surface tension. These may be particles of dust or spray mist, either from the coating itself or from a foreign substance. The liquid coating is then unable to wet these particles completely and reduces the surface area of contact by forming a crater.

Another cause of cratering can be the presence in the coating of scarcely wettable, insoluble particles, e.g. gel particles or additives such as defoaming agents added in the incorrect quantity.

In such cases too, surface-active additives bring the surface tension of the coating into line with that of air-borne particles or of particles present at the substrate, thus preventing cratering.

Another form of cratering is known as *popping*. This occurs in stoving coatings, for example, in the presence of highly volatile solvents, which swiftly evaporate from the upper layers of the coating film during drying, causing the viscosity at the surface to increase rapidly and the surface to close quickly. The evaporating solvents in the lower layers then form bubbles, which burst and cannot be closed by the already highly viscous coating at the surface. This results in pinholes or craters in the dry film. This problem can normally be remedied by the addition of a highly solvent high boiler or of a suitable blend of high boilers.

Ghosting

The causes of the phenomenon known as *ghosting* are similar to those of cratering. Ghosting refers to the appearance of grinding marks or smears in the subsequent coating during refinishing work. Grinding creates areas of differing surface tension (ground – unground) which can then generate flows in different directions in the applied coating. The result is the formation of clearly visible small swellings or bars.

The counter-measures are again the same as those for similar phenomena: surface-active additives lead to uniform wetting of the entire substrate surface, thus preventing cratering or ghosting.

Slip additives

In addition to remedying the surface defects described above, surface-active substances can also be used for the selective modification of other properties of the applied coating or of the surface of the cured coating system.

The sliding resistance of a coating surface is largely dependent on the interactions between surface and sliding body. Substances characterised by slight interactions between one another and with other substances are thus particularly suitable for use as additives to improve the surface smoothness of a coating film (*lubricants or slip additives*). Apart from fluorinated compounds, the main substances used are silicones and modified siloxanes with a relatively high polydimethyl siloxane content. These accumulate at the surface of the coating during drying and form a lubricating film.

In solvent-containing systems the formation of this lubricating film is based on the flows arising during evaporation of the solvents and occurs relatively quickly. In solvent-free systems the slip additives slowly become insoluble in the system - with the progressive curing of the binder - and as a result are pushed to the surface. These systems do usually require somewhat larger amounts of additive than the solvent-containing systems. In aqueous systems the efficiency of slip additives is extremely dependent on the compatibility of the additive with the binder. Their efficiency increases as the cure progresses. Emulsifiers can trap the slip additives in the coating, however, thus reducing their efficiency.

In addition to surface smoothness, slip additives improve

- the blocking resistance of a coating and
- generally also the scratch resistance of a coating thanks to its enhanced surface smoothness.

Scratching is a form of linear damage to the surface caused by sharp objects that either irreversibly deform the surface or actually perforate it. Slip additives form what is effectively a lubricating film, making it more difficult for the object to perforate the surface. To do so, however, the additive film must remain closed despite the pressure from the sharp object; it must therefore demonstrate high load-bearing capacity, which can only be achieved by intensive interactions amongst the additive molecules and between the additive molecules and the binder.

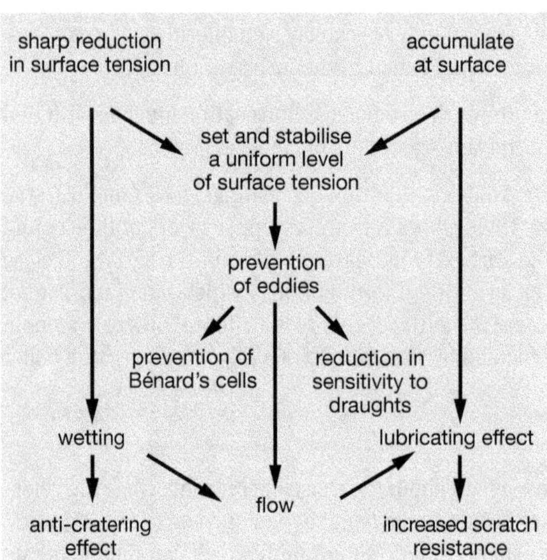

Figure 2.58: Overview of the made of action of surface-active additives, according to [40, Tego Chemie Service GmbH]

Many surface-active additives reduce the dust dry time of an applied coating film, since they accumulate at the liquid/air interface and form a closed film there, reducing the tackiness of the coating. Figure 2.58 provides an overview of the mode of action of surface-active additives in coatings and their influence on various coating properties. In order to be able to work effectively, these additives must be homogeneously incorporated into the coating. To this end it is often helpful to pre-dissolve the various additives in the given solvent; at the same time this ensures an accurate metering of the often small quantities.

2.4.3 *Rheological additives*

2.4.3.1 *General introduction*

The rheological properties of a coating material are of prime importance for optimum performance during production, transport and application. However, the requirements of each process stage are often contradictory. During dispersion, for example, an appropriate level of viscosity is needed for transferring the highest-possible shear forces onto the pigment agglomerates, yet the mill base must remain workable. A high viscosity is usually preferred for storage and transport as it prevents settling of pigments and fillers. The viscosity needed for application depends on how the coating is applied, but it normally has to be low. After application at the desired thickness, the coating should level out well, without sagging.

The shear rates which arise vary considerably. Dripping, flow and sedimentation all occur in the low shear rate range, between around 0.001 and 0.1 s^{-1} (see Figure 2.59). By contrast, the coating is manufactured and generally also applied at a very much higher shear rate of around 10^4 s^{-1}.

Influence of shear rate

At *high shear rates* the flow characteristics of a coating material are principally influenced by the binders, solvents, pigments and fillers. In this shear rate range, for example, the viscosity at a given concentration rises in the same way with the increasing molecular weight of the film former as with rising PVC. The influence of the solvent is less clear, however. *Inter alia*, the inherent viscosity of the solvent (blend) and its capacity to form hydrogen bridge bonds play an important role here. Accordingly, the viscosity of a coating material is principally adjusted for the high shear rate range by providing a suitable balance between its components, e.g. by diluting with solvent at a given composition of the cured film.

At *low shear rates* the viscosity is naturally also influenced by these three main coating components - very strongly in respect of the film former, for example, by thixotropic alkyd resins. In this range, however, any necessary adjustment is made using special rheological additives (thixotropic agents, thickeners) , which only change the flow behaviour in certain shear-rate ranges (e.g. high-shear and low-shear thickeners).

Rheological additives are roughly classified as thickeners and thixotropes in line with their primary effect. Often, however, additives that raise the viscosity of a coating material more or less uniformly across the entire shear rate range are termed thickeners too, whereas those which influence the shape and position of the flow curve only are termed rheological additives. Thixotropic agents, by contrast, are those rheological additives which primarily influence the temporal change in viscosity under altered shear conditions. It is not always possible, moreover, to unequivocally classify the various additives under one of these headings and sometimes other

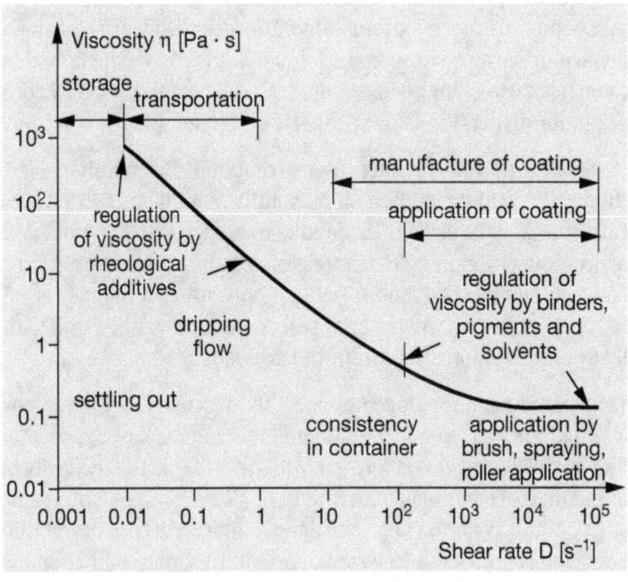

Figure 2.59: Overview of the shear rates occurring during manufacture, transportation, storage and application of coatings

Table 2.35: Influence of modified cellulose on the properties of emulsion paints

Property	Methyl cellulose	Methylhydroxy-propyl cellulose	Methylhydroxy-ethyl cellulose	Hydroxyethyl cellulose
thickening effect	⇐	decreases		⇐
spraying tendency	⇒	increases		⇒
water retention	⇒	increases		⇒
pigment flocculation	⇐	decreases		⇐
foam formation	⇐	decreases		⇐
wash resistance	⇐	decreases		⇐
susceptibility to microbial attack	⇒	increases		⇒

criteria are used for this. Additives that are used to specifically adjust pseudoplasticity can be found in all these groups, for example.

2.4.3.2 Thickeners

Thickeners are designed primarily to prevent the settling out of pigments and fillers in coating materials during storage and transportation. Their most important area of application is in emulsion paints. Cellulose derivatives (e.g. methyl cellulose, hydroxyalkyl cellulose) as well as derivatives of heteropolysaccharides (e.g. xanthane), polyacrylates, polyether polyols and polyurethane derivatives are all used here. Their efficiency can be controlled within certain limits by their chemical structure. In line with their mode of action, they are variously classified as hydrocolloids, water-soluble polymers and associative or system thickeners.

Types of thickeners

Most thickeners in the first group, e.g. the cellulose derivatives and certain poly(meth)acrylic acid/poly(meth)acrylate copolymers, work by increasing the hydrodynamic volume (see section 2.2.3.5) via the aggregation of water molecules, the formation of hydrogen bonds along the polymer chain (swelling) and the association of thickener molecules (molecular enlargement). Since they thicken the aqueous phase, they are often called phase thickeners. They have the advantage of being system-independent and thus predictable. However, they often make a high elastic contribution to the flow behaviour of the coating material which they have thickened (see section 8.1.4), and this precludes them from use in high-gloss paints. Furthermore, they may contribute to flocculation if the pigments or binder dispersion is not properly stabilised (see Table 2.36).

Associative thickeners, such as hydrophobically modified PU thickeners or products of the HASE (**h**ydrophobically modified **a**lkali **s**oluble **e**mulsions (acrylate copolymers)) or HMHEC (**h**ydrophobically **m**odified **h**ydroxy**e**thyl**c**ellulose) type, consist of molecules with water-soluble hydrophilic and hydrophobic regions. Their mode of action can be expressed in simplified terms as an interaction between their hydrophobic regions and other components of the coating materials, such as film formers, pigments, surfactants and coalescing agents, due to the formation of complexes, whilst the hydrophilic regions remain in the aqueous phase.

The water-soluble region increases the viscosity of the aqueous phase, and micelle formation and aggregation lead to a three-dimensional network of aggregates, which can be a three-dimensional network of aggregates is formed, which can be relatively easily broken down again, however. Similarly to the increase in molecular weight of a dissolved polymer, a network of this type increases the viscosity of the system. The precise molecular structure of the thickeners determines whether the flow behaviour tends to be Newtonian or whether they will be more likely to act in the lower to medium or the upper shear-rate range. As a result, the shape of the flow curves changes. Often, the flow behaviour is optimised by combining associative thickeners with each other, with phase thickeners or with thixotropic agents. Associative thickeners generally also have a thixotropic action. Since the strength of the interaction during aggregation depends on the interfacial tension between the aqueous phase and the hydrophobic thickener regions or other hydrophobic constituents of the coating material, the thickening action is influenced by any materials that alter this interfacial tension. That is why paints which have been adjusted with associative thickeners often do not respond predictably to recipe changes.

Associative thickeners generally also have a thixotropic action. Table 2.34 provides a summary of the action of the various types of thickeners.

Table 2.36: Comparison of the efficiency of various thickeners (by basic grade) [40]

Property or effect	Thickener	
	non-associative	associative
low-shear thickening	+	+/o/–
high-shear thickening	–	+/o
resistance to flocculation	–	+/o
sensitivity to co-solvents	+/o	–
sensitivity to surfactants	+/o	–/o
reduction in spraying tendency	–	+
maturing period	+	o
dripping	+	o/–
flow	–	+/o
gloss	–	+
blocking resistance	+	o/+
price	+/o	–
+ = favourable, positive influence o = scorcely any influence, neutral – = unfavourable, negative influence		

2.4.3.3 Thixotropic agents

Mode of action

Good flow requires the viscosity to be kept sufficiently low for long enough to allow the surface of the coating to form an even, smooth surface under the influence of surface tension. When the coating remains "open" in this way, however, dripping may occur on vertical surfaces due to the influence of shear force. To prevent dripping, the viscosity must be as high as possible to stop the thin coating from flowing. This means that the viscosity of the coating must be relatively low during and shortly after application - to ensure good flow - and relatively high thereafter - to prevent dripping.

In order to obtain good flow without dripping, a thixotropic agent is added to the coating material. Because of the high shear rates during application, a material that has been thixotropically adjusted displays the low viscosity immediately after application that is required for good flow. Since the shear forces in the coating film after application are only low, the viscosity rises again, preventing dripping.

The speed of this rise in viscosity is critical for the dripping and flow behaviour of the coating material. If the increase in viscosity occurs relatively slowly, the coating film displays good flow but may also have a strong tendency to drip. Too rapid an increase in viscosity, however, whilst preventing dripping, would only result in poor flow. The optimum viscosity behaviour must be found by adding the appropriate quantity of rheological additive. The time-related progression of the processes associated with thixotropy is shown schematically in Figure 2.60.

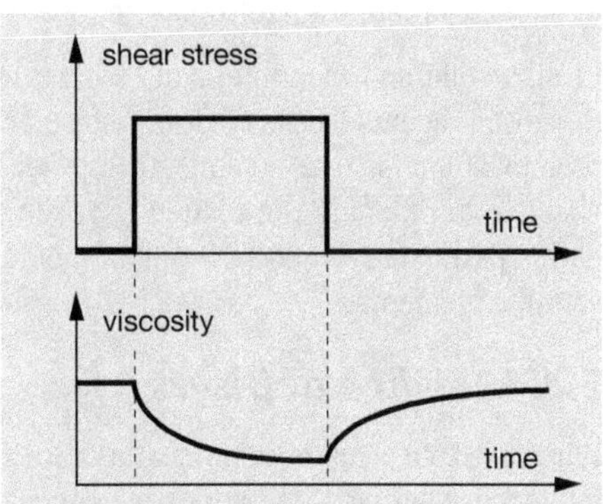

Figure 2.60: Progression over time of shear stress and viscosity in a thixotropic system, according to [40]

Types of thixotropic agents

The thixotropic agents employed are substances which build up framework structures in the coating material which are gradually destroyed again by the action of

Table 2.37: Comparison between various types of thixopropic agents

Thixotropic agent	Rise in viscosity	Drip control	Antisettling effect	Difficulty of incorporation into	
				solvent-containing systems	aqueous systems
phyllosilicates	+	+	+	moderate	somewhat difficult
pyrogenic silicic acid	+	+	o	moderate	moderate
hydrogenated castor oil	+	+	+	fairly great	–
high-molecular polyolefins	+	+	o	somewhat difficult	–
polyamides	+	+	–	easy	–
cellulose derivatives	+	+	+	–	easy
polyacrylates	+	+	+	–	easy
+ strong o moderate - low					

shear forces, e.g. modified phyllosilicates (e.g. hectorite, bentonite), pyrogenic silica, derivatives of polyurea and of castor oil. The attainable effects are presented in Table 2.37.

- *Phyllosilicates* form a three-dimensional network - similar to a house of cards - by means of hydrogen bridge bonds at the edges of the phyllosilicate platelets; this builds up a high viscosity. The network is destroyed by shear and the viscosity falls. When the shear is removed, the house of cards rebuilds itself. In aqueous systems a similar network is held together by forces of attraction between the differently charged surfaces and edges of the silicate platelets.
- *Pyrogenic silicic acids* consist of spherical SiO_2 particles, at the surface of which are located silanol groups (-SiOH). These congregate by means of hydrogen bridge bonds to form a three-dimensional framework of silicic acid particles.
- Castor oil derivatives contain *inter alia* 12-hydroxystearic acid by way of fatty acid and use its hydroxyl groups to form hydrogen bonds to adjacent, similar molecules or to suitable binder molecules, a fact which also leads to the formation of a framework structure.

Since thixotropic agents can set a high viscosity in a resting coating material, they act as antisettling agents at the same time. Furthermore, in some of the thickeners already discussed, the thickening effect diminishes under shear stress, enabling them to be used as thixotropic agents too.

Thixotropic agents are naturally just as important in water-thinnable coating materials as they are in solvent-containing types. Because of the polarity of water and its capacity to build hydrogen bridge bonds, however, the same substances cannot always be used in the former as in the latter. Nevertheless, many different thixotropic agents are also available for use in aqueous systems.

In high solid and medium solid coatings, SCAs (sag control agents) are sometimes used now as thixotropic agents. These are low-molecular, semi-crystalline, urea-based organic compounds, whose principal function is as an antisettling agent. *Microgels* of acrylate copolymer microparticles are also effective in reducing the tendency to run, especially in one-component high solid systems. They have a particularly important role in metallic coating systems, however, where they are used to fix metallic pigments.

2.4.4 Light stabilisers

When used out of doors, many coatings are gradually degraded and destroyed under the simultaneous influence of light (UV radiation), oxygen, moisture and often also of air pollution. Various changes in the coating can then be detected, such as gloss reduction, change in hue, yellowing, embrittlement, cracking, blistering or delamination, which not only detract from the appearance of the coating but also impair its function. For this reason particularly high-grade industrial coatings and topcoat paints, e.g. for automotive applications, are protected by light stabilisers from the consequences of the action of harmful UV light. In other cases, the substrate or deeper-lying layers of the coating structure need to be protected against UV light. For pigmented topcoats, the UV protection can be provided by the pigments therein. In clearcoats, by contrast, this role has to be played by photostabilisers.

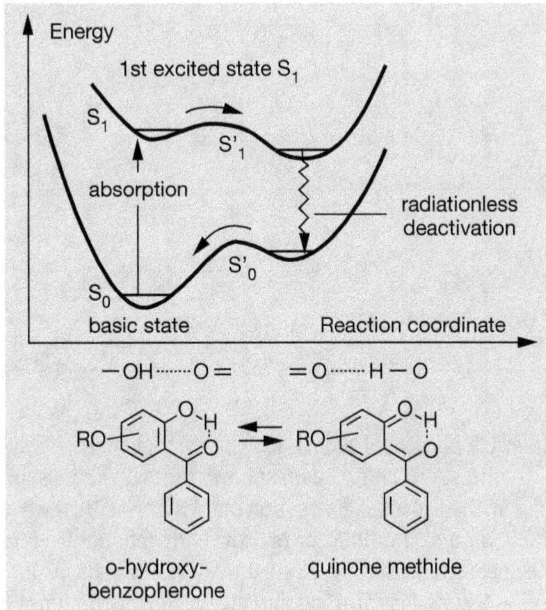

Figure 2.61: Mechanism of energy conversion by phenolic UV absorbers, taking o-hydroxybenzophenones as an example, according to [40, BASF AG]

Mode of action

Light stabilisers can be divided according to their mode of action into four classes, of which only the first two are used to any notable extent in coatings, however.

- UV absorbers
- radical interceptors
- quenching agents
- hydroperoxide decomposers

Platelet-shaped aluminium pigments have been used for many years in highly weather resistant masonry coatings. The "mirror effect" of these pigments limits photochemical film degradation to a surface zone of only a few hundred nanometres in thickness. These are covered in section 2.3.5.2 together with other special pigments. In addition to their protective action, light stabilisers are also required to exhibit high stability, a weak self-colour, ready incorporability, low volatility, good long-term performance and good compatibility with the binder.

UV absorbers

The action of UV absorbers is essentially based on the fact that harmful UV radiation in the wavelength range from 290 to 350 nm is absorbed and converted into harmless heat energy by means of radiationless deactivation (Figure 2.61). (Although UV radiation with an even shorter wavelength is even more harmful, it is filtered out of sunlight completely as it passes through the ozone layer of the earth's atmosphere). To this end UV absorbers must have as high an absorption coefficient as possible in this wavelength range together with a steep absorption edge in the near UV region to keep their self-colour weak.

Four classes of substances are mainly used today as UV absorbers:

- 2-hydroxybenzophenones

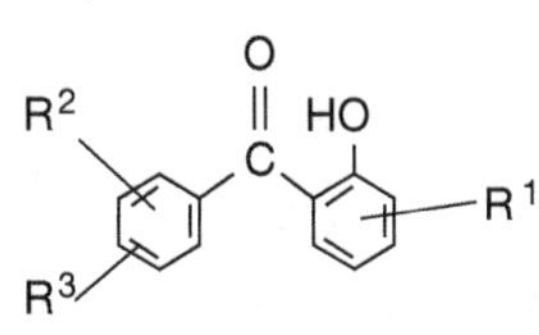

- 2-hydroxyphenylbenzotriazoles

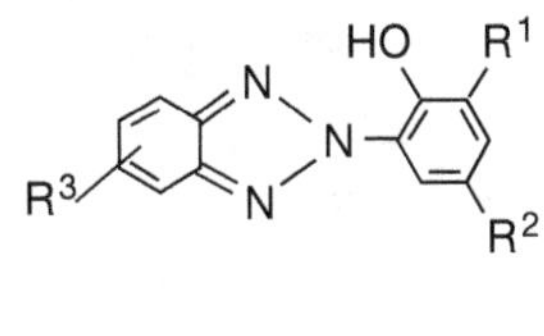

• 2-hydroxyphenyltriazines

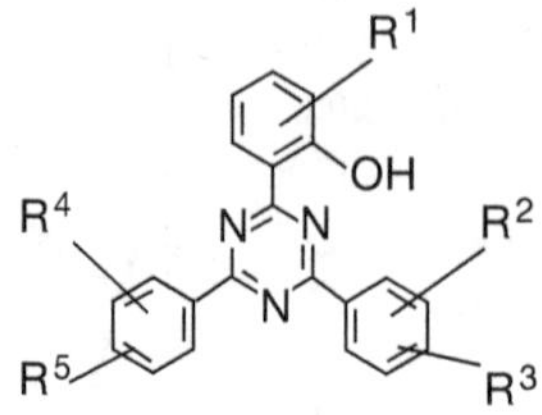

• oxalanilides

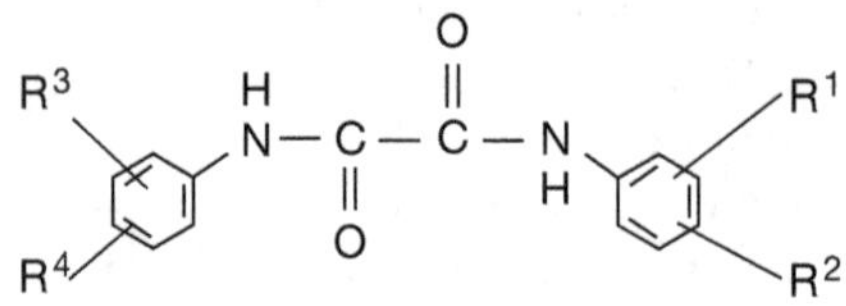

Other UV absorbers such as cyanacrylic acid derivatives play only a minor part in the UV stabilisation of coatings. Probably the most important substances are the 2-hydroxyphenylbenzotriazoles, which display the broadest absorption band in the critical wavelength range and are more photochemically stable than oxalanilides or hydroxybenzophenones. Figure 2.62 shows the transmission spectra for the specified absorbers. According to *Lambert-Beer's law*, the protective action is dependent not only on absorber concentration and absorption coefficient but also on the coating thickness, which must have a certain minimum value. The uppermost layer of coating, in which the UV absorbers cannot yet fully absorb the harmful UV radiation, remains unprotected however.

Radical interceptors

The above-mentioned gap can be closed by radical interceptors. Although they do not prevent the formation of reactive radicals under UV radiation, they intercept these radicals and convert them into stable compounds. This interrupts the radical reaction chain of photochemical degradation of the coating binder. The most commonly used radical interceptors in modern coatings are the HALS type products (**H**indered **A**mine **L**ight **S**tabiliser), which without exception are derived from 2,2,6,6-tetramethyl piperidine.

The *stabilising mechanism* can be represented as follows (Figure 2.63): The basicity of piperidine can be regulated by means of the R^1 substituents on the N atom. Where R^1 = H or CH_3, a pK_B[1)] of around 5 is obtained. Such products cannot be used in coating systems in which curing is catalysed by strong acids, however. In such cases an acyl derivative is used, where R^1 = CO-R and pK_B is still only around 10. As the diagram in Figure 2.63 shows, the stabilising effect of HALS systems is based on the continuous formation of nitroxyl radicals during weathering and is not derived directly from the HALS itself.

Sterically hindered phenols are not used as radical interceptors to any great extent in coatings. *Quenching agents* and *hydroperoxide decomposers* are very rarely used for light stabilisation in coatings. Quenching agents are usually chelating complexes of transition metals, which deactivate groups that are excited in the coating binder, e.g. carbonyl groups. They are almost always coloured. Peroxide decomposers are metal complexes of sulphur-containing compounds, which are oxidised by peroxides.

Selection criteria

The choice of a suitable light stabiliser is governed by the type of coating system.

In one-coat paint systems UV absorbers alone have only a moderate effect since they cannot protect the surface area of the coating (see above). This light stabilisation, already limited in clear coatings, is even further reduced in opaque coatings if the binder is photo-oxidatively decomposed by the

1) pK_B: defined as the negative decimal logarithm of the base constant K_B.
The higher the value of pK_B, the weaker the base.

pigment (with absorption of UV radiation) (→ 2.3.3.2), which can very quickly lead to damage in the coating system.

A relatively large addition of HALS (approx. 3 %) is required to achieve an effective stabilisation in such systems. This type of combination of UV absorber and HALS often exhibits synergistic effects. Transparent one-coat paint systems with UV absorbers act as a UV filter for the substrate below but often need to be protected themselves from the effects of UV radiation by the addition of HALS. Although this UV protection is often adequate in less sensitive substrates, only a limited protection can be achieved with economically realistic quantities of UV absorber alone in materials more susceptible to photo-oxidative degradation - such as wood - although the protection can be improved in combination with HALS.

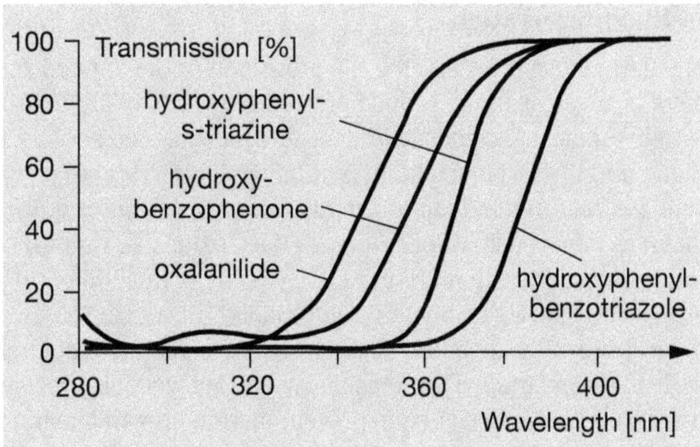

Figure 2.62: Transmission spectra for various classes of UV absorbers vc = 1.4 · 10^{-4} Mol/l in $CHCl_3$ (1 cm cell), according to [33]

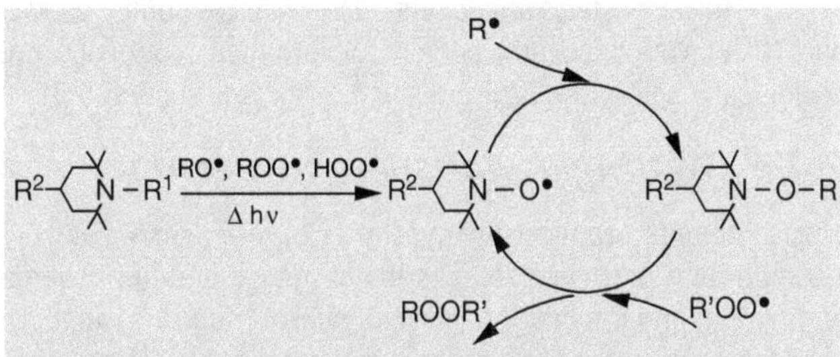

Figure 2.63: Schematic view of the stabilisation mechanism of a HALS compound, taking a tetramethylpiperidine system as an example

Two-layer topcoat systems generally consist of one colour-imparting, opaque layer, on top of which a clear, transparent topcoat is applied for optical reasons (gloss) and/or for protection. In this case the pigmented layer is comparable to the substrate below a transparent one-coat paint system. Stabilisation by UV absorbers alone may often be sufficient here, although enhanced stabilisation effects are observed with a combination of UV absorbers/HALS due to synergistic effects.

2.4.5 Biocides

As systems in which organic substances are finely dispersed in water, aqueous coating materials represent an ideal nutrient medium for fungi, algae and bacteria. As the content of residual monomers and organic solvents - which often have an anti-microbial action - is reduced, the risk of microbial contamination increases. In storage containers this can lead to fouling or discoloration of the surface of the product, to changes in rheological behaviour and to variations in pH, coagulation phenomena, breaks in dispersion, putrid odours or gas evolution. Applied coatings can also show signs of microbial attack, however, such as visible fouling with algae or fungi, green or grey discoloration or cracking.

The unchecked growth of such microorganisms in the coating material or coating system can be reduced or even prevented by the use of chemical biocides.

In-can preservatives

These substances protect against attack during production, transportation and storage. The most commonly used substances today are formaldehyde and various reaction products of formaldehyde with alcohols, amides and amines, as well as N, S-heterocyclics such as isothiazolinones, and also chloroacetamide.

In-film preservation

This term is used to describe the protection of the applied coating film against attack by bacteria, moulds, algae or mosses. Such protection is principally required if building materials and coating systems come into contact with moisture for extended periods, e.g. when exposed to condensation or high humidity and inadequate ventilation in interior applications, and when situated on the shaded side of structures in exterior applications. The list of requirements for such in-film preservatives includes items such as long-lasting effect, extensive water-insolubility and low environmental pollution. Substances that are used today include, in addition to some ecologically unsafe substances such as organic Hg compounds and trialkyl tin compounds, a number of S- and N-containing, often cyclical, organic compounds such as dithiocarbamates, thiophthalimide derivatives and benzimidazole derivatives and trialkyl tin compounds. This category also encompasses *antifouling additives* in marine paints, which are designed to prevent marine growth on ships' hulls and port installations.

Wood preservatives

These substances represent a special case, being intended to prevent the biological degradation of wood by fungi, bacteria and insects. The once commonly used chlorophenols such as pentachlorophenol (PCP), which has now been banned in many countries, are increasingly being replaced now by ecologically safer, less harmful substances (see section 5.5).

2.4.6 Wetting and dispersing agents

Most pigments produced by synthesis have a primary particle size of between 0.05 and 0.5 μm. The optimum particle size in terms of optical and application-oriented properties, and depending on pigment and application, normally lies within this range. During drying these primary particles congregate more or less strongly to form agglomerates with particle sizes of up to 100 μm and above. Since this means that during packing and transportation the pigments are in the form of easily manageable powders rather than dusts (with a particle size of below 20 μm), agglomeration is actually desirable by this stage. In the finished coating material, however, the pigments again need to be sufficiently finely divided to enable their optical and application-oriented properties to develop to the full.

Mode of action

The complex process of wetting the pigments as completely as possible with binder (solution) and distributing them evenly is often referred to simply as *dispersion*. It can be broken down into three *individual steps*.

1. Thoroughly moistening and wetting of the pigment agglomerates and aggregates
2. Breaking down the agglomerates
3. Stabilising the pigment dispersion.

These steps can be characterised as follows:

1. The thorough moistening and wetting of the pigment agglomerates and aggregates by the binder solution is substantially influenced by the nature of the pigment and of the binder solution. The initial stage of the wetting process can be described by the *Washburn equation*:

$$\dot{V} = \frac{\pi \cdot r^3 \cdot \sigma_L \cdot \cos \Theta}{2 \cdot \pi \cdot L}$$

where V = transported liquid flow, r = capillary radius, σ_L = surface tension of the liquid, Θ = angle of contact between the liquid and the surface of the pigment, $\sigma_L \cdot \cos \Theta$ = wetting tension, η = viscosity and l = length of capillaries in the pigment agglomerate.

Since the moistening rate, i.e. V/t, depends on the cube of the capillary radius, pigments should be in the form of loose, uncompacted powders with a large pore volume.

The coating manufacturer can influence the moistening rate by the viscosity (→ 3.9) and the surface tension of the grinding solution (and hence the wetting tension). A low viscosity can be produced in the binder solution not only by means of a correspondingly low concentration of film former but also by raising the temperature or selecting a solvent (blend) in which the polymer is only poorly solvated. In the last case the low affinity of the solvent to the polymer encourages polymer adsorption on the pigment, improving the stability of the pigment dispersion (see below).

Wetting additives can accelerate the moistening and wetting of the pigment powder by reducing the interfacial tension between the binder solution and the pigment surface and hence increasing the wetting tension. Such wetting additives display a surfactant structure and are interfacially active. The non-polar end generally consists of hydrocarbon chains, the polar end is either anionic or non-ionic.

2. The breaking down of agglomerates - essentially by shear forces - is discussed in section 4.9.2. Breaking down the pigment agglomerates into primary particles and aggregates introduces energy into the system to overcome the forces of attraction between the pigment particles. The pigment dispersion generally endeavours to change back into a low-energy state by congregating the finely divided particles. *Flocculates* may be formed, which are very similar in structure to *agglomerates* except that the space between the pigment particles is filled not with air, as in agglomerates, but with binder solution.

Stabilising possibilities

3. The third step in dispersion is therefore to stabilise the resulting pigment suspension against flocculation with the aid of *dispersion additives*. These bond to the surface of the pigment and prevent the pigment particles from coming close enough to one another to allow the van der Waals forces acting between them to lead to flocculation.

One possible means of preventing this is to produce such a high viscosity in the binder solution that the particles are no longer able to move together. This route is used in offset printing inks, for example. Other similar mechanisms are electrostatic and steric stabilisation, or a combination of the two.

Electrostatic stabilisation

Electrostatic stabilisation is based on the self-repulsion of like electrical charges. Although solids dispersed in aqueous solution generally already carry surface charges, this charge can be increased even further by adsorption of polyelectrolytes. These surface charges together lead to the formation of a mobile ion cloud around the particle. When two particles approach each other, provided that the electrostatic repulsion forces are greater than the van der Waals attraction forces, flocculation does not occur (with increasing ionic charge and ionic concentration in the dispersion, the range of the electrostatic forces decreases however, leading ultimately to flocculation in such cases). This type of stabilisation is predominantly employed in aqueous media. Polyphosphates and polyacrylates are widely used here as dispersing agents.

Steric stabilisation

In solvent-containing coatings, by contrast, steric stabilisation is more commonly used (Figure 2.64). Effective dispersing agents in these systems contain both binder-compatible segments and one or more (mainly in the case of polymeric wetting agents) pigment-affinitive groups, which are critical for a stable and durable bond between the additive and the pigment surface.

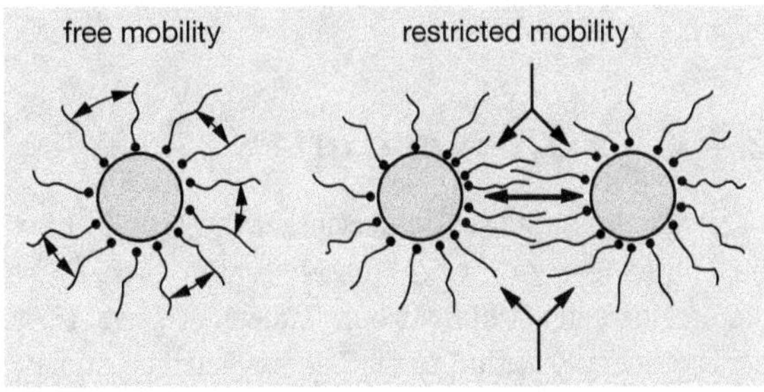

Figure 2.64: Schematic model of steric stabilisation, according to[22]

The chains of additives bonded to a pigment particle should be as distant as possible from the pigment and should extend into the binder solution like "bristles" or ties. If two particles approach one another, the result is deformation of this polymer shell or mutual perforation by these "bristles", restricting their mobility. This equates to a loss of entropy, which is why this type of stabilisation is also known as entropic stabilisation. This loss of entropy can only be counterbalanced by expenditure of energy, which means that the pigment particles are separated from one another by an energy barrier.

Due to their structure consisting of polar anchor groups and non-polar chains, sterically active dispersing agents, like wetting agents, also have distinct surfactant properties. In solvent-containing systems it is frequently possible to use a single dispersion additive as both wetting and dispersing agent. Owing to the necessary charge separation, electrostatic stabilisation in solventborne coating materials is not possible. In water-thinnable coating systems, both mechanisms - steric and electrostatic stabilisation - are possible.

Floating and flooding

Dispersing and wetting agents can also act as *anti-flotation* or *antisettling agents*. A distinction is made between horizontal *flooding* and vertical *floating*. "Flooding" refers to a uniform change of colour in the still wet coating film, which can be observed by means of a *rub-out test*. "Floating", on the other hand, is the uneven distribution of several pigments contained in the coating at the surface of the coloured film, which can be seen as colour cells and/or stripes. Both phenomena are caused by flows in the drying coating and by different pigment mobilities (see surface-active additives, Bénard's cells → 2.4.2.2). The differing pigment mobilities are closely linked to the particle size. As the difference in size between individual particles can be considerable, especially between inorganic and organic pigments, floating is a major problem in the coatings industry. Ideally these size differences can be largely balanced out by adsorption of dispersion additive molecules.

Co-flocculation (or *controlled flocculation*) represents one means of suppressing these pigment separations. In additives that support controlled flocculation, the pigment-affinitive groups are not restricted to a severely limited area of the molecule as they are in the dispersion additives discussed above, but instead are distributed throughout the entire molecule. They can therefore act as a bridge between various pigment particles and ultimately form a three-dimensional network, or *controlled flocculate*, in the dispersion. The size and stability of the network is determined by the properties of the additive. The critical difference from normal flocculation lies in the fact that in controlled flocculation similar pigment particles mostly are not in direct contact.

Since in controlled flocculation the different pigments are linked in *one* flocculate, they are no longer able to separate. This prevents floating. At the same time, controlled flocculation also alters the rheological properties of a coating. The three-dimensional network structure creates thixotropic flow characteristics (see rheological additives) and, frequently, also a yield point, thus reducing the tendency to settle out. Co-flocculating additives are often used in combination with other rheological additives, such as pyrogenic silicic acid, hydrated castor oil, etc., as a result of which synergistic effects can often be seen.

2.4.7 Catalysts and driers

Catalysts are supposed to accelerate the crosslinking reaction in the applied coating film - and hence curing. For historical reasons, catalysts that accelerate the crosslinking of oxidatively drying oleo-resins and alkyd resins are known as *driers*. They consist of the metal salts of natural organic acids, generally with 8 to 11 C atoms, such as linoleic, abietic or naphthenic acid, or of synthetic organic acids.

The catalytically active component in such driers is the metal atom. The decomposition of hydroperoxides formed by the action of atmospheric oxygen on unsaturated organic compounds is catalysed according to the diagram below:

$$Co^{2+} + ROOH \rightarrow RO^\bullet + OH^- + Co^{3+}$$

$$Co^{3+} + ROOH \rightarrow ROO^\bullet + H^+ + Co^{2+}$$

H_2O and reactive free radicals are ultimately formed, which initiate a radical polymerisation of the C=C double bonds.

Classification basis

Co, Mn, Pb, Zr, Ca and Ba are the metals predominantly used in driers; others, such as Ce or other rare earths, are less commonly used. Many metals are excluded because of the colour of their salts. Driers are divided into

- primary driers and
- auxiliary driers (secondary driers).

Primary driers

These siccatives have an inherent catalytic action. Co and Mn salts catalyse crosslinking at the surface of the coating in particular, and are therefore known as *surface driers*. Pb driers, which instead promote the complete drying of the film, are known as *hard driers*.

Auxiliary driers

Salts of Ca, Ba and other metals are only slightly active, if at all. In combination with primary driers, however, they can provide substantial support to their action.

Combination driers

Combinations of driers, e.g. Co, Zr and Ca driers, are almost always used in order to achieve optimum drying characteristics. The use of Pb and also co-driers is now avoided wherever possible, however, for ecological and toxicological reasons.

Skinning

Driers can cause a skin to form on the coating during storage, particularly if the container has already been opened. For this reason ketoximes, substituted phenols, aldoximes, etc., are added to the coating as *anti-skinning agents*; together with the co-salts in particular these form reversible inactive or at least less active complexes. After the coating has been applied, these anti-skinning agents evaporate to allow the drier to fulfil its function.

Other catalysts

Other classes of catalysts include

- acid catalysts,
- amines and
- organic tin compounds.

Their mode of action naturally depends on the nature of the catalysed crosslinking reaction and hence on the binder. It is therefore discussed together with the corresponding binders.

After driers, the most widely used catalysts are the *acid catalysts*. These are used for a large number of stoving coatings, quick-drying coatings and also acid-curing wood coatings. They accelerate the crosslinking of coating systems based on phenolic/formaldehyde and amido-/formaldehyde resins. Various sulphonic acid derivatives are generally used here. Blocked acids, whose acid H atom is inactivated by reacting the acid with an amine, are used to extend the pot life of the finished coating. The ammonium salts produced in this reaction then decompose at elevated temperature, the amine evaporates and the free acid is able to catalyse the crosslinking reaction. If a catalyst is used, the stoving temperature or time can be reduced, conserving energy and also enabling temperature-sensitive substrates to be coated.

The curing of two-component polyurethane coatings can inter alia be accelerated by means of *organotin compounds*, e.g. DBT(D)L (dibutyl tin dilaurate), or *tertiary amines* such as DABCO (diazabicyclooctane). The latter are also effective in a number of epoxy resin systems.

Initiators are substances that initiate polymerisation reactions, usually by means of radical formation. Strictly speaking they are not catalysts at all, since they participate themselves in the reaction once they have decomposed into radicals. The usual initiators for unsaturated polyester resins are organic peroxides, e.g. benzoyl peroxide or 2-butanone peroxide, etc. Depending on their splittability (thermally and/or by UV radiation), initiators must be protected from heat and/or light.

2.4.8 Flatting agents

The gloss of a coating is largely determined by its surface roughness. In a given binder system this in turn is strongly dependent on the PVC and particle size distribution. Consequently the gloss of a pigmented coating can be controlled by the PVC level and by the particle size of the pigments and fillers used.

In a number of cases, including some furniture varnishes and certain automotive and industrial coatings, low gloss and high transparency are required. This is achieved by the addition of small quantities of finely divided substances, such as PE waxes or precipitated silicic acids. During the drying process, they prevent the formation of a smooth surface and thus give rise to the desired flatting effect. Provided that they are present in only small concentrations, their prevailing combination of low refractive index and small particle size causes only minimal light scattering and so guarantees good transparency.

Sources and references for Chapter 2

[1] G. Benzing: Pigmente in der Lackindustrie, Expert Verlag GmbH, Ehningen, 1992

[2] J. Bielemann: Lackadditive, Wiley-VCH Verlag GmbH, Weinheim 1998

[3] E. Brandau: Duroplastwerkstoffe, VCH Verlagsgesellschaft mbH, Weinheim, 1993

[4] D. Braun, H. Cherdron, W. Kern: Praktikum der makromolekularen organischen Chemie, 3. Aufl., Hüthig, 1979

[5] G. Buxbaum, Pfaff, G. (ed.): Industrial Inorganic Pigments, Wiley-VCH Verlag GmbH & Co. KGoA, Weinheim 2005

[6] K. Dören, W. Freitag, D. Stoye: Wasserlacke: Umweltschonende Alternative für Beschichtungen, TÜV Rheinland, 1992

[7] H. Endriß: Aktuelle anorganische Bunt-Pigmente, Curt C. Vincentz Verlag, Hannover 1997

[8] H. Ferch: Pigmentruße, Curt R. Vincentz Verlag, Hannover 1995

[9] R. Gächter, H. Müller: Plastic-Additives, Carl Hanser Verlag, München 1990

[10] G. Pfaff, Spezial Effekt Pigments, Vincentz Network, Hannover 2007

[11] A. Goldschmidt, H.-J. Streitberger, BASF-Handbook Basics of Coating Technology, 2nd edition, Vincentz: Hannover (2007)

[12] C. H. Hare: Protective Coatings, Technology Publishing Company, Pittsburgh 1994

[13] W. Herbst, K. Hunger: Industrial Organic Pigments; 3rd edition,, Wiley-VCH: Weinheim (2004)

[14] H. F. Huber: Dauerhaft kleben, Curt R. Vincentz Verlag, Hannover 1994

[15] C. Jentsch: Angewandte Chemie für Ingenieure, BI-Wiss.-Verl., 1990

[16] E. Karsten: Lackrohstoff-Tabellen, 10th edition, Verlag Vincentz Network, Hannover 2000

[17] Kittel, H.; Ortelt, M. (Hrsg.), Lehrbuch der Lacke und Beschichtungen, 2. Aufl., Bd. 1 bis 5.; Hirzel: Stuttgart (1999 bis 2007)

[18] R. Laible et al.: Umweltfreundliche Lacksysteme für die industrielle Lackierung, Expert Verlag 1989

[19] O. Lückert: Pigment + Füllstoff Tabellen, Vincentz Verlag, Hannover 2002

[20] J. Möckel, U. Fuhrmann: Epoxidharze, Die Bibliothek der Technik, Bd. 51, Verlag moderne Industrie, Landsberg/Lech, 1990

[21] P. Nanetti: Lackrohstoffkunde, 3rd edition, Vincentz Network, Hannover 2008

[22] P. Oldring, G. Hayward (Hrsg.): Resins for Surface Coatings, Vol. 1-3, SITA-Technol., London, 1987

[23] J.M. Oyarzún: Handbuch der Pigmentverarbeitung, Vincentz Verlag, Hannover 1998

[24] J. Ruf: Organischer Metallschutz, Curt R. Vincentz Verlag, Hannover 1993

[25] H. Saechtling: Kunststoff-Taschenbuch, 24. Ausgabe, Carl Hanser Verlag, München 1989

[26] E. V. Schmid, Farbe + Lack 96 (1990), 108 („Gedanken zur Glasumwandlung von Anstrichfilmen")

[27] M. J. Schwuger: Lehrbuch der Grenzflächenchemie, Georg Thieme Verlag, Stuttgart 1996

[28] K. Sponsel, W. O. Wallenfang, I. Waldau: Lexikon der Anstrichtechnik 1 - Grundlagen, 9. Aufl., Verlag Georg D. W. Callwey, München 1992

[29] D. Stoye (Hrsg.): Paints, Coatings and Solvents, VCH Verlagsgesellschaft, Weinheim 1993

[30] D. Stoye, W. Freitag (Hrsg.): Lackharze, Carl Hanser Verlag, München 1996

[31] D. Stoye: Lösemittel, Hüls Publikationen 1993

[32] J. Bentley, G. P. A. Turner: Introduction to Paint Chemistry and Principles of Paint Technology, 4th Edition, Chapman and Hall, 1998

[33] Ullmann's Encyclopedia of Industrial Chemistry, 6. Auflage, CD-ROM, VCH, 2001

[34] Ullmann (6. Auflage) CD-ROM Februar 2004

[35] A. Valet: Lichtschutzmittel für Lacke, Curt R. Vincentz Verlag, Hannover 1996

[36] Verband der Lackindustrie: Betriebliche Fachkunde, 4. Aufl., 1996

[37] Vincentz Verlag: „Lehrgang Lacktechnologie", Modul 2 (November 1996); „Epoxidharze I + II" (K. Hoffmann, Dow Deutschland), „Alkydharze für Metalle und Holz" (H.-J. Luthardt, Synthopol Chemie, „Dispersionen und Emulsionen" (R. Kuropka, Clariant GmbH)

[38] G. Walz (Hoechst AG), Kunstharz-Nachrichten 33 (1974), Heft 6, S. 30 („Die Auswahl von Lösemitteln für „High Solids"-Systeme mit Hilfe des Löslichkeitsparametermodells")

[39] G. Walz (Hoechst AG), Kunstharz-Nachrichten 34 (1975), Heft 8, S. 19 („Löslichkeitsparameter-Diagramme ausgewählter Alkyd- und Melaminharze")

[40] Z. W. Wicks, Jr., F. N. Jones, S. P. Pappas: Organic Coatings, Science and Technology, 2nd ed., J. Wiley & Sons, Inc., New York 1999

[41] A. Zosel: Lack- und Polymerfilme, Curt R. Vincentz Verlag, Hannover 1996

[42] Firmenschriften der Firmen BASF SE, Bayer AG, Borchers GmbH, Byk Chemie GmbH, Dow Corning GmbH, Kronos, Du Pont, Rheox GmbH, Tego Chemie Service GmbH

[43] D. Gysau, Fillers for Paints, Vincentz: Hannover (2006)

[44] K. Hunger (Hrsg.), Industrial Dyes, Chemistry, Properties, Applications, Wiley-VCH: Weinheim; (2003)

[45] S. Sepeur: Nanotechnology, Vincentz Network GmbH & Co.KG, Hannover 2008

[46] H. M. Smith (Hrsg.), In High Performance Pigments, Wiley-VCH: Weinheim; (2002)

[47] J. Winkler, Titanium Dioxide, U. (Hrsg.), Vincentz: Hannover (2004)

[48] R. Baumstark, M. Schwartz: Dispersionen für Bautenfarben. Vincentz Network, Hannover 2001

[49] J. Bentley and G. P. A. Turner: Introduction to Paint Chemistry and principles of paint technololgy. 4th Ed., Chapman and Hall, London 1998

[50] Encyclopädia of Polymer Science and Technology: Coating Methods, Powder Technolgy. John Wiley & Sons, Inc. (www. plasticdecoratingdata.com (August 2008))

[51] H. Friedrich, WELT DER FARBEN, 5/2006, S. 8 („Vorteile der Dual-Cure-Vernetzung bei wässrigen UV-Dispersionen")

[52] W. Förster, G. Wagner, Phänomen Farbe, 12/2001-1/2002, S. 34 („Sol-Gel-Beschichtungen als dünne Korrosionsschutzschichten")

[53] „HighChem hautnah". Broschüre der GDCh (Fachgruppe Lackchemie), Frankfurt 2008

[54] J. Leuninger, F. Tiarks, H. Wiese, B. Schuler, FARBE&LACK, 110, 10/2004, S. 30 („Wässrige Nanaokomposite")

[55] A. R. Marrion (Edr.): The Chemistry and Physics of Coatings. 2nd Ed., The Royal Soc. of Chemistry 2004

[56] U. Meier-Westhues: Polyurethane - Lacke, Kleb- und Dichtstoffe. Vincentz Network, Hannover 2007

[57] P. Mischke: Filmbildung in modernen Lacksystemen. Vincentz Network, Hannover 2007

[58] B. Müller und U. Poth: Lackformulierung und Lackrezeptur. 2. Aufl., Vincentz Network, Hannover 2005

[59] D. Ondratschek (Red.): besser Lackieren! Jahrbuch 2008. Vincentz Network, Hannover 2007

[60] Römpp-Lexikon: Lacke und Druckfarben. Georg Thieme Verlag, Stuttgart 1998

[61] H. K. Schmidt, Chemie in unserer Zeit, 35, 3/2001, S. 176 („Das Sol-Gel-Verfahren")

[62] R. Schwalm et al., WELT DER FARBEN, 5/2005, S. 10 („UV-härtbare Lacke - kratzfeste Beschichtungen für Automobile")

[63] VILF-Vorträge, Schriftenreihe, Bd. 4: Neue Bindemittelkonzepte für moderne Beschichtungen. Neu-Isenburg 2001

[64] W. Weigt und F. Auer-Kanellopoulos, FARBE&LACK, 110, 10/2004, S. 20 („Nichthaften erwünscht")

[65] U. Wienhold und G. Wagner, FARBE&LACK, 109, 1/2003, S. 82 („Billige Alternative")

[66] Winnacker-Küchler: Chemische Technik. 5. Aufl., Bd. 7: Industrieprodukte. Wiley-VCH, Weinheim 2004

[67] R. D. Athey, farbe + lack, 96, 7/1990, S. 523 („Das Molekulargewicht von Lackharzen - Bestimmungsmethoden und ihre Anwendung")

[68] H. Breuer: dtv-Atlas zur Chemie. Bd. 2: Organische Chemie und Kunststoffe. 11. Aufl., DTV, München 2006

[69] G. Meichsner, Th. G. Mezger, J. Schröder: Lackeigenschaften messen und steuern. Vincentz Network, Hannover 2003

[70] J. H. Meyer, FARBE UND LACK, 78, 9/1972, S. 813 („Löslichkeitsparameter und ihre Anwendung in der Praxis")

[71] B. Tieke: Makromolekulare Chemie. 2. Aufl., Wiley-VCH, Weinheim 2005

[72] Z. W. Wicks, F. N. Jones, S. P. Pappas, D. A. Wicks: Organic Coatings. 3rd ed., J. Wiley & Sons, Hoboken/New Jersey 2007

3 Coating systems, formulation, film-forming

3.1 Composition of coating materials

Having discussed the most important raw materials for coatings from a predominantly chemical perspective in Chapter 2, in this next chapter we will cover some of the fundamental aspects involved in the formulation of coating materials - in other words, the problems of coating technology. Let us first review the key principles of composition.

According to DIN EN ISO 4618,
a **coating material** is a product in liquid, paste or powder form that, when applied to a substrate, forms a film possessing protective, decorative and/or other specific properties.

This generic term is accompanied by a series of subsidiary terms by which coating materials can be classified. Nomenclature is conventionally based on

- *film former system*, e.g. alkyd resin coating, PU coating,
- *function* in a coating formulation, e.g. primer, filler, topcoat or
- *solvent content*, e.g. low solid coating, high solid coating, powder coating, water-borne coating.

The terms coating, coating material and paint are commonly used as synonyms for one another (see section 1.3).

Classification

The components of coating materials can be divided according to their function into four groups, namely film formers (see below), pigments and fillers, solvents, and additives.

As can be seen from Figure 3.1, every coating material must contain a film former. This makes up the binder together with the non-volatile matter in the additives. The film former alone is also often described - not entirely accurately - as the binder.

According to DIN 55 945,
the **film former** is the part of the binder that is essential for the formation of the film. Film formers are generally macromolecular organic substances or substances from which macromolecules are produced during film-forming, such as alkyd resins, polyacrylates or a combination of polyester and polyisocyanates. Examples of inorganic film formers include water glass or alkyl silicates.

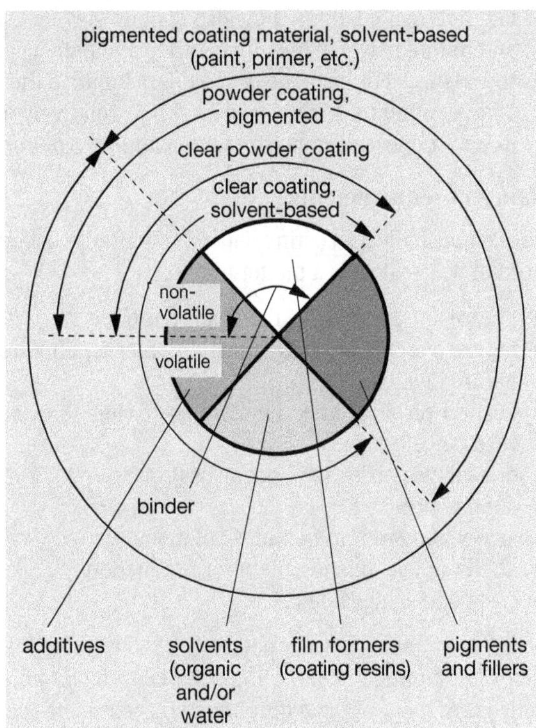

Figure 3.1: Basic composition of a coating material, according to[1]

Brock, Groteklaes, Mischke: European Coatings Handbook

ISBN 978-3-86630-849-7

According to DIN EN ISO 4618,
a **binder** is the non-volatile part of a medium. Plasticisers, driers and other non-volatile additives are also components of the binder. Reactive, inherently volatile substances belong to the binder too if they become part of the coating by chemical reaction.

Functions

Let us first review the key facts relating to the coating components mentioned earlier in preparation for our introduction to the formulating aspects:

- Most of the chemical and physical properties of a coating material are decided by the film former (→ 2.1). There are even a few glazes and clear coatings that consist solely of the film former but, as we have already said, there are no coating materials that do not contain a film former. It holds the coating together by means of *cohesion* and bonds it to the substrate by means of *adhesion*.
- In addition to the film former, a clear coating contains additives and in the vast majority of cases also solvents (the exception being clear powder coatings and some systems with reactive thinners). The purpose of additives (→ 2.4) is to impart certain properties to the coating material or final coating, or to eliminate any undesirable properties that may exist. They are usually added in concentrations of below 5 %, sometimes as low as 0.01 to 1 %.
- Without solvents (→ 2.2), most coating materials could not be applied at all. In addition to adjusting the viscosity, in chemically-curing coating materials they also influence the reactivity of the system both before and during application. Examples of solvents include low-molecular organic substances such as benzines, alcohols or esters, often as blends with different boiling points and dissolving properties, and, increasingly these days, water. They influence the flow properties of a coating material during drying and hence also the drying time and the properties of the final coating through the speed and sequence of their evaporation.
- Coloured and opaque coating systems contain, in addition to film formers and additives, pigments (→ 2.3.3 to 2.3.8) and optionally fillers (→ 2.3.9); with the exception of powder coatings and hot-melt coatings they also contain solvents. The pigments provide a coating with colour and hiding power, whilst fillers are principally used to obtain certain mechanical properties in the coating. For instance, some fillers improve the barrier effect, others the substrate adhesion and yet others the film strength. Their relatively low price means that fillers can also be used as a cost-effective means of increasing the volume ("body")[1] of a coating.

Range of requirements

The composition of a coating depends on numerous factors, in relation to which the following points need to be decided as a matter of priority:

- nature and function of the desired product (e.g. stopper, primer, topcoat)
- nature of application (application process, film-forming)
- nature of substrate
- desired physical and chemical properties (e.g. weather resistance, scratch resistance, alkali resistance)
- optical properties (e.g. hue, gloss)
- supply form
- safety and environmental regulations
- ability to use existing plant for production
- costs and achievable price.

Given the number of such points it is clear that no one solution can ideally satisfy every expectation and requirement of coating materials. The formulator has to find a reasonable compromise that satisfies as many of the requirements as possible, especially those of more fundamental importance,

1) Body = filling property (DIN 55 945): the ability of a coating material to give a uniform film over an irregular substrate.

without giving rise to any serious disadvantages or defects. This means that for even seemingly small changes in the range of requirements, a whole series of tests has to be performed before the best solution can be discovered.

Since coatings generally consist of mixtures of various raw materials - most of them technical products that are not usually chemically pure - the implications of even small formulation changes can never be fully predicted by theory. The formulator will constantly be surprised by unexpected consequences of a change in formulation, which in some cases may make it easier to find a solution, but usually make it even more difficult. A successful formulator therefore not only needs a good knowledge of the raw materials on the market - substance class, purity, chemical and physical properties, etc. - but must also have a thorough grounding in chemistry and engineering to enable him or her to take into account at least the majority of the possible interactions between the coating components in the formulation.

In the sections that follow we have chosen not to list and describe starting formulations for coating systems. This would go beyond the scope of this handbook. Anyone with an interest in these should consult the appropriate literature or obtain starting formulations from raw materials manufacturers; these are an extremely valuable source of information.

3.2 Basic formulating parameters

Let us consider some of the important formulating parameters by reference to a simplified formulation for a white alkyd resin coating (Table 3.1).

Of all the components in this coating, only the "non-volatile matter" remains as the coating on the substrate after film-forming. This is often also referred to as "solids content" or "solids".

Characteristic values

According to DIN EN ISO 4618,
the **non-volatile matter** is the residue by mass obtained by evaporation under specified conditions.
When establishing the content of non-volatile matter it is principally the drying temperature, drying time and dry coating thickness that are determined. If we make the value up to 100 % we obtain the proportion of the coating material that is released into the atmosphere during processing or film-forming and that may consequently be harmful to the environment. The yield of a coating is dependent on its content of non-volatile matter.

In our sample formulation (Table 3.1) the non-volatile matter is calculated as follows:
non-volatile matter = content of alkyd resin (without solvent) + content of pigment
non-volatile matter = 60 % · 0.75 + 27 % = 72 %

Table 3.1: Simplified formulation for a white alkyd resin coating

Item	Function in coating	Substance	Proportion by weight (%)
1	filmformer/solvent	alkyd resin, 75% in white spirit (solids density: 1.04 g · cm^{-3})	60.0
2	pigment	titanium dioxide (density: 4.1 g · cm^{-3}, oil absorption value 20)	27.0
3	drier	cobalt octoate (10% Co)	0.2
4	drier	zirconium complex (6% Zr)	0.5
5	drier/dispersing agent	calcium octoate (5% Ca)	1.7
6	solvent	white spirit	10.6
			Σ = 100.0

Of course the non-volatile matter contained in the additives, e.g. the metal salts in the driers, also remains in the coating. As in the example above, however, their non-volatile content is often ignored; this generally leads to only a slight error but would make the calculation more comprehensive if included. The information needed in order to include it is frequently not available, however.

According to DIN EN ISO 4618,
the **spreading rate** is the surface area that can be covered by a given quantity of coating material to yield a dried film of requisite thickness.
It is expressed in m^2/l or m^2/kg.

Pigment volume concentration

One mathematical quantity that is fundamental to an understanding of the many correlations between the composition of a coating and its properties is the **p**igment **v**olume **c**oncentration (PVC) or the latter's ratio to the "**c**ritical" **p**igment **v**olume **c**oncentration (CPVC). These correlations are discussed in section 3.3.

According to DIN EN ISO 4618,
the **PVC** is the ratio, expressed as a percentage, of the total volume of the pigments and/or extenders and/or other non-film-forming solid particles in a product to the total volume of the non-volatile matter.

According to the same standard,
the **CPVC** is the value of the **pigment volume concentration** at which the voids between the solid particles which are nominally touching are just filled with **binder** and above which certain properties of the **film** are markedly changed.

In order to be able to calculate the PVC of a coating from a formulation expressed in proportions by weight, the non-volatile content and density of every component of the coating material is required. From these can be calculated the approximate volume of the individual components in the coating film. The following equation is then valid for the PVC:

$$\text{PVK [\%]} = \frac{\Sigma V_{\text{pigments}} + \Sigma V_{\text{fillers}}}{\Sigma V_{\text{pigments}} + \Sigma V_{\text{fillers}} + \Sigma V_{\text{binders}}} \cdot 100 = \frac{\Sigma V_{\text{pigments}} + \Sigma V_{\text{fillers}}}{\Sigma V_{\text{total}}} \cdot 100$$

Here again, in the interests of simplifying the calculation, the volume of non-volatile matter in the additives is often disregarded. In our example the PVC is thus:

$$\text{PVK [\%]} = \frac{\left(\frac{27.0}{4.1}\right)}{\left(\frac{27.0}{4.1}\right) + \left(\frac{60.0 \cdot 0.75}{1.04}\right)} \cdot 100\ \% = 13.2\ \%$$

Possible methods of determination

The CPVC can be approximately determined from the oil absorption value (→ 3.3) for the pigment contained in the coating material. If it contains a mixture of several pigments and fillers, an average oil absorption value must be used, although this can lead to quite a substantial degree of inaccuracy - especially where differently shaped and sized pigment and filler particles are involved. Determining the oil absorption value for the pigment/filler blend is a more accurate route in such a case.

Other methods for determining the CPVC may also be considered (→ 3.3). In general the following relationship applies:

$$\mathrm{CPVC} = \frac{100\ \%}{1 + \frac{\rho_{pig}}{\rho_B} \cdot \frac{\mathrm{OAV}}{100}}$$

where OAV = oil absorption value, ρ_{pig} = density of the pigment and ρ_B = density of the binder ($\rho_{\text{linseed oil}}$ = 0.935 g/cm^3).

In our example (Table 3.1) the CPVC is then:

$$\mathrm{CPVC} = \frac{100\ \%}{1 + \frac{4.1}{0.935} \cdot \frac{20}{100}} = 53.3\ \%$$

Pigment content

Simply stating the PVC or the CPVC of a coating is not usually very meaningful. The position of the PVC relative to the CPVC, generally expressed as the *Q-value* (→ 3.3), is much more important:

$$Q = \frac{\mathrm{PVC}}{\mathrm{CPVC}} \cdot 100\ \%$$

In our example the Q-value is:

$$Q = \frac{13.2}{53.3} \cdot 100\ \% = 24.8\ \%$$

which is a typical value for a glossy topcoat (see section 3.3). Another parameter that is often stated as a measure of the pigment content is the pigment-binder ratio, defined in terms of the masses (m) of the two components as

$$\text{pigment-binder ratio} = \frac{m_{\text{pigment}}}{m_{\text{binder}}}$$

However, this disregards the fact that the densities of different pigments (and hence their volumes) can differ enormously from one another, particularly when organically pigmented and inorganically pigmented coatings are compared. In our example the pigment-binder ratio is:

$$\text{pigment-binder ratio} = \frac{27.0}{60.0 \cdot 0.75} = 0.6\ (= 3 : 5)$$

3.3 Pigment volume concentration and film properties

The defining equation on page 190 introduced the PVC as a fundamental value that is critical to an understanding of many of the correlations in a coating. In order to be able to apply this value, the following information is also required.

Packing density

The total volume V_{total} of a dry coating film consists of the volume of binder and that of pigments and fillers. For the sake of convenience, pigments and fillers are referred to below simply as "pigments" and their volume as "pigment volume" V_{pig}.

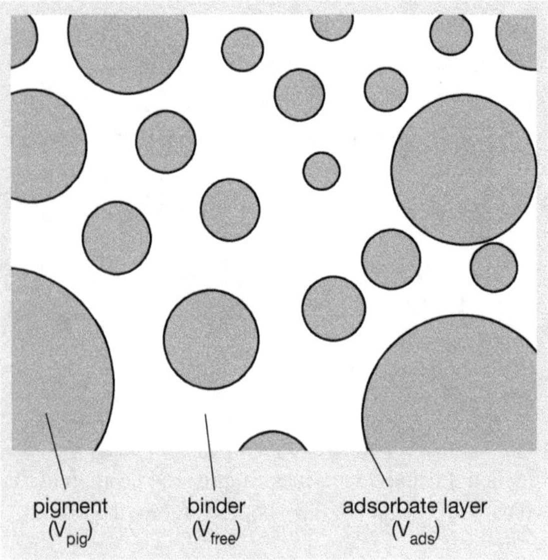

Figure 3.2: Morphological appearance of a coating below the CPVC

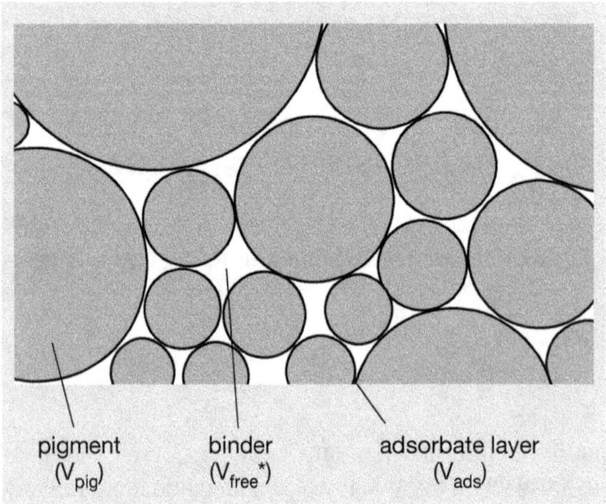

Figure 3.3: Morphological appearance of a coating at the CPVC

We have to assume that in a coating film every pigment particle is enveloped by a thin layer of adsorbed non-volatile binder components. These components consist mainly of wetting or dispersion additives or film former molecules (Figure 3.2). The quantity of adsorbed binder components is dependent on the specific surface area of the pigment and on its chemical properties. The larger the specific surface area, the greater the volume of adsorbed binder components V_{ads}. The total volume of binder V_B is normally greater than V_{ads}. The difference between the two is termed the *free volume* of binder V_{free}; this makes up the remaining volume of the coating film that is not occupied by pigment particles with their layer of adsorbate. In other words, V_{ads} and V_{free} together make up V_B. The expression for the PVC can then be written as follows:

$$PVC = \frac{V_{pig}}{V_{pig} + V_{ads} + V_{free}}$$

The size of V_{free} depends on the PVC.

In a clear coating the PVC = 0. The properties of the coating are determined solely by the binder properties. At low PVC values (see Figure 3.2), isolated pigment particles with their adsorbate layers, which are not in contact with one another, are intercalated into the binder film. Here again the properties of the binder dominate, and the film correspondingly appears glossy and free from pores.

As the PVC increases, the solids particles move closer and closer together and the volume of free binder V_{free} decreases. Above a certain PVC, the volume of pigment is so great that the individual particles are touching. They are now separated only by the thin adsorbate layers. Nevertheless, there is still just enough free binder to completely fill up the spaces between the pigment particles. The volume of free binder is exactly equal to that of the hollow spaces in the pigment packing (Figure 3.3). The film is matt but still closed. This PVC is in fact the critical pigment volume concentration:

$$CPVC = \frac{V_{pig}}{V_{pig} + V_{ads} + V_{free}^*}$$

except that here

V_{free}^* = volume of the spaces between the touching pigment particles.

As the PVC continues to increase, the volume of free binder V_{free} is no longer sufficient to fill up the hollow spaces between the pigment particles, and some of these are instead filled with air. In this case the PVC is supercritical. The film is porous and is now only weakly held together by the binder (Figure 3.4).

Effect of the CPVC

The properties of a coating thus evolve - as the PVC increases - from the high gloss of a dense, impermeable film (low PVC) through a silk-matt stage to the matt appearance of a permeable film - which also has a reduced tendency to blister. As early as 1949, *Asbek* and *van Loo* demonstrated that these and a number of other physical properties, such as resistance to underbody rusting, can be altered considerably as a function of the PVC within a small range around the CPVC (Figure 3.5).

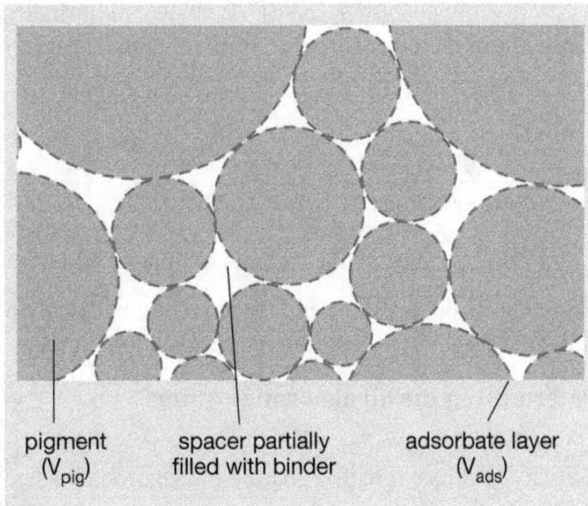

Figure 3.4: Morphological appearance of a coating above the CPVC

The location of the CPVC is governed by several factors, including particle size, particle size distribution and particle shape as well as by the chemical properties of the particle surface. The first three quantities decide the specific surface area of a pigment. This in turn, together with the chemical properties of the pigment surface, determines the quantity of binder components that is fixed there.

Filling effects

Particle size and particle size distribution are the critical quantities influencing the packing density of the pigment particles. Spherical particles of uniform particle size can make up a maximum of around 74 % of the available space when densely packed. Taking into consideration the layer of adsorbate around each particle, the CPVC for such particles would in any event be below 74 %.

Real pigments do not have a uniform particle size, however, but instead display a particle size distribution. Smaller particles are thus able to fill the spaces between the larger particles to some extent. This reduces the volume of the space not filled by pigment particles, which corresponds to the variable $V_{free}{}^*$ indicated in the defining equation for the CPVC (page 192). As $V_{free}{}^*$ falls, the CPVC increases accordingly.

In anti-corrosive primers, fillers and stoppers, in particular, the objective is to achieve an optimum space-filling behaviour by using a blend of fillers or pigments with different particle sizes.

Platelet- or needle-shaped pigments would likewise exhibit good space-filling properties when arranged in an optimum parallel orientation. When mixed with spherical particles, however, they can act as spacers, thus reducing the CPVC. In practice a maximum CPVC is obtained using a blend of differently sized, similarly shaped particles with broad particle size distributions.

Determining the CPVC

In principle, the CPVC can be determined by means of any measurable coating property that alters significantly at the CPVC. It has been found that the values obtained by measuring different properties - e.g. adhesive strength, gloss or binder absorption - generally differ very little from one another. In the case of binder adsorption, however, it must be remembered that chemically different film formers are absorbed in varying degrees at the pigment surface, so the CPVC depends not only on the pigment properties but also on the binder. The influence of the chemical nature of the film former can be reduced by the use of dispersion additives, however. Dispersing agents themselves influence the prevailing particle size distribution in a coating and hence naturally also the CPVC. Whilst it is theoretically possible to determine the CPVC by calculating the geometrical packing density, in practice this is virtually impossible.

As we mentioned earlier, the CPVC value can normally be calculated to an acceptable degree of accuracy from the oil absorption value (→ 2.3.2.4) and the density of a pigment or pigment blend.

The oil absorption is (usually) expressed as follows in terms of the quantities (m) of the substances involved:

$$\text{ÖZ} = \frac{m_{\text{linseed oil}}}{m_{\text{(pigments + fillers)}}} \quad [\text{g}/100\ \text{g}]$$

It can also be expressed as volume of linseed oil per weight of pigment, however.

The oil absorption value can be used along with the densities of the linseed oil and the pigment to calculate the CPVC according to the equation on page 191.

Determining the oil absorption value

Linseed oil of a specified grade is incorporated into a pigment or filler sample on a glass or marble plate by grinding it with a spatula (as vigorously as possible). The addition of oil is halted when a soft paste has been produced that can just about be dispersed without tearing or crumbling and that just about adheres to the plate. The result depends very much on the intensity of incorporation. The values obtained by different testers can thus differ somewhat from one another. Any pigment agglomerates that may have been present are broken down by grinding during incorporation of the linseed oil. On completion of the process to determine the oil absorption value, the linseed oil has wetted the pigment particles and just fills the spaces between the pigment particles. These are precisely the conditions that prevail at the CPVC.

Since the oil absorption value (OAV) is determined without the use of dispersion additives, the CPVC value calculated from the OAV cannot be directly transferred to other binder systems. For a binder that wets less effectively than linseed oil, the real CPVC would be lower than that calculated from the OAV.

Pigment and filler blends

Highly-filled coating materials in particular frequently contain more than one pigment or filler. The oil absorption value for these pigment blends cannot generally be calculated from the oil absorption values of the individual components and their proportion by weight. If we calculate the oil absorption value for a blend of a pigment and a filler at different pigment/filler ratios, we often find that it passes through a minimum (Figure 3.6). The packing density of the pigment/filler blend is at its greatest at this minimum.

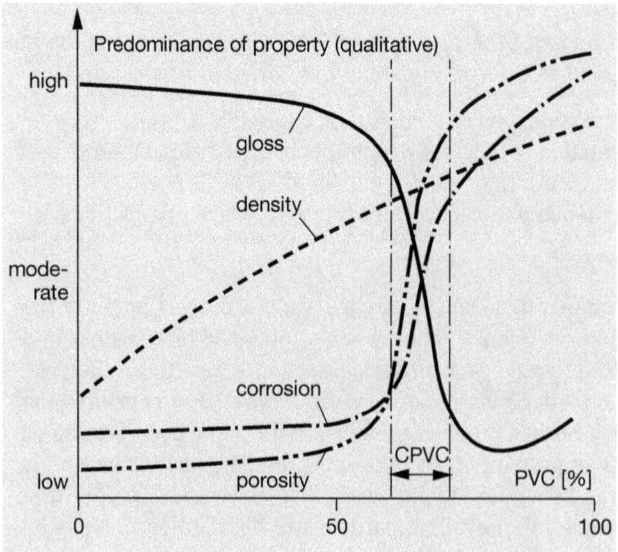

Figure 3.5: Change in various coating properties as a function of the PVC (with realistic CPVC)

With blends of more than two components, it is clearly not so easy to determine the optimum mixing ratio - with the greatest packing density. In this case too, however, by determining the oil absorption value we can tell whether a particular change in the mixing ratio is useful or not.

In primers and stoppers in particular, the PVC must be kept constant even if a pigment or filler is substituted. Provided that the particle shapes are roughly the same, the corresponding quantity of the substitute substance can be calculated approximately from the oil absorption values of the two substances concerned. The following equation is used:

$$m_{new} = m_{old} \cdot \frac{OAV_{old}}{OAV_{new}}$$

where m_{new} = weight of the pigment to be used and m_{old} = weight of the pigment to be replaced.

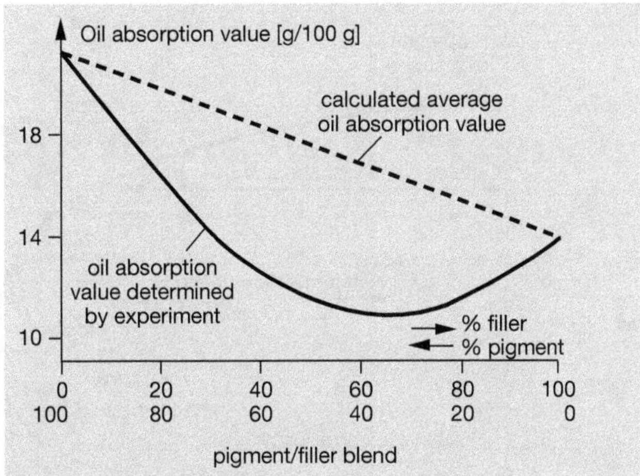

Figure 3.6: Dependence of the oil absorption value of a pigment and filler blend on the composition

Significance of the PVC and Q-values

With the exception of the values at the upper and lower ends of the PVC scale (from 0 to 100 %), merely stating the PVC of a coating is not particularly meaningful. A PVC of 65 % may be below the CPVC, very close to the CPVC or even above the CPVC. Without knowing its position relative to the CPVC, no conclusions can therefore be drawn from the PVC alone in respect of the properties of a coating. This is possible, however, if the Q-value is known (as defined in section 3.2, page 189).

In Figure 3.7 various Q-value ranges are allocated to typical coating materials. High-gloss topcoat paints and also emulsion paints display a low Q-value (below 0.5). Their PVC is usually below 25 %. More highly filled coatings with a Q-value between 0.5 and 0.8 - corresponding to a PVC of between 30 and 40 % - are matt. Intermediate coatings and primers cover a broad range of Q-values, from around 0.4 to 0.8 (PVC between around 20 and 55 %).

Active anti-corrosive primers must be as highly filled as possible in order to have a good barrier effect. At the same time, however, they still need to display a closed binder film. Their Q-values are therefore between around 0.7 and 0.95, which corresponds to a PVC range between 40 and 60 %. Emulsion paints for exterior use are also found in this range.

Fillers and stoppers are applied in relatively high coating thicknesses (from around 40 µm to >> 1 mm). To prevent cracking during the drying process due to volume shrinkage in the binder, the pigment and filler particles must be densely packed. These coating materials therefore have Q-values of between 80 and 100 % or higher, corresponding to PVC values of between 50 and 80 %. Since fillers need to retain a certain elasticity, their Q-values tend to be at the lower end of this range.

Zinc dust primers are typical of coatings with supercritical PVC. With PVC values of up to 80 %, their Q-value is accordingly over 100 %. Emulsion paints for interior use need to display good water vapour permeability. They too are supercritical, with a Q-value above 100 %. Water vapour can diffuse through the air-filled pores between the pigment and filler particles more easily than it can through a closed binder film.

3.4 Solvent-based coating materials

3.4.1 Low solid and medium solid systems

Classification principles

The terms "low solid" (LS) and "medium solid" (MS) are used here to mean the opposite of the term "high solid" (HS). They refer to those solvent-based coating materials that do not satisfy the conditions for high solid coating materials. Medium solid systems, whose solvent content is between that of low solid and high solid systems, have currently already taken over the status of low solid

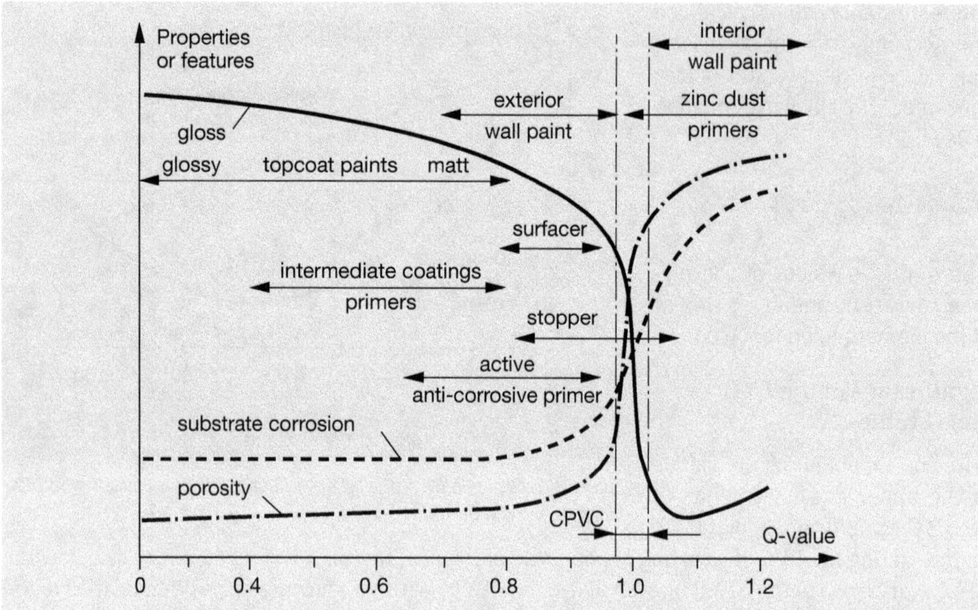

Figure 3.7: Typical coating materials and their decisive Q-value ranges

systems as the most commonly used coating materials in many applications. Although located in many respects between the extremely low solid and high solid systems, the raw materials used in their manufacture and the phenomena that occur within them are more comparable to those of the low solid systems. At the same time, however, medium solid systems containing only slightly more solvent than is permissible for equivalent high solid systems naturally have more in common with the latter.

Virtually all known film former systems as discussed in Chapter 2 are used in the manufacture of solvent-based coating materials. They can be used either alone or in a blend, according to need. In the latter case a distinction is made between the *main film former*, which essentially determines the properties of the coating, and the *additional resins*, with which effects can be produced that cannot be achieved with the main film former alone. Examples include adding epoxy and polyvinyl butyral resins to primers in order to enhance their adhesive strength on metallic substrates.

Film-forming possibilities

As we have already explained elsewhere, coating materials can be divided according to drying method into

- physically drying coating materials,
- room-temperature curing coating materials,
- stoving systems and
- radiation-curing coating materials.

In terms of the film-forming process the following features should be emphasised:

Physically drying systems form a film unaided, by evaporation of the solvents and, in the case of dispersions, by coalescence of the polymer droplets or particles.

Film formers that do not achieve the required technological properties after physical drying have to be modified by chemical reactions - either on their own or with other components of the coating material - in order to meet these requirements. A distinction is made between room-temperature curing coating materials and stoving systems, depending on the temperature at which these curing reactions occur.

Room-temperature curing coating materials are often used in the form of two-component coatings, 2K coatings for short, or multi-component systems, because this avoids the risk of a curing reaction taking place during storage. The components are added together shortly before use. Thereafter only a limited period of time is available during which the coating remains workable. The maximum time within which the coating material must be used after mixing is known as the *pot life*. The substances most commonly used as film formers here are OH group-containing polymers such as OH-functional polyesters, polyethers or acrylic resins crosslinked with isocyanates, or epoxy resins reacted with amines. Some stoving systems can likewise be converted to room-temperature curing coating materials by the addition of a catalyst as the second component, shortly before application. Acid-curing coatings are an example of these. If drying is accelerated by raising the temperature (up to 80 °C), we refer to *forced drying*.

Room-temperature curing one-component systems, 1K coatings for short, are naturally less common. They include alkyd resin coatings, which cure with atmospheric oxygen, as well as moisture-curing polyurethane systems. In principle, however, these are actually two-component systems since the component needed for curing, i.e. oxygen or water, is contained in the ambient air which can diffuse into the coating material either when the container is opened (which is in fact undesirable) or - as intended - only after application.

Stoving systems by contrast are often manufactured as one-component coatings. This prevents the risk of blending errors on the part of the processor, as can occur in multi-component systems. It also removes the need for the complex technology required for two-component systems. Some of the most important film formers used here are OH group-containing polymers crosslinked with phenolic resins, melamine- or urea-formaldehyde resins or blocked isocyanates. The curing reactions can be accelerated using catalysts. To prevent the catalyst from becoming active during storage, it is often added in blocked form. However, two-pack coatings are often stoved in order to achieve better durability or mechanical properties through denser crosslinking, relative to curing at room temperature.

In recent years a number of new crosslinking reactions have been investigated for their suitability in coating materials, but such systems are restricted to isolated applications since generally they offer no great advantage over existing materials.

3.4.2 High solid systems

According to DIN EN ISO 4618,
high solids is a term applied to coating materials in which the content of volatile matter is kept to a minimum, consistent with the maintenance of satisfactory application properties. As a result, their VOC emissions (in the processed state) are often reduced by roughly half or more relative to the corresponding low solid materials.

In order to be able to compare those coating materials containing differing amounts of pigment too, the solids content for high solid systems should be given not as a percentage by weight as is usual for the content of non-volatile matter, but as a percentage by volume. The calculation is based on the densities of the liquid coating and the solvent-free film, according to the following equation:

$$\text{solids [vol.\%]} = \text{nvm } [\eta_o] \cdot \frac{\rho_L}{\rho_{solid}}$$

where ρ_L = density of the liquid coating and ρ_{solid} = density of the solvent-free coating film.

In modern high solid systems the content of non-volatile matter is generally around 60 to 80 %, which corresponds to a volume-related solids content of around 50 to 65 vol.%. These high concentrations of film former cannot be achieved with low solid film formers. Possibilities for increasing the non-volatile content without altering the processing viscosity include:

- reducing the average molecular weight of the film former (generally between 800 and 1500 g/mol)
- narrowing the molecular weight distribution
- using highly effective viscosity-reducing solvents
- using reactive thinners
- high-temperature application.

The relation between the viscosity of a polymer solution, its concentration and the molecular weight of the polymer can be obtained in approximate terms from the following equation (see also section 2.1.1.8):

$$\lg \eta = nvm \cdot K \cdot \sqrt{M}$$ [1)]

where η = viscosity of the polymer solution, M = molecular weight of the polymer, nvm = non-volatile matter and K = constant (dependent on dissolved polymer, solvent, temperature).

To double the content of non-volatile matter whilst keeping the viscosity unchanged, the average molecular weight of the film former must be reduced to around one quarter. Depending on the breadth of the molecular weight distribution, the film former contains a varying degree of higher-molecular matter, which influences viscosity much more strongly than the low-molecular matter and thus raises it sharply even when present in only a small proportion. For that reason film formers for high solid systems usually have a restricted molecular weight distribution, with number averages $\overline{M}_n$ of between 800 and 1500.

The solvents used in high solid systems must demonstrate the greatest possible viscosity-reducing effect. The best mixture is generally found by means of practical experiments, but the viscosity of such high solid solutions can be selectively influenced by using solvents with particular molecular structures and physical properties. Thus, for example, solvents that can react only as H acceptors in hydrogen bridge bonds are more able to suppress the mutual association of film former molecules via OH groups than are solvents that act as both H acceptors and H donors. The latter can form hydrogen bridge bonds with two film former molecules, which seemingly more than doubles the molecular weight of the film former, whereas the former can only ever form a hydrogen bond with one film former molecule.

Viscosity reduction can also be achieved using *reactive thinners*. This term refers to low-molecular, low-viscosity compounds with functional groups via which they can be chemically bonded into the film network. They include low-volatility diols such as 2-ethylhexane-1,3-diol as well as numerous other (even monofunctional) compounds. The most important of these today is still styrene.

In *high-temperature application*, the viscosity is reduced to the level required for the application method by raising the temperature (→ 2.1.1.8).

When selecting components and designing the formulation, a number of special features need to be taken into consideration, as described below.

Film formers

Like low solid coatings, high solid systems too can be used both as one-component and – more frequently in this case – as two-component systems. Common film formers include long oil, oxidatively drying alkyd resins containing around 85 % non-volatile matter, and saturated polyester resins

1) Strictly speaking, the quantities in this equation should be divided by the units. For the sake of clarity, however, this step has been dispensed with here.

crosslinked with highly etherified melamine-formaldehyde resins or with blocked (in the case of one-component coatings) or free isocyanates. They can also be combined with hydroxy-functional acrylic or alkyd resins. Aliphatic or cycloaliphatic polyisocyanates are frequently used as hardeners for exterior applications. Epoxy resins are used as hardeners (predominantly as two-component systems) with amines or other compounds with active H atoms. PVC plastisols are also used for specialist applications, such as underbody protection in cars.

Additives

In comparison with low solid systems, high solid systems are more prone to the development of problems that have to be resolved using additives; examples include problems in

- the wetting of critical substrates
- flow and drip behaviour and
- pigment stabilisation (risk of flocculation and/or influence of pigment content on viscosity).

Compared with low solid coating systems, high solid systems have a higher surface tension, as a consequence of which wetting defects - such as cratering - are more likely to occur on conta-minated and inadequately cleaned metal surfaces. In high solid systems these problems generally cannot be resolved by the use of suitable solvents, since they are subject to many more restrictions than lower solid coatings in regard to choice of solvent. For that reason it is often necessary to use *flow-control agents* or other *surface-active additives*, e.g. anti-cratering additives.

In most cases solvents have a lower surface tension than the binder, which means that the surface tension of high solid systems is already higher than that of low solid systems due to their lower solvent content. In addition, however, the solvents used in high solid coatings generally have a higher surface tension themselves. Since the film formers in high solid systems display a lower equivalent weight than similar low solid film formers - which equates to a higher content of mainly polar functional groups - they too have a higher surface tension.

The lower molecular weight of the high solid film formers leads to a more frequent occurrence of flocculation, the reasons being as follows. Of primary importance for the stabilisation of pigment dispersions is the thickness of the layer of film former adsorbate on the pigment surface, which is naturally lower in the case of a low-molecular film former than it is with a higher-molecular type. This means that a dispersion additive is almost always needed in pigmented high solid systems.

In contrast to low solid coating systems, drip behaviour in high solid coating systems generally cannot be influenced so easily, as can occur by varying the evaporation rate of the solvents or by altering the distance from spray gun to substrate, for example. Instead it is often necessary to adjust the required thixotropic flow behaviour by means of suitable *rheological additives*. Since the viscosity in high solid coatings falls more quickly with rising temperature than it does in low solid coatings, and since the onset of crosslinking delays its subsequent rise, running can occur during stoving even if no other problems arise during application.

Pigments

The same pigments can essentially be used in high solid systems as in low solid and medium solid systems. The influence of pigment content on viscosity is much greater in high solid systems, however, which means that the achievable PVC values are lower.

In pigmented low solid coatings the influence of pigment content on viscosity is relatively low provided that the pigments are not flocculated. However, as the proportion of non-volatile matter increases for a given PVC, the volume of the disperse phase - including the adsorbate layer on the surface of the pigments - likewise increases and then influences the viscosity of the coating material in turn (→ 3.2), particularly if the PVC is relatively high (as in the case of primers, for example).

Figure 3.8 illustrates the dependence of viscosity on non-volatile matter (in vol.%) based on model calculations for various examples, e.g. a clear coating, a coating with a PVC of 20 % and a primer with a PVC of 45 %. The growing influence of pigment content on viscosity as the non-volatile content increases can be clearly seen. For anti-corrosive primers or fillers with a PVC close to the CPVC, a non-volatile content of around 60 % should therefore be regarded as the maximum feasible value at present.

3.5 Aqueous coating materials

3.5.1 Water-soluble and emulsifiable systems (water-borne coatings)

The principles of formulation are essentially the same for water-soluble systems as they are for conventional solvent-containing systems. Nevertheless, there are a number of notable differences in terms of the suitability of the individual coating components.

Film formers

Only a few highly polar organic film formers – such as polyvinyl alcohols, polyacrylamides, polyethylene glycols, some cellulose derivatives as well as acrylates and polyesters with a very high acid value – are soluble in water. Such film formers naturally then remain water-soluble in the dried film as well. For that reason alternative routes had to be found for the development of water-soluble or water-thinnable film formers (→ 2.1.1.9).

In *synthetic resin dispersions* the polymer is not dissolved in water; instead the discontinuous phase forms a dispersion in water. Coating materials based on these film former systems are covered in section 3.5.2.

Another route has led to *water-thinnable* but not strictly water-soluble *film formers*, consisting of relatively short-chain polymers (M_r < 10,000) with acid or basic functional groups capable of salt formation incorporated into the side chains. They are neutralised with suitable bases or acids, which prevents precipitation of the film former as a macroscopic phase (→ 2.1.1.9). Following application, the neutralising agents evaporate and the resulting coating film is no longer water-soluble.

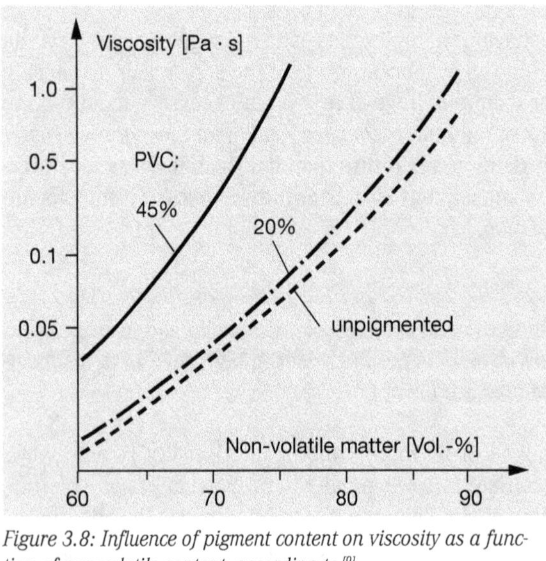

Figure 3.8: Influence of pigment content on viscosity as a function of non-volatile content, according to [9]

The nature and quantity of amines used as basic *neutralising agents* in anionically dissolved film formers have a substantial influence on the viscosity anomaly ("water mountain") that arises when such systems are diluted with water. Whereas the water mountain (→ page 38) is usually extremely pronounced in the case of neutralisation with ammonia in the absence of co-solvents, organic amines such as dimethyl-ethanol amine (DMEA) or triethylamine (TEA) exhibit more of a co-solvent behaviour and thus reduce the water mountain.

The water mountain is much more apparent when neutralisation is incomplete than when it is complete. In the interests of safety, neutralisation is therefore often performed with a small excess of amine. Figure 3.9 shows examples of viscosities adjusted in this way. Even when amines with a good co-solvent characteristic are used, however, acceptable viscosity behaviour and good film-forming capacity can frequently only be achieved with elevated solvent contents of up to 15 %. The non-volatile content of such water-soluble systems is relatively low (usually 20 to 40 %) because of this viscosity anomaly.

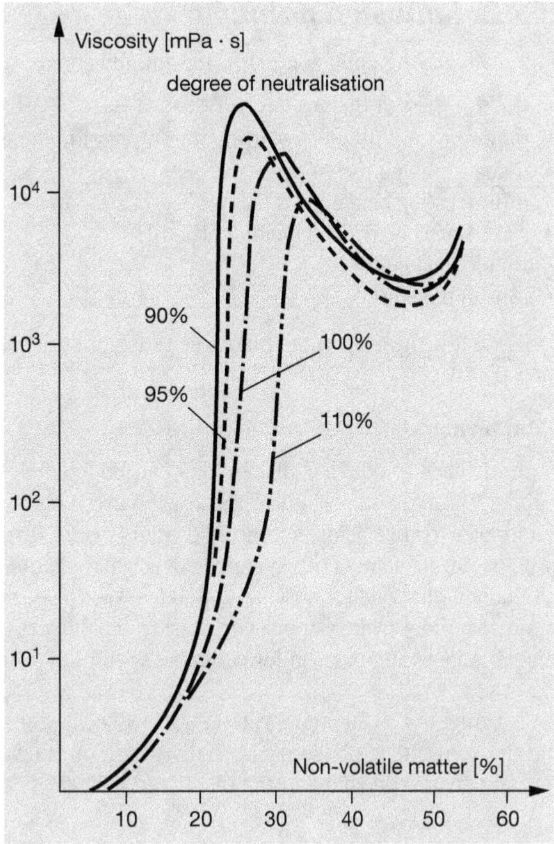

Figure 3.9: Schematic view of the dependence of the viscosity of a polymer solution in water on the degree of neutralisation

In addition to such film formers, *hybrid systems* comprising water-soluble resins and resins dispersed in water are also used. Water-emulsified film formers have recently been developed, which enable a higher non-volatile content to be obtained than is possible with water-soluble film-formers. Prime examples of these include PU and polyacrylate dispersions, which are used in base coats, clear varnishes and plain coloured topcoats, for example.

Water-soluble systems may be entirely physically drying or also chemically drying, either at room temperature or by forced drying. The principal substances used as *hardeners* are melamine resins and suitably modified polyisocyanates. As we have already mentioned, epoxy resins can also be crosslinked with amines. Since the film formers for water-soluble systems generally need to have a relatively hydrophilic structure, coatings made from such coating materials often display a certain tendency towards moisture sensitivity.

Pigments

Apart from a few exceptions - those with limited alkali resistance or sensitivity to hydrolysis, for example - most pigments are suitable for coating materials with water-soluble film formers. However, too high a content of water-soluble components in the pigment can lead to binder coagulation during storage and to blistering if the coating is exposed to moisture or water. Incidentally, the pigment dispersion is already largely stabilised (in contrast to emulsions) by the water-soluble film formers.

Additives

Because they are surface-active substances, water-thinnable film formers and any *dispersing agents* that may be present can increase the tendency towards foam formation, which means that *defoaming agents* may need to be added to the mill base. The surface tension of coating materials containing water-soluble film formers is high itself - firstly because of the high content of polar groups in the film former molecules and secondly because of the high surface tension of the water - often too high even to enable the substrate to be thoroughly wetted. Whilst the surface tension can be sharply reduced by the use of suitable *co-solvents*, it may need to be reduced still further by the addition of *flow-control agents*. *Biocides* are generally not necessary, since the content of organic solvents is usually still sufficiently high to prevent attack by microorganisms.

3.5.2 Emulsion paints

Physically drying emulsion paints essentially consist of

- the film former dispersion
- pigments and fillers (+ wetting and dispersing agents), if required
- defoaming agents
- (often) coalescing agents
- NH_3, NaOH or similar to adjust the pH
- rheological additives
- (often) biocides.

Individually, these components must satisfy a series of important requirements specific to emulsion paints.

Film formers

Several types of polymer dispersions can be used as film formers in physically drying emulsion paints. The most important of these are vinyl acetate copolymers with dibutyl maleinates, vinyl esters of Versatic® acids, acrylic acid esters or terpolymers with ethylene and vinyl chloride, vinyl propionates (usually as copolymers with acrylic acid esters), pure acrylates consisting of polyacrylates and polymethacrylates, acrylate copolymers with styrene and styrene-butadiene copolymers. Their main advantage over systems with dissolved polymers lies in their low viscosity with relatively high non-volatile contents, even for polymers with high average molecular weights.

According to the following relation, the viscosity of a dispersion is dependent on the viscosity of the continuous phase η_{con}, usually water, on the volume of the discontinuous phase V_{dis} and on the geometric shape of the dispersed particle as expressed by a form factor:

$$\ln \eta = \ln \eta_{con} \cdot \frac{K \cdot V_{dis}}{1 - \frac{V_{dis}}{\Phi}}$$ [1]

where η_{con} = viscosity of the continuous phase, K = form factor (2.5 for spherical particles), V_{dis} = volume of the discontinuous phase and Φ = packing factor.

It is not dependent, however, on the molecular weight of the polymer. This means that molecular weights up to and over 10^6 g/mol are possible. The packing factor Φ corresponds to the maximum volume of the discontinuous phase that is reached when the polymer particles are touching one another and the continuous phase fills up the spaces.

Depending on their glass transition temperature T_g and resistance to UV radiation or alkalis, polymer dispersions are used in coating materials for all sorts of substrates, both for mineral substrates and for metals, paper, wood and even plastics, in both interior and exterior applications. In terms of volume, the most important of these are probably *wall paints* for internal and external use, *synthetic resin plasters* and *latex paints*. No particular safety precautions are generally required for their use. Most polymer dispersions dry physically at normal temperatures of above +5 °C.

Fine to medium-fine dispersions with particle diameters of between 0.1 and 2 µm are used for most applications. If a good penetrating power or good surface appearance is required, very fine dispersions with particle sizes of between 0.04 and 0.1 µm are used. Polymer dispersions are stabilised against coagulation by means of appropriate additives.

[1] Strictly speaking, the quantities in this equation should be divided by the units. For the sake of clarity, however, this step has been dispensed with here.

Pigments and fillers

Almost all common pigments (with the exceptions mentioned in section 3.5.1) can be used in combination with polymer dispersions. By far the most important of these is titanium dioxide, which usually produces coatings with good hiding power and - with appropriately surface-treated grades - good weather resistance for exterior applications. Fillers are predominantly used in stoppers, filling compounds and wall paints. They have a considerable influence on the permeability of the coating in respect of water, water vapour, sulphur dioxide and other gases and also affect the coating thickness; some also influence the adhesive strength of the coating on the substrate or its grindability.

Examples of the substances most commonly used as fillers (→ 2.3.9) include natural inorganic minerals, particularly chalks, but also dolomites, kaolin, talc, mica or barytes. Synthetic fillers such as precipitated calcium carbonate, blanc fixe or silicic acids, being more expensive, are used only for specialist applications. The precise composition of the pigment/filler blend (type and particle size of the individual pigments and fillers) has a critical influence on the CPVC and hence on the porosity of the final coating.

Additives

Since polymer dispersions do not stabilise pigments and fillers, *wetting and dispersion additives* such as polyphosphates or low-molecular polyacrylic acids are required in pigmented emulsion paints. As they increase the water sensitivity of the coating, an excess of such additives must be avoided.

As a consequence of their content of polymerisation additives, polymer dispersions have a tendency to foam, which means that the use of *defoaming agents* is usually essential. These should be added straight to the mill base in the presence of wetting agents.

Coalescing agents (→ 3.10.2.2) are frequently also recommended as a means of enhancing film formation. Dispersions are stable only at a particular pH, usually in the slightly basic range, and so bases are often also added to emulsion paints.

The viscosity of most dispersions is too low for application, so the necessary rheological behaviour has to be adjusted by means of rheological additives. In addition to the traditional cellulose derivatives, fully synthetic species and inorganic products are also used in such cases. With their polymer particles finely dispersed in water, dispersions make an ideal nutrient base for microorganisms. They therefore often need to be protected against microbial attack in containers by means of biocides. The coating film itself can also be liable to attack under correspondingly damp conditions, however, and must therefore be treated with a suitable anti-microbial product.

3.6 Radically-curing coating materials

This section covers coating materials that crosslink in a chain reaction triggered by various - mostly radical - initiation mechanisms. Specifically, the chain reaction may be initiated

- by high-energy radiation
- thermally or
- catalytically.

Aside from their initiation process, during which reactive radicals are formed which then trigger the polymerisation reaction, the crosslinking mechanisms of the coating materials discussed here are identical. The great advantage of such materials is that they can be stored almost indefinitely provided that the crosslinking reaction is not initiated, e.g. in the absence of high-energy radiation, but then cure very rapidly after initiation - in fractions of a second or within a few hours at most, depending on the system. Epoxy resin systems capable of cationic polymerisation are characterised by an essentially similar initiation and by an equally rapid curing reaction and so are also covered here.

A number of special features have to be taken into consideration in the composition and processing of radically-curing coatings.

Film formers

The most common substances used as film formers in coating materials that cure either by radiation or by thermal or catalytic means are as follows:

- unsaturated polyester-styrene blends (→ 2.1.4.2),
- oligomeric polyester, polyether, urethane or epoxy resins bearing acrylate functions and with polymerisable double bonds at the ends of the molecules (UV-curable acrylates, → 2.1.4.3),
- epoxy resins (→ 2.1.4.8).

In addition to the film former resin they generally also contain

- an initiator system (with the exception of electron beam curing),
- flow-control agents,
- reactive thinners,
- defoaming agents,
- (often) rheological additives and
- (sometimes) solvents or water

and, if necessary,

- pigments together with the additives required for incorporation and
- flatting agents.

Solvents

A great advantage of all of these systems is that with solvents that are incorporated by polymerisation, known as *reactive thinners* (→ 2.2.1), the VOC content (see page 364) can be reduced almost to 0 %. In film formers based on unsaturated polyester resins (UP resins), styrene is still widely used as a solvent although it is clearly volatile and is also suspected of being carcinogenic. Monomers are used as such thinners in acrylate systems. They may contain just one acrylate double bond, as in the case of isobornyl acrylate, or up to as many as four, as with pentaerythritol tetracrylate. The MAK value for such systems is generally < 20 ppm. As a consequence of the incorporation of reactive thinners into the coating film, they not only initially reduce the viscosity of the coating material but subsequently also strongly influence the physical and chemical properties of the finished coating. In principle, therefore, they are classed as film formers.

Initiator systems

The composition of an initiator system is naturally governed by the means by which the crosslinking reaction is initiated. UV and electron beam-curing coating systems are mostly one-component systems. If polymerisation is initiated thermally or catalytically, two-component systems are used.

Electron beam curing

If polymerisation is triggered by high-energy electron beams, no initiator system is required. The high-energy radiation excites the film former molecules or ionises. The initiation comprises the following individual stages:

$$\text{polymer} + \text{radiation energy} \rightarrow \text{polymer}^* + \text{polymer}^{\bullet +} + e^-$$

$$\text{polymer}^* + \text{polymer}^{\bullet +} + e^- \rightarrow \text{polymer}^*$$

$$\text{polymer}^* \rightarrow 2\ R^\bullet$$

Excited polymer or reactive thinner molecules (polymer*), radical cations (polymer$^{\bullet +}$) and secondary electrons e^- are thus produced. As the radical cations are recombined with electrons, excited film former molecules are formed which then, like the directly excited molecules, decompose into two radicals via a homolytic bond breakage. These then initiate the polymerisation.

UV curing

UV curing systems polymerise radically or cationically. In the first case, under the influence of UV radiation, a photoinitiator system releases reactive radicals which then initiate the radical polymerisation of the film former system. Substances such as bezoin ethers are split into two radicals by UV light, for example, as follows:

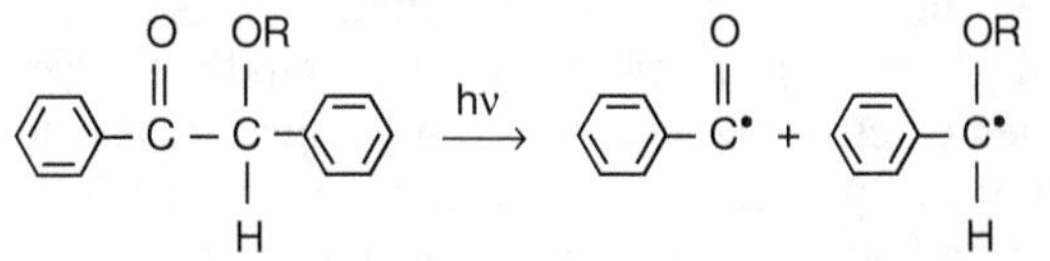

In other systems, such as a combination of benzophenone with tertiary amines, the radicals are formed by intermolecular hydrogen abstraction according to the following scheme:

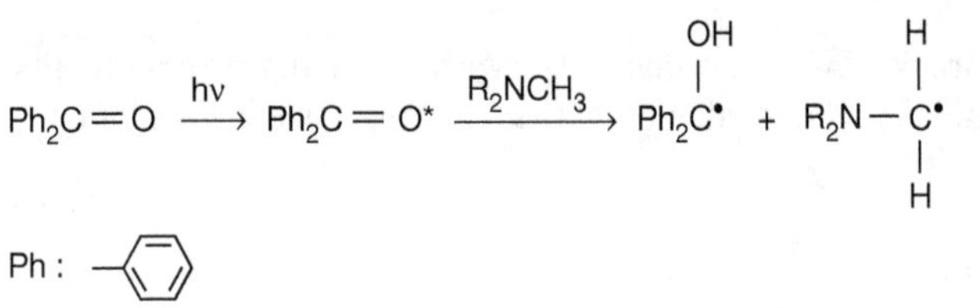

In mercaptans the sulphur-hydrogen bond is very quickly split under irradiation with the aid of bimolecular photoinitiator systems. The resulting thiyl radicals can add not only to relatively reactive double bonds, such as are found in acrylates, but also to more inert double bonds like those in allyl derivatives, which then homopolymerise.

Thermal or catalytic curing

Radicals can also be generated *thermally* or *catalytically* from appropriate *precursors*, generally peroxides. Typical examples of such catalysts (a better term for which is *initiators*), are ketone peroxides such as cyclohexanone or methylethyl ketone peroxide, or acyl peroxides such as benzoyl peroxide, perbenzoates or peroctoates. They are mainly used in the form of 2 % desensitised[1)] solutions and are added to the coating material before it is processed. In other cases they form part of what is known as an *active base*, which is applied as the first layer on top of the substrate. At elevated temperature these peroxides decompose into radicals, which initiate polymerisation of the film former molecules.

Acceleration

In order to initiate an adequate degree of decomposition even at room temperature, substances known as *accelerators* can be added. In this way ketone peroxides are conventionally split by heavy metal salts, generally with 0.02 to 0.05 % cobalt in the form of octoate or naphthenate. Peroxides and accelerators must always be added separately to the coating material. To further increase acceleration, 1 to 5 % of *promoters* such as acetyl acetone or acetoacetic ester can be added; although they do not themselves accelerate the decomposition of the peroxides, they strengthen the action of the cobalt salts. Tertiary amines such as dimethyl aniline, etc., have proven effective for acyl peroxides.

Stabilisation

Occasionally the stabilisation provided by the film former in use is inadequate, for example because pigments and fillers have adsorbed part of the *stabiliser* (generally substituted hydroquinones) in the system. In such cases secondary stabilisation must be performed, e.g. with hydroquinone ether, to prevent premature polymerisation.

1) Desensitisation: reduction in the sensitivity of an explosive substance to mechanical influences.

Inhibition

Since radical reactions are inhibited by oxygen, this must be excluded during curing (see below for exceptions). One method, albeit rather expensive, is to work in an inert gas atmosphere. An alternative is to add *waxes* to the system, which after application of the coating form an oxygen barrier on its surface. However, they impair its appearance and make the application of a subsequent coat more difficult. Another option is to take advantage of the fact that when UV light sources of very high intensity are used, free radicals are formed in such large numbers that they trap the oxygen that has diffused into the coating. The coating then cures before new oxygen molecules are able to diffuse into it. Since the reaction with oxygen proceeds particularly quickly in the case of radicals with nitrogen atoms at the α-C atom, appropriate amines are also added in small quantities to the monomolecular photoinitiator systems in order to reduce the oxygen content in the coating.

Cationic initiators

Epoxy resin systems on the other hand are polymerised cationically. Typical initiators are onium salts of very strong acids, which release cations under irradiation in the following system, for example:

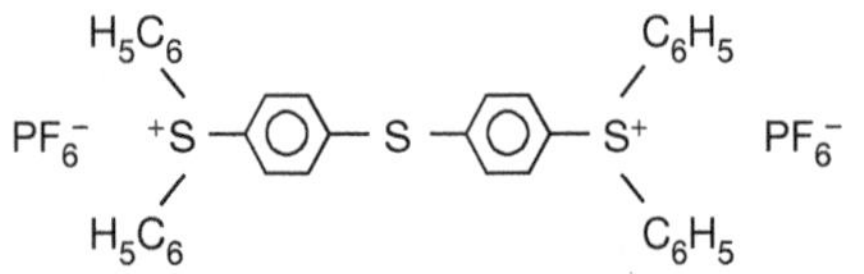

The cations in turn attack the oxiran rings in the film former molecules, initiating cationic polymerisation. The acid anions must be only very slightly nucleophilic, since otherwise the acids would simply be added by means of ring opening. Since polymerisation does not proceed radically, oxygen presents no problem in this case.

Pigments and fillers

A number of special features must be taken into account when using pigments and fillers. Since radiation-curing coating materials are generally free from solvents, the volume of pigments and fillers approximately corresponds to the PVC of the cured film. As a consequence of the high pigment content right from the start, the viscosity tends to be higher than is desirable for good flow in the coating material during the short period following application, particularly as the viscosity rises noticeably above a pigment volume concentration of only around 20 %. It continues to rise in line with the PVC and is extremely high by the time the CPVC is reached. At the same time, due to the low molecular weight of the film former, stabilisation against flocculation is difficult. This applies in principle to all of the systems covered here.

Radiation absorption

Opaque pigmentation is generally problematic in UV curing systems since it requires a higher proportion of light-scattering pigments, which prevent the UV light from penetrating to the lower layers of the coating. An adequate level of crosslinking is therefore unable to take place there, so only coatings of a few μm in thickness can be successfully cured. The same effect occurs with absorption of UV light, i.e. even under the most favourable conditions coating thicknesses of up to roughly 30 μm can be achieved in practice, although attempts are currently being made to substantially increase the curable coating thickness by modifying the initiator system. By contrast, electron beam-curing systems cure completely even when they are pigmented, but the pigmentation has a negative influence on flow, so here too they are of limited use.

If polymerisation is initiated thermally or catalytically using peroxides, only pigments that have neither an accelerating nor a retarding action can be used. Pigments that release radicals from the peroxides by redox reactions, such as metal powders, carbon blacks or some organic pigments, may act as accelerators, reducing the storage life.

Talc, barytes, kaolin and silica flour are the main fillers used.

Use

Radiation-curing systems enable curing to take place very rapidly at room temperature. They are typically used as clear coatings for paper and plastics - for both interior and exterior use - and especially for timber coating. The rapid cure means that the accompanying shrinkage in volume of about 5 to 10 % cannot be offset, however. The resulting stresses between coating film and substrate may then lead to adhesion problems. This volume shrinkage in epoxy systems is of course significantly lower, at only around 3 %, making them suitable for use as primers on metals too.

In addition to their use in timber coating, thermally or catalytically curing systems are also found in stoppers and fillers.

3.7 Powder coatings

Replacing organic solvents with water is one means of reducing the proportion of VOCs in coatings. Another is the reduction in solvent content *per se*, from low solid through medium to high solid systems. In this context, powder coatings - which contain no solvent whatsoever - represent the final destination along the route of VOC reduction. This is probably the fastest growing market in the coatings sector.

The composition of powder coatings is generally relatively simple. In addition to the

- film former system

they usually only contain
- flow control agents,
- degassing agents and
- optionally pigments (and rarely fillers).

These components are pre-mixed, homogenised in an *extruder* and then ground. This means that every particle of powder coating contains all components in the same proportions. Another method of producing powder coatings is the *dry blend process*, in which the components are mixed in a high-speed mixer (without being melted), although the homogeneity of products manufactured in this way is often inadequate. Both *thermoplastic* and *thermosetting* (after subsequent curing) systems are used as film formers, the latter group being by far the more important at present. Thermoplastics are principally used in fluidised bed coatings, whereas thermoset products predominate in coating powders applied by electrostatic processes. Powder slurries, which can be used with conventional spray technology in automotive painting, for example, are a recent development.

The selection of components for a powder coating is governed by a number of key criteria.

3.7.1 Film formers

Table 3.2 (page 208) provides an overview of the most important film former systems currently used for powder coatings. In chemically curing film formers, a fine balance must be found between glass transition temperature T_g, average molecular weight $\overline{M}_n$, average functionality $\overline{f}_n$ and reactivity. Specifically, the pre-mixed powder coating must be able to be melted in the extruder without noticeably crosslinking. Furthermore, the powder coating material must not sinter too much during storage, but then during stoving it must melt in such a way that it flows into the desired film before the expected film properties are set by crosslinking. The resins are therefore usually amorphous polymers with a sufficiently high T_g (at least 40 to 50 °C) to ensure that sintering is prevented during storage, and with a $\overline{M}_n$ of several thousand g/mol.

Table 3.2: Common film former systems for powder coating

Conventional abbreviation	Film former		Use
Thermoplastic systems			
PE	polyethylene (LDPE, LLDPE, HDPE)		interior
PA	polyamide 11 or 12		interior
SP	polyester		interior
EVOH	ethylene-vinyl alcohol copolymer		interior and exterior
PVC	polyvinyl chloride (and copolymer)		interior and exterior
PVDF	polyvinylidene fluoride (and copolymers)		interior and exterior
Thermoset systems			
	Resin	**Hardener**	
EP	epoxy resin	phenolic hardeners imidazoline derivatives anhydride adducts	interior
EP-DCD	epoxy resin	modified dicyandiamide	interior
EP-SP (hybrid system)	COOH polyester resin	epoxy resin	interior
SP-TGIC	COOH polyester resin	TGIC , glycidyl ester arom. carboxylic acids	exterior
SP-HAA	COOH polyester resin	hydroxyalkyl amide hardener	interior and exterior
SP-PU	OH polyester resin	isocyanate adduct	interior and exterior
AC-PU	OH acrylate resin	isocyanate adduct	interior and exterior
AC-DDA	glycidyl acrylate resin	dodecanedicarboxylic acid	interior and exterior

Thermoset powder coating types

Epoxy resins

The most important powder coating systems in terms of quantity are those based on bisphenol-A type *epoxy resins.* Resins having n values from 3 to 5 (→ 2.1.4.8) are most commonly used, although the trend is towards lower-molecular types - down to n = 2.5 - which allow better flow in thin films. For greater film thicknesses, n values of up to 8 are preferred, however. The most important hardener is accelerated or modified dicyandiamide (DCD)[1]. The modification improves the solubility of DCD in the epoxy resin and hence also enhances the film quality. Phenolic hardeners lead to greater chemical resistance and improved corrosion protection, whereas powder coatings hardened with carboxylic acid anhydrides have greater resistance to yellowing, acids and solvents. Imidazole derivatives are only used as hardeners for matt coatings.

Hybrid systems

COOH-functional polyester resins with a $\overline{M}_n$ of some thousand g/mol are used in *SP-EP* systems *(hybrid systems).* Since their $\overline{M}_n$ is greater than that of the epoxy resins used, the SP-EP system is termed the parent resin here. These powder coatings display improved colour retention and UV resistance in comparison to the other EP systems, but like them have only poor weather resistance, as a consequence of which all of these products are used primarily for interior applications.

1) also DICY

TGIC systems

SP-TGIC systems - usually called *TGIC systems* - used to be the leading products for outdoor applications. TGIC (triglycidyl isocyanurate) is suspected of being mutagenic and TGIC-containing preparations must be labelled as toxic. In central Europe and Scandinavia, the use of TGIC coating powders has declined sharply. A successor product contains a mixture of aromatic polycarboxylic acid glycidyl esters instead of TGIC.

Primid systems

In central Europe nowadays, TGIC powder coatings are frequently being abandoned in favour of powder coatings based on acid polyesters and hydroxyalkyl amides as curing agents (*SP-HAA systems, Primid coatings*). Since water is released during curing, the coating thickness is restricted to at most 120 µm. On the other hand, there is the risk of pinhole formation. However, since there are efforts underway to generally decrease the thickness of powder coatings to well below 100 µm, this restriction rarely poses a problem in practice.

Polyurethane systems

Other film former systems for weather-resistant powder coatings include the *SP-PU* and *AC-PU* systems - together known as *polyurethane powder coatings* - consisting of OH-functional polyesters or acrylate resins which can be crosslinked with isocyanate adducts. Blocked derivatives of isophorone diisocyanate are most commonly used here. One disadvantage of these blocked systems is that the blocking reagent is separated off during stoving. The quantity of VOCs in powder coatings blocked with e-caprolactam can therefore easily be 2 to 4 %. However, isocyanate prepolymers dimerised to uret diones are available that naturally cease to release separation products during stoving. In order to lower the stoving temperature, these systems are catalysed with dibutyl tin dilaurate (DBTDL) or tertiary amines. Since species formulated with OH-functional polyesters flow better than those formulated with COOH-functional variants, SP-PU systems generally display better flow than the other powder coatings. Finally, AC-PU powder coatings are characterised by their outstanding weather resistance.

Acrylate systems

AC-DDA powder coatings are predominantly used in Japan. They consist of an epoxy-functional acrylate resin crosslinked with dodecanedicarboxylic acid. A disadvantage of powder coatings based on polyacrylates is their pronounced incompatibility with other powder coating systems, and this necessitates strict separation of these powder coatings from the others, during both production and processing.

More recent developments

There are increasing attempts being made to lower the stoving temperatures of powder coatings not only to save energy but also to be able to powder-coat heat-sensitive substrates as well. Since this means that the glass transition temperature of the powder coating resins has to be reduced as well, the temperature of powder coatings that have relatively low stoving temperatures often needs to be controlled during transport and storage. Already, timber derived materials, such as medium density fibreboard (MDF) can be powder coated.

One way to cure powder coatings without damaging the substrate is to use a combination of near-infrared (NIR) radiation for stoving and special pigments that strongly absorb this radiation. The outcome is curing times of substantially less than one minute.

To obtain good levelling despite low stoving temperatures, attempts are being made to melt the coating powder (NIR) and to use UV-curable systems to uncouple the start of the crosslinking reaction. Some powder coatings of this kind are already being used for MDF boards.

Thermoplastic powder coatings

The very first powder coatings to be used were *thermoplastic*. In order for these to be able to produce sufficiently resistant films, high-molecular film formers must be used. However, this means that their melts are relatively highly viscous even with high stoving temperatures, and thus products of this type display poor flow or require elevated temperatures (300 to 450 °C) to flow. Themoplastic powder coatings are now designed primarily for interior use. Those formulated with *polyamide 11* and *polyamide 12* produce films with outstanding abrasion and weather resistance. Their good adhesion also makes them suitable for use as primers. *Polyethylene* coatings (PE coatings), by contrast, display only poor adhesion without a primer and are not weather resistant unless they have been stabilised. In copolymers with *vinyl alcohol* (EVOH systems), adhesion is significantly better. These powder types are suitable for flexible anti-corrosive coatings with good weather resistance. *Polyvinyl chloride* (PVC) intended for use in coatings is treated with plasticisers. It also requires stabilisation, for which organotin compounds or metal soaps, e.g. of calcium or zinc, are generally used. Given their extremely high price, *fluoropolymer powders* are only used in applications requiring exceptionally high weather resistance or resistance to aggressive media, e.g. in chemical plants.

3.7.2 Additives

In the molten state powder coatings display a relatively high surface tension due to the absence of solvents. On inadequately cleaned or pretreated metal surfaces this may lead to defective wetting, edge effects and poor adhesion. For that reason powder coatings often contain 1 to 2 % of *flow-control agents*, usually polyacrylates, which at the same time reduce the tendency towards cratering, although if added in too high a quantity they may lead to orange peel effect. Many powder coatings also contain 0.1 to 1 % benzoin, which acts as *degassing agent* and likewise counters cratering.

3.7.3 Pigments

The following requirements apply to pigments used in powder coatings :

- thermal stability at stoving temperature,
- minimum possible rise in melt viscosity,
- no reaction with other powder coating components and
- stability in respect of shear forces arising during extrusion and grinding.

These criteria can best be met by inorganic pigments. In the case of organic pigments, relatively high grades often have to be used. Lustre pigments by nature cannot be incorporated in an extruder because their platelets would become buckled or broken. Thus, they are either physically incorporated only or, in order to avoid signs of separation, are fixed to the powder coating particles in a bonding process.

Since powder coatings contain no volatile components, their PVC corresponds to the volume of pigments in the powder. Above a volume content of 20 %, the viscosity of pigment dispersions - here in the solvent-free molten binder - rises sharply. Matt powder coatings therefore cannot be obtained by means of a PVC close to the CPVC, as is possible with liquid coating systems, because the high pigment content would lead to poor flow and poor mechanical properties due to the high viscosity of the melt. Appropriate additives, e.g. based on silica or wax must then be used if flatting effects are required.

3.8 Inorganic coating materials

3.8.1 Water glass paints (silicate paints)

Water glass coatings generally consist of

- water glass as the film former, also known as *fixative* in this context, and
- water glass-resistant pigments and fillers, which support *silicification*.

They can be formulated as both one-component and two-component systems.

When formulating water glass paints, the following aspects apply in respect of their components.

Film formers

The film formers in water glass paints largely consist of potassium or sodium water glass and, more rarely, lithium water glass. These film formers dry both physically, by evaporation of water, and chemically, whereby several reactions run simultaneously. During silicification the alkaline silicates react with atmospheric carbon dioxide, forming the corresponding carbonates and silica gel, approximately according to the following reaction scheme.

$$m\,K_2O \cdot n\,SiO_2 \cdot x\,H_2O + m\,CO_2 \rightarrow n\,SiO_2 \cdot x\,H_2O\uparrow + m\,K_2CO_3$$

If such paints are applied to mineral substrates, they react with the calcium hydroxide contained within the substrate, forming calcium metasilicates, whereupon an insoluble bond between coating and substrate is created. A distinction is made between the following formulation types:

• **One-component silicate emulsion paints.** They may also contain *hydrophobing additives* and up to 5 % of *organic film formers*, normally the same as those used in emulsion paints. The latter stabilise the system and improve water resistance. Depending on the water absorption coefficient and storage temperature, they can have a shelf life of over a year.

• **Two-component systems.** In these products the fixative on the one hand is thoroughly mixed in a ratio of 1:1 with fillers, pigments and any additives on the other hand and homogenised for a number of hours. The quality of the product depends on the accuracy of measurement, the mixing time, the degree of homogenisation and the temperature.

Pigments

Pigments used in water glass paints must be resistant to alkalis because of the strong basicity of the film former. As the film formers set, the inorganic pigments are chemically incorporated into the binder matrix through the formation of sparingly soluble silicates on their surface.

Use

Water glass coatings are very hard and are permeable to water vapour and carbon dioxide. They do not tend to detach and, unlike mortars or plasters, they do not prevent carbon dioxide exchange between atmosphere and substrates. Since they are chemically related to mineral substrates, water glass coatings have similar coefficients of thermal expansion. This means that no stresses occur between substrate and coating even under widely fluctuating temperatures. Since the binder is inorganic, self-cleaning coatings can be made with photocatalytic titanium dioxide nanoparticles.

3.8.2 Alkyl silicate paints

The following formulation principles are the starting point for this class of coating materials:

Film formers

The use of silicon esters as film formers is based on the hydrolytic splittability of the Si-O bond. Selective hydrolysis ultimately leads to silicon dioxide via variously condensed silicic acids:

$$Si(OR)_4 + 4\,H_2O \longrightarrow Si(OH)_4 + 4\,ROH \longrightarrow SiO_2 + 4\,ROH + 2\,H_2O \qquad R = \text{Alkyl, meist } C_2H_5$$

The reaction rates of the two reactions running in parallel can be influenced independently of each another.

The manufacturing process involves firstly mixing the esters with a defined quantity of water and solvents, generally alcohols, organic esters or ketones. 2 moles of water for 1 mole of silicon ester are required for full hydrolysis. The addition of acid, usually hydrochloric acid, sets a low pH, around 2 for ethyl ester, which means that a mixture of this type, with a degree of hydrolysis of 80 to 90 % and around 20 % SiO_2 content, is stable for about 1 year. If a further 10 % of water is added to a mixture containing around 80 % water, the shelf life is reduced to below 25 % of the original value. Reducing the proportion of water produces a more inert paint that takes too long to dry.

Pigments

Alkyl silicate paints can in principle be pigmented with all of the pigments and fillers that are conventionally used in other aqueous systems. Pigments such as titanium dioxide or chromium oxide green, and fillers such as talc, mica or kaolin, may also improve the processability and properties of the paint. *Zinc dust paints* have to be formulated as two-component systems, however, since the finely divided zinc reacts with hydrochloric acid to form zinc chloride, thus raising the pH. This then initiates condensation and the paint solidifies within a few hours.

Additives

The addition of *antisettling agents* such as bentonites or highly disperse silicic acids to the film former may stabilise the pigment dispersion.

Use

After application the paint dries quickly by evaporation of the alcohols and solvent. The silanol groups formed by reaction with atmospheric humidity condense or react with the zinc or the substrate to form a scarcely soluble SiO_2 network. Coatings of this type are extremely thermally stable, highly chemically resistant and are also resistant to weathering, ageing and UV radiation. They are used predominantly in high-temperature applications and corrosion protection.

3.9 Formulating the mill base

3.9.1 General introduction

Pigments and fillers are supplied in more or less *agglomerate form* (→ 2.3.2). In the finished coating material these agglomerates must have disappeared to as great an extent as possible. Therefore the pigment and filler agglomerates must be broken down during the manufacture of the coating. This is generally done in a separate stage of coating production, known as *dispersion*. The success of dispersion in a given apparatus is very much dependent on the composition of the blend - of pigments, fillers, binders and solvent - in which the pigments and fillers are to be dispersed, i.e. the *mill base*. Since the dispersion processes are often the most expensive processes involved in coating manufacture, the quantity of pigments that can be dispersed in such an apparatus per unit of time must be optimised. The prerequisites for rapid dispersion are

- high shear forces and,
- to ensure rapid pigment wetting, a low viscosity in the (pigment paste) binder solution and hence a high pigmenting capacity.

Mill base viscosity

The mill base viscosity must be adjusted to the individual dispersion unit. The pigment content of the mill base can be maximised by starting from a binder solution with the lowest possible viscosity. Whilst a solvent used on its own would most easily enable low viscosity, rapid pigment wetting and hence a high pigment content to be achieved, it would not be able to stabilise the dispersion against flocculation. For that reason the mill base always contains a small quantity of film former and/or wetting and dispersion additives in addition to pigments and solvent. Since the viscosity of the mill base remains necessarily low even with the maximum pigment content, the quantity of film former should be restricted to the minimum required. Finally, the risk of *shock effects* (→ 4.7) during manufacture of the coating limits both the maximum and minimum pigment and film former concentrations.

Table 3.3: Orientation values for mill base composition

Pigments and fillers	Dispersing unit	Proportion in mill base in %
inorganic (relatively readily dispersible)	high-speed mixer	≥ 60
	bead mill (also sand or ball mill)	45 to 60
inorganic (relatively scarcely dispersible) and uncritical organic	agitator/triple roll mill	10 to 30
	bead mill (also sand or ball mill)	5 to 25
finely divided organic and carbon blacks	agitator/triple roll mill	≤ 5
	bead mill (also sand or ball mill)	≤ 4

Mill base formulation

The mill base can be formulated in various different ways.
For *organic pigments and carbon blacks*, orientation values as set out in Table 3.3 can be used as a guideline. Since the pigments and fillers in the individual groups differ from one another both in their specific surface area and in the chemical nature of their surfaces and can exhibit very diverse interactions with the binder, the orientation values in the table are set within very broad limits. Starting from these values a series of dispersion experiments using various mill base compositions needs to be performed in order to find the best possible formulation from the above perspectives.

For *inorganic pigments*, the orientation data in Table 3.3 still apply, and the optimum mill base composition can also be obtained from experimentally determined starting points. In this determination, also known as *Daniel titration*, the *flow value* for the proposed pigment blend with film former solutions of varying concentrations can also be established.

The flow value is defined as the volume of a film former solution that is needed to transform 20 g of pigment in a beaker by homogenisation with a glass rod into a rheological state whereby the mixture runs off the glass rod in just cohesive threads. The flow value depends on the particle size distribution of the pigment, its surface properties and any surface treatment, as well as on the properties and concentration of the film former. The properties of the solvent naturally also play a part.

Optimum mill base composition

If the flow values for various film former concentrations are plotted in a coordinate system, with the concentrations on the abscissa and the flow values on the ordinate, the curve obtained normally shows a clear minimum. This is where the mixture contains the greatest proportion of pigment and from the above perspectives represents the *optimum mill base composition*. An example is shown in Figure 3.10.

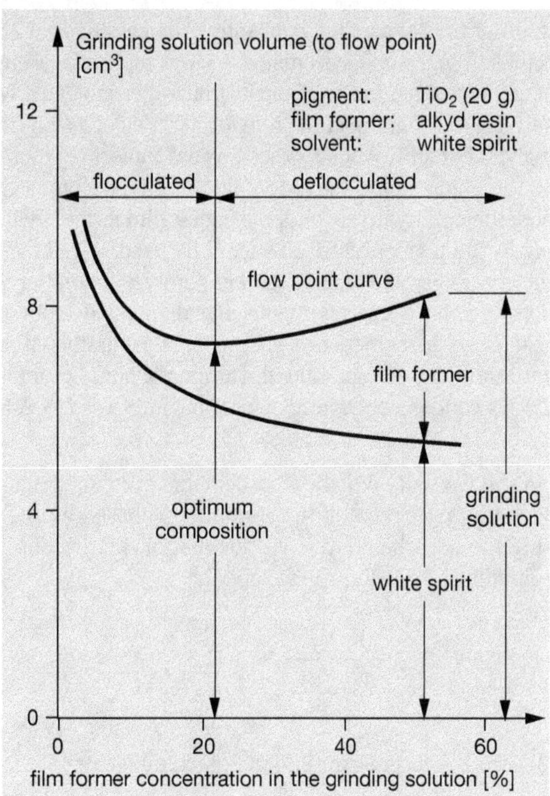

Figure 3.10: Dependence of the flow value of a TiO_2 pigment on the concentration of an alkyd resin solution in white spirit, according to [9]

In dispersion units that require a more highly viscous mill base, the starting point is not the flow point or flow value but instead the *smear point*, which is determined in a similar way. Generally, however, neither the determination of the flow point nor that of the smear point is particularly accurate. Since the mill base blend determined in this way contains only the minimum quantity of film former necessary, even a slight reduction in film former can lead to flocculation. Such highly pigmented systems often also display flow anomalies, and there is a risk that signs of shock may occur during manufacture of the coating. It is therefore recommended that dispersion be performed with a larger proportion of film former; the excess should be around 30 %.

Controlling the mill base

More modern methods of mill base formulation are oriented instead towards the result of the dispersion. A number of different evaluation constants can be drawn upon in this respect, such as granularity, colour intensity, gloss or hiding power; not all of these are suitable for every pigment.

In the case of inorganic pigments *grind gauge measurement* (→ page 369) is a quick method; for organic pigments, although frequently used, it is not particularly satisfactory, however. A more suitable criterion for these pigments is colour intensity. The *colour intensity evolution* of the mill base with varying PVC is tracked over time. When this is plotted as a three-dimensional coordinate system a concave area is produced, in which the "valley" at a given dispersion time indicates the optimum mill base composition. Since the dispersion time needs to be kept as short as possible to reduce costs, this method can also be used to establish the optimum mill base composition for a given grind gauge value. With the introduction of computers into the laboratory, optimisation by means of statistical experimental planning is also becoming increasingly important, however.

3.9.2 High solid systems

In order to achieve high concentrations combined with a workable viscosity, the film former molecules in high solid systems have lower molecular weights than in low solid systems. This reduces at any rate the number of those groups per molecule which are capable of bonding to the pigment surface; the bonding of the film former molecules on the pigment surface becomes weaker. Accordingly, the quantity of adsorbed solvent molecules increases. Thinner adsorption layers are formed on the pigment and the risk of flocculation becomes greater. To obtain stable pigment dispersions nonetheless, relatively high-molecular film formers with a large number of groups capable of bonding to the pigment surface should be used. A relatively low-molecular film former which affords the possibility of a high content of non-volatile matter is then used for letting down. It is inevitable, however, that some high-molecular dispersion resin remains in the solution in equilibrium, and that its viscosity increases sharply. It is generally not possible to obtain stable pigment dispersions with solutions of high solid film formers alone. The mill base for high solid systems therefore almost always contains polymeric dispersion additives (→ 2.4.6).

In the viscosity equation for disperse systems (page 197), if we use the viscosity of the binder solution η_B for that of the continuous phase and the volume of the dispersed pigment, including the adsorbate layer, V_{pig} for that of the discontinuous phase, then for the viscosity of the pigment dispersion η_{pig} we obtain:

$$\ln \eta_{pig} = \ln \eta_B + \frac{K \cdot V_{pig}}{1 - \frac{V_{pig}}{\Phi}}$$

where K = form factor and Φ = packing factor.

The observations made in respect of the cited viscosity equation can also be applied to the above.

In low solid systems the volume of pigments including the adsorbate layer is fairly low relative to the total volume. Differences in the thickness of the adsorbate layer therefore have little effect on the rheological behaviour of the coating material. In highly-pigmented high solid systems, however, this volume makes up quite a considerable proportion of the whole, and even small changes can have a significant influence on the viscosity.

3.9.3 Aqueous systems

A distinction must be made between film former solutions in water and in dispersions for pigment dispersion in *aqueous systems* too. Two properties of water then play an important part in both systems, however. Because of the high surface tension of water, some pigments are only poorly wetted. Sometimes the interactions between the water molecules and the pigment surface are so great, however, that a very strong affinity must already exist between the latter and the anchor groups of a dispersing agent to enable these groups to compete successfully with water for binding sites on the pigment surface. In any event, given the relatively high surface tension of the aqueous systems, thorough wetting of the pigment agglomerates occurs only slowly.

Behaviour of inorganic pigments

Most *inorganic pigments* and fillers, too, are wetted by water very easily because of their polar surface. However, the adsorbed water layer can in no way stabilise a pigment dispersion against flocculation. Unlike dispersions in organic media, aqueous pigment dispersions are usually stabilised electrostatically. Their stability depends on the surface charge of the pigment (*zeta potential*), which in turn depends on the pH. For any combination of water, dispersing agent and pigment there is a pH at which the surface charge is zero, known as the *isoelectric point* (IEP). At lower pH values the surface charge is positive due to the adsorption of protons or separation of hydroxide ions, at higher pH values it is negative. The IEP is pigment-dependent and can cover a broad range, e.g. pH 4.8 for kaolin or pH 9 for calcite. It can be shifted quite considerably for a given pigment by means of surface treatment. The range of titanium dioxide pigments includes grades designed especially for aqueous systems, for example.

Most emulsion paints contain more than one pigment or filler. In combinations of this type care must be taken to ensure that at the adjusted pH value the surface charge is the same for all of the pigment types - either positive for all or negative for all. In any event, dispersing agents level out the differences in IEP between the individual pigments. To compound matters, the polymer dispersions too are only stable in certain pH ranges.

Stabilisation possibilities

For *stabilisation* of the pigment dispersion, anionic substances such as salts of acrylate copolymers with acrylic acid as comonomer or also potassium polyphosphates (usually potassium tripolyphosphate) are conventionally used. These form a strong adsorptive bond on the pigment surface via their ionic groups. Their compatibility with the aqueous phase is due to their hydroxyl groups. By virtue of the non-ionic sequences between the ionic groups, electrostatic stabilisation is accompanied by steric stabilisation too. Stabilised *pigment slurries* are frequently also used now in the manufacture of emulsion paints. They save pigment manufacturers the expense of drying the pigments and paint manufacturers the costs of dispersion.

Behaviour of organic pigments

With the exception of a few surface-treated types, organic pigments have a significantly lower surface tension than water. The surface tension of the latter must be reduced sufficiently by means of surface-active substances - in the form of *wetting and dispersing agents* - to allow wetting to take place. Depending on whether anionic or long-chain non-ionic molecules are used, either electrostatic or entropic repulsion will dominate during stabilisation.

Other aspects of formulation

Depending on the surface activity of the dispersing agents, the mill base has a strong tendency to foam in aqueous systems. It is therefore recommended that part of the *defoaming agent* is added straight to the mill base. Some *rheological additives* also have to be added as early on as at the dispersion stage.

Even today, formulation of the mill base is still largely conducted on an empirical basis. The pigments and fillers are measured out with a part of the additives. Then water is added until the viscosity has reached the optimum value for the dispersion unit.

3.10 Film-forming

3.10.1 General introduction

Most coating materials are liquids, whose viscosity at high shear rates - depending on the application method - lies in the range from 0.05 to 1 Pa·s. Following application, the coating materials should be transformed into a dry, solid film. Powder coatings represent an exception to this rule. They are solid at the time of application and their particles are melted, run together and after cooling form a solid film.

Since the film formers and the resultant films are usually amorphous, the term "solid" is not to be understood as meaning "crystallised", but merely only as "solidified" (and possibly crosslinked). Amorphous substances, however, exhibit is no clear boundary between the solid and liquid states, unlike the case for crystalline substances. For our purposes, a substance is usually said to be solid if it does not flow visibly under defined shear stress. Thus, the ***degree of film dryness*** can be expressed indirectly via its dynamic viscosity: it is "dust dry" (or "sand dry") at a viscosity of around 10^3 Pa·s; "block-resistant" at around 10^7 Pa·s, and "through-dry" at around 10^{12} Pa·s. Drying can be measured or monitored by various methods (section 8.4.3).

The viscosity η of an amorphous substance depends on the distance of the temperature T from the glass transition temperature T_g according to the following equation of Williams, Landel and Ferry (WLF equation):

$$\ln \eta = \ln \eta_{Tg} - \frac{A\,(T - T_g)}{B + (T - T_g)}$$

where A and B are semi-empirical constants relating to the frozen free volume (see Figure 2.19) and the coefficient of thermal expansion. η_{T_g} is the viscosity at T_g. The equation is sufficiently accurate for amorphous polymers in the range T_g to T_g + 100 K.

For a given drying temperature, the above equation can be used to provide a rough estimate of how high T_g needs to be in order for a film material to display the required viscosity or required degree of dryiness. If we use the "universal constants" (A = 40.2 and B = 51.6 K) in this equation, then for a material with an η_{T_g} of 10^{12} Pa·s, we can calculate that the drying temperature may be at most 55 °C above the T_g in order for the film to reach the "dust dry" degree of dryness. In order for drying to occur at room temperature (20 °C), the value of T_g would have to be above -35 °C. For "block-resistance", the minimum value for T_g is -1 °C.

3.10.2 Physical drying

Coating films are normally produced by application of a binder solution or dispersion, optionally containing dispersed pigments and fillers, and subsequent evaporation of the solvent or dispersant. Chemically curing films additionally undergo crosslinking by means of chemical reactions. Let us first consider the processes involved in purely physical drying.

3.10.2.1 Drying of dissolved binders

Influence of viscosity

The viscosity of a coating material for spray application is around 10^{-1} Pa·s. Even during the flight of the coating droplets from the spray gun to the substrate, a substantial part of the solvent evaporates due to the large surface area of the droplets. This means that the viscosity of the coating material depositing on the substrate is already significantly higher than 10^{-1} Pa·s. After deposition, the "mountain of droplets" (see Figure 6.20) levels out to form a coherent film, from which the solvent continues to evaporate, causing the viscosity to rise also. Initially the evaporation rate of the solvent is largely dependent on its vapour pressure and hence also on the interactions between solvent and binder molecules (→ 2.2.3.1), as well as on the surface-to-volume ratio of the film and the solvent concentration in the air immediately above the film, i.e. ultimately on the circulation of air. The influence of the dissolved polymer, however, is still slight. The following deliberations are based on Fick's 1st law of diffusion:

$$\dot{n} = DA\frac{\Delta c}{\Delta x}$$

$\dot{n}$: mole flux
D: Diffusion coefficient
A: area plane across which diffusion takes place
$\Delta c/\Delta x$: molar concentration gradient

The mass flow is given by $\dot{m} = \dot{n} \cdot M$; M: molar mass of the diffusing substance

Influence of the glass transition temperature

As the solvent content falls, the T_g of the binder-solvent mixture rises, as does the viscosity of the film in accordance with the WLF equation above. If the drying temperature is (still) high enough above the T_g, then the rate at which the solvent diffuses out of the film is greater than the rate at which it evaporates, and so the diffusion rate has hardly any influence on the drying speed. As drying continues, the viscosity or T_g of the film then becomes so high, and hence the diffusion coefficient (D) so low, that the solvent can no longer diffuse to the surface of the film as quickly as it evaporates from there. The ongoing drying rate is then increasingly controlled by diffusion. Since the diffusion rate is proportional to the concentration gradient between the surface - where the solvent concentration is very low - and the more solvent-rich interior of the film, the drying rate decreases as the solvent concentration in the film falls. In short, the dryer the film already is, the more slowly it continues to dry. Fick's law can also be used to explain the empirical observation that the rate of physical drying where drying is controlled by diffusion decreases with the square of the film thickness.

The transition from an evaporation rate controlled by the vapour pressure to one controlled by diffusion frequently occurs in the range from 40 to 60 %, but naturally depends on the system. As the solvent content in the film decreases, the T_g rises further and hence, at the drying temperature T, the difference ($T-T_g$) becomes smaller, which means a further increase in viscosity according to the WLF equation above. If the T_g of the remaining binder-solvent mixture reaches and exceeds the drying temperature, the evaporation rate drops sharply. This means that a film composed of a binder whose Tg is much higher than the drying temperature (e.g. PMMA with a T_g of 105 °C) still contains a small percentage of solvent, even when it appears hard and dry. This is known as solvent retention. At worst, this small percentage will take several years to completely evaporate). If the solvent needs to be completely eliminated from the coating film within a reasonable time period, the film has to be heated to a temperature above the T_g of the pure binder.

Special features of high solid systems

In general, the amount of solvent that evaporates on the way from the spraying nozzle to the substrate during the spray application of high solid systems is much less than in low solid and medium-solid systems. This means that the viscosity of the newly applied coating has also risen less rapidly, and increases slowly during the initial drying phase on the substrate, too. Although a lesser wet film thickness is required in high solid systems to produce the same dry film thickness, prevention of sagging is consequently more difficult than it is in low solid systems. One of the reasons for the lower solvent loss is the much lower molar solvent concentration and the resultant lower saturation vapour pressure. For example, in a low solid clear coat with a non-volatiles content of 30 %, an average molecular weight of 20,000 g/mol for the binder and of 100 g/mol for the solvent blend, the molar fraction of the solvent blend is 0.998. In a high solid clear coat with a non-volatiles content of 70 %, an average molecular weight of 800 g/mol for the binder and of 100 g/mol for the solvent blend, it is just 0.774. This difference alone cannot explain the differences in solvent loss during spraying, however. A further reason could be that many high solid systems exhibit such high binder concentrations that the release of solvent is controlled by diffusion even during spraying.

In contrast to the case for most low solid and medium solid coatings, solvent flash-off of their high solid counterparts can hardly be described as a physical drying process. Whereas in low solid and medium-solid systems the release of solvent in itself is enough to produce a more or less dry (uncrosslinked) film, in high solid systems the T_g is so far below the ambient temperature throughout flash-off that film-forming in the conventional sense barely occurs at all. The applied high solid coating thus remains "open" for longer and the film-forming process, which in this case consists almost solely in crosslinking, is delayed both in respect of time and, in the case of stoving systems, of temperature.

Special features of water-dissolved binders

The evaporation rate of water in coating systems based on aqueous binders depends heavily on the relative humidity. At 70 % relative humidity, it is only very slow while at 100 %, the water does not evaporate at all. Reducing the relative humidity by either heating the air or condensing the moisture is expensive. Since it is cheaper to wet air rather than to dry it, the coating is best formulated in such a way that its optimum application properties are obtained at a comparatively high relative humidity of around 55 %.

Given the high heat of evaporation and the low rate of evaporation, water loss during spraying is low. For this reason and because of the generally low content of non-volatile matter, the wet film thickness needed in waterborne coatings to obtain a particular coating thickness is greater than in the case of solventborne systems. The high film thickness and the sluggish initial drying in turn are a reason for the greater sagging tendency of waterborne coatings.

Following application, too, the water evaporates more slowly than most organic solvents. Evaporation can be accelerated by means of co-solvents that form an azeotrope with water (mixture of both liquids that has a lower boiling point). Until the volatile neutralising agents (amines, ammonia) have been evaporated, the salt-like structures of the binder also retain the water. When heated in the oven, water initially evaporates predominantly from the surface of the coating, causing the viscosity to rise. Then, when the water contained in the lower layers subsequently evaporates also, the water vapour bubbles may not necessarily be able to break through this more viscous layer. Bubbles are formed. If they burst, the upper layer of coating may already be too viscous to level back into a smooth surface again. In that event, craters (pinholes) remain. The tendency to form bubbles or craters ("popping") can be countered by a longer flash-off time and slower oven heating or by infra-red (pre-)drying, The scale of this problems naturally depends very much on the coating thickness. The formulator can reduce the risk of popping by using a resin of low T_g or by including less volatile co-solvents, which keep the viscosity of the uppermost coating layer sufficiently low even after the water has evaporated.

Once the water and solvent have been eliminated, the curing sequence is similar to that of equivalent conventional systems.

3.10.2.2 *Drying of primary dispersions*

Even though most dispersion-based coating materials, especially emulsion paints, contain pigments, fillers and additives in addition to the polymer dispersion, and these naturally have a significant influence on the film-forming processes, in the interests of clarity, we will consider only the role of the polymer dispersion in describing these processes.

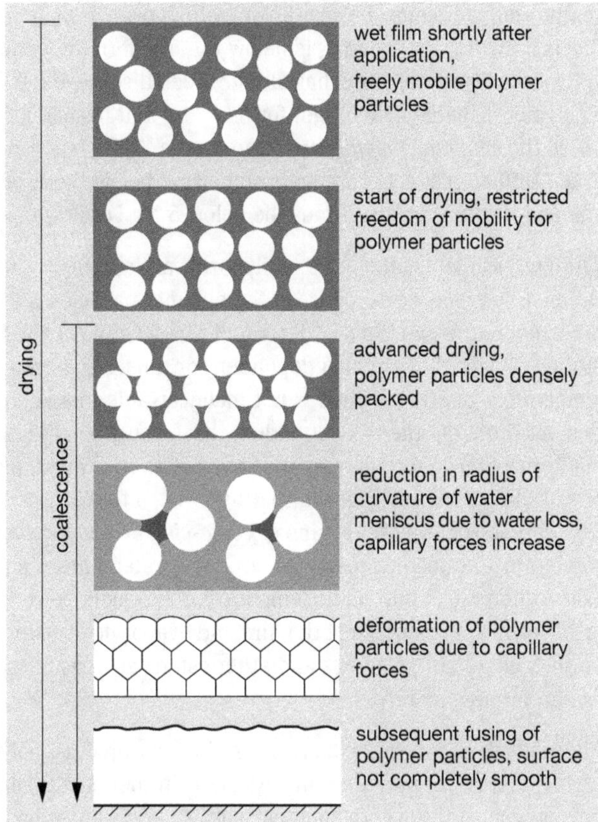

Figure 3.11: Model representation of film-forming in a polymer dispersion

Coalescence mechanism

The mechanism is shown schematically in Figure 3.11.

When the water evaporates from the dispersion, the polymer particles move closer and closer together.

As soon as they touch one another, the space between them, which is still filled with water, becomes so small that it can be considered to be a tangle of capillaries. Capillary forces composed of interfacial compressive forces in the interior and of surface suction forces from the meniscii in the ends of the capillaries cause the polymer particles to congregate even more closely together and to become deformed. (The physical causes of the resultant driving force for this sub-process are still the subject of scientific debate). Increasingly large forces thus develop in the capillaries, which likewise become narrower. Under the resulting pressure in the order of magnitude of 10 MPa (100 bar) and due to the interdiffusion of the polymer molecules on contact between the particles, the polymer particles more or less completely fuse together. The fusion of the individual polymer particles to a coherent film is called coalescence.

The surface area of the resulting polymer film is only a tiny fraction of the surface area (interface) of all the polymer particles before coalescence. Thus, a great expanse of interface disappears during film-forming and this can be seen as the macroscopic thermodynamic driving force behind coalescence. It might therefore be expected that more finely divided dispersions would coalesce much more readily than coarser particles of the same polymer - this would manifest itself in a lowering of the minimum film-forming temperature (MFT; see below). And this relationship does in fact tend to be observed, but the effect is so slight that a desirable lowering of the MFT via a reduction in particle size cannot be achieved to the technical degree needed.

Minimum film-forming temperature

A prerequisite for coalescence is that

- the polymer particles are highly deformed by the forces generated, and
- polymer molecules - or at least along most of their length - rapidly diffuse into adjacent particles.

To this end, there must be a sufficiently high proportion of free volume in the polymer particles, yet this is only the case when T_g is below the film-forming temperature. The dependence of the T_g of a polymer on its composition has already been discussed (→ 2.1.1.12). The temperature below which a polymer dispersion no longer forms a film, but instead forms a cracked or even powdery layer, is called the *minimum film-forming temperature (MFT)*. It is generally between -5 and 50 °C and, owing to the softening effect of water, is somewhat below the T_g of the dry film. For emulsion wall paints and architectural paints, it must be below 5 °C. For drying in heat, it can be higher.

Although an emulsion paint usually forms a film quickly during drying (too quickly in some applications), complete coalescence of the polymer particles is a slow process. Its speed correlates with the temperature gap from T_g. For rapid film-forming, a low T_g is desirable. Several other properties of the polymer film are also dependent on this gap, however, including blocking resistance. At a temperature of 50 °C, which is not inconceivable in a sunny spot on a summer's day, this means that, according to the WLF equation, the T_g must be at least 2 °C. A polymer of that nature would not form a film at normal temperatures, however. T_g must therefore change during the film-forming process in such a way that the coating forms a film at 5 °C and yet is still block-resistant at 50 °C. This can be achieved by means of *coalescing* or *film-forming agents*, which have to be moderately soluble in the polymer particles and the water phase, and should have a low, yet appreciable evaporation rate. They act as temporary plasticisers, lowering the T_g. After film-forming, they diffuse slowly to the surface of the film and evaporate, causing the T_g to increase to its original value. Various glycol ethers, esters and ether esters, such as butyl diglycol (acetate) are commonly used as film-forming additives.

Alternatively, a two-step copolymerisation can produce polymer particles that display a relatively high T_g inside, yet have a relatively low T_g in their shell. Thanks to the soft outer shells of the particle polymers, these core-shell particles fuse readily at low temperatures. Over time, the T_g in the coalesced film more or less equalises across the film. The use of coalescing agents is also possible here, of course. It should not go unmentioned that, even after a dispersion has undergone complete physical drying, the contact surfaces and gussets between the (original) particles can still be seen under the electron microscope after contrasting. In other words, areas of inhomogeneity that weaken the film remain.

Nowadays, various kinds of dispersions based on core-shell technology and/or specific monomer compositions (e.g. vinyl acetate-ethylene) are available that enable emission-free and solvent-free paints to be made.

3.10.2.3 Drying of polyurethane dispersions

Polyurethane dispersions are classified as secondary dispersions, even though the final molecular chain length is not obtained until after emulsification in the water phase. They differ in many respects from primary dispersions:

- the average molecular weights are lower (3 to $11 \cdot 10^4$ g/mol)
- the molecules are present in the form of extensively hydrated molecular associations
- the dispersions are just about thermodynamically stable before drying.
- the dry films have a high resilience on account of the molecular structure: soft segments connect hard domains from packed urethane groups.

As a result of this property combination, PU dispersions have an MFT whose lower limit is governed by the freezing point, but which dry to yield tenacious tack-free films. In summary, film-forming of PU dispersions can be viewed as intermediate between the drying of a solution and the coalescence of a primary dispersion.

3.10.3 *Curing (crosslinking) of liquid coating materials*

3.10.3.1 General principles

Network formation

A major disadvantage of physically curing coating materials is that, because the physical film properties require the dissolved film formers to have high molecular weights and because the application viscosity has to be low, they usually have a relatively low content of non-volatile matter.

Furthermore, the thermoplastic films are not durable enough for many applications. By contrast, film formers of a much lower molecular weight and hence a higher concentration at the application viscosity can be used in chemically curing coating materials. The high molecular weight, and ultimately the polymer network, required for good strength and durability only develops in the course of the chemical curing process(es).

The individual chemical reactions leading to chemical crosslinking in a coating film have already been discussed (→ 2.1.2 and 2.1.4), and basic information on network formation and characterisation has been provided in section 2.1.1.4. Only general, supplementary aspects will be provided below. Crosslinking is shown roughly in Figure 3.12.

Hardening by one of the established poly-reactions involving continuous molecular enlargement (lengthening and branching) leads to crosslinking as soon as a certain amount o f reactive groups have been converted. This called the gel point and it manifests itself as a gel-like state. From this point on, the ability to flow freely gradually declines as elastic resilience increases, which explains why the film can now no longer level out completely. Any residual solvent present has no influence over this.

Once the gel point has been reached, network formation proceeds quickly at first, but curing itself becomes progressively slower since

- the concentration of the reactive groups decreases, and
- the network increasingly restricts the mobility of the hardener molecules or the reactive parts of the molecules

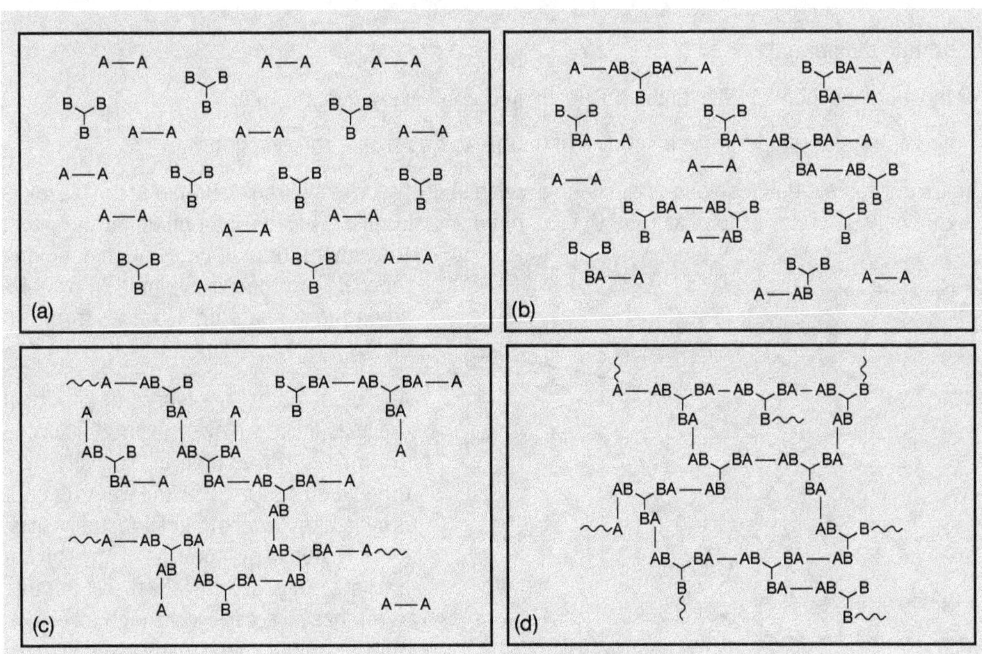

Figure 3.12: Formation of a network from bifunctional and trifunctional molecules bearing reactive groups A and B. a) prior to reaction; b) formation of branched chains; c) gel point exceeded; d) conversion/crosslinking complete

Both of these causes lead to a reduction in the number of collisions between reactive groups per unit of time.

Just as solvent evaporation in the advanced stage is increasingly controlled by diffusion due to the rise in T_g, the same thing happens with curing. The rise here in T_g, however, is caused not only by any ongoing solvent evaporation, but also by molecular enlargement and network formation. The larger the molecules are initially and the denser the network is, the higher is the T_g. As a rule of thumb:

- if the T_g exceeds T - 10 K (T: instantaneous curing temperature), diffusion control sets in; from T_g = T + 35 to 50 K and above, the curing rate of step reactions (polyaddition, polycondensation) tends towards zero. In the case of free-radical polymerisation (e.g. UV curing), the T_g threshold is somewhat higher.

The practical outcome is that the final degree of conversion and thus the crosslink density also decreases with decrease in curing temperature. Thus, final conversion rates for room-temperature-curing two-pack epoxy systems are typically 90 to 95 %, provided that the curing agent is not present in excess; for UV curing, double-bond conversion is just 60 to 90 %, the exact level depending inter alia on the film temperature during polymerisation. Solvent retention should not be forgotten here - it occurs especially in the case of films that cure at room temperature.

Film formation is generally boosted by temperature increase, since

- solvent evaporation proceeds faster,
- chemical crosslinking is boosted,
- the retarding effect of diffusion control sets in later, if at all.

The Arrhenius equation governing the rate of a chemical reaction is

$$k = A \cdot \exp\,[-(E_a/RT)]$$

k: reaction rate constant
A: factor (frequency factor)
E_a: molar activation energy of the reaction
R: general gas constant
T: absolute temperature

For most organochemical reactions, this leads to the following general rule:

- increasing the temperature by 10 K will double to quadruple the reaction rate

Thus, a curing reaction for a one-pack coating which takes five years in the container at 25 °C, takes roughly only 3^{-11} times as long at 135 °C, i.e. around 25 minutes. It has been found empirically that the stoving time of conventional stoving topcoats is reduced by some 37 to 44 % when the temperature is raised from 130 to 140 °C.

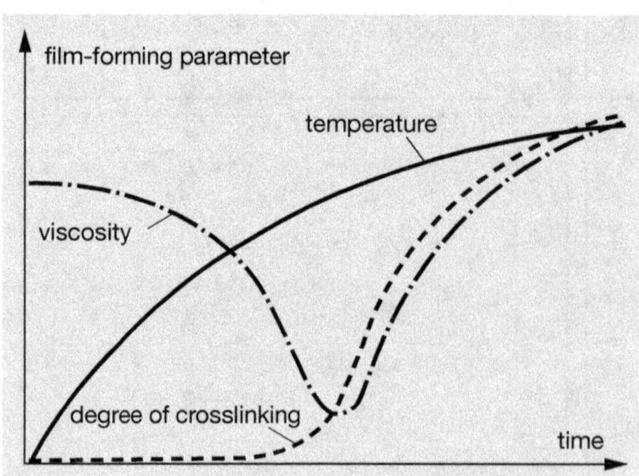

Figure 3.13: Evolution of degree of crosslinking, viscosity and temperature as a function of stoving time

A crucial point in respect of practical thermal curing ("oven drying", "forced drying") is the change in viscosity over time when an article or the wet film on it is heated up. After application, the viscosity rises due to evaporation of any solvent present, but generally falls thereafter due to the heating, passes through a relative minimum and then finally rises sharply again due to molecular enlargement (curing).

Figure 3.13 schematically illustrates how a number of film parameters change during oven drying (after physical setting). The viscosity minimum harbours the risk of sagging, which can be counteracted by optimising the heating curves and the paint formulation (solvent composition, anti-sagging additives).

Network structures

Three ways of producing networks have been described in section 2.1.1.4:

1. crosslinking of long, linear molecules
2. linking together of large, highly branched molecules
3. polyreaction of low-molecular (monomeric or oligomeric) binder molecules

These yield characteristic network structures that exhibit different mechanical behaviour. Generally, it may be said that:

- networks having meshes of the same size (homogeneous networks) but no long, free chain ends or only singly bonded loops (rings) yield the best combination of hardness (or tensile strength) and flexibility and thus are the desired objective.
- the higher the crosslink density, the higher are the hardness (or tensile strength), media resistance, and internal stresses in the film, and therefore the lower is the elongation at break (flexibility).
- molecules not incorporated into the network, such as thermoplastic binder fractions, additives or residual solvent increase the flexibility at the expense of strength and durability.

The crosslink density increases with increase in mass fraction (e.g. OH content in %) or the number of reacting functional groups per unit of mass (e.g. hydroxyl number) of the binder; note that the concentration of this group is inversely proportional to the equivalent weight (e.g. the epoxy equivalent weight).

Crosslinking is always accompanied by volume shrinkage of the film former as the molecules or molecular segments bond more tightly to each other. On a one-dimensional basis, this manifests itself as "film shrinkage", which, on rigid substrates, leads to substantial shear stress if the film has a high T_g. The outcome of this can be dehesion and/or subsequent cracking.

3.10.3.2 High solids

The lack of physical setting and drying on the part of high solid systems and possible problems arising therefrom have already been covered (→3.10.2.1). The influence of the higher binder concentration on chemical curing is complex. One reason is that solvent evaporation affects the rate of curing, i.e. the physical and the chemical components of film formation cannot be viewed in isolation from each other.

Since the film formers for high solid systems are relatively low-molecular, it takes many linking steps to make macromolecules and to reach the point at which crosslinking can start (the gel point). Not only is physical setting hindered or even completely suppressed but actual curing is delayed and this needs to be factored into the technical design of the drying process, e.g. in the form of a higher heating rate.

In the case of two-pack materials, the high concentration of functional groups (low equivalent weight of the film formers and less dilution by solvent) accelerates the chemical reaction in the mixture. However, since the average molecular weight in the pot remains relatively low at first despite the increased rate of reaction, the increase in viscosity is also slow. Overall, though, the pot-life is shortened relative to coatings with a higher solvent content.

3.10.3.3 Crosslinking of waterborne film formers

Most of the general principles underlying crosslinking of film formers dissolved in organic solvents apply to waterborne materials. However, extra chemical features or additional aspects need to be taken into account, which can only be discussed in connection with the respective crosslinking chemistries and formulation ingredients.

Generally:

- The crosslinking functional groups and any reaction accelerators must be stable in water long enough so as not to be destroyed by water prior to curing. (Isocyanates, silanes and metal catalysts, for example, are critical).
- The rise in the pH (especially at the surface) of the wet film due to evaporation of amine often exerts an accelerating influence on the crosslinking reactions.

Dispersions, too, are frequently crosslinked to meet the needs of more demanding applications, even though polymer dispersions (primary dispersions), as opposed to binder solutions, have high molecular weights despite their low viscosity and so will yield stable films simply by physical drying. Crosslinking can take the form of

- crosslinking of appropriately functionalised dispersions by stoving, in two-pack form, oxidatively or by UV irradiation (→2.1.4)
- interparticulate, relatively weak supplementary crosslinking with substances such as polycarbodiimides, polyaziridines, zinc or zirconium complexes, dihydrazides or silanes functionalised with organic groups.

The second group of substances are usually added just before application because, as they are 1-pack materials whose storage stability ranges only from a few days to several weeks in accordance with their crosslinking mechanism.

What was said about crosslinking of primary dispersions more or less applies in full to PU dispersions.

3.10.3.4 Radiation curing

The chemistry of film formers and curing is covered in sections 2.1.4.2, 2.1.4.3 and 3.6.

Radiation-curable coatings - with the exception of aqueous products - are mostly 100 % systems, i.e. virtually free of volatile substances. Curing may follow a free-radical or ionic mechanism and be initiated by UV radiation or electron beams.

In free-radical UV curing, free-radicals generated photochemically from the photoinitiator trigger intermolecular polymerisation to produce a network very quickly. The network is severely impaired by numerous free chain ends and loops, and the double-bond conversion rate is only 60 to 90%. The choice of photoinitiators and their concentration should reflect the Lambert-Beer's law ($I(x) = I_0 \cdot 10^{-E(\lambda)}$, $I(x)$: intensity at depth x, I_0: intensity at the surface, $E(\lambda)$: decadic extinction = absorption strength) to ensure that too much radiation is not absorbed in the film surface and too little is present at deeper levels. As a rule of thumb:

- free-radicals are only formed wherever radiation goes
- free-radicals can trigger polymerisation only over a distance of a few nm
- this means that the polymerisation does not propagate over appreciable distances within the film.

Polymerisation occurs so quickly that atmospheric oxygen is unable to inhibit it at the film surface. There is therefore usually no need to use inert gas (N_2, CO_2). Nonetheless, UV curing proceeds much more readily in CO_2 than in air, and so allows the concentration of photoinitiator and intensity of irradiation to be reduced. In particular, it greatly facilitates curing on 3D articles.

Electron beam curing, which is much faster overall than UV curing, requires the use of nitrogen to exclude air because the electrons have a lower density per unit area than the UV quanta.

Cationic UV curing is much slower than its free-radical counterpart. However, dark curing continues for up to 24 hours, and inhibition by air, and attendant blocking resistance, does not occur here.

3.10.4 Curing (crosslinking) of powder coatings

Powder coatings consist of finely divided, discrete solids particles. They must not sinter together during transportation or storage. The T_g of the uncrosslinked binder must therefore not be too low; it is usually around 55 °C. Powder coatings are often applied by electrostatic spraying, especially on metallic substrates. On heating, the particles coalesce on these substrates to form a coherent film; crosslinking commences at the same time. Non-crosslinking, thermoplastic powder coatings (sinter powders) coalesce on the pre-heated article to yield thick layers, without undergoing a chemical reaction. They will not be discussed any further here.

Stoving process

A key technological goal of processing powder coatings is short stoving times at low oven temperatures. This can be achieved by highly reactive film formers (or highly effective catalysts), for example. However, there is then the risk that the curing reaction will start even during extrusion and that the melt viscosity during curing will not fall far enough to ensure uniform substrate wetting and good levelling. Furthermore, their average molecular weight may rise during protracted storage if they are not adequately cooled. A compromise must therefore always be sought between the requirements for good storage stability on one hand and the most cost-effective and gentle stoving conditions on the other. As a rule of thumb, the stoving temperature should be about 50 °C above the temperature of the melt during extrusion and around 70 to 80 °C above the T_g, i.e. at least around 130 °C, although this can be reduced in the case of powder coatings that are refrigerated during transportation and only stored briefly ("low-temperature powders").

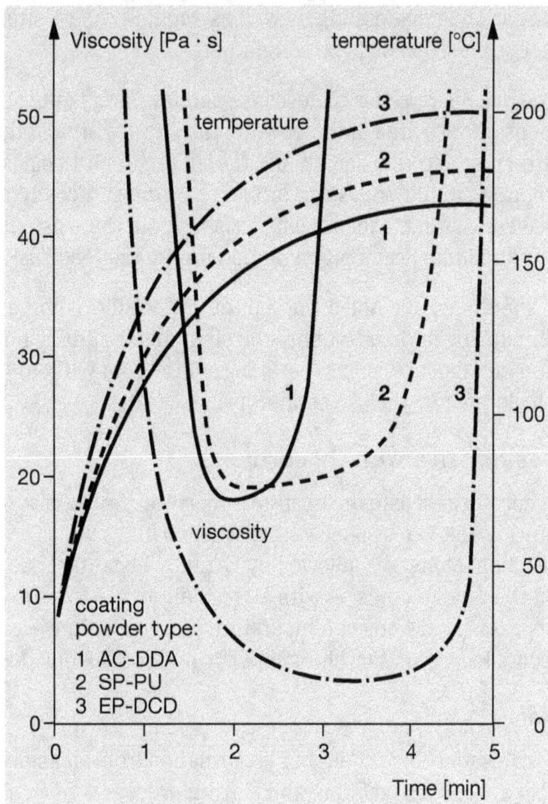

Figure 3.14: Change in temperature and viscosity of a powder coating over time during stoving, according to [10]

Melt viscosity and surface tension

Figure 3.14 provides a schematic illustration of the change in viscosity as a function of stoving time. The viscosity initially falls as the temperature rises. At the same time, as the temperature rises, the crosslinking reaction becomes increasingly rapid, until finally it is so fast that the drop in viscosity due to rising temperature is more than offset that the influence of the increasing molecular weight of the film former. At this point, the viscosity passes through a minimum at around 5 to 20 Pa·s.

The crosslinking reaction and hence also the rise in viscosity come to a stop either when T_g is around 50 °C above the stoving temperature or the film former has completely reacted. The more quickly

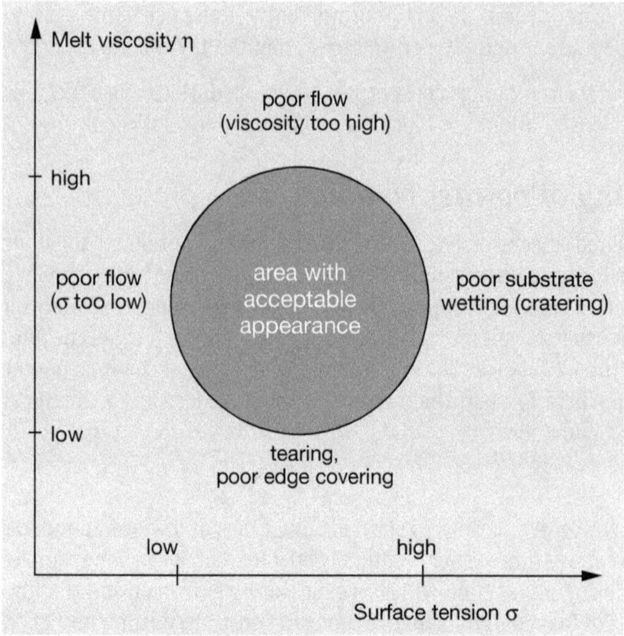

Figure 3.15: Dependence of the surface quality of a powder coating on the surface tension and melt viscosity

the powder coating is heated, the deeper and narrower is the value of the viscosity minimum and the better is the levelling. As a general rule, levelling is possible only below a melt viscosity of 1000 Pa·s. Accordingly, the quality of the expected levelling and wetting can be gauged by

$$\int_{t_1}^{t_2} \frac{dt}{\eta(t)}$$

t_1, t_2: times at which 1000 Pa·s are not met or are exceeded respectively. Note that this integral, which has incorrectly been termed the "levelling number" does not represent the area of the viscosity curve under the 1000 Pa·s line.

Nor must the melt viscosity drop too far, either, and/or hover for too long at the minimum because, in that event, while the appearance (levelling) may be good, the protective effect is poor owing to corner and edge pulling as well as sagging. As so often in coatings technology, a golden mean or compromise needs to be sought here.

The driving force for coalescence and levelling is the surface tension. In comparison with the enormous surface area of the powder, the surface area of the coating after coalescence is insignificant. A high surface tension should therefore support coalescence and smooth the molten coating film. Figure 3.15 illustrates the influence of melt viscosity and surface tension on the appearance of a powder coating. The problems arising from the - usually high - surface tension of powder coatings and the influence of pigmentation on the melt viscosity have already been covered (→3.7).

It should also be noted that optimum film formation also requires deaerating or degassing of the film during curing. Not only the air trapped initially but also any cleavage products (water of reaction etc) must be able to escape from the film without causing lasting problems. Degassing can be accelerated by the use of appropriate additives (→3.2.7).

Combination with UV curing

Where heat-sensitive substrates (derived timber materials, plastics) need to be given a clearcoat quickly, this can be done with powder coatings which level out well at 100 to 120 °C, without thermal reaction. Melting is followed by virtually instantaneous curing - free-radical or cationic, depending on the type of binder - with UV radiation; electron beam curing may also be used. Aside from the necessarily low object temperature (especially in the case of IR heating), the key advantage of this technology is that levelling and curing are uncoupled with regard to time.

Note:

The described principles of film formation from an amorphous compound that has no initial cohesion also apply to the melting and stoving processes of an electrodeposition coating.

Sources and references for Chapter 3

[1] R. Baumstark, M. Schwartz: Dispersionen für Bautenfarben. Vincentz Network, Hannover 2001

[2] A. Goldschmidt, H.-J. Streitberger: BASF-Handbook: Basics of Coating Technology, 2nd edition, Vincentz: Hannover (2007)

[3] C. H. Hare, Protective Coatings: Technology Publishing Company, Pittsburg 1994

[4] „HighChem hautnah". Broschüre der GDCh (Fachgruppe Lackchemie), Frankfurt 2008

[5] H. Kittel: Lehrbuch der Lacke und Beschichtungen. 2. Aufl., Bd. 3, S. Hirzel Verlag, Stuttgart 2001

[6] R. Laible u. a.: Umweltfreundliche Lackiersysteme für die industrielle Praxis, Expert Verlag, Ehningen bei Böblingen 1989

[7] A. R. Marrion (Edr.): The Chemistry and Physics of Coatings. 2nd Ed., The Royal Soc. of Chemistry 2004

[8] Meier-Westhuis, U.; Polyurethanes, Vincentz Network, 2007

[9] P. Mischke: Filmbildung in modernen Lacksystemen. Vincentz Network, Hannover 2007 (und dort zitierte Literatur)

[10] B. Müller und U. Poth: Coatings Formulation. 2nd edition, Vincentz Network, Hannover 2007

[11] D. Ondratschek (Red.): besser Lackieren! Jahrbuch 2008. Vincentz Network, Hannover 2007

[12] Poth, U.; Polyester und Alkydharze, Vincentz Network, 2005

[13] J. Ruf: Organischer Metallschutz, Curt R. Vincentz Verlag, Hannover 1993

[14] D. Stoye (ed.): Paints, Coatings and Solvents, VCH Verlagsgesellschaft mbH, Weinheim 1993

[15] Ullmann (6.) [CD-ROM Februar 2004]

[16] Z. W. Wicks, F. N. Jones, S. P. Pappas, D. A. Wicks: Organic Coatings. 3rd ed., J. Wiley & Sons, Hoboken/New Jersey 2007

[17] A. Zosel: Lack- und Polymerfilme. Vincentz Verlag, Hannover 1996

[18] Ullmann's Encyclopedia of Industrial Chemistry, 5. Auflage, Vol. A20, Paints, VCH Verlagsgesellschaft mbH

4 Manufacture of paints and coatings

4.1 Preliminary comment

Unlike the standard manufacturing methods used in primary chemistry for substances such as ammonia, isoprene, phenol and many others, in the case of coating materials and other "preparations" (mixtures of substances), there are no universally applicable, clearly defined manufacturing instructions or systems. The reasons for this can be summarised as follows:

Coating manufacture (in narrow terms) takes place without any noticeable chemical reaction(s), so "reaction conditions" regulated by set parameters (temperature, pressure, reaction time) do not apply here. Instead it is the product of an optimum combination of metering, mixing, dispersing and testing operations matched to the individual product (formulation), specific batch size and available production equipment. The individual stages of coating manufacture can to a limited extent be combined in a variety of ways, and a choice of equipment is available for certain stages.

Consequently, this and other books do not contain descriptions along the lines of "How to produce an automotive assembly line coating" or "How to make a stopper". Nevertheless, it is useful to examine the characteristics of various production sequences (production "strategies") and to investigate the physico-chemical and engineering principles behind the individual operations and *equipment.*

4.2 General introduction to the manufacture of paints and coatings - layout of a coating

Types of operation

Production processes can be divided into continuous and discontinuous or *batch* operations. Transitional forms are known as "*semi-batch*" or "*semi-continuous*" processes.

The working method used in the coatings industry is that of batch operation, possibly with some continuous sub-processes, such as dispersion in an attrition mill or in an extruder. Given the usually wide variety of products that have to be made available in irregular sequences and quantities and often produced to order at short notice (just in time production), continuous operation would be impossible, or at least impractical.

Production features

The characteristics or problems of coating manufacture can be summarised as follows:

- large number of raw materials
  → high costs for acquisition (purchasing), inspection, storage and distribution
- versatile yet also heterogeneous production plants
  → high operating, cleaning and maintenance costs
- huge range of finished products (generally thousands of "live" formulations)
  → high costs for inspection, storage and delivery.

In addition to these production-related matters, other necessities also arise within a coating factory, such as

- high costs for development, product stewardship and customer support (the last including "on site" support at the customer's premises)

Brock, Groteklaes, Mischke: European Coatings Handbook

ISBN 978-3-86630-849-7

- IT-based management of vast collections of formulation data and manufacturing instructions
- compliance with numerous safety and environmental regulations relating to raw materials, production processes and final products (key terms: regulations covering hazardous substances and products, safety data sheets, accident prevention regulations[1] and many others).

These demands are compounded by the now virtually obligatory requirement to introduce a *quality management system* (QM system), usually to the international standards ISO 9000 to 9004 (and particularly ISO 9001 in this case), and to have it certified by an independent accredited body or company.

Manufacturing layout

Figure 4.1 shows the layout of a coating factory with an in-house film former production capability. Figure 4.2 provides a schematic view of the material flow within a coating plant. Both designs will naturally need some alteration according to the specifications of the individual manufacturer.

Coating manufacturers buy their raw materials from the *coating raw materials industry* which, like the coatings industry itself, features some medium-sized companies but is largely owned by the major chemicals groups. Some, mainly the larger coating manufacturers, produce at least a proportion of their film formers ("binders") in-house, however, and are therefore able to use customised raw materials that are not commercially available.

4.3 Process stages in the manufacture of coatings

Organisation of production

The processing sequences involved in the production of paints and coatings are illustrated schematically in Figure 4.3. The manufacturing processes for the main types of coatings can be described as follows, with reference to the diagram:

- Pigmented paint ("from scratch"):
  Raw materials placed in storage (1) - raw materials weighed out for the dispersion batch (2) - pre-mixing, i.e. coarse homogenisation (3) - pre-dispersion (4) - fine dispersion (main dispersion) (5) - letting down cutback and adjustment (6) - filtration (7) - packaging (8) - finished products placed in storage (9).

The two schemes described below are characteristic of other types of coating products (with reference to the same numbered stages):

- anti-corrosive coatings and undercoats, primers, emulsion paints:
  1 2 3 4 6 7 8 9
  (fine dispersion (5) is unnecessary in many (but not all!) cases)
- clear coatings:
  1 2 3 6 7 8 9 (or 1 2 6 7 8 9)
  (the two dispersion stages (4 and 5) are omitted).

The main production stages

Before the filtration and packaging stages, various *release inspections* are carried out, the number and complexity of which varies according to the particular area of application and quality requirements.

Where no pigment dispersion or fillers are required, as in clear coats, relatively simple inspections (from the point of view of manufacturing technology) are sufficient.

1) In Germany, for example, the German trade association regulations 86a ("Manufacture of coatings") must be complied with in coating manufacture.

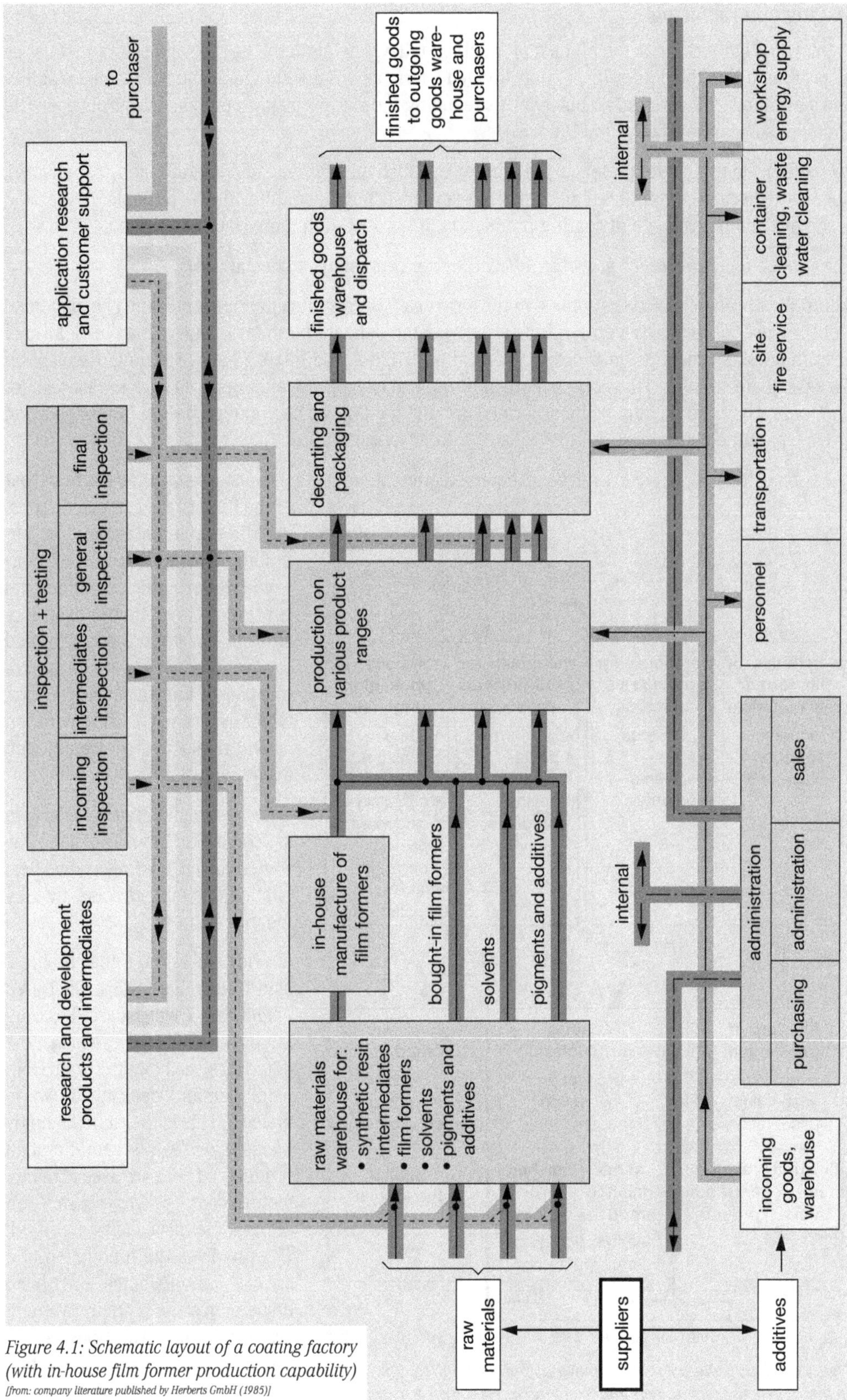

Figure 4.1: Schematic layout of a coating factory (with in-house film former production capability)
[from: company literature published by Herberts GmbH (1985)]

Mixing and metering

Pigmented coatings incur the highest production outlay. The pigment, as the solid colorant, must be incorporated into the binder in as finely divided a form as possible to yield a hue of defined stability and constancy ("dispersing"). This involves reducing the size of the pigment particles from over 100 μm to, in some cases, far below 1 μm.

"*Dispersion*" in this context refers to the process by which the pigment agglomerates and, if applicable, filler particles are comminuted as completely and distributed as uniformly as possible. Dispersion is the most complex stage of coating production; it is covered in more detail in section 4.9.

The production of *metallics and other effect coatings* represents a special case:

Aluminium bronze, which is supplied in the form of a paste, or mica pigments etc. must be dispersed ("*opened up*") particularly gently, since otherwise the delicate pigment platelets may be deformed or even crushed, causing "*greying*" of the coating. "High-speed mixers", slow-running dissolvers if necessary, are suitable. This is even more applicable to layer-substrate pigments such as pearlescent pigments. They should only be incorporated into the medium to be pigmented (film former solution or transparent coloured coat-ing) by means of intensive stirring.

The composition of the *paste* (mill base) obtained after dispersion is almost always different from that of the final coating, since dispersing the pigments in the final formulation (complete formulation) would be too inefficient. Additional substances must therefore be added to the paste in a stage termed *cutting back*. One element of cutting back is known as "*making up*", which refers to the incorporation of additional (concentrated) film former solution.

Adjustment, sometimes also known as "correction", serves to fine-tune the required coating properties, particularly the precise hue by means of "*tinting*".

In addition to the linear progression from stages 1 to 9 illustrated in Figure 4.3, which describes coating production "*from scratch*", the diagram also shows the jump from 5 (dispersion) back to the warehouse 1. This is the route taken by *pigment pastes* (as semi-finished products). These are used to manufacture coatings "*from pastes*", an alternative processing method. This involves the production of a special coating batch - with no dispersion stage - simply by intensively mixing several pastes and other components, as described in the following section.

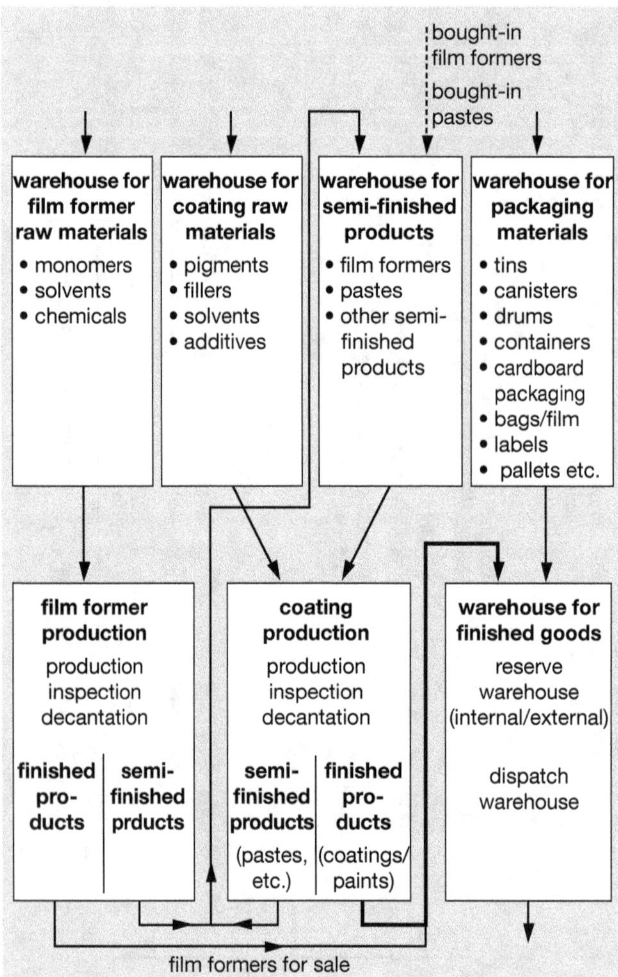

Figure 4.2: Block diagram showing material flow for paint/coating production with in-house film former production (no recycling/waste recycling)

4.4 Production "from scratch" and from pastes - formulation example

Alternative production methods and comparison

If the **goal** of batch production is a certain **colour**, e.g. a RAL colour or a "fashionable purple" car colour, etc., three different production strategies are possible for making and letting down the mill base as illustrated in Table 4.1 (page 234), which shows the essential effects and application areas.

Production "from scratch", as we have already explained, means that the pigment or blend of pigments is dispersed for every batch of coating; in the latter case we also refer to the *co-dispersion* of pigments.

In production from pastes, also known as "mixed production", no dispersion is required for production of a colour batch. Instead, the dispersion stage is separate in that every pigment required is stocked in disperse form as a pigment paste for use in several coating batches, often of different colours. Alternatively, rather than producing the pigment pastes themselves, coating manufacturers may buy them from *paste manufacturers* or have them produced externally by means of "*toll dispersion*".

A further production variant consists in cutting back each pigment paste to form a *mixer paint* and then combining the appropriate blend of mixer paints to produce the desired coating in the specific hue. This method is particularly suitable for preparing coatings on a customer's own premises (where there are no coating production facilities) quickly and in almost any hue, in the small quantities required in each case.

Table 4.1 mainly shows how the **batch sizes** to be produced and the number of **colours** needed dictate the procedure to be followed.

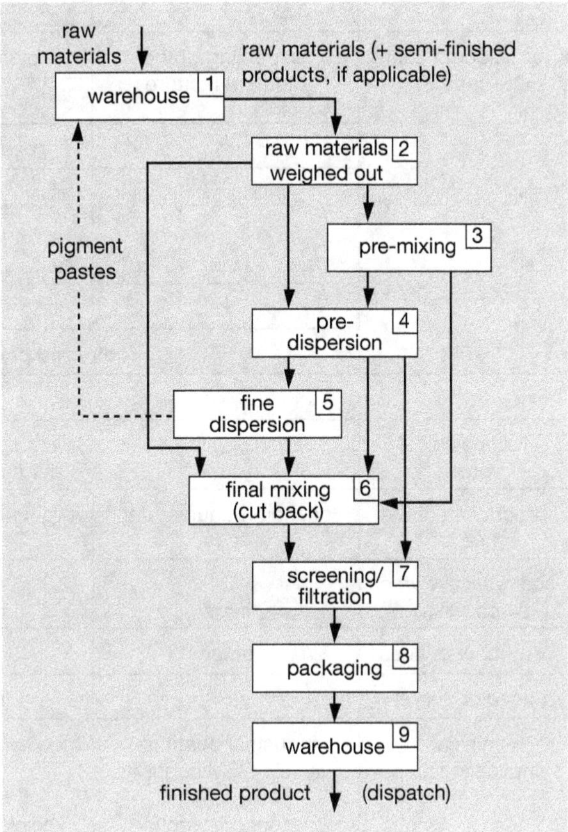

Figure 4.3: General flowchart for coating production

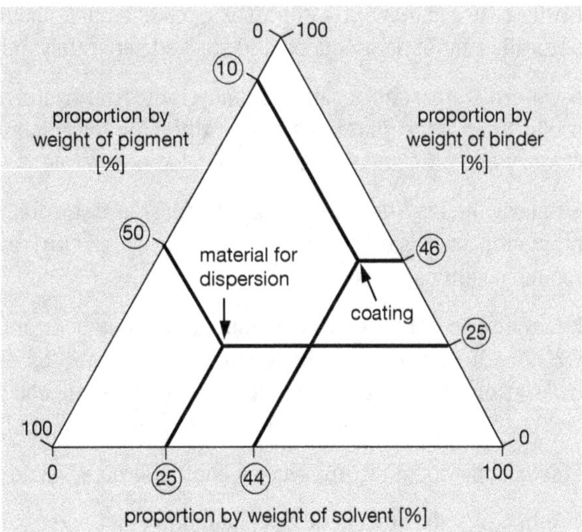

Figure 4.4: Automotive topcoat, plain-coloured. Composition of material for dispersion (paste) and coating as a triangular diagram

Table 4.1: Production strategies

Production strategies	**Production from scratch**	**Via pigment pastes**	**Via mixing coatings**
	raw materials		
	↓	↓	↓
for coloured coatings	dispersing of the pigment mixture	pigment pastes	pigment pastes
		↓	↓
	↓	paste mixtures	let down to mixing paint
	let down	↓	↓
		let down	mixing to produce the hue
	↓	↓	↓
	coloured paint (ready mix hue)		

technological grade	equivalent (for same final hue)		
production of the hue for	coating manufacturer	producer or processor	usually at the processor
cost effective when batch sizes:	large	medium	small
number of hues	small	small-large	large
speed of delivery	slow	medium	fast
typical application areas	industrial coatings vehicle OEM coating, house paints. masonry protection	modular industrial coating systems, vehicle OEM coating	mixing systems: automotive refinishes, industrial, DIY finishes

In the case of automotive refinishes, for example, about 15,000 colours must be available, which naturally cannot be produced and stocked separately and can only be handled by mixing systems.

By contrast, masonry coatings, many industrial coatings etc. are required in fewer hues, but are produced in large quantities: predestined for "production from scratch". In addition, available machinery and utilisation in production also play a role.

Similarly, depending on the manufacturer's assortment, consideration is accorded to whether a pigment paste system can be used as a so-called **"universal paste system"** for various chemically related coating systems.

It may also be worthwhile for smaller paint manufacturers or where hue batches are small to medium-sized not to do one's own dispersing, but to obtain the necessary pigment pastes (as "finished mill base") from **toll** dispersers or from appropriately specialized paste manufacturers. These rules out any possible overcapacity in terms of expensive and bulky dispersing equipment.

Finally some manufacturers use a combination of mixer coatings and pastes:

The hue of a starting coating is shaded to a defined extent using pigment pastes ("tinting pastes", "full-tone paints"). A familiar example of this from the decorating sector is the use of one or more coloured pastes to tint a white emulsion wall paint.

To illustrate the differences between the two basic strategies of "production from scratch" and "production from pastes", Table 4.2 provides formulation details for a conventional, plain-coloured automotive paint using the two alternative manufacturing methods. Figure 4.4 illustrates the composition of the pastes and coatings by means of a *triangular diagram*, the principle behind which, if not already familiar to the reader, should be evident from the example itself.

Table 4.2: Formulation details for a conventional, plain-coloured automotive paint [2]

	Proportion by weight in %	
	Production from scratch	Mixed production
material for dispersion (mill base)		
solvents	25	
film formers (100%)	23	
pigments 1 to 5	50	
additives	2	
paste mix		
paste 1		60
paste 2		30
paste 3		7
paste 4		2
paste 5		1
total	100	100
desired overall formulation for both productions		
solvents	44	
film formers (100%)	45	
pigments	10	
additives	1	
total	100	
to be mixed in order to obtain the complete formulation:		
paste[1])	20.0	
solvents	39.0	
film formers (100%)	40.4	
additives	0.6	
total	100.0	

[1]) The proportion of paste in the complete formulation is calculated from the following observation: 100 g finished coating contain 10 g pigment; these 10 g of pigment are contained in exactly 20 g paste comprising 50 g pigment.

The most important finding from the data in Table 4.2 is that, in principle, five different pigments would have to be co-dispersed in the "from scratch" method. However, since each pigment requires highly specific dispersing conditions to achieve optimum dispersion, i.e. to make maximum use of its coloristic potential, the co-dispersion of five in some cases very different pigments would be technologically (and economically) unworkable. On balance, the coating can only be produced from pastes.

Comparison of production methods

In conclusion, the advantages and disadvantages of the two "strategies" are listed below; (+) stands for advantage or favourable, (-) for disadvantage or unfavourable.

- Production "from scratch"
  - (+) dispersing conditions for each batch can be selected or influenced as required
  - (+) no storage facilities for pastes required
  - (+) no problems arising from instability (flocculation, loss of colour intensity) of pastes
  - (+) no foreign substan-ces, e.g. special paste resins, in the coating
  - (-) each batch requires individual dispersion (mills in use, have to be cleaned)
  - (-) in co-dispersion none of the optimum dispersing conditions can be chosen for any of the pigments
  - (-) the deviations in hue are often too great in co-dispersion (separate correcting pastes are needed)
  - (-) production time for one batch is longer than for mixed production.

- Production from pastes ("mixed production"):
  (+) optimum dispersion possible for each pigment
  (+) where many hues or batches are involved: total dispersion/cleaning costs are significantly lower
  (+) greater flexibility in terms of range of hues and batch size
  (+) if all pastes are bought in, no dispersion on own premises is required; no pigment dust generated
  (+) correction of coating hue is generally straight-forward
  (+) production time for one batch shorter than in production from scratch
  (-) pastes have to be kept in stock
  (-) if pastes are bought in, foreign paste components may cause detrimental effects
  (-) bought pastes are more expensive than the equivalent raw materials
  (-) sometimes the converging pigments co-flocculate.

4.5 *Aspects of equipment for the manufacture of coatings*

Figure 4.5 provides a pictorial view of the processing sequences involved in coating production, according to the commonly used *vertical process flow*. The advantage of this production method is that for the most part the substances pass along the production route without the need for pumping or other conveying processes. It does of course mean that almost all of the raw materials have to be transported up to the top storey, and the individual floors consequently need to have a high load-bearing capacity.

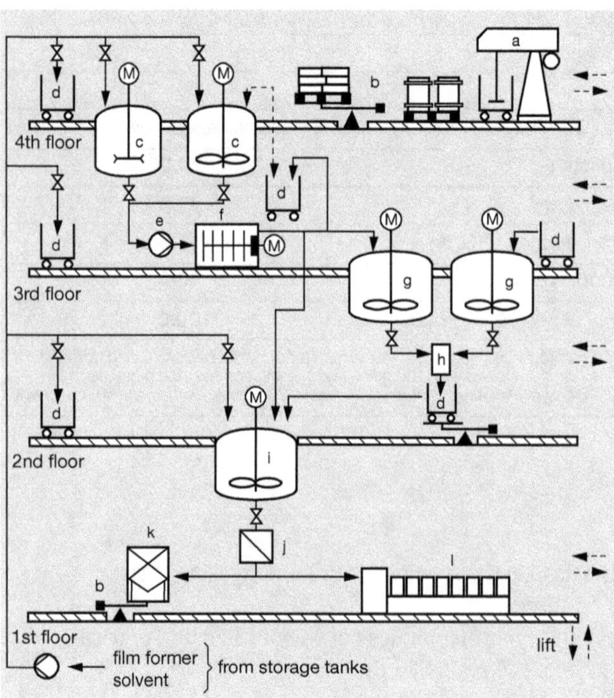

Figure 4.5: Flow chart for coating production with vertical process flow, according to [8]
a) stationary dissolver, b) scales, c) pre-mixer or dissolver, d) portable vessels, e) pump, f) attrition mill, g) paste mixer, h) paste metering station, i) final mixer, j)screen, filter, k) containers, l) packaging line for small drums

Although the details in Figure 4.5 should be self-explanatory, some additional notes follow:

a: stationary dissolver for pre-dissolving, pre-mixing or pre-dispersing in portable vessels
b: recording floor scales with tare setting (plus automatic supply cut-off for liquid components)
c (left): dissolver with fixed, closed vessel
c (right): pre-mixer for pre-dissolving and pre-mixing
d: portable vessels, capacity up to around 1000 l
e: generally a gear pump
f: attrition mill (pearl mill) for fine dispersion
g: paste mixer for temporary storage of pigment pastes
j: filtration is almost always carried out using depth filters such as filter bags or cartridges (tubes), frequently together with surface filters such as screen bags or vibrating screens

k: various reusable vessels used as containers (for major customers)

l: packaging line for filling, sealing, labelling

The fixed vessels are often placed on load cells so that they also function as scales, allowing automatic filling or metered addition.

Liquid components are frequently admixed volumetrically nowadays. This naturally entails constant density monitoring, because of the temperature fluctuations experienced by raw materials stored outdoors.

Figure 4.6 shows small-scale coating production on one level only.

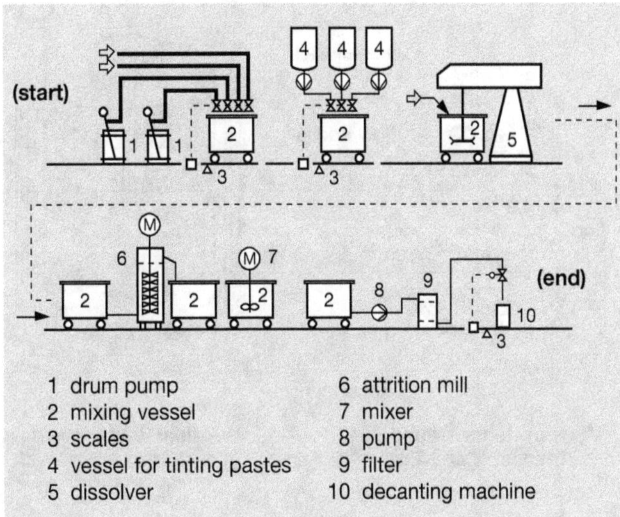

Figure 4.6: Small-scale coating production [from: company literature published by RHE Händel Engineering GmbH]

4.6 Manufacture of powder coatings

The processing sequences involved in powder coating manufacture differ by nature quite considerably in some respects from those in liquid coating manufacture. The method illustrated in Figure 4.7 has become established as the standard operating procedure - with one or two minor variations. The central step in the manufacturing method, namely dispersion of the pigments in the molten binder, takes place under strong shear forces in a *screw compounder (extruder)*; see section 4.9.6 for further details.

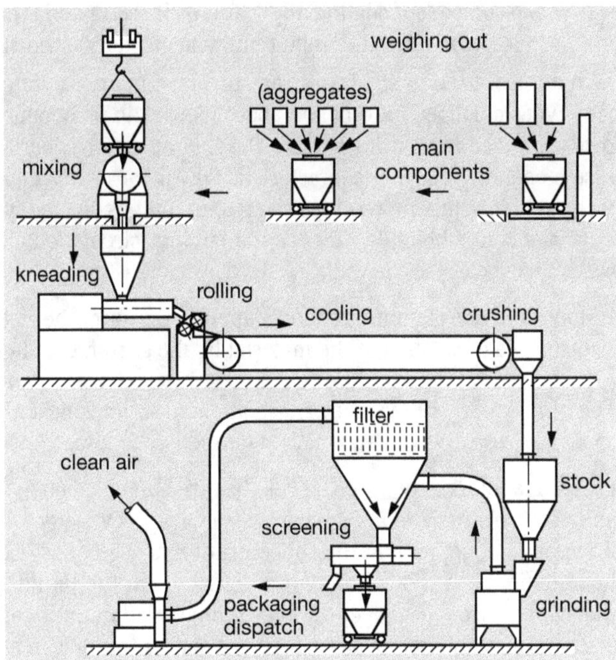

Figure 4.7: Powder coating manufacture, according to [5]

4.7 Further information about mixing and dissolving

Homogenising processes

The purpose of mixing is to obtain an everywhere same composition of initially separate then blended individual components throughout the mix. This process is also generally known as *homogenising* (even if the outcome is often only heterogeneous and visually uniform mixing), or specifically - according to the final state obtained - *dissolving, dispersing, emulsifying* or *suspending*.

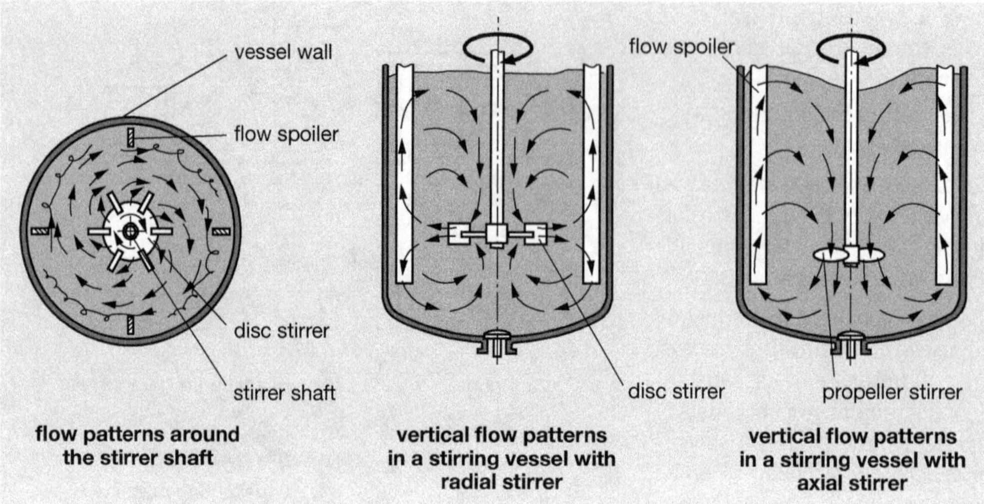

Figure 4.8: Flow patterns in the stirring vessel

Mixing of substances can be based on various physical and technological principles. Mixing by *stirring* and *static mixing* (flow mixing) are particularly important in coating technology; the latter is mainly used in two-component coating applications. If the material to be stirred is doughy or viscous, the term used is *kneading*; if it is solid (lumpy to powdery) then stirring is a variant of *dry mixing*.

The main tool used in stirring is the stirrer or, more accurately, the "stirring element". There are many types of stirrer, information about which can be obtained from the literature on chemical or process engineering. During stirring, the super-position of horizontal and vertical flows in the stirring vessel generally produces characteristic flow patterns, examples of which are shown in Figure 4.8. Since rapid intermixing requires the maximum possible eddy formation and circulation, in highly liquid media it is helpful to reduce the laminar circulation of the liquid by means of vertical *flow spoil-ers* fixed close to the wall.

The following rule of thumb generally applies: the lower the viscosity of the material to be stirred and the faster the circulation of the material by the stirrer, the shorter the mixing (stirring) period.

The quality of the inter-mixing or homogenisation is described by a statistical quantity known as the *degree of mixing* M, which is defined as follows:

Let us imagine that two substances that are homogeneously miscible with each other, e.g. water and acetone, are to be mixed together in a ratio by weight of 1:1. The two liquids are therefore first poured together and the concentration of acetone is then immediately measured (in one's imagination) at various points (i = 1, 2, ... n) in the liquid, producing the results w_1, w_2, ... n; the average acetone concentration in the overall quantity of liquid is always $\overline{w} = 0.5$ (50 %), of course. The standard deviation s for the concentration can now be obtained as a measure of the inhomogeneity of the mixture:

$$s = \sqrt{\frac{1}{n-1} \cdot \sum_{i=1}^{n} (w_i - \overline{w})^2}$$

The standard deviation s can then be normalised to its initial value s_0, which existed immediately after the liquids were poured together, i.e. with no (additional) mixing; this gives us the relative standard deviation:

$$s_{rel} = \frac{s}{s_o}$$

The degree of mixing is then simply

$$M = 1 - s_{rel}$$

It is thus easy to see that:

$M = 0$ at the start of mixing or before mixing (time $t = 0$)
$0 < M < 1$ after a finite mixing time ($t > 0$)
$M \rightarrow 1$ for a very extended mixing period ($t \rightarrow \infty$)

Let us then make the following "reasonable" assumption: the better the existing level of mixing, i.e. the greater the value of M, the lower the mixing speed dM/dt becomes:

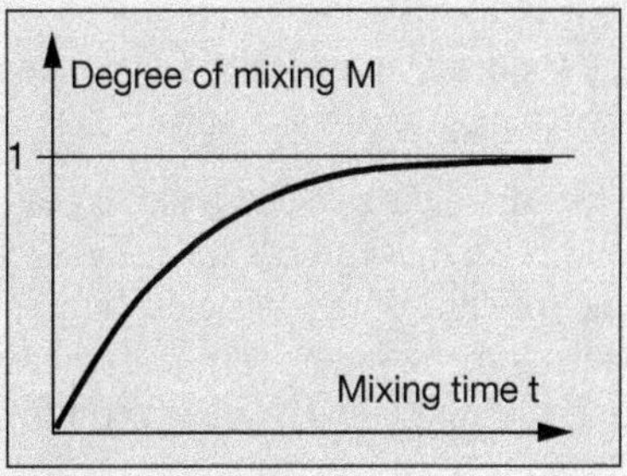

Figure 4.9: Dependence of degree of mixing on time

$$\frac{dM}{dt} = A \cdot (1 - M)$$

A is a mixing constant (proportionality factor dependent on the equipment and substance).

The solution to this simple differential equation gives us

$$M = 1 - e^{-A \cdot t}$$

the graph for which is shown in Figure 4.9.

The rate of increase in M is typical for a stirring process. What is interesting is that complete mixing ($M \rightarrow 1$) is never achieved in certain cases, no matter how long the mixture is stirred! This is always the case when the *mixing process* is superposed with a *demixing process* - e.g. due to sedimentation. This phenomenon frequently occurs in dry mixing processes in particular, where in many cases M passes through a maximum and then approaches its lower final value (limiting value) asymptotically. (To illustrate this, an extreme example would be like trying to distribute lead pellets evenly amongst polystyrene chips by stirring.) The fact that the degree of mixing passes through a maximum during dry mixing means in practice that there is an optimum length for the mixing period, the value for which has to be determined by experiment.

Shock phenomena

Shock phenomena during the course of the mixing process in coating manufacture are of great importance. The term predominantly refers to a partial *flocculation of the pigment* (agglomeration of pigment crystallites), which can occur within a pigment-containing mixture due to too pronounced differences in film former concentration. For example, too rapid an addition of concentrated binder solution to the pigment paste, i.e. during "let down", can lead to what is known as "*pigment shock*".

The theoretical explanation for this is that solvent diffuses out of the paste phase, with its low binder content, into the "make up" solution, which is rich in binder (to balance out the concentration); this causes the paste phase to virtually dry out and the pigment particles contained within it to "cake". The risk of shock can be reduced if the concentration of film former chosen in the material to be dispersed is as high as possible and if the concentrated film former solution is added particularly slowly during "make up", whilst stirring vigorously.

Conversely, flocculation can also occur during cutback with pure solvent ("*solvent shock*", "*dilution shock*"). The cause here is the diffusion of solvent into the paste phase, which is now richer in binder, thus reducing the stabilising effect of the film former molecules.[1)]

1) A third type of shock is known as "binder shock". This is caused by phase separations (precipitations) arising from incompatibilities between solvent and film former or between different film formers.

Sequence effects

The sequence in which the components are added during mixing or dissolution is often extremely important. The two examples below illustrate this:

When dissolving a solid resin, e.g. an epoxy resin in flake form, in a blend of organic solvents, the solid resin must always be absorbed first in a weaker solvent or a non-solvent to prevent clogging due to caking of the particles. Only then can the stronger solvent be added, in batches if necessary, whilst stirring intensively. (Alternatively it may be sufficient to add the resin slowly or in batches to the solvent blend, whilst stirring vigorously.)

If the solution of a water-soluble unpreneutralised resin in butyl glycol is to be diluted with water, for example, the amine must naturally be added and completely stirred in before the water is added, otherwise the resin would precipitate and coagulate.

4.8 Kneading

Highly viscous, doughy or tenacious mixtures, such as dispersion plasters, stoppers or adhesives, are prepared by *kneading*. Kneading is both a mixing or stirring process and also - given the relatively high shear forces in the material - a dispersing process. Since kneading plays only a minor part in coating technology - with the exception of screw compounding in powder coating production - we will not discuss it in any further detail here.

4.9 Dispersion, dispersing units

4.9.1 General introduction to dispersion

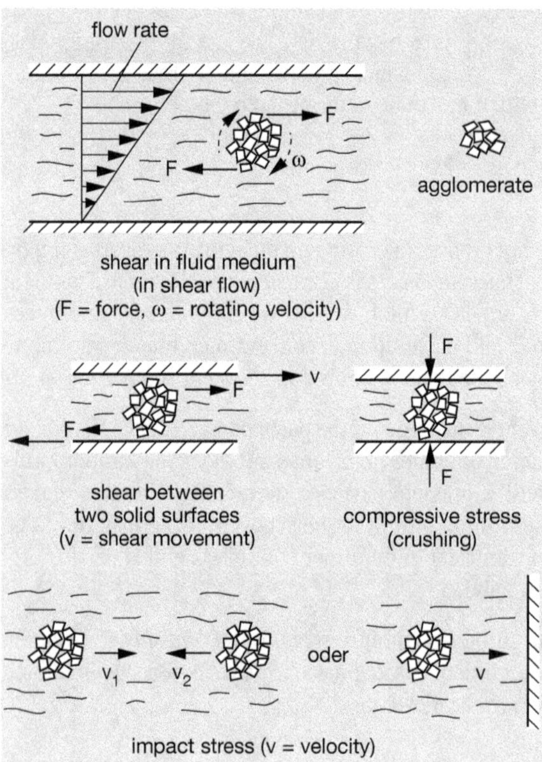

Figure 4.10: Stress mechanisms involved in the breaking down of pigment agglomerates

Dispersion takes place in the manufacturing process for paints and coatings whenever pigments and/or fillers have to be incorporated. *Dispersion* here means the breaking down of pigment agglomerates (agglomerations of crystallites or primary particles, or aggregates → 2.3.2.1) and their distribution in a liquid phase, usually a film former solution or - in coating powder production - a polymer melt.

Dispersion is often also known as "*grinding*"; it is the most energy-intensive and technologically demanding stage in the manufacture of pigmented coatings.

"Grinding" in the sense of size-reduction of primary particles may at best take place at the raw materials manufacturer's premises, where the coloristic and other properties are adjusted precisely to specifications. If further size reduction occurs due to excessively high energy input during coating manufacture, important properties such as hiding power, colour, durability etc suffer.

To ensure that dispersion proceeds as ef-

Table 4.3: Stress mechanisms in various types of dispersion equipment

	Equipment				
Stress	**Dissolver**	**Stator-rotor disperser, colloid mill**	**Triple roll mill**	**Attrition/ ball mill**	**Extruder/ kneader**
shear in fluid medium	● (t)	● (t)	● (l)	❍	● (h)
shear between two solid surfaces (mixed friction)			❍	●	❍
compressive stress (crushing)				●	
impact stress	❍	❍		❍	

● highly effective
❍ somewhat effective
(t) turbulent (in field of fine-structure eddies, viscosity relatively low)
(l) laminar (in laminar flow field, viscosity relatively high
(h) highly viscous (viscous resin melt or highly viscous material to be kneaded)

ficiently as possible in the given equipment (unit, apparatus), the composition of the material to be dispersed, comprising the three raw material groups - pigment(s)/filler(s), film formers ("grinding resin") and solvents, together with dispersing agents if applicable - must be adjusted by means of a process known as "*mill base optimisation*". (In aqueous pastes the grinding resin can be entirely replaced by wetting and dispersing agents.) See section 3.9 for more information about mill base optimisation.

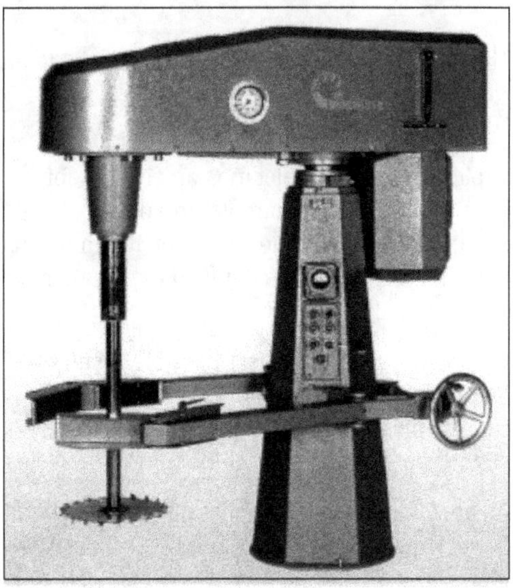

Figure 4.11: Simple production dissolver [from: company literature published by Wilhelm Niemann GmbH & Co.KG]

4.9.2 Stress mechanisms during dispersion

The application of a mechanical force on a particle, e.g. an agglomerate, is also known as mechanical stress. Figure 4.10 illustrates the *stress mechanisms* to which the particles (pigment agglomerates) may be subjected during dispersion.

Table 4.3 lists the principal stress mechanisms that occur in the various types of dispersing equipment.

4.9.3 Dispersion using dissolvers

An *dissolver*, which can be defined as a high-speed *toothed disc mill*, is used for premixing, predispersion and - if a particularly fine dispersion is not required or if the pigments are readily dispersible - also for complete dispersion. A simple production dissolver (without vessel) is shown in Figure 4.11; Figure 4.12 illustrates a *toothed disc.*

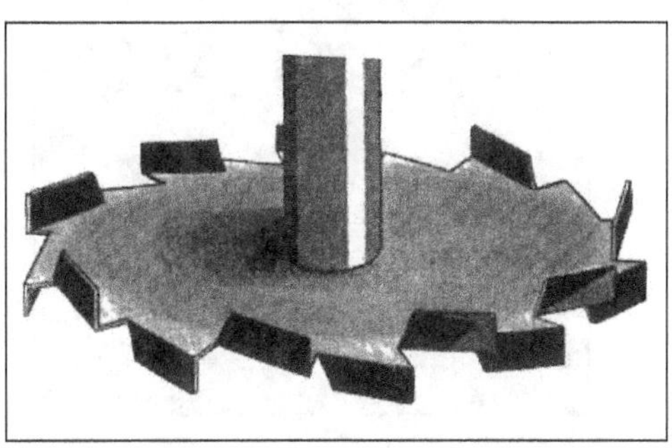

Figure 4.12: Toothed disk from a laboratory dissolver (schematic view)

Working mechanism

What is the basis of the dispersing effect in an dissolver?

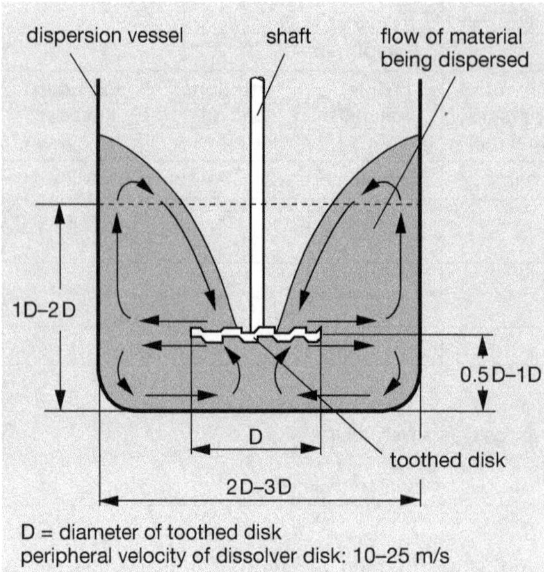

Figure 4.13: Optimum geometric proportions for dispersion in a dissolver (general recommendation according to [9])

The teeth on the disc are oriented relative to the circumference in such a way that an excess pressure is generated at their outer side and a reduced pressure at their inner side. Material to be dispersed that comes within the range of action of the teeth is thus exposed to considerable variations in pressure. Moreover, numerous small but strong eddies develop in the vicinity of the teeth, shearing and hence breaking down the agglomerates (see page 244 for a more detailed explanation of the processes involved).

Since it is the toothed edge of the disc that exerts the dispersion effect, the *peripheral velocity of the disc* v_p has to be regarded as the most important value for the operating status of an dissolver. This is calculated from the speed n and the disc diameter D as follows:

$$v_p = \pi \cdot n \cdot D$$

The energy P introduced into the material to be dispersed at torque M is

$$P = 2 \cdot \pi \cdot n \cdot M$$

Typical peripheral velocities are 10 m/s for matt interior emulsion paints and 25 m/s for solvent-based coatings, for example. In spite of these high speeds, the average particle sizes that can be obtained are still in the region of 10 µm or higher, which is insufficient for many topcoat applications, for example. (See detailed explanation on page 244 for the reasons for the limited dispersing effect of an dissolver).

In order to obtain the best possible results with an dissolver, the geometric proportions shown in Figure 4.13 must be complied with. In particular, setting the correct flow pattern is absolutely critical. The material to be dispersed must move in a spiral pattern from the side towards the disc - with slight swelling and smooth (laminar) flow - leaving the entire shaft clear (see Figure 4.14).

As preconditions for establishing this flow pattern (also known as "*donut effect*"), the viscosity must be set to its optimum level and the composition of the material to be dispersed adjusted accordingly.

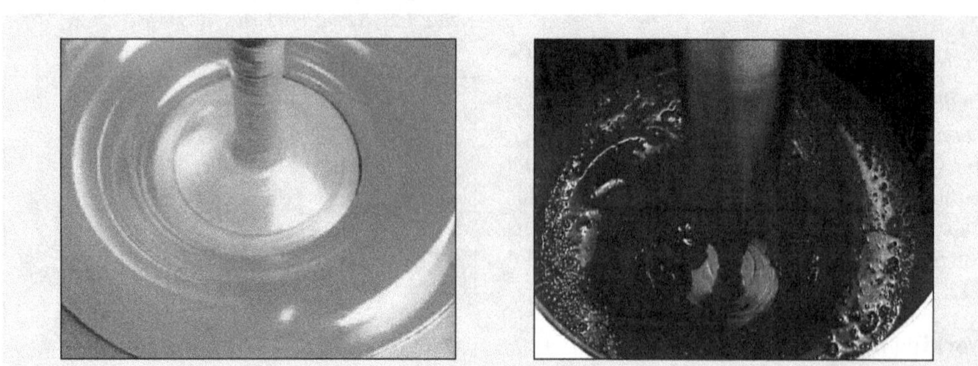

Figure 4.14: Optimum mill base with "donut effect" (left) and unfavourable ("lumpy") mill base with "dissolving hole"
[from: company literature published by VMA-Getzmann GmbH]

Too low a viscosity leads to spraying and foaming, whilst if the viscosity is too high or the material is lumpy, the disc rotates in a "*dispersing hole*", since the flow of material to the disc is interrupted.

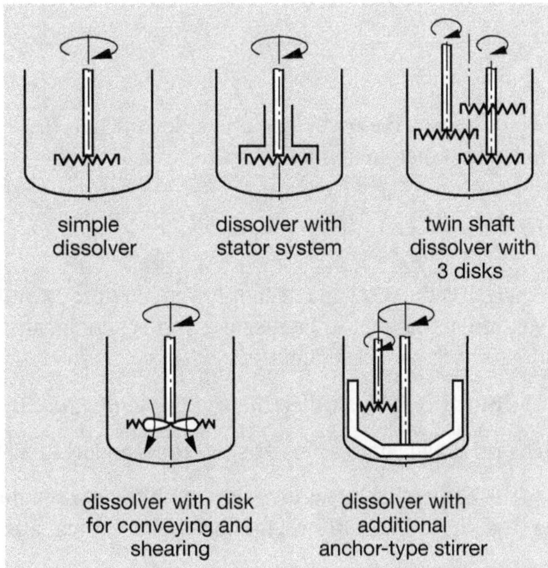

Figure 4.15: A number of design forms for dissolvers, according to [2]

Energy aspects

Since only a small percentage of the energy P (see above) introduced into the material to be dispersed is consumed in breaking down the agglomerates, i.e. the actual dispersion process, as a consequence of which the overwhelming proportion of the energy is converted into heat, the paste gradually heats up to a temperature of typically around 60 °C or even higher. This heating effect has an essentially positive influence on pigment wetting and hence on the dispersion process. However, if the temperature threatens to rise above a permissible value, the material must be counter-cooled by means of a cooling jacket fitted to the dissolver vessel or dispersion in the dissolver is deemed to have ended..

Modern production dissolvers automatically adjust their speed to the changing viscosity during the dispersion process in such a way that the motor power up to a given maximum speed is fully utilised. Under optimum conditions the dispersion time for a batch is from 10 to 30 minutes. On the basis of information supplied by one dissolver manufacturer, for example, 60 t per day of emulsion paint or 30 t per day of white solvent-based coating can be produced using a 100 kW dissolver under optimum operating conditions.

Special-purpose designs

To enable dispersing conditions to be adapted to the sometimes extremely diverse dispersion materials, many special-purpose designs and accessories are available for dissolvers. The following special forms are encountered relatively frequently:

- *vacuum dissolvers* (to prevent inviscous milling bases and to remove air bubbles)
- dissolvers with coolable or heatable vessels
- dissolvers designed as stator-rotor systems
- dissolvers with multiple stirring elements or toothed discs
- dissolvers fitted with wall scrapers or anchor-type stirrers.

Figure 4.15 illustrates a number of dissolver models.

The mechanical energy introduced by the dissolver disc into the material to be dispersed (mill base), i.e. a mixture of pigment powder and liquid (binder solution, water with wetting and dispersion additives), initially causes a coarse turbulence to develop in the vicinity of the disc; this is then converted into a fine turbulence consisting of fine-structure eddies and ultimately into molecular movement, i.e. heat. This "dissolution" of work, whose end point is heat, is called *dissipation*.

A shear stress acts on the particles or agglomerates located in the vicinity of the fine turbulence, and this leads to the breakdown of the agglomerates. This shear stress τ_t is defined as the shear force F_s relative to a surface section A of the agglomerate,

$$\tau_t = \frac{F_s}{A}$$

which causes the particle to break down when the agglomerate bond forces are exceeded. The following relation applies:

$$\tau_t = \text{const.} \cdot \rho^{\frac{1}{3}} \cdot \left(\frac{P}{V}\right)^{\frac{2}{3}} \cdot x_P^{\frac{2}{3}}$$

where P is the mechanical energy applied (torque multiplied by angular velocity), V is the given volume of material to be dispersed, ρ is the density of the material to be dispersed and x_P is the current particle diameter.

The quantity P/V is called the *energy dissipation density*.

The principal conclusions arising from the formula above are as follows:

- The dispersing effect increases with the energy dissipation density.
- The shear stress destroying the agglomerates falls as the particle size decreases.

The second statement is extremely important. From this we can further conclude that below a certain particle size the shear forces are no longer sufficient to overcome the agglomerate bond force and hence to act dispersively.

To summarise, we can state that:

At a given energy dissipation density and given agglomerate bond forces, there is a certain value below which the particle size cannot fall, no matter how long dispersion is continued in the dissolver.

In the most favourable case, i.e. at the maximum achievable (local) energy dissipation densities in the dissolver of around 10^4 kW/m^3 and at low agglomerate bond forces ("dispersing hardnesses"), the smallest achievable particle size is around 10 μm.

4.9.4 Dispersion using triple roll mills

Triple-roll type roll mills are rarely found in coating production now, although they are still used to some extent in the printing inks sector (especially for offset printing inks based on non-volatile mineral oils), for pigment concentrates and in specialist sectors of the coating materials industry. The single and double roll mills have practically disappeared.

Design principle

The basic design of a triple roll mill – also known as a *triple roller* – is illustrated in Figure 4.16. The rolls operate within the following speed ranges: $40 < n_1 < 100$ rpm, $140 < n_2 < 200$ rpm, $300 < n_3 < 600$ rpm (rpm = revolutions per minute); these are roughly in the proportion of 1 : 3 : 9, sometimes also 1 : 2 : 4.

The rolls are made from case-hardened steel; they are ground smooth but have a certain grip. To compensate for deflection under load (up to around 100 bar surface pressure), the diameter of the rolls is slightly greater in the middle than at the ends; this is known as *roll camber*.

In order to be able to vary or to set an upper limit for the temperature of the paste in the nip – despite substantial evolution of heat due to shearing – the rolls can be cooled or heated with water from inside. The rolls are arranged horizontally in smaller mills and obliquely in larger models. The operating conditions for modern triple rollers (speeds, contact pressure, nip width, roll temperature) can of course be measured, regulated and logged electronically. In large triple roll mills the rolls can be up to around 0.5 m in diameter and up to approx. 2 m in length.

The dispersing effect of a triple roll mill arises from the fact that the three rolls rotate at differing speeds and thus shear the paste containing the agglomerates in the two nips. Since the pigment particles are constantly enveloped in liquid during this process, the dispersion is very "gentle"; there is virtually no dry friction between the pigment particles and the surface of the roll.

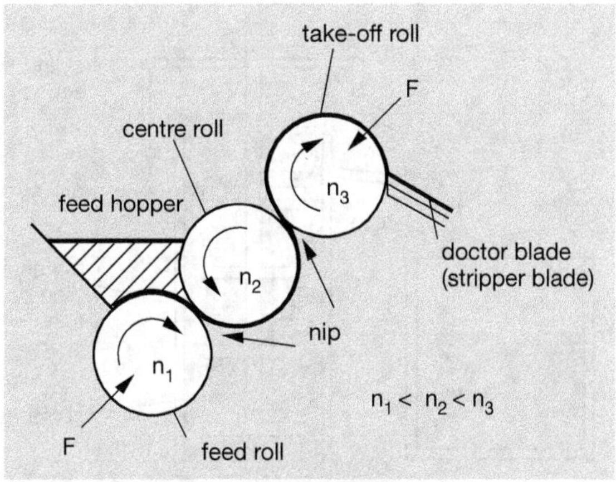

Figure 4.16: Triple roll mill (triple roller)
(F = contact pressure, n = rotating velocity)

Operating notes

The theory behind roll mill dispersion gives rise to the following general rules for the most efficient operation of a triple roll mill:

- nip not too narrow (otherwise throughput will be too low and risk of dry-running)
- speed no higher than necessary (for a cost-effective throughput)
- viscosity as high as possible (to obtain high shear forces).

Let us conclude with a look at the characteristic strengths and weaknesses of the triple roll mill.

- Strengths:
  - gentle dispersion, both mechanically (important for abrasion-sensitive pigments, e.g. coated types) and in respect of temperature
  - very pasty materials can be dispersed
  - mechanism is straight-forward and easily adjustable
- Weaknesses:
  - open system: substantial, uncontrolled solvent loss when volatile solvents are used (extraction necessary; solvent content of paste unknown)
  - relatively complex monitoring systems required
  - relatively low throughput rate, expensive

4.9.5 Dispersion using attrition mills

4.9.5.1 Dispersion mechanism in the presence of grinding media

Whilst dissolvers and triple roll mills are examples of dispersing units that operate *without* grinding media, *ball mills* (rarely used now) and *attrition mills* contain balls that act as grinding media. The dispersing effect arises from the balls being set in translational and rotational movement, as a result of which they impact both against one another and against the walls and the other surfaces in the grinding compartment. The stress mechanisms exerted on the agglomerates by this movement are principally compressive stress (crushing) and shear - through both laminar shear flow and mixed friction[1].

The following simple illustration is a very helpful means of obtaining a basic understanding of the processes involved in dispersing with grinding media. In order for agglomerates to be able to be broken down, they must

- firstly pass between shearing or impacting surfaces (grinding medium/grinding medium, grinding medium/wall, etc.) and
- secondly be subjected to sufficiently strong stresses.

1) Mixed friction refers to the intermediate state between friction of dry surfaces against one another and friction with a liquid film (lubricating film) located between the shearing surfaces.

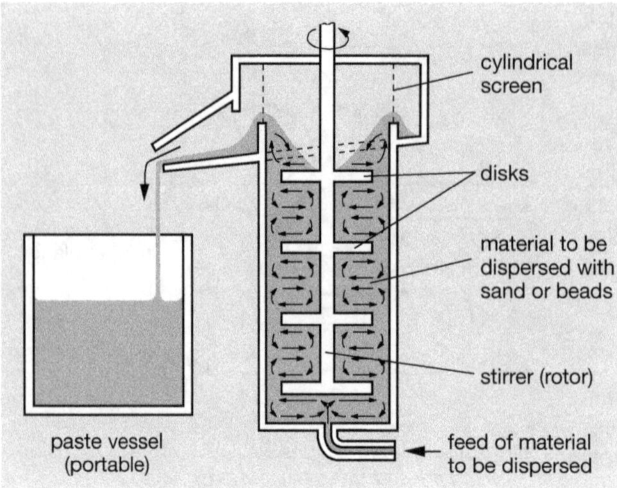

Figure 4.17: Vertical attrition mill, open at top ("sand mill")
[from: company literature published by Hüttenes-Albertus GmbH (1987)]

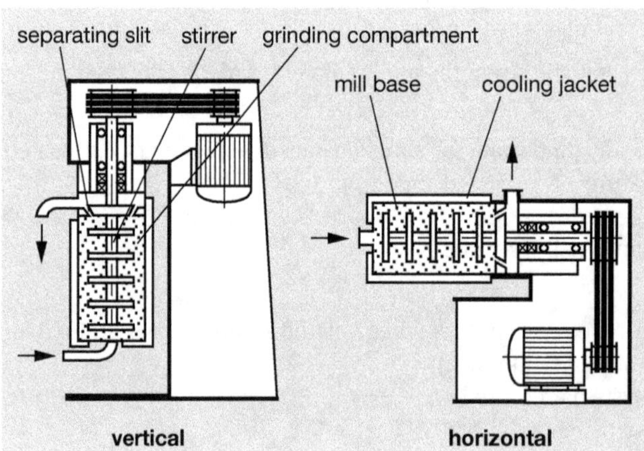

Figure 4.18: Basic design of closed attrition mills, according to [2]

Successful dispersion is therefore subject to both spatial and energetic preconditions. Both conditions obey statistical regularities due to the chaotic movement of the grinding media, but we will not discuss these in any further detail here.

4.9.5.2 Design and operating parameters for attrition mills

Whereas the traditional *ball mill* comprises a rotating horizontal closed cylinder partially filled with large balls and material to be dispersed, in an *attrition mill* the vessel is at rest and the mixture, consisting of relatively small balls and material to be dispersed, is kept moving by a rapidly rotating stirring element.

Design principle

The first attrition mills essentially consisted of a vertical, coolable "*grinding compartment*", open at the top, in which *Ottawa sand* (silica sand, round, grain size 0.8 to 1.0 mm) was set in rapid motion by means of a multi-disc stirring element. The material to be dispersed was pumped in from below and left the mill flowing freely (without pressure) through a cylindrical screen (see Figure 4.17). This type of mill was known as a "*sand mill*" even if beads were used in place of sand as the grinding medium.

The characteristic weaknesses of the sand mill - narrow viscosity range for the mill base, low throughput, risk of overflow, development of sand spouts due to drawing in of air, solvent emission - were overcome by the transition to closed mills. The Ottawa sand was replaced by grinding beads, giving rise to the name "*bead mill*".

Attrition mills (bead mills) can have a gross capacity of up to around 1000 l and a drive power of 300 kW. The machines found in coating production are always smaller than that, however, with a maximum capacity of around 250 l. Figure 4.18 illustrates the basic design of vertical and horizontal closed attrition mills.

Grinding media

The beads used in attrition mills are made from materials such as steel, zirconium oxide, aluminium oxide, Si/Al/Zr mixed oxide (SAZ beads), steatite (modification of talc), glass and plastics; their diameter mostly lies in the range from 0.1 to 3 mm. The harder the beads, the greater the intensity of dispersion, but the wear on the mill increases at the same time. The density of the beads has also influence on the dispersion result. Steel beads cannot be used for dispersing light-coloured pastes because of the dark grinding dust they produce. Developments are leaning strongly towards the use of ZrO_2 beads, owing to the hardness, smoothness, stability (service life) and density of the beads. They contain yttrium or cerium oxide to stabilise the crystal lattice. Dispersion of nanoparticles requires novel particle diameters, especially of around 0.01 mm. Larger beads would have too low a "success rate".

The beads are accelerated to a velocity in the order of 10 m/s and also set in rotation by the *rotor* (stirring element) or by the discs or pins attached to it. Since the probability of an agglomerate passing between two beads increases with the number of beads in the mill, the efficiency of the machine rises as the diameter (d) of the beads falls. The number of beads is proportional to $1/d^3$, i.e. halving the diameter of the beads causes the number of beads to increase eight-fold. The aim is therefore to use the smallest beads possible.

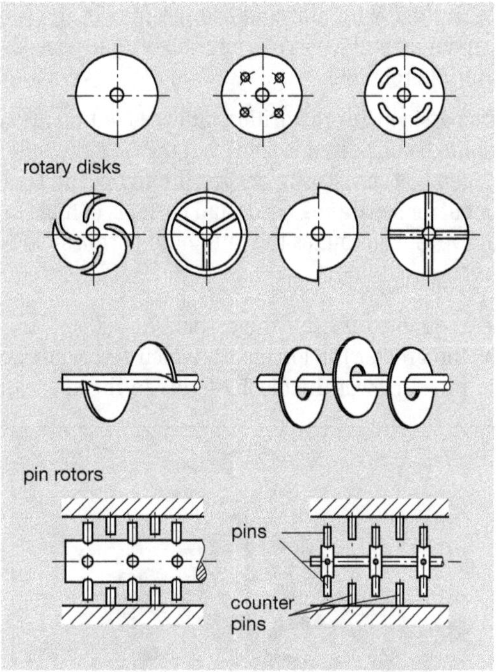

Figure 4.19: Disk and rotor designs for attrition mills, according to [2]

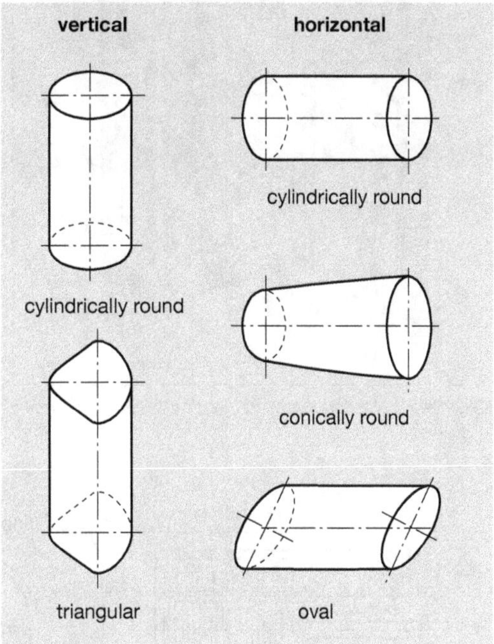

Figure 4.20: Grinding vessel designs for attrition mills [2]

Figs. 4.19 and 4.20 show examples of a number of different design variants of conventional attrition mills. Basket mills are also becoming more widespread - stator-rotor mills filled with beads and boasting an additional dissolver blade which is immersed in the material to be dispersed.

Operating notes

Unfortunately we do not have room here to discuss in detail the advantages and disadvantages and the attributes of the various mill designs. We can make a few general comments, however.

There is a certain trend away from vertical towards horizontal mills. This is due to the following advantages of the horizontal arrangement over the vertical design:

- smaller grinding beads can be used (below 0.1 mm)
- the grinding media compartment can be filled to a capacity of about 85 %
- the machine is easier to start up after shutdown
- there is a more uniform dispersing effect across the entire length of the mill

A better dispersing effect is attributed to non-circular grinding vessels (grinding compartments) due to the formation of compressions in the grinding media bed. These are not very widespread, though.

Mills fitted with pins and counter-pins in the walls of the rotor and vessel are suitable for the dispersion of highly viscous and thixotropic pastes, e.g. for printing inks, and represent an alternative to triple roll mills.

Dispersion with an attrition mill is a continuous sub-process within coating manufacture. The material to be dispersed is conveyed through the mill at a varying level of excess pressure (up to 6 bar), depending on viscosity, desired throughput and mill design. The design of the *separating system* (screen or slit for separating beads and paste) is of huge importance to the operational efficiency of a mill and above all determines the minimum permissible bead size. In general, the follow-ing applies:

The smaller the beads,

- the more effective the dispersion
- the greater the force with which the beads are carried to the mill outlet or separating system
- the narrower the mesh or slits in the separating system must be.

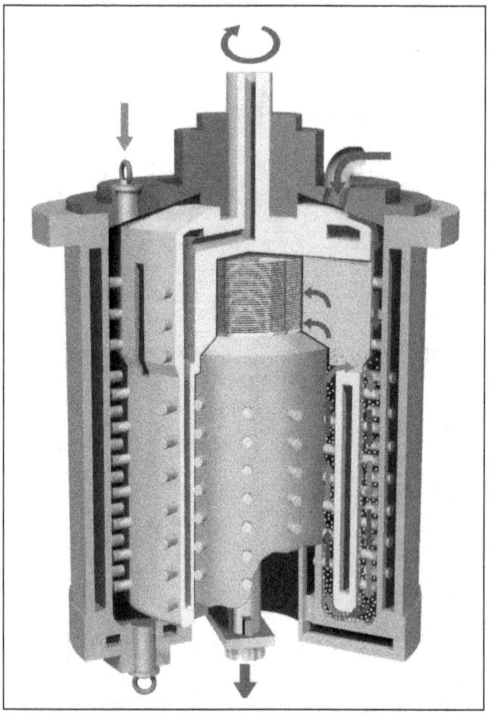

Figure 4.21: Nip-ball mill with circulating beads, Source: Draiswerke

Optimising the dispersion process

It is vital that the beads are prevented from piling up ahead of the separator, blocking it and subjecting it to unacceptably high mechanical stress; this phenomenon is known as (hydraulic) *bead jam*. The smaller the beads, the greater the risk of this. An empirical rule suggests that a bead jam occurs when the increase in the mill base throughput causes the temperature of the emerging paste to rise. At the sign of a jam,

- the paste throughput should be reduced,
- and/or the paste viscosity reduced (by adding solvent),
- and/or the rotor speed in-creased.

The principal objectives of modern mill designs are aimed at minimising or ideally eliminating altogether the development of a jam ahead of the separating system. As examples of the various technical solutions for jam prevention, Figs. 4.21 and 4.22 show two common mill types: the *nip-ball mill* also annular passage mill with circulating beads and the *centrifugal fluidised bed mill*. In the first of these jamming is prevented by forced bead circulation in the narrow grinding nip. In the second, the centrifugal force of the beads rotating around the mill shaft prevents them from jamming ahead of the centrically rotating screen system.

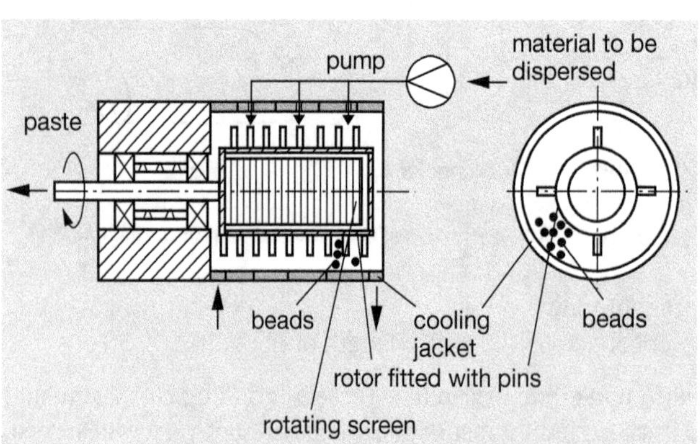

Figure 4.22: Centrifugal fluidised bed mill, according to [8]

Another important aspect of mill design is the means of cooling the mill base. The considerable energy dissipation density (energy per unit volume) introduced into an attrition mill means that cooling is definitely necessary. In larger, modern mills, not only the grinding compartment walls but also the rotor are cooled.

Mill bases for bead mills must be thoroughly pre-dispersed in an dissolver; otherwise the less demanding pre-dispersion is transferred into the complex bead mill dispersion, which ultimately leads to higher costs.

Dispersion control

In practice it is often necessary to check the progress and/or the quality of the dispersion - in the dispersing equipment itself - quickly and as simply as possible. A grind gauge (according to *Hegman*) is often used for rough estimates; this provides an approximate guide to the upper limit of the particle size distribution , i.e. not the size distribution itself. An indirect, more complicated yet in return significantly more accurate and - in terms of coating properties - more meaningful method involves determining the colour intensity of the paste or, e.g. the gloss or haze of the coating (see section 8.4.1 for more information about these methods of measurement).

Table 4.4 provides a qualitative summary of the dependence of the dispersion quality on the main operating parameters in an attrition mill.

4.9.5.3 Residence time distribution in an attrition mill

An agglomerate entering the mill may be carried - by chance - relatively quickly to the outlet by the turbulent conditions in the grinding compartment or it may remain in the mill for a long time. There is therefore no uniform *residence time* for agglomerates in the mill, but only a *residence time distribution.* Since a short residence time leads to poor dispersion and a long residence time to good dispersion, a broad residence time distribution also results in a broad *particle size distribution*, which is generally undesirable. By way of example, Figure 4.23 illustrates the residence time distribution for the dispersion of a talc suspension in a horizontal laboratory-scale bead mill. It can be seen that short residence times lead to large particle sizes ("Hegman values"), i.e. incomplete dispersion[1].

Mean residence time

Whilst the residence time distribution can only be determined by experiment, the *mean* (average) *residence time* $\bar{t}$ can be calculated from the paste throughput (volume flow rate) $\dot{V}_P$ and the free grinding compartment volume V_f:

$$\bar{t} = \frac{V_f}{\dot{V}_P}$$

Table 4.4: Dependence of the dispersion quality on the operating parameters in an attrition mill

change in parameter	Dispersion quality	Note
grinding media fill level increases	rises ⇒ optimum ⇒ falls	limited by separating systems risk of jamming increases
grinding media size falls	rises	restricted by separating system; risk of jamming increases
grinding media hardness increases	rises	wear on mill increases
grinding media density	increase	
temperature rises	rises ⇒ optimum ⇒ falls	take temperature sensitivity of mill base into consideration
mill base viscosity increases	rises ⇒ optimum ⇒ falls	risk of jamming increases (reduce throughput)
throughput increases	falls	risk of jamming increases
speed increases	rises	temperature rises or greater cooling required; risk of "overdispersion"

1) The residence time distribution is determined by injecting a test substance into the feed directly ahead of the mill inlet for a short time and then analysing the time dependence of the concentration of the substance beyond the mill outlet.

The free grinding compartment volume is equal to the total grinding compartment volume minus the total volume of the beads (approx. 0.5 times the volume of the bead bed), the rotor and other fixtures. In practice the mean residence times are in the region of a few minutes.

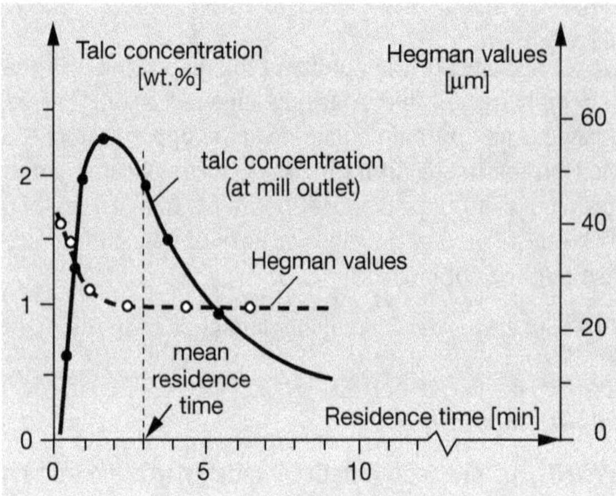

Figure 4.23: Residence time distribution and Hegman values after injection of talc into a bead mill, according to [6]

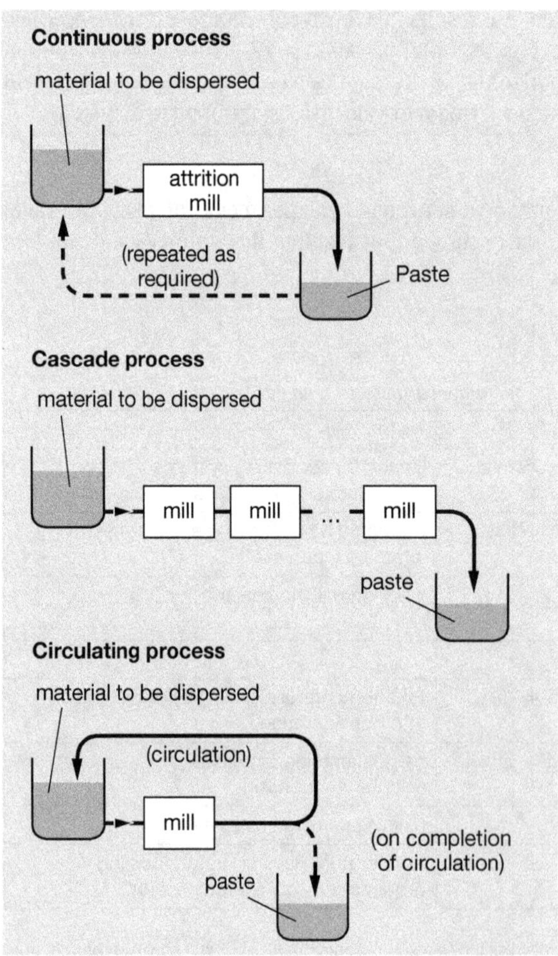

Figure 4.24: Basic process variants for dispersion in attrition mills

Influences on residence time

In order to obtain the narrowest possible residence time distribution, at a given throughput the axial flow rate of the material to be dispersed must be as high as possible to ensure that the agglomerates are continuously forced to move in the direction of the mill outlet. A narrower residence time distribution can thus be obtained by means of

- a longer, relatively narrow grinding compartment
- larger, less heavily perforated discs
- a larger number of discs.

All three of these measures increase the risk of jamming (see above), however, so a compromise has to be found here.

4.9.5.4 Continuous and circulating processes

Process variants

A single passage of the paste through an attrition mill is frequently insufficient for complete dispersion. Multiple dispersion then becomes necessary, for which two essentially different processes exist (see Figure 4.24).

- Continuous/cascade processes:
  The mill base is dispersed several times in succession in a single mill or passed through a cascade (series) of mills.
- Circulating processes:
  The mill base is conveyed from a receiving vessel through the mill and continuously from the mill outlet back into the receiving vessel.

The two processes differ in regard to the breadth of the residence time distribution. The mean residence time for one cycle through the mill is calculated - as we have already said - as

$$\bar{t} = \frac{V_f}{\dot{V}_P}$$

where V_f is the free grinding compartment volume and $\dot{V}_P$ the paste flow rate (volume flow rate). For n complete (discrete) cycles through the mill in the continuous or cascade process, the mean residence time is n times greater.

For the circulating process, only a mean number of cycles can be stated. This is

$$\bar{z} = \frac{\dot{V}_P \cdot t}{V_P}$$

where $\dot{V}_P$ is the paste flow rate (volume flow rate), t the machine running time (throughput time) and V_P the paste volume.

Above a mean number of cycles ($\bar{n}$) of five, the mean residence time for the circulating process can be estimated according to the following equation:

$$\bar{t} = \frac{V_f \cdot t}{V_P}$$

in this case V_f representing free grinding compartment volume, V_P paste volume and t the machine running time (throughput time).

Process comparison

If we now compare the breadth of the residence time distribution for the continuous process of 5, 10 or 20 cycles through the mill with that for the circulation process based on a *mean* cycle number of 5, 10 or 20, we obtain the graph shown in Figure 4.25; the circulation process results in a visibly broader residence time distribution (the reasons for which are made clear in the detailed explanation below).

The theoretical explanation given above can be summarised by the following qualitative statements.

- Circulating process:
  Lower operating and investment costs, but broader residence time distribution (less uniform dispersion)
- Continuous/cascade process:
  Higher operating costs for continuous process (single mill) or higher plant costs for cascade process (several mills), but narrower residence time distribution (more uniform dispersion).

The greater breadth of the residence time distribution in the circulating process can be explained as follows:

An agglomerate that leaves the mill and returns to the receiving vessel may remain there by chance for a relatively long time before (possibly) reaching the mill again, or it may pass straight to the outlet and thus back into the mill. This considerable divergence in residence times in the receiving vessel leads to a correspondingly wide divergence in the total residence time in the mill.

A model calculation shows that even after a mean cycle number of $\bar{n} = 5$, 0.7 % of the mill base still has not passed through the mill. In a continuous process comprising 5 cycles, by contrast, the entire paste would have passed through the mill five times.

4.9.6 *Dispersion in the extruder in the manufacture of powder coatings*

The dispersion of pigments in the context of powder coatings manufacture takes place almost exclusively in extruders (screw compounders). So far trials with other devices have not proved successful.

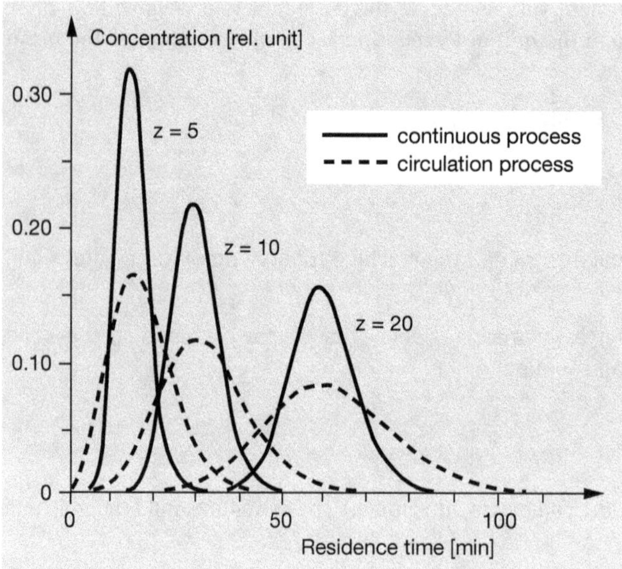

Figure 4.25: Residence time distribution for continuous and circulation processes, according to [6]

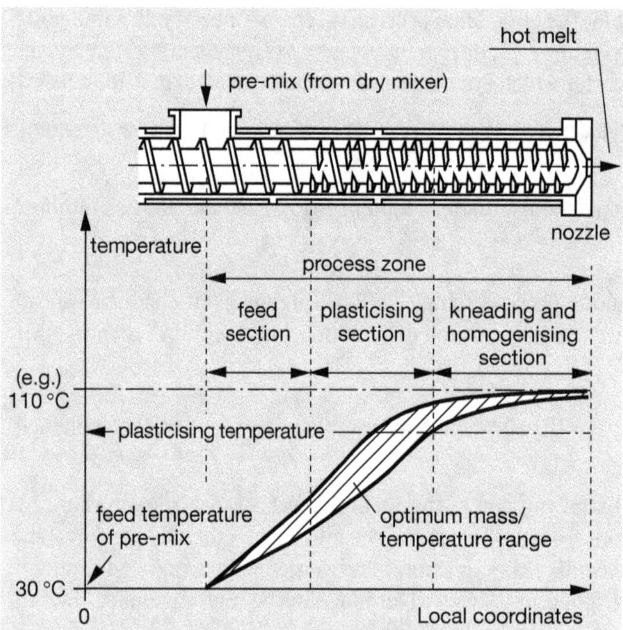

Figure 4.26: Temperature profile diagram for powder coating dispersion in a single-screw extruder ("co-compounder")

Design principle

The central component of an extruder is the screw, which rotates in a cylindrical runner; the screw draws the resin-pigment-additive mixture out of the feed hopper and then passes it through the runner, melting and kneading (shearing) it as it goes. Two rival types of extruder have become established in the powder coatings sector, the first being the *twin-screw extruder*, with two intermeshing screws rotating in the same direction, and the second being the *single-screw extruder*, in which the screw is periodically moved back-wards and forwards (known as a "co-compounder").

The screws in the twin-screw extruder are fitted with kneading discs in addition to their conveying elements. In the single-screw extruder the kneading effect derives from the complex shape and movement of the screw combined with the kneading cogs fixed to the inner wall of the barrel. In contrast to plastics extrusion, the powder coating melt leaves the extruder virtually without pressure and then passes into the cooling device.

Mode of action

Figure 4.26 shows a temperature profile diagram for a single-screw extruder used for powder coatings. The plasticisation or melting of the compound derives exclusively (except in the start-up state) from the mechanical shearing and kneading operations. The temperature profile along the screw is adjusted by means of a barrel cooling system, which can be regulated on a zone-by-zone basis, and optionally by an internal screw cooling system.

Due to the high shear forces in the extruder, the pigment agglomerates are broken down in a peeling action from the outside inwards. The shear forces and hence also the energy dissipation density increase as the speed of the screw increases (up to around 400 rpm). At the same time, however - if operating at 100 % capacity - the throughput (mass flow rate) also increases, and the mean residence time falls. Where technically feasible, the mean residence time can be increased by extending the kneading zone.

Operating parameters

The most critical parameter in powder coating extrusion is the temperature of the melt. Since the material to be kneaded already contains all of the components of the powder coating, it is reactive to heat. If local temperatures in the extruder are too high, partial fusion (curing) may occur, which must be avoided at all costs. The maximum temperature of the melt must be at least 20 K below the stoving temperature for the powder coating; at the same time the mean residence time should be no longer than is absolutely necessary for dispersion and the residence time distribution should be as narrow as possible. Typical data are given below:

- temperature: 80 to 100 °C (locally and for short periods may also be higher)
- mean residence time 15 s.

In general the operating conditions should be adjusted such that

- shear effect (speed, torque)
- mean residence time
- throughput
- temperature
- temperature profile
- melt viscosity

all lie in the optimum range. Since the above parameters influence one another, this is a very complicated task.

Typical data for a large powder coating single-screw extruder are as follows: screw diameter 140 mm, screw length 7 m, drive power 139 kW, maximum throughput 2500 kg/h.

4.10 Filtration

After cutback, coatings always contain a certain amount of undesirable particles, e.g. dust particles, grinding dust from grinding media, oversize pigment particles and gel particles. Before being decanted, these components must be removed from the coating as fully as possible by means of filtration.

Types of filter

We essentially distinguish here between surface and depth filtration. Figure 4.27 illustrates the method of operation of the two variants.

Surface filters

These filters, which are also known as *screens*, consist of a layer of wire or synthetic fibre mesh in the form of a cloth or bag ("screen bag"). By fixing (e.g. heat setting in the case of synthetic fibres), the screen is given a defined, practically uniform mesh size, usually in the range from 5 to 800 µm. In conjunction with the fibre thickness, the mesh size also determines the proportion of free surface area in the total surface area, known as *free screen area*. The principle behind surface filtration ("screening") is that all particles that do not pass through the mesh are separated off as *screening residue*. Screens are used in coating production - and also in the application of coatings - whenever the passage of oversize solid particles needs to be prevented ("policing function").

If there is a risk of the screen becom-ing clogged by separated particles, vibrating screen machines can be used.

Depth filters

Depth filters (or, more accurately, filter elements) are porous plates, pipes (cartridges, tubes) or bags of varying thicknesses. A diverse range of materials is used; examples include cellulose fibre/kieselguhr press mouldings for plates, synthetic resin-bonded filament windings for cartridges

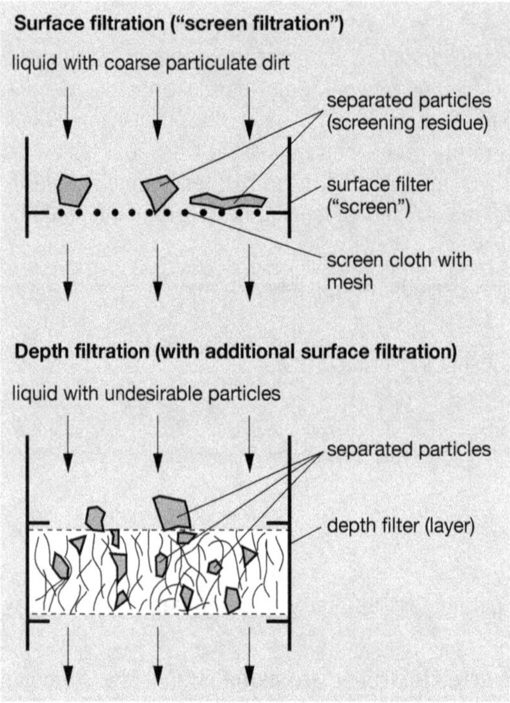

Figure 4.27: Principle behind surface and depth filtration

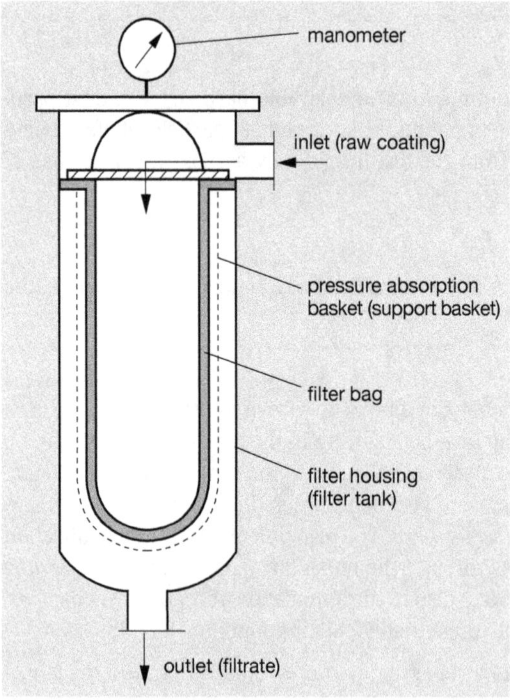

Figure 4.28: Bag filter apparatus

and needle felt for filter bags. The mechanism underlying depth filters involves the particles that are to be separated passing into the pores and being deposited adsorptively or trapped mechanically in the chaotic labyrinth. Particles that do not pass into the pores are retained by surface filtration (screen-ing effect). In contrast to the constant mesh size, which is the central characteristic for screens, the average pore size in filters - due to the entirely different separating mechanism and the wide variation in pore sizes - is a less useful specification.

The characteristic value for depth filters is instead the *retention rate*, which can be determined by test filtrations. The retention rate depends not only on the particular filter element but also on the test impurity and on the physical filtration conditions (flow rate, differential pressure and temperature).

Operating parameters

By (absolute) *retention rate* we mean a particle size (in µm), for which the following qualitative rule applies:

- particles larger than the retention rate are almost completely separated; most of the particles smaller than the retention rate pass through.

The retention rate must not be confused with the *degree of separation*. The latter describes the proportion by weight of impurities that is separated by filtration.

An important operational quantity in filtration is the differential pressure between the upper and lower side of the filter layer. It should be as low as possible and should not undergo any sharp rise during filtration.

In practical operation it is particularly important to note that all depth filter elements not only separate out particles but also release particulate components themselves, e.g. pieces of fibre. This problem can be alleviated by installing a surface filtration stage (screen) downstream.

As an example of a coating filtration unit, Figure 4.28 illustrates the cross-section of a bag filter apparatus. Filtering devices are generally combined with a pump (e.g. gear pump) and connecting tubes to make a portable unit, which can then be quickly moved to different parts of the manufacturing facility.

4.11 *Further information about the manufacture of water-borne paints and coatings*

The manufacture of water-borne paints and coatings is fundamentally no different from that of solvent-based products. It involves a number of special features, however, which are outlined below:

- risk of plant corrosion and hence damage to product (discoloration, coagulation) from corrosion products
  → use stainless steel, plastics or protective coatings; avoid non-ferrous metals in particular
- risk of binder coagulation due to ionic impurities (especially tap water containing lime, regeneration residues from the deionisation system)
- importance of the pH with regard to the stability of the waterborne coating
- increased foaming
  → use defoaming agents
- risk of bacterial and fungal attack (particularly prevalent in emulsion paints and similar coatings)
  → use sterile production methods; add biocides if necessary
- risk of incrustation of production equipment due to irreversible surface drying
  → work under clean conditions, immediately wash away spillages or product residues from the equipment
- dispersions and colloidal film formers frequently not thermally stable or shear resistant
  → use gentle pigment dispersion methods, e.g. in an dissolver only, at max. approximately 50 °C, or (if possible and necessary) without film formers in a blend of water and dispersion additive only
- risk of freezing of water-containing raw materials, semi-finished products and finished products
  → keep storage temperatures above 5 °C
- risk of precipitation and/or coagulation of film formers under the influence of atmospheric CO_2 (drop in pH)
  → minimise air contact during production and storage; optimise formulation
- risk of shocks greater than in solvent-based formulations
  → perform "make up", dilution and neutralisation, if required, slowly and with stirring; avoid large differences in concentration; pre-dilute strong solvents and neutralising agents.

Sources and references for Chapter 4

[1] G. Benzing: Pigmente in der Lackindustrie. Expert Verlag GmbH, Ehningen 1992

[2] A. Goldschmidt and H.-J. Streitberger, BASF Handbook: Basics of Coating Technology, 2nd edition, Vincentz Network Hanover, 2007

[3] W. Hemming: Verfahrenstechnik. 7th edition, Vogel Buchverlag, 1993

[4] E. Ignatowitz: Chemietechnik. 4th edition, Verlag Europa-Lehrmittel, 1992

[5] H. Kittel: Lehrbuch der Lacke und Beschichtungen, Vol. 8, S. Hirzel Verlag, Stuttgart, 2nd edition, 2004

[6] R. Laible et al.: Umweltfreundliche Lackiersysteme für die industrielle Lackierung. Expert Verlag GmbH, Ehningen 1989

[7] B. Müller and U. Poth, Coatings Formulation, Vincentz Network, Hanover 2006

[8] J.M. Oyarzún, Pigment Processing, Vincentz Verlag, Hanover 2000

[9] J. Prieto and J. Kiene, Holzbeschichtung, Vincentz Network, Hanover 2007

[10] D. P. Roelofsen, farbe + lack 97 (1991), 235 ("Entwicklungen auf dem Gebiet des Pigmentdispergierens mit Rührwerkskugelmühlen")

[11] K. Sponsel, W. O. Wallenfang, I. Waldau: Lexikon der Anstrichtechnik, 1 - Grundlagen. 9th edition, Verlag Georg D. W. Callwey, Munich 1992

[12] Vincentz Network: "Lehrgang Lacktechnologie", Module 2 (November 2008); "Lackherstellung" (T. Brock, Hochschule Niederrhein)

[13] F. Vock: Verfahrenstechnik und Prozeßkette Lackherstellung. CC Press AG, Termen (CH), 1997

[14] RömppOnline, with updates, see www.roempp.com

[15] Ullmann's Encyclopedia of Industrial Chemistry, 6th edition, CD-ROM, VCH: Weinheim, (January 2001)

[16] Literature from the firms: Bernd Schwegmann GmbH & Co. KG, Buss AG, Fryma-Maschinen AG, Hüttenes-Albertus GmbH, Netzsch Feinmahltechnik GmbH, RHE Händel Engineering GmbH, Seitz-Filterwerke GmbH & Co., VMA-Getzmann GmbH, Wilhelm Niemann GmbH & Co. KG

5 Substrates and pretreatment

5.1 General introduction

Strictly speaking, coating materials are not end products but intermediate products. The true end product is the finished, cured coating. This in turn is likewise only part of a composite consisting of the coating itself and the coated substrate. Within this composite a coating can serve various functions, e.g. protecting the coated object against damaging environmental influences, enhancing its appearance or – by means of special physical or functional effects – improving its surface finish. These tasks can only be completed if the coating adheres strongly to the object. If it were to become even partially detached from the substrate, it would not fulfil its purpose.

The *strength of adhesion* to the substrate is not determined by the coating material alone, however, but substantially also by the properties of the coated substrate. As prerequisites for optimum adhesion, firstly the object to be coated must be given an appropriate pretreatment or preparation, and secondly the coating material must be matched to the substrate in terms of specific film properties, e.g. hardness or thermal expansion behaviour. In addition to this, the substrate governs the method of application and the possible curing methods, which are determined most importantly by the heat resistance of the substrate and by its dimensions.

All solid substrates can be coated, from inorganic materials such as metals or concrete, to organic materials such as paper, wood or leather. The following sections cover the most significant proper-

Table 5.1: Surface pretreatment of various substrates, according to [23]

Substrate	Contaminants	Cleaning		Preparation/pretreatment [1)]
metals	metal filings, oils, greases, scale, rust, oxides, rust film, dust, silicones, lapping paste, paint residues, old coatings	mechanical:	wiping grinding polishing brushing blasting	activating, pickling (Al), phosphatising, chromating, chromate-free processes, secondary passivating
		thermal:	flame treatment	
		chemical: (possibly with mechanical support)	pickling, cleaning with solvents or aqueous cleaning agents	
plastics	greases, oils, release agents, dust, perspiration from hands	mechanical:	abrading, blowing off, rinsing	flame treatment, plasma treatment, corona discharge, fluorination, chromic acid process, satinising, benzophenone/UV
		chemical:	solvents or aqueous cleaning agents	
wood	sanding residues, dust, moisture, wood constituents, paint residues	mechanical only:	sanding, polishing, brushing	impregnating, sealing, priming
mineral substrates	dust, salts, greases, tyre marks, old coatings	mechanical:	blasting	impregnating, sealing, hydrophobing
		chemical:	solvents or aqueous cleaning agents	

1) Strictly speaking, pretreatment also includes cleaning

Brock, Groteklaes, Mischke: European Coatings Handbook

ISBN 978-3-86630-849-7

ties for a coating and the pretreatment of the most important industrial metals, plastics, woods and wood products, and mineral substrates. Table 5.1 lists the *cleaning and pretreatment methods* for the main substrates.

5.2 Principles of adhesion

We have already said that adhesion of a coating to the substrate is an absolute prerequisite for the serviceability of the coating. Some of the main factors influencing this include the surface topography of the substrate and - most importantly - the interactive forces at the interface between substrate and coating. Apart from chemical bonds, adhesion has the greatest influence on adhesive strength here.

- *Adhesion* is the term generally used in this case for the adhesive effect at the interface between a solid substance and a second phase consisting either of molecules, particles, droplets or powders or of a continuous liquid or solid phase, in this case the coating, the latter being of particular relevance to coating technology.
- By *cohesion*, on the other hand, we mean the strength of the internal coherence of a homogeneous substance, i.e. of the coating film as well as that of the substrate in this case.
- *Adhesive strength* refers here to the bonding force per unit area, calculated from the sum of all of the individual forces acting between substrate and coating film.

Nature and range of adhesive forces

As yet we still do not have an adequate scientific understanding of the basic relations between the multitude of variables that influence adhesion and the resulting adhesive strength. A whole range of theories has been developed, but these theories still deal with only partial aspects of the adhesion of coatings to substrates. Nevertheless, two preconditions for a perfect bond do exist:

1. The surface of the substrate must be sufficiently solid to form a workable basis for the coating. A loose surface cannot provide a good anchorage for the coating.
2. The coating material must come close enough to the surface of the substrate to enable it to be held in place on the substrate by the forces of attraction acting at the interface.

The nature of the forces of attraction is naturally dependent on the type of substrate and coating material. *Secondary valency forces* are always involved, such as dipole-dipole forces, induction forces or dispersion forces (→ 2.1.1.3). In order for true bonds or hydrogen bridge bonds to form, the appropriate reactive groups must be present both on the surface of the substrate and in the film former.

If necessary, additional *adhesion promoters* containing such reactive groups matched to the substrate can be added to the coating material, or special adhesion promoting layers applied, either on the substrate side (e.g. phosphatising in the case of metals or plasma treatment for plastics) or on the coating side in the form of *primers*. The range of the forces of attraction between substrate and coating lies in the atomic range between around 0.1 and 0.3 nm (see Table 5.2).

Table 5.2: Interactive forces between boundary layers

Type of forces	**Bond forces**			**Dipole-dipole forces**	**Induction forces**	**Dispersion forces**
	covalent	ionic	H bridge bonds			
range (nm)	0.1 to 0.	0.1 to 0.2	0.3 to 0.5	0.3 to 0.5	0.3 to 0.5	0.3 to 0.5
bond energies [kJ/mol]	60 to 700	600 to 1000	< 50	< 20 Keesom energy	< 2 Debye energy	0.1 to 40 London energy
dependence of bond energy on distance d	–	$\sim d^{-1}$	–	$\sim d^{-3}$	$\sim d^{-6}$	$\sim d^{-6}$

Such small distances between coating material and substrate can only be achieved if the surface of the substrate

- is firstly completely clean and
- secondly is completely wetted by the coating material.

Given the short ranges of the forces of attraction at the interface, even a monomolecular oil film would sharply reduce the bonding strength. The first stage in preparing the substrate must therefore be a thorough cleaning. Any components that impede wetting or are not permanently attached to the surface, whether corrosion products such as rust, drawing compounds on a metal substrate, or release agents in the case of plastics, must be completely removed before it is coated. The methods used to achieve this naturally depend on the nature of the substrate and are therefore discussed in the appropriate section.

Substrate wetting

Another important and even limiting factor for good adhesion is the wetting of the substrate surface by the coating material. If the latter cannot spread on the substrate, no intermolecular interactions and hence also no adhesion can develop. The conditions under which spreading occurs on a surface were discussed in connection with surface-active additives (→ 2.4.2.2).

As a rule of thumb, in order for the coating to spread on the substrate, the surface tension of the coating material must be lower than that of the substrate surface. By way of clarification, typical surface tension values for a number of solids and liquids are given in Table 2.32. In practice the wetting properties can be checked by two simple manual tests.

- The spreading behaviour of a drop of the coating material on the substrate surface is observed. If it remains as a small drop without spreading, i.e. it does not break up, poor wetting and poor adhesion can be anticipated.
- Secondly, a little coating material is mechanically distributed on the surface. If this quantity of coating material subsequently contracts back into drops, poor wetting can likewise be expected.

Strictly speaking, these tests ought to be carried out in an atmosphere saturated with the solvent blend used in the coating material, in order to prevent the solvents from evaporating during the test. This is very difficult to do in practice, however. Normally a material that spontaneously spreads in the first test also wets the substrate well in the second test. In the presence of surfactant-type additives with only a single polar group and a long non-polar radical, it can sometimes happen, however, that although the coating material spontaneously spreads, it later contracts back to drops. Such substances produce a low surface tension in the coating material, as a result of which it initially spreads well and during which process these additives cover the substrate with an oriented monomolecular film. Their polar groups bond to the substrate surface and the non-polar hydrocarbon chains form a new interface with very low surface tension.

A classical example of this type of behaviour is the occasionally poor adhesion of coating materials containing dodecyl benzene sulphonic acid as catalyst. It is clear that the surface to be coated must first be entirely free from such impurities which have a low surface tension.

Influences of roughness

The strength of adhesion of a coating to a substrate is also influenced by the size (see below) and the roughness of the substrate surface. On a completely smooth surface, the coating adheres with the substrate only by means of mutual forces of attraction. If on the other hand the surface exhibits microscopic structures, two further factors influence the adhesive strength.

- ***Surface expansion***

Under equal macroscopic expansion, a microscopically rough surface displays a much larger boundary layer to the coating than a smooth surface. If we assume that the number of binding sites is proportional to the interface between substrate and coating, then the adhesive strength thus increases accordingly.

- ***Surface structure***

If the rough surface also displays cavity-like structures, the coating can hook into them like a zip. The coating can then only be detached in the event of cohesive failure in either the substrate or the coating. Improved adhesion is thus generally observed on roughened surfaces.
The roughness of a surface can be tested by various methods. The most important measured value in practice for the roughness of a surface is the roughness height (page 381).

Penetrating power

As a prerequisite for improved adhesion, however, the entire surface of the substrate must be completely wetted. The influencing quantities involved here can be taken from the *Washburn equation*, which can be used to describe the penetration of a liquid, in this case a coating, into capillary-type surface structures:

$$L = 2{,}24\sqrt{\frac{\sigma}{\eta}\cdot(r\cdot\cos\Theta)\cdot t}$$

where L is the depth of penetration of the liquid into a capillary, σ is the surface tension of the liquid, η is the viscosity of the liquid, r is the capillary radius, Θ is the angle of contact and t is time.

Accordingly, the penetrating power of the liquid is at its best when its surface tension is high. Unfortunately, as the surface tension rises, cos Θ falls, which means that there is an upper limit to the surface tension. The variable that is most easily influenced by the coating formulator is the viscosity η. Low-viscosity coating materials that release their solvent only slowly and likewise cure only slowly generally provide the best adhesion. Apart from the aspects already mentioned, it must be borne in mind that as the surface becomes rougher, the roughness of the coating increases as well, extending into the topcoat region even when several coatings are applied. Consequently, it is important not to ignore the desired surface quality of the topcoat when specifying the substrate. Thus, sheet-steel parts which have been pre-treated by blasting and which have a roughness height of 15 to 30 µm have a very rough surface, whereas bodywork panels for this reason only have roughness heights of a few micrometers.

Thus if the roughened surface of the substrate is not com-pletely wetted, e.g. because the coating material is too highly viscous to be able to penetrate into the pores, the beneficial "zipping" action does not occur. The actual interface between the substrate and the coating may then be even smaller than in the case of a completely smooth substrate. Even when complete wetting takes place, however, stress concentrations may occur at the tips and corners of the coating/substrate interface (see Figure 5.1), causing the system to be more readily destroyed under mechanical stress. This can often reverse the basically positive effect of surface expansion.

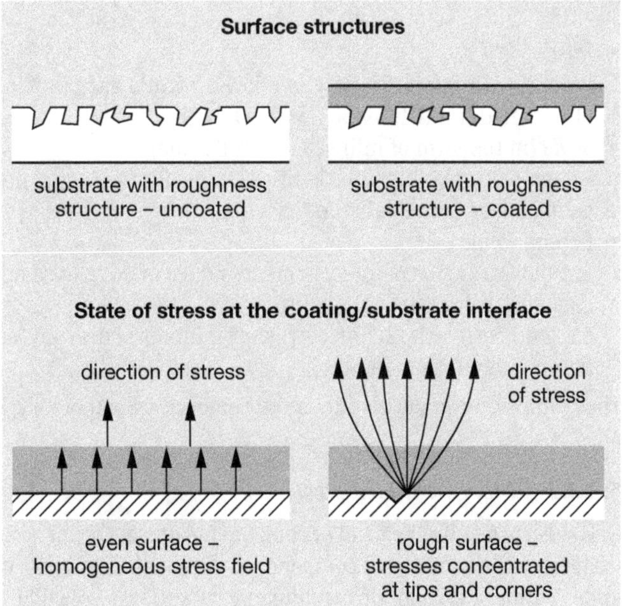

Figure 5.1: Influence of surface roughness on adhesive strength, according to [12]

5.3 Metal substrates

5.3.1 Metals and their surfaces

Perhaps the most striking material property of most metals is their *ductility*, i.e. their capacity to undergo plastic deformation whilst retaining great strength. Metals may be intentionally or unintentionally bent, leading to either elastic recovery or permanent deformation. A coating must survive these stresses without the least possible tearing or delaminating (detaching from the substrate).

In terms of the application of a coating, the nature of the metal surface is naturally of primary importance. If it is covered even only in places with poorly adhering and/or inadequately wettable, i.e. coating-rejecting, foreign layers, a permanent and visually perfect coating system will not be obtainable. As we have already explained, however, the roughness of the surface also has a major influence on the adhesion of the coating and - in the case of thinner coating films - on the quality of the coating surface. In their crude state, metals commonly used in industry are typically always covered with varyingly thick layers of oxides or corrosion products and are generally also dirty. In one respect, however, metals are also "coating-friendly": they immediately cover themselves with a thin oxide layer when exposed to air. As a result, cleaned metal surfaces have such a high surface tension that they are wetted well by most coatings. This is in stark contrast to most plastics, for example.

5.3.2 The most important metal substrates

5.3.2.1 Steel

Conventional *plain carbon* or *low alloy steel* consists essentially of the chemical element iron; it also always contains some carbon together with small quantities of foreign metals and traces of other chemical elements (P, S, Si, etc.). Steels containing high levels of foreign metals are known as *high alloy steels*; if they contain more than 12 % chromium they are resistant to rust. High alloy steels are rarely used as coating substrates.

The only material properties of steels that are of any real interest to coating technologists are effectively their chemical surface condition and corrosion behaviour. Depending on the origin or method of production and shape of a given steel component, various problematic films may be found on its surface.

- *Rust*
  Porous, "coating-unfriendly", yellow-brown to dark brown layer or crust consisting essentially of hydrated iron(III) oxide ($Fe_2O_3 \cdot x\ H_2O$).
- *Scale* (in the form of mill scale or roll scale)
  Brown-black to bluish black, brittle oxide film found on hot-rolled sheet steel or annealed steel components; principally Fe_3O_4.
- Foreign films
  e.g. old coating or paint systems, residues of adhesives and sealants, lubricating films.
- Fats, oils, salts, particulate dirt
  e.g. anti-corrosive oils or corrosion inhibitors, drawing compounds, shop dust, chalk marks, perspiration from hands.

These films and impurities must be removed as completely as possible before the coating is applied.

5.3.2.2 Zinc, galvanised steel

Galvanised steel (in the form of components or steel strip) occurs as a coating substrate much more frequently than solid zinc components. Although zinc is far more resistant to corrosion than steel under normal atmospheric conditions, zinc surfaces are still frequently coated, either for decorative reasons or to provide additional protection. Examples include partially or fully galvanised automotive bodies as well as "white goods" (especially washing machines).

In regard to galvanised steel a basic distinction is made between

- *hot-dip galvanised* and
- *electrogalvanised* steel.

The former is more widespread. The hot-dip zinc coating, which can be up to approx. 80 μm thick, is generally characterised by a more or less fine bloom (marbling of the surface). A special method of hot-dip galvanising is known as the "*Sendzimir* process", by which particularly thin and above all flexible zinc coatings are produced on steel strip in a continuous process.

"Hot-dip zinc", particularly when applied using the Sendzimir process, is not a "coating-friendly" substrate; the heavily bloomed types are frequently left uncoated.

The thinner electrogalvanised zinc coating forms a uniform matt grey film and responds much more favourably to being coated.

Aged and weathered zinc may exhibit films composed of various substances, depending on atmospheric conditions. The adherent, anti-corrosive layer of zinc oxide and "basic zinc carbonate" (2 $ZnCO_3 \cdot 3\ Zn(OH)_2 \cdot H_2O$) usually converts in practice into less protective hydroxide sulphates and other basic salts due to acid air pollution. In the long term these layers of corrosion known as "*white rust*" generally impair the adhesion of the coating.

5.3.2.3 Aluminium

The principal forms of aluminium found as coating substrates are sheets, profiles and low to higher diecasting alloys with magnesium, silicon, copper and manganese. Since these alloying additives are not homogeneously distributed throughout the metal microstructure, local cells can be formed, e.g. in the case of aluminium alloys containing copper, which greatly increase susceptibility to corrosion. Wherever high corrosion resistance needs to be achieved nonetheless, e.g. in aircraft construction, panels made from such alloys are protected by a type of sandwich structure having thin layers of pure aluminium on the outside (this is called plating). Aluminium and aluminium materials are characteristically covered by a thin, colourless oxide layer which protects the aluminium (in neutral medium) against corrosion. The porous, outer section of this oxide layer can be fairly thick and will prevent paint adhesion, which is why it has to be removed prior to coating. Because the oxide skin is electrically insulating, aluminium has to be pretreated before it can be coated by electrodeposition.

Aluminium is an *amphoteric* metal, which means that unlike iron, for example, it is attacked strongly not only by acids but also by alkalis. (Zinc is also classed as an amphoteric metal, but its liability to attack by alkalis is not particularly pronounced.)

5.3.2.4 Other metal materials

To save weight, it has become increasingly common to use materials containing magnesium, especially in auto-motive construction. Like aluminium, magnesium is auto-passivating and also unstable in acid and alkaline mi-lieu. It is therefore alloyed with other metals to increase its strength. Current pre-treatment, like that of alumin-ium and its alloys, may or may not involve chromate. However, the resultant surface properties are not as con-sistent as those obtained in the case of aluminium.

For the sake of completeness we should point out that many other metal materials occasionally serve as coating substrates. In many cases they are alloys or composites of the three base metals iron, zinc and aluminium in the form of sheet metal, such as zinc-aluminium alloys or steel strip coated with zinc dust (trade names: "Galfan", "Galvalume" or "Zincrometal", etc). *Tin* in the form of tinplate (tinned sheet steel) is important for can coating. *Non-ferrous metals* and alloys thereof are relatively rarely coated.

We should also mention that electrically conductive composites of various metals are always prone to corrosion problems - even underneath a coating. For electrochemical reasons the contact zones are particularly at risk.

5.3.3 Removal of adherent coatings

Layers of corrosion products such as rust and scale as well as "coating-unfriendly" and/or uneven foreign layers must be removed as completely as possible. (Layers of old coatings that are still firmly attached can be overcoated after rubbing down and filling where necessary.) Depending on the size and accessibility of the surfaces to be treated (the complexity of the components), the nature of the layers (strength, thickness, chemical composition), surface or component throughput (manual treatment of individual components or assembly line/mass production), different mechanical or chemical processes can be used.

The degree to which the foreign coatings are removed is expressed by the degree of cleanliness (DIN EN ISO 8501, parts 1 to 3). The type of cleaning is quoted in the form of a combination of letters (*Sa* for blast-cleaning, *St* for manual or power-tool cleaning, *Fl* for flame cleaning, *Be* for pickling and *Ma* for mechanical grinding). In the case of *Sa* and *St*, the cleanliness of the surfaces is additionally ranked on a scale of 1 to 3, where 3 represents the best. Complete removal of layers of foreign matter by blast-cleaning of the pure, bright metal surface corresponds to the degree of cleanliness *Sa 3*; the degree of cleanliness *Sa 2½* is generally required for coatings. Where foreign layers are only partially removed (because intact coatings and overcoats remain behind), a P (for partial) precedes the symbol for degree of cleanliness (e.g. P *Sa 2½*).

5.3.3.1 Mechanical processes, abrasive blasting

Rust removal

Manual rust removal is carried out using simple tools such as a pick hammer, scraper, wire brush or sandpaper or sanding block. Very solid films such as mill scale cannot readily be removed in this way.

Mechanical rust removal with rotating wire brushes, rotating or vibrating grinding discs, pneumatic chisels or needle guns, etc., is more effective.

Abrasive blasting

By far the most effective method of foreign coating removal is abrasive blasting. This involves directing *reusable abrasives*, e.g. steel grit (angular), steel shot (round), cut wire pellets, corundum, silicon carbide, glass beads, thermoset powder, or *disposable abrasives* (slag) at high speeds onto or over the surface. "*Sand blasting*" with silica sand is banned in the UK with just a few exceptions, because of the risk of silicosis.

A distinction can basically be made between the following *blasting processes*:

- *Dry blasting*
  The abrasive is blasted onto the object from guns using compressed air or is hurled at the object by a bladed wheel ("*centrifugal blasting*").
- *Damp blasting*
  The abrasive is moistened to reduce the formation of dust.
- *Wet blasting*
  Water - with or without the addition of abrasives - is sprayed onto the object by means of compressed air or without air at a pressure of up to around 2000 bar ("wet compressed air blasting" or "ultra-high pressure jet blasting").

Blast cleaning plants

Modern blast cleaning plants recirculate the abrasive in a continuous cleaning process.

Abrasive blast cleaning is carried out in closed housings, booths or shops; in the case of large objects such as bridges, the blasting location has to be enclosed and a means of extraction provided for the abrasives. Figures 5.2 and 5.3 illustrate the principles behind *jet blasting* and *centrifugal blasting*.

In regard to the choice or efficiency of abrasive, the following rule of thumb applies:
The more coarsely grained, sharp edged and harder the abrasive and the higher its speed of impact, the more abrasive the action. In the case of thin-wall components or filigree structures, it should be borne in mind that too vigorous abrasive blasting can lead to damage.

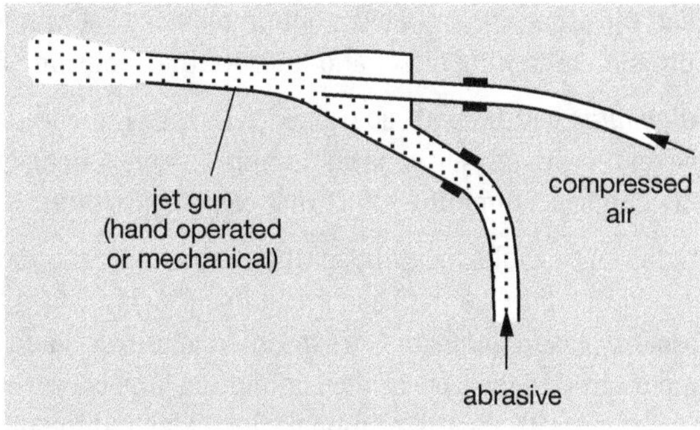

Figure 5.2: Principle behind jet blasting

Surface quality

Clean, metal-bright metal surfaces with a defined roughness height of between 30 and 100 µm are obtained by abrasive blast cleaning. Freshly blasted metal surfaces are highly reactive, tend to corrode rapidly and must therefore immediately be chemically passivated or primed. Coating adhesion on blast-cleaned surfaces can generally be classed as good.

We should also mention that abrasive blasting (with spherical or soft abrasives, frequently in the form of wet blasting) is also used for deflashing components as well as for smoothing, dimming and hardening surfaces and also for non-metallic materials.

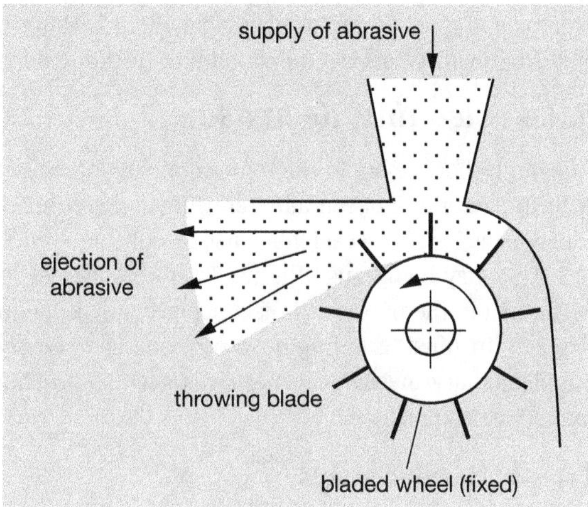

Figure 5.3: Principle behind centrifugal blasting

5.3.3.2 Flame cleaning

Rust removal on large steel struc-tures can also be carried out by means of flame cleaning. This involves passing a reducing oxy-acetylene flame slowly over the surface, as a result of which organic coatings and impurities carbonise or combust and the rust loosens and detaches as it disintegrates. The residues of the coating are then brushed or scraped off. As with mechanical blasting, the surface in this case too must be primed immediately; alternatively a preser-vative adhesive film can be produced by burning a phosphoric acid solution ("flame phosphatising"). The disadvantage of flame cleaning is that dangerous chloride and sulphate salts are often incompletely removed, remaining on the surface as chloride and sulphate pockets that promote rapid corrosion.

5.3.3.3 Pickling

"Pickling" refers to the chemical dissolution of oxidic coatings or corrosion products found on metals. The simplified chemical reaction equation for the removal of scale by pickling with hydrochloric acid is as follows:

$$\underset{\text{scale}}{Fe_3O_4} + \underset{\text{hydrochloric acid}}{8\,HCl} \rightarrow \underbrace{FeCl_2 + 2\,FeCl_3}_{\text{dissolved}} + 4\,H_2O$$

Both strip or wire and individual components can be pickled using a continuous or intermittent process. Dilute sulphuric or hydrochloric acids are the principal *pickles* used; phosphoric acid is used more rarely. The pickling process can be accelerated by raising the temperature (in closed plants). As a moderately strong acid, phosphoric acid has a significantly weaker pickling effect than the other mineral acids, but on metal surfaces it forms very thin phosphate films that have a corrosion-inhibiting effect and improve the adhesion of a subsequent coating.

Aluminium can be both acid pickled - with the addition of hydrofluoric acid (HF) - and alkaline pickled - with caustic soda containing complexing agent or with alkaline cleaning agents. Alkaline pickling has a dimming effect that can be eliminated by means of acid secondary pickling.

As far as possible, pickling solutions should only remove oxidic and other inorganic coatings and should not dissolve the base metal itself. Acid pickles therefore generally contain additives in the form of *pickling inhibitors* ("restraining additives") and are thus called "pickling restrainers". Most importantly, in the case of steel, inhibition largely prevents most importantly the hydrogen evolution associated with the dissolution of the metal and the hazardous *hydrogen embrittlement* that results therefrom. Organic films such as greases, mineral oils and old coating systems disrupt the pickling process and must be removed completely in advance by degreasing or stripping. If the surface is only slightly dirty, surfactant-containing "pickling degreasers" can also be used.

5.3.4 Cleaning, degreasing

"Cleaning" here refers to the removal of any foreign substances from the surface of the substrate, with the exception of the adherent films covered in section 5.3.3. In practice, however, the term "degreasing" is still occasionally used for the sub-process of removing organic, fatty or oily impurities. This may be because the treatment of surfaces with organic solvents, which was frequently performed in the past, can in fact achieve only a "degreasing" effect for the most part, whilst particulate dirt (dust, grinding dust, etc.) and sometimes also salts mostly remain on the substrate. Supplementary mechanical measures such as brushing, spraying or ultrasound may offer some limited assistance here.

Types of cleaning agents

Although organic solvents - especially chlorohydrocarbons - dissolve fats and oils very effectively, degreasing with solvents, apart from a few exceptions such as the manual cleaning of small and/ or complexly shaped parts, has been superseded by cleaning with *aqueous solutions*. This is partly because of environmental and safety issues and partly because of the incomplete removal of particulate dirt, as mentioned above.

The products used for aqueous cleaning (*industrial cleaning agents*) take the form of soluble powders or aqueous concentrates. The best way to categorise cleaning agents is by pH value: As can be seen from Table 5.3, the boundaries between pickling, cleaning and the application of conversion layers are often fluid (section 5.3.5).

Immersion and spray cleaners

A further distinction is made between immersion and spray cleaners. Whilst shorter treatment times and lower temperatures and concentrations are possible with spraying, immersion offers better access to interior and undercut surfaces. Spray cleaners must be "foam-inhibited". Highly alkaline cleaning agents are not suitable for spraying due to the absorption of atmospheric CO_2. Multi-stage cleaning processes or additional brushing are utilised for heavy-duty cleaning requirements. Surfaces usually need to be rewashed with water after aqueous cleaning. Directly before applying the coating, the surface should additionally be rinsed with *demineralised water* and then air dried. (Residues of electrolytes (salts) remaining on the metal surface can later trigger corrosion below the coating and lead to osmotic processes such as delamination with blistering).

Special process variants (cascade principle, etc.) and equipment (oil traps, ultrafiltration cycles, etc.) mean that cleaning plants can use less water and chemicals, making them both cost-effective and environmentally friendly.

Table 5.3: Aqueous cleaning and pre-treatment methods

pH	Method	Important ingredients	Substrate	Application
>12	strong alkaline cleaning	sodium hydroxide, silicates, sodium carbonate, phosphates, surfactants	iron	heavy duty cleaning or degreasing tasks
10 to 12	(mild) alkaline cleaning	silicates, sodium carbonate, phosphates, surfactants	iron, zinc, aluminium, plastics	suitable for most applications, good results without chemically attacking the metal
7 to 9	neutral cleaning, passivation	surfactants, phosphates, emulsifiers	iron, zinc, aluminium	for less demanding cleaning tasks; when used for metals, usually combined with temporary corrosion protection; leave virtually no residue on drying
4.5 to 5.5	degreasing, acid alkali phosphate cleaning, iron phosphatising	alkali phosphates, polyphosphates, phosphoric acid, accelerators, surfactants	iron, zinc, aluminium, plastics	produce very thin phosphate films on steel and zinc which substantially improve the adhesion of coatings
approx. 4	chromating	Cr (VI), zirconium and titanium compounds	zinc, aluminium, (magnesium)	passivating
approx. 2	zinc phosphatising	Zn(II), Mn(II), Ni(II), Cu(II) salts, oxidising agents, phosphoric acid	iron, zinc, aluminium	produce phosphate films for very good adhesion of coatings
1 to 2	pickling, derusting, descaling	mineral acids	iron	complete derusting

5.3.5 *Application of conversion coatings*

In many instances, the metal surface can be coated directly after being cleaned. However, where more stringent requirements exist in respect of coating adhesion or resistance to corrosive migration beneath the coating, e.g. in industrial automotive assembly line production, domestic appliances, metal furniture, etc., as well as in continu-ous coating of metal strip (coil coating), cleaning is followed, where possible, by the formation of a *conversion coating*, which creates an ideal bond between the metal phase and the organic coating. Such conversion coatings are produced by selective chemical transformations of the substrate surface to form a highly adherent, more or less rough, inorganic protective coating.

Types of coating/process variants

The most important processes for producing conversion coatings are outlined below in brief.

- *Zinc phosphatising*

Finely crystalline zinc or (on a steel substrate) zinc iron phosphate coatings of a few μm in thickness are deposited principally on steel or zinc (but also on aluminium by means of special processes) from a phosphoric acid solution containing zinc and other metal ions together with various additives. When this is combined with a passivating *rewash*, outstanding corrosion or submigration resistance is obtained. The formation of the coating on steel is illustrated in simplified and summarised form by the following equation:

$$\underbrace{3\,Zn^{2+} + 6\,H_2PO_4^- + 6\,OX}_{\text{in the phosphatising solution}} + \underset{\text{steel substrate}}{4\,Fe} \rightarrow \underset{\text{zinc phosphate film deposited on the surface}}{Zn_3(PO_4)_2 \cdot 4\,H_2O} + \underset{\text{phosphatising slurry}}{4\,FePO_4} + 6\,X + 2\,H_2O$$

OX = oxidising agent (depolariser)

(Depending on the composition of the phosphatising solution and the application conditions, the phosphatising films on steel also contain varying quantities of iron. In addition to iron phosphate, other compounds - predominantly zinc phosphate - are also found in the phosphatising slurry.)

Figure 5.4 shows scanning electron micrographs of a bright and a (manganese-modified) zinc-phosphatised steel surface. It should be clear from these that because of its fissured structure a film of this type, as well as having a protective function, also represents an ideal adherend surface for the coating system.

Zinc phosphatising is the most common method of pretreatment for automotive underbodies and other high-grade mass-produced goods subject to high corrosive stress.

- *Alkaline phosphatising*

Alkaline phosphatising ("iron phosphatising") of steel and zinc is generally performed as a combined cleaning and phosphatising process. In comparison to zinc phosphatising, this method is much more attractive in terms of plant investment and control costs and is consequently widely used. Alkaline phosphatising produces only relatively thin films, however, and the corrosion resistance that can be achieved after application of the coating system is significantly lower. For that reason alkaline phosphatising is generally not adequate for coated surfaces subject to weathering and permanently exposed to damp conditions.

- *Chromating*

Aluminium and occasionally also zinc and magnesium alloys are still frequently chromated before being coated. In this case the treatment is carried out with acid chromium(VI) (i.e. chromate-containing) solutions. Depending on the composition of the chromating agent, differently coloured films are produced, a distinction being made above all between *green* and *yellow chromating*.

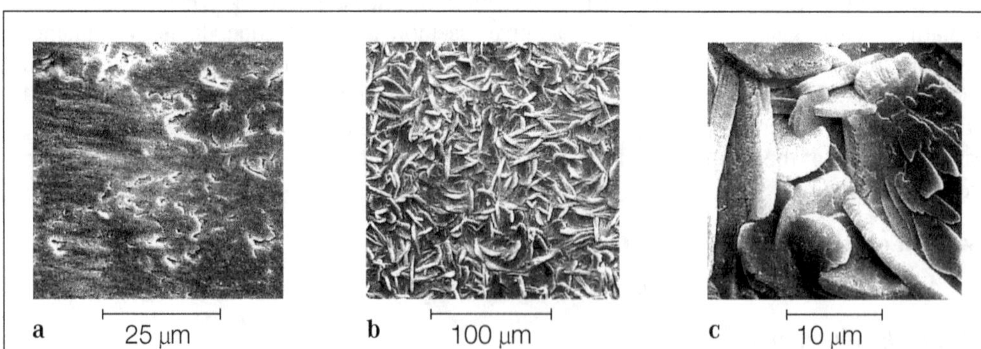

Figure 5.4: Bright (a) and (manganese-modified) zinc-phosphatised (b, c) steel surfaces viewed through a scanning electron microscope (SEM) [from: "Bonder-Technik" 23, 1987 (Chemetall)]

• *Chromium-free processes* (as a substitute for chromating)

Owing to the toxicity of chromium(VI) compounds and the general problems surrounding heavy metals, chro-mium (VI)-based pre-treatment methods for zinc and aluminium (and increasingly for magnesium) are gradually being replaced by chromium-free methods. Various methods have become established for different substrates, e.g.:

- In alkaline passivation (which is suitable for galvanised substrates), a very thin layer (0.05 to 0.5 g/m^2) of complex metal oxide is formed on the surface and protects the underlying metal against corrosive attack.
- Layers containing titanium, zinc or zirconium form when zinc or aluminium substrates are attacked by an acid pickling solution containing phosphoric acid and fluoride. The titanium and zirconium are de-posited as phosphates and/or oxides and are incorporated into the oxide layer as it forms. The thickness of the resultant coatings varies with the method employed but can be very thin (0.01 to 0.02 g/m^2). However, combinations with organic polymers can yield thicknesses of 20 to 30 g/m^2. In the cerate method, cerium-containing oxide layers form in highly acidic milieu.
- Methods based on organofunctionalised silanes dispense with transition metals, being on one hand firmly bound to the metallic substrate by silicon-oxygen bridges and, on the other, incorporated into the binder matrix of the organic coating via the functional group on the organic silane group.
- The method of self-assembling molecules (SAM) consists in the attachment of organophosphonic acids to the metal surface via the acid group to form a layer which possesses anticorrosive properties due to the alignment of certain functional groups.

• *Oxidation*

The above process of anodic film-forming on aluminium is generally intended more as a means of "bright corrosion protection" rather than as a pretreatment method, however. A special alkaline oxidation process has been developed for the pretreatment of galvanised steel strip. The alkaline oxidation of steel is known as "black oxidation" or "black finishing"; this is a decorative bright corrosion protection rather than a pretreatment.

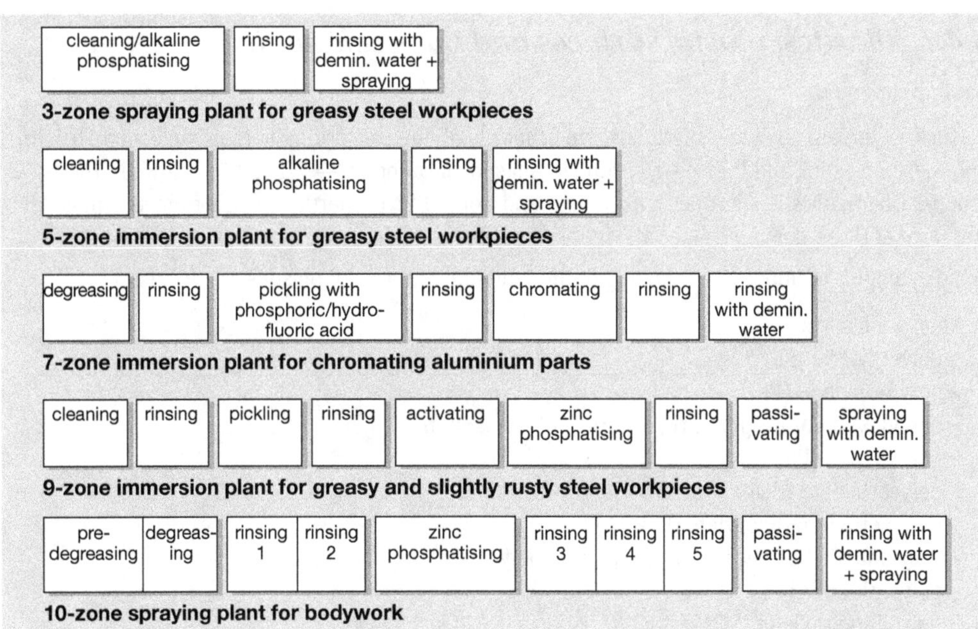

Figure 5.5: Overview of pretreatment processes for metal frequently used in practice

Finally, Figure 5.5 illustrates a number of typical processing sequences for the pretreatment of metal surfaces, as practised in industrial manufacturing.

5.3.6 Manual preparation of metal substrates

Any rust present on steel surfaces must first be removed using the methods listed in 5.3.3.1. (Inaccessible areas can be treated with "rust converters" after degreasing.) Suitable processes for the subsequent degreasing/cleaning stage include steam jet cleaning. The surface is then primed using a wash primer or two-component epoxy primer, for example.

Zinc surfaces must be washed down thoroughly (using steel wool) with ammonia liquor to which a small amount of detergent (neutral cleaner) has been added. Any white rust must be ground off. (Caution! Avoid grinding through the zinc film on galvanised steel!) Special zinc primers are available for priming.

The following procedure can be used for aluminium: clean - dry grind with "Scotch Brite®", for example - remove dust - scour with solvent. Smooth "Duralumin" is preferably "washed" with phosphoric acid solutions or steam jet cleaned.

5.4 Plastic substrates

Plastic components too are frequently coated, particularly when high optical standards are required. Application of a coating can provide

- a decorative appearance,
- colour matching with surrounding surfaces, if required, e.g. coated bumpers
- improved surface quality (especially when combined with stoppers and fillers)
- protection of the plastic surface against the direct influence of aggressive agents and weathering
- particular surface textures or effects, e.g. "soft-feel" effect
- particular technical surface properties, if required, e.g. electrical conductivity (by application of conductive coatings).

5.4.1 Plastics, plastic surfaces and their coatability

Basic properties

Plastics, i.e. organic polymer materials, are primarily divided according to their molecular structure into thermoplastics and thermosets; rubbery-elastic elastomers (→ 2.1.1.1) fall between these two groups. Characteristic thermal, mechanical and solubility properties derive from the molecular structure of these polymer materials; these were discussed in detail in section 2.1.

Some examples of important members of the individual groups of plastics are listed below:

- Thermoplastics
  - polyethylene (PE)
  - polypropylene (PP)
  - polyvinyl chloride (PVC, rigid: uPVC, plasticised: pPVC)
  - polystyrene (PS)
  - polycarbonate (PC)
  - polymethyl methacrylate (PMMA)
  - acrylonitrile-butadiene-styrene graft copolymer (ABS)
  - polyamide (PA)
  - polyoxymethylene (POM)
  - polytetrafluoroethylene (PTFE)

- Thermosets (often filled and/or fibre-reinforced)
  - unsaturated polyesters (UP)
  - epoxy resins (EP)
  - melamine resins (MF)
  - phenolic resins (PF)

- Elastomers
  - polyurethane (PU)
  - silicone (SI)
  - ethylene-propylene-diene rubber (EPDM)

Polyurethane is also found as a coating substrate in the form of *integral PU foam* of varying hardnesses. It has a cellular core and a closed surface.

The properties of the plastic surface, such as purity, wettability, affinity to the coating (adhesive strength) and etchability by coating solvents, are of primary importance in regard to the application of coatings to plastics. Secondly, the mechanical and thermal behaviour of the plastic as well as its electrical conductivity must also be taken into consideration. Thermosets, for example, are usually extremely resistant to high temperatures, which means that coatings can be stoved for short periods at up to around 140 °C. Many thermoplastics, by contrast, lose their dimensional stability at temperatures of only between 60 and 100 °C.

Surface condition, coating problems

Plastics are not "pure" in the chemical sense; like coating materials, they always contain *foreign components* generated either by secondary reactions during synthesis , are intended to modify certain properties in the finished part, e.g. fillers, pigments, reinforcing fibres, or incorporated in the form of additives during processing. Where they influence the surface properties of plastics, a basic distinction is drawn between

- additives such as plasticisers, flow improvers, slip additives, blowing agents (in foams), dyes, stabilisers, antioxidants and also decomposition products, e.g. water and formaldehyde, that migrate to the surface from inside the plastic
  and
- processing aids applied to the surface, such as release agents (waxes, greases, oils, etc.).

All of these substances are (also) found to a greater or lesser extent on the surface of the unmachined part, and also after surface cleaning they can migrate to the surface in varying degrees from inside the plastic. The significance of this phenomenon lies in the fact that these substances generally have a negative influence on coatability.

Even if the surface of the plastic is adequately freed from the above foreign substances, *dust contamination* of the surface is often the next problem to be tackled. The absence of surface conductivity means that electrostatic charges cannot be discharged from the surface of the plastic; air-borne dust particles are then trapped and retained. (By way of comparison, think of the phenomenon whereby scraps of paper can be attracted to a plastic comb that has just been used to comb dry hair.) Even if the dust issue is then resolved, a third problem may potentially emerge: the coating does not readily wet the surface and does not adhere adequately after drying.

If the presence of residual foreign substances on the surface is not the cause of this final problem, the reason lies in the plastic itself: a material with a non-polar molecular structure likewise has a non-polar surface and hence too low a *surface tension*. The major bulk plastics PE and PP in particular, as well as POM and EPDM, are all difficult to coat in this sense.

5.4.2 *Pretreatment of plastics*

Elimination of surface defects

If the surface displays production-related defects such as pores, voids (depressions), blisters and weld lines or even more serious damage, these must first be ground or blasted and even filled if necessary. (Gentle) grinding or blasting also gives the surface a defined roughness, which has a positive influence on the adhesion of the coating.

Cleaning

Cleaning of plastic surfaces (including removal of the additives mentioned in 5.4.1) is generally now performed - as with metals - using aqueous detergent solutions. However, it is often more difficult to clean plastic because, as mentioned earlier, the surface is frequently diffi-cult to wet and because the cleaning agent is usually at a lower temperature. This is followed by rinsing with water, finally with demineralised water, and by blow drying. Cleaning with solvents by wiping, spraying, dipping and/or "vapour degreasing" is still used to a limited extent (e.g. if release agent residues cannot be completely removed by other means), if necessary in conjunction with preliminary or intermediate annealing. It can be difficult choosing suitable solvents that will remove impurities on one hand, but will neither attack nor migrate excessively into the plastic on the other.

Removal of dust and surface charges

It is usually impossible to prevent cleaned and dried plastic components from attracting dust en route from the cleaning area to the coating area. This must be blown off with *ionised* air at the entrance to the spray booth. Alternatively an "*antistatic liquid*" can be applied after aqueous cleaning, which prevents electrostatic charging during and after blow drying, without subsequently impairing the adhesion of the coating.

Improving wettability and coating adhesion

The non-polar bulk plastics PE and PP in particular display inadequate wetting by coatings, especially water-borne coatings. Furthermore, the adhesion of the dry coating film to the former and also to a number of other plastics is relatively poor. These problems, as we have already indicated, are due to too low a surface tension or polarity in the plastic surface. To increase the polarity, the plastic surface is oxidised or fluorinated by various methods, causing non-polar C-H-bonds to convert to polar C-O- or C-F-bonds.

The principal methods are briefly described below.

• *Flame treatment*

A gas flame burning with excess air, i.e. a non-luminous flame, is passed manually or automatically over the surface of the plastic (exposure time usually less than 1 second). This leads to surface oxidation and hence to a rise in polarity. The most commonly used method.

- Advantages: continuous, simple and effective.
- Disadvantages: thermal stress on components, very short-term effect.

• *Corona treatment*

An A.C. voltage electrode takes the place of the flame used in flame treatment; this electrode forms high-energy oxygen ions by means of luminous discharge (corona discharge), which oxidise the plastic surface.

- Advantages: continuous, no thermal stress on components.
- Disadvantages: limited reach, only short-term effect, can only be used to treat (practically) plane surfaces and film.

- *Plasma treatment*

The components are exposed to a plasma[1)] of a reaction gas (oxygen, possibly also other gases) in a vacuum chamber. Exposure time 1 to 5 minutes.

- Advantages: can be used in-line, universal, more effective than corona treatment, virtually independent of component geometry, highly robust process, effect lasts up to several weeks, also suit-able for materials which are difficult to activate.
- Disadvantages: technically complex, component size restricted.

- *Fluorination*

The components are exposed to a fluorine-nitrogen mixture in a vacuum chamber for a few minutes.

- Advantages: very effective (substantial and permanent increase in surface tension), components of any shape can be uniformly treated, effective up to several months.
- Disadvantages: discontinuous, technically complex. Risk of stress cracking in the substrate, component size restricted.

- *Silicatisation*

Similar to flame treatment. The combustion gas additionally contains organosilicon compounds that deposit a layer of amorphous silica 20 to 40 nm thick on the substrate.

- Advantages: continuous, simple and effective, an easy upgrade from flame treatment, coating becomes very moisture resistant
- Disadvantages: thermal stress on the parts, short-term effect (but longer than that of flame treatment)

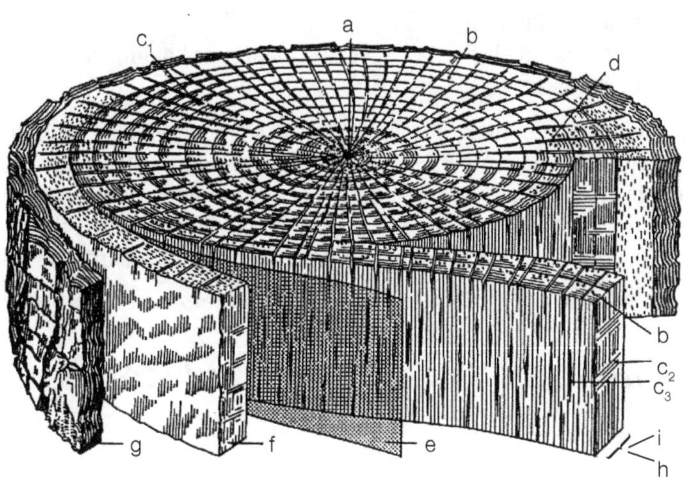

Figure 5.6: Structure of a conifer trunk, according to [15]
a) pith (usually only approx. 2 mm in diameter); b) growth rings. The older (inner) rings form the heartwood, the outer rings, with softer walls, form the sapwood; c) rays: c1 viewed from above, c2 viewed from the side as end fibres, c3 viewed from the front; d) resin canals; e) cambium (growth layer), forms sapwood on the inside and bark or phloem on the outside; f) inner bark or phloem (with broader rays from here on and intercalated sieve tubes; g) bark; h) late wood and i) early wood (both shown in two growth rings).

1) A plasma is a gas that, in addition to neutral molecules and atoms, also contains positive and negative ions produced from these particles.

• *UV irradiation*

The parts to be coated are transported past UV lamps of very high power density. A free-radical reaction with circulating oxygen causes the surface to be activated.

- Advantages: simpler than plasma treatment, continuous, low process temperatures
- Disadvantages: high ozone out-put

Adhesion promoters

The adhesion of coatings to plastics can sometimes be improved by the use of adhesion promoters (in the coating) or primers - possibly combined with abrading (roughening) the plastic surface - to such an extent that oxidative pretreatment can be dispensed with. This is of particular significance for the manual application of coatings.

Annealing

Certain rigid plastics, such as PMMA, still often contain "frozen-in" stresses arising from the manufacturing process; these can lead to cracking under the influence of organic solvents or coatings. Such stresses can be broken down by annealing the components (before surface pretreatment), i.e. by storing them for an extended period at a temperature slightly below the flow or melting range, and then cooling them slowly.

Foam components with a closed outer skin (SFM and RIM components[2)]) are also occasionally tempered before coating, in order to drive out residual gases or blowing agents that can lead to blistering and delamination of the coating during oven drying.

Increasing the electrical surface conductivity

Surface conductivity (cf. DIN 53 482), e.g. for electrostatic coating, can be increased by means of the following, fundamentally different methods:

- reduction of the plastic resistance in the compound
- application of conductive films
- technical equipment (special electrodes, special supports or holding devices for parts) which allow the charge to be discharged.

Let us briefly consider just one of these: the application of conductive films. These films can consist of the "antistatic liquids" mentioned in 5.4.2; alternatively, evaporating (i.e. only temporarily active) conductive liquids (hydrophilic solvents) can be used. A third option is the conventional application of a special electrically conductive coating ("conductive coating"), which then becomes part of the structure of the coating and must therefore be matched both to the particular plastic and to the subsequent coating films.

5.5 Wood and wood products as substrates

5.5.1 Wood

Composition

Wood, as we know, is a natural substance derived from the "structural material" of trees. Chemically speaking it is an inhomogeneous, organic material whose solid component, depending on the type of wood, consists of 30 to 50 % cellulose, 15 to 35 % *hemicelluloses* and 20 to 35 % *lignin*. Cellulose is poly-β-D-glucose (→ 2.1.3.3).

1) SFM = structural foam moulding, RIM = reaction injection moulding

2) The % content of moisture in wood can be regarded as the % by weight of water, relative to the weight of kiln-dried wood.

Hemicelluloses - unlike the linear-structured cellulose - are branched polysaccharides. Lignin is an aromatic-aliphatic, irregularly structured, branched polyether; it acts as a kind of cement gluing the cellulose fibres together. Other components of raw wood include fats, waxes, proteins, sugars, terpenes, resins, tannins, dyes, alkaloids and mineral substances.

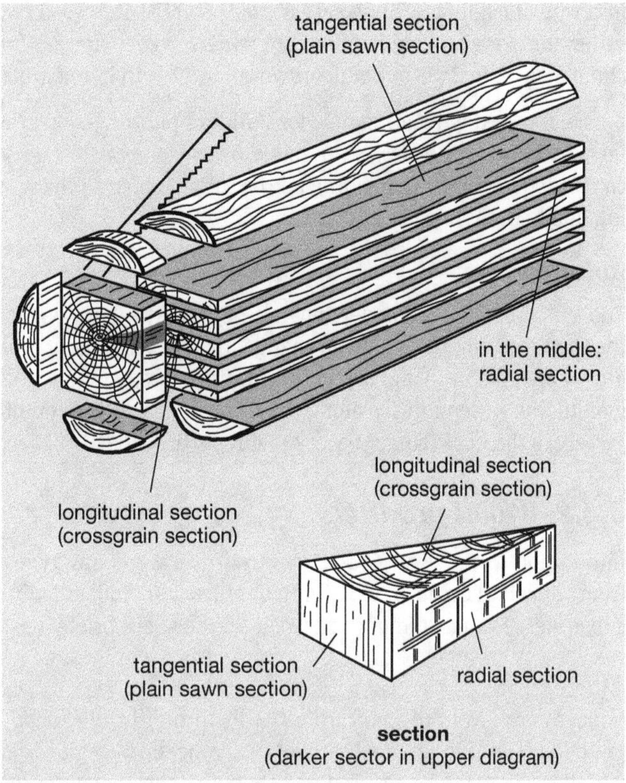

Figure 5.7: Types of cuts for timber, according to [15]

Wood hardnesses

Wood is divided into various hardness categories [15]:

- very soft: aspen, spruce, fir, poplar, willow
- soft: sycamore, lime, larch, alder, birch
- moderately hard: elm, pine, pitch pine, chestnut
- hard: oak, ash, pear wood, teak
- very hard: beech, hickory, rosewood, yew, walnut
- bone hard: quebracho, lignum vitae, ebony

The bulk density of the wood increases in line with its hardness; in the kiln-dry state it is between 0.1 g/cm^3 (balsa) and 1.2 g/cm^3 (lignum vitae).

Features of wood

In order to be able to surface treat and coat wood correctly (usually by painting), it is important and helpful to know the major characteristics of wood as a substrate.

The macroscopic structure of wood is as complex as its chemical structure. Figure 5.6 illustrates a section from the trunk of a conifer. Most striking are the *growth rings* with the inner layer of early wood consisting of larger cells and the outer layer of late wood composed of smaller cells. Other features include the *rays* (radial strands of tissue running from the pith to the bark) and the *pores* or resin canals (chopped vertical transport ducts for the sap).

A tree trunk can essentially be sliced (sawn through) in three different ways to produce *tangential* (plain-sawn), *radial* (rift-cut) and *longitudinal* (crossgrain) sections, as shown in Figure 5.7.

In tangential sections the growth rings display the characteristic *grain* of the wood. The surface of the longitudinal (crossgrain) section contains the severed canals and is therefore especially highly porous or absorbent.

In many types of trees, the canals in the inner growth rings gradually block up - from the pith outwards - forming the resistant, often darker-coloured *heartwood*. The *sapwood* that extends out to the bark is softer, absorbent and vulnerable to attack by fungi and insects.

After damage to the wood by organisms, change in water content combined with an increase in volume due to *swelling* or a reduction in volume due to *contraction*, is the greatest technical problem in terms of coating. (The wood "warps"). Whilst the wood of a living tree can contain up to 80 % water, the average residual moisture (*moisture content equilibrium*) of wood, depending on type and climate, is 12 to 25 % out of doors and 6 to 12 % in living areas[1)].

Before processing (coating, use for building purposes, etc.) the moisture content of the wood must be adjusted as accurately as possible by means of selective *seasoning* to the anticipated moisture content equilibrium, depending on the intended use of the wood, to prevent deformation and cracking due to excessive contraction.

Anisotropy

The anisotropy (direction-dependence of properties) of wood is revealed when it swells or contracts: the change in length for a given change in moisture content is around 10 times greater radially (in the direction of the rays) and even up to 20 times greater tangentially (along the growth rings) than it is longitudinally (lengthwise along the trunk). As a rule of thumb, the average change in length transversely to the fibre (tangentially, radially) is 0.2 % per 1 % change in moisture content of the wood.

5.5.2 Wood products

In addition to the "*solid wood*" covered in 5.5.1, many types of wood products, which are made from wood broken down to varying degrees and binders (adhesive resins, etc.), are used as coating substrates. The most important wood products are briefly described below.

- *Veneers*

Veneers are layers of wood from 0.3 to 10 mm in thickness, cut from solid wood by a variety of methods. The thinner, decorative "face veneers" are glued onto the surface of the blank material (chipboard, etc.) to be covered and thus give the visual impression of solid wood.

- *Plywood*

In the broader sense plywood covers veneer plywood ("plywood" in the narrow sense) and wood core plywood. The first consists of several (at least three) layers of veneer glued crosswise on top of one another; in the second a layer of top veneer is glued to either side of a core layer made from small bars or strips glued together.

- *Chipboard* ("particle board")

Chips of resin-coated wood are compression-moulded into relatively rough and porous sheets. These cannot be coated until they have been filled and sanded.

- *Fibreboard*

Fibreboard consists of compressed wood fibres with or without the addition of binder. A distinction is made between the absorbent *softboard* (insulating board) and the thin brown *hardboard*. Hardboard is mainly used for rear and separating panels in furniture and doors; it has a rough "wire side" and a very smooth, paraffined top side, which should not be sanded before coating.

- *Medium-density fibreboard* (MDF)

This board can be roughly pictured as particularly dense, smooth and homogeneous chipboard even though - strictly speaking - it is not chipboard. The surface of MDF can generally be coated directly with no need for surface treatment. The absorbent edges, however, require insulation or impregnation.

5.5.3 Pretreatment of wood and wood products

The pretreatment - or preparation - of wood, depending on its initial state or origin, conversion process and intended use, involves a selection of the following individual measures:

- *seasoning* the freshly sawn timber in seasoning kilns to obtain the anticipated moisture content equilibrium
- *facing* and *smoothing* by fine-planing, sanding, hot dressing and filling
- *graining* the surface, i.e. incorporating visual textures
- *soaking in water* then *drying* and *fine-grinding* to "top" the erect fibres (beneficial if it is to be treated with aqueous stains and/or coatings)
- *deresinifying* and removal of marks (if not already achieved by planing and grinding)
- *staining*, i.e. colouring the surface of the wood (possibly in combination with bleaching)
- *impregnating* with a wood preservative (only necessary if constantly or repeatedly exposed to moisture)

The selection and preparation of solid wood and wood products for various applications is covered by DIN 18 355 VOB Part C (ATV).

The section below describes in a little more detail only those facing methods that are particularly relevant for coating purposes. It is followed by some notes on wood preservation. Staining is covered in section 7.9, together with the coating of wood.

5.5.3.1 Facing and smoothing

Processing method

The purpose of facing and smoothing is to obtain an even or smooth material surface that is free from dust, marks, problematic foreign substances and protruding fibres and has a reduced, uniform absorbency.

In the manual sector, general facing of raw, solid wood components such as boards or profiles frequently begins with *planing*. This involves using a blade to slice off relatively thick chips. Then the surface is sanded.

Fine planing (*hydraulic planing*) has become the standard practice in industry. Here the raw boards are passed under cutter heads fitted with numerous cutting edges. The truth of running of the cutters is 0.003 mm, which means that a very smooth surface is obtained even at this stage, and subsequent sanding is often unnecessary.

The core process in the preparation of wood and wood products with raw or uneven and possibly marked surfaces is *sanding*. Practitioners say that the secret of a flawless coating lies in the sanding, underlying the importance of this technique.

Abrasives and sanding tools

With the exception of some special cases, *abrasive papers* or *cloths* in the form of strips, discs or sheets are used for sanding wood and where necessary also coatings. The abrasive generally consists of corundum grains fixed to the substrate with synthetic resins. (Silicon carbide (SiC) is more suitable for sanding coatings, however).

An abrasive is characterised by the parameters of *grit* and *dispersion*.

- Grit

Grit is expressed as a figure - properly preceded by "P" (for paper) - whereby the fineness increases as the figure rises. The figure indicates the number of meshes per linear inch in a sieve through which the grains will just pass. Grit ranges from P 60 (or coarser) for calibration sanding (rough and thick sanding) through around P 220 for fine sanding of wood and intermediate sanding of coatings, up to a maximum of P 1200 for the final sanding of coatings. These are only rough guidelines; individually they depend on the given material, machines and operating parameters such as belt speed and sanding pressure.

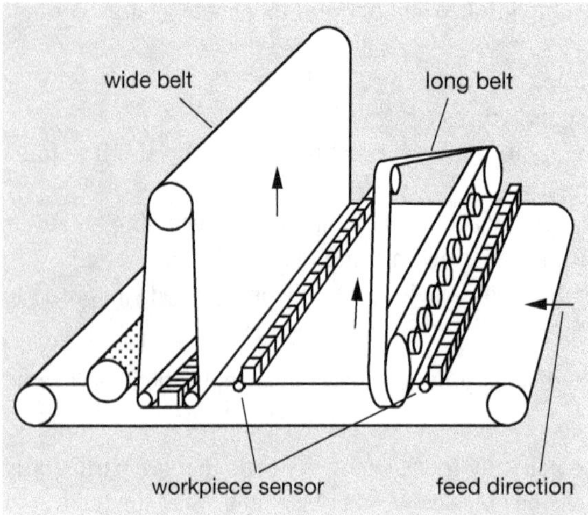

Figure 5.8: Two-belt cross-sanding machine [from: Taschenbuch für Lackierbetriebe 1995]

- Dispersion

An "open dispersion" exists if up to 60 % of the substrate surface is covered with grains; a degree of coverage of over 60 % is referred to as a "closed dispersion". The open dispersion is recommended for soft wood and a high level of sanding, because the abrasive dust is removed more effectively than with a closed dispersion. If an optically flawless coating is to be achieved, it is essential that the abrasive dust be completely removed from the material surface by brushing and vacuuming.

Figure 5.8 shows a schematic view of the design of a modern sanding machine with workpiece sensor, crossgrain-sanding long belt and parallel/counter-sanding wide belt. It may also include a hard sanding roll (contact drum) as the initial sanding tool for calibration sanding and a "Fladder" (a lamellar, rapidly rotating spindle covered with abrasive paper) as the final tool for topping any remaining fibres.

Figure 5.9 reproduces scanning electron micrographs of sanded and fine-planed wood surfaces suitable for possible use as a coating substrate.

5.5.3.2 Notes on wood preservation

The requirements for wood preservation are defined *inter alia* in DIN 68 800, Part 2: *Structural wood preservation*, and Parts 3 + 5: *Pre-ventive chemical wood preservation.* This standard identifies 5 hazard classes, ranging from class 0 (no wood preservative required) to class 4 (insect-repellent, fungicidal, weath-er-resistant, soft rot-repellent wood preservation). In summary, preven-tive chemical wood preservation is generally only necessary if the long-term application-related moisture level exceeds 20 % for solid wood and 18 % for wood products. Under no circumstances should furniture be treated with wood preservatives.

If wood is subject to acute infestation, *controlling wood preservation* is necessary, although this need not necessarily be chemical in nature. The damaged timber component or structure can be heated with hot air, for example, since insects cannot survive a temperature of at least 55 °C for more than one hour.

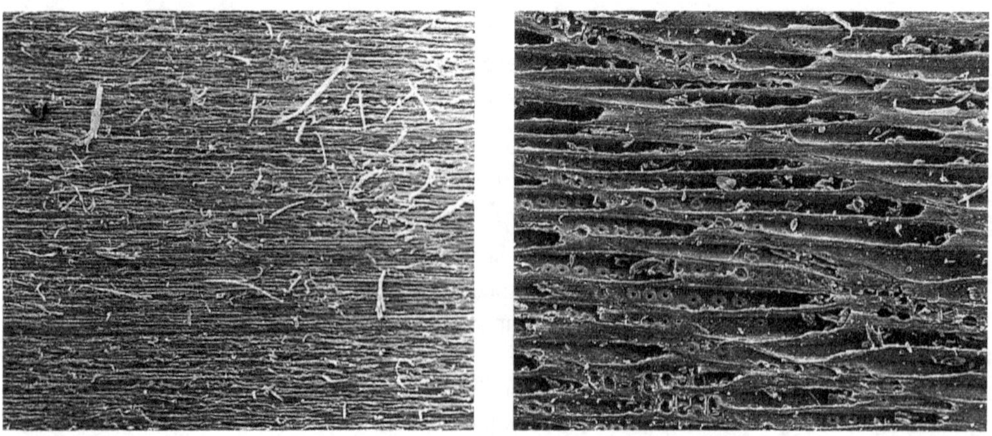

Figure 5.9: Wood surfaces as substrate; left: sanded; right: fine-planed (REM shots)

Wood preservatives are applied in liquid form, ideally by immersion, flow coating or forced impregnation (pressure process). Wood preservatives for load-bearing components (structural timber) in Germany must bear the official mark of conformity of the *Institut für Bautechnik* (institute of structural engineering). For all other applications the *RAL quality symbol* (quality assurance arm of the German standards institute) is proof of effectiveness and safety.

Since the biocidal action of wood preservatives is generally often harmful to plants, animals and humans and in some cases still contain organic solvents too, the rule is:

- use as much wood preservative as necessary, but as little as possible!

Since February 1998 the use of biocides and hence also wood preservatives has been the subject of the *EU Biocides Directive*; this has now passed into national legislation.

5.6 Mineral substrates

5.6.1 Composition and properties

Even though mineral substrates are predominantly oxidic in nature, like plastics they still come in a wide variety of forms, e.g. stone, brick, concrete or plaster.

Various compounds of the element calcium form the principal component of mineral compounds, e.g. in inorganic building materials and mortars, etc., as listed in Table 5.4.

Since some of the interactions that can occur between these substrate materials and coatings, such as solvent-based coatings or paints, are manufacture-related, it seems advisable to obtain an idea of their properties.

- ***Lime and gypsum***

Calcium oxide (burnt lime) is obtained from limestone ($CaCO_3$) by calcining at around 1000 °C. Calcium hydroxide (slaked lime) is obtained by the addition of water. Both substances undergo a highly alkaline reaction with atmospheric CO_2 to form calcium carbonate in the presence of moisture or calcium hydrogen-carbonate in the presence of larger amounts of water. This "carbonatisation"

Table 5.4: Some of the main Ca compounds in building materials and mortars

Chemical composition	Name
CaO	calcium oxide (burnt lime)
$Ca(OH)_2$	calcium hydroxide (slaked lime)
$CaCO_3$	calcium carbonate (chalk)
$Ca(HCO_3)_2$	calcium hydrogen-carbonate (calcium bicarbonate)
$CaSO_4$	calcium sulphate (anhydrite, screed plaster)
$CaSO_4 \cdot 1/2\ H_2O$	calcium sulphate monohydrate (hemihydrate, mortar gypsum)
$CaCl_2$	calcium chloride
$3\ CaO \cdot Al_2O_3 \cdot CaSO_4 \cdot 32\ H_2O$	calcium aluminium sulphate hydrate (ettringite, cement bacillus)
$CaO \cdot SiO_2 \cdot H_2O$	calcium silicate hydrate (component of concrete and sand lime bricks)
$3\ CaO \cdot SiO_2$	tricalcium silicate (component of cement)
$3\ CaO \cdot Al_2O_3$	tricalcium aluminate (component of cement)
$CaO \cdot Al_2O_3 \cdot H_2O$	monocalcium aluminate (component of cement)
$CaCO_3 \cdot MgCO_3$	calcium magnesium carbonate (dolomite)

is of key importance in the hardening of calcium binders. Depending on the moisture content, an equilibrium is set here between calcium carbonate and calcium hydrogen-carbonate, as follows:

$$CaCO_3 + H_2O + CO_2 \rightleftarrows Ca(HCO_3)_2$$

In this continuous changeover between carbonate and hydrogen-carbonate, the calcite crystals form that are fundamental for the final consolidation of lime mortar plaster or whitewash.

Another reaction of calcium (hydr)oxide that is extremely important in "**hydraulic lime**" and also in cement is the formation of scarcely soluble silicates and aluminates with various aluminas. The calcination of alumina-containing limestone at temperatures of over 1200 °C produces, in addition to calcium oxide, the calcium salts of the "alumina acids" SiO_2, Al_2O_3 or Fe_2O_3, e.g.:

$$3\ CaCO_3 + SiO_2 \cdot H_2O \rightarrow 3\ CaO \cdot SiO_2 + 3\ CO_2 + H_2O$$

After calcination, true hydraulic lime is slaked with water, whereupon the calcium oxide converts into the hydroxide. The content of free, unbonded lime depends on the lime content of the raw materials used and is between around 8 and 29 % calcium hydroxide. An excess of water must be avoided since otherwise the hydraulic components react too early. Setting and hardening are based on the formation of various water-containing compounds, e.g. the fibrous calcium silicate hydrate $CaO \cdot SiO_2 \cdot H_2O$. The setting rate can be controlled by the addition of gypsum.

Gypsum consists of the dihydrate of calcium sulphate. When it is calcined, the water of crystallisation is eliminated to a varying extent according to the calcining temperature.

$$2\ CaSO_4 \cdot 2\ H_2O \overset{>120\ °C}{\rightleftarrows} 2\ CaSO_4 \cdot 0.5\ H_2O + 3\ H_2O \overset{>290\ °C}{\rightleftarrows} 2\ CaSO_4 + 4\ H_2O$$

The setting process proceeds in the opposite direction. The water content of the various types of gypsum is theoretically 0 % for anhydrite, 6.21 % for hemihydrate and 20.9 % for set gypsum.

In addition to these binders, most mineral construction materials also contain **aggregates**, usually sand. The composition and density of sand differ according to the type of stone, e.g. basalt, quartzite or limestone, from which it is made. The quantity of aggregates is very much governed by their binder requirement, which depends on their volume of pore space and their specific surface area. Both variables can be attributed to the grain size distribution of the sand. The ratios here are similar to those relating to the binder requirement or oil absorption value for pigments and fillers in coating materials.

Table 5.5: Properties of various coating substrates in the construction industry

Substrate	Thermal expansion ($mm \cdot m^{-1} \cdot K^{-1}$)	Water vapour diffusion resistance value μ (m)	Capillary water absorption ($g \cdot cm^{-2} \cdot h^{-1}$)	pH
bricks	0.036 to 0.0058	6 to 12	0.4 to 1.7	neutral to alkaline
sand lime bricks	0.0078	12 to 20	1.5 to 3.0	neutral to alkaline
concrete	0.007 to 0.015	10 to 25	0.2 to 0.3	alkaline
gas concrete	0.008	0.2 to 0.3	0.8 to 0.9	alkaline
lime rendering mix	0.008 to 0.009	7 to 14	approx. 0.5	alkaline
hydraulic lime plaster	–	9 to 11	approx. 1.1	alkaline
machine-mixed gypsum plaster	–	5 to 6	–	neutral
plaster of Paris	–	7 to 8	–	neutral

A large proportion of mineral construction materials is produced from these raw materials. Table 5.5 provides an overview of their properties.

- *Masonry*

Bricks are baked from loam, clay or clayey materials, possibly with the addition of sand, brick dust or similar substances. Red bricks contain iron-rich brick clays. Water-soluble salts, e.g. sulphates or chlorides, may cause efflorescence on bricks that is usually harmless but in the case of impregnation or hydrophobing may force the coating away from the substrate. Frost-resistant bricks, also known as face bricks, are used for visible brickwork and facing work. Lining bricks are used for brickwork that is to be faced, plastered or provided with a weather-resistant coating. They do not need to be frost-resistant. Klinker bricks are brick-like stones that are baked until they sinter. They are usually only slightly absorbent and are extremely chemically and mechanically resistant.

Sand lime bricks are obtained from lime and quartz-containing aggregates. During the hardening process the lime sets with the silicon dioxide at the surface of the grains of sand, forming calcium silicate hydrates. Complete conversion of the lime is often achieved to a depth of only 3 to 4 cm, which means that the core of such bricks may still contain free water-soluble calcium hydroxide.

Natural stone, from a coatings point of view, is a problematic substrate because the strength and tendency to weather can fluctuate even within a masonry structure. It can be protected against weathering by the use of con-solidating primers based on silicates or of transparent water-repellent solutions based on silicone. Hiding coat-ings employed include those which permit diffusion of air, such as silicone resin emulsion paints and silicate emulsion paints.

- *Types of concrete*

Concrete is an artificial stone made from a mixture of cement, concrete aggregate and water. Residues of forming oil may be left on the surface of the concrete from the manufacturing process, and these must be removed before applying a coating. Concrete is generally categorised according to its bulk density:

- breeze concrete: dry bulk density $\leq 2.0\ g/cm^3$
- standard concrete: dry bulk density 2.0 to 2.8 g/cm^3
- high-density concrete: dry bulk density $> 2.8\ g/cm^3$

Depending on the type of cement, quality of cement, temperature and moisture content, concrete dries at varying rates; the afterbake may take up to a year to dry out in air and several years if stored in a humid atmosphere. Porous concrete surfaces can absorb moisture, which in some cases can lead to frost cracking. If concrete is attacked by aggressive liquids, the process is decisively influenced not only by their quantity but also by flow rate, pressure and naturally also temperature. Rainwater enriched with industrial gases causes surface destruction of concrete, and gypsum-containing water destroys the structure of the concrete. Infiltration and subsequent drying out of salt solutions can give rise to osmotic processes that ultimately also lead to destruction of the concrete. To a certain extent concrete is also sensitive to organic acids, vegetable fats and polyhydric alcohols.

Reinforced concrete and **prestressed concrete** are reinforced with steel armouring. This steel armouring is normally protected against corrosion due to concrete's prevailing pH of 12 to 13. With progressive carbonatisation, the pH value falls – as the depth and age of the concrete increase – to below 9, which is no longer sufficient to passivate the steel. Steel armouring must therefore be covered by a sufficiently thick layer of concrete. Any coating applied to concrete must further provide a high diffusion resistance for carbon dioxide, which suppresses carbonatisation.

The diffusion resistance s_d (m) is calculated from the diffusion resistance value[1] μ for the gas under consideration and the coating thickness s (m) as

$$s_d = \mu \cdot s$$

It indicates the thickness of a layer of air that would oppose the diffusion of the gas with the same resistance as the given covering. The diffusion resistance for carbon dioxide of a 1 cm thick layer of concrete is around 3.5 m. A 300 µm thick polymer coating with a diffusion resistance value of 80,000 has a diffusion resistance of 24 m. It therefore opposes carbon dioxide with a diffusion resistance corresponding to a concrete covering of approx. 7 cm.

Gas concrete *or* pores is made from an aqueous slurry of cement, calcium oxide and finely ground aggregates. Aluminium grit, which reacts to form aluminates and hydrogen under the highly alkaline conditions during curing, is generally used as blowing agent. It produces the characteristic porous structure of gas concrete.

- ***Other construction materials***

Cement-bonded fibreboard contains reinforcing fibres as well as cement. In the case of cement-bonded chipboard, the expansion in thickness of up to 2 %, which for mineral substrates is quite considerable, must be taken into account. Under the influence of moisture a pH of 11 to 13 is set. In contrast, the expansion and contraction behaviour of cement-bonded plastic or fibreglass board is only in the region of 0.15 %. Both are substitute materials for asbestos cement, which is no longer produced in the UK.

Cement-bonded fibreboard should first be treated with a primer sealer to avoid lime efflorescence. The topcoat can be a standard house paint.

Plaster blocks are produced from rapidly stiffening types of gypsum with the addition of suitable accelerators. Arising from the production process, release agents may remain on their surface which, in the case of phenol-containing mineral oils for example, can cause yellowish specking.

Gypsum plasterboard also consists essentially of gypsum board encased in thick paper.

Plasters are single- or multi-layer coatings of rendering mix applied to exterior and interior wall surfaces. They form hard-wearing surfaces and improve moisture-proofing, flame resistance, thermal insulation and sound-proofing. Interior plasters must demonstrate good absorbency and water vapour permeability, whereas exterior plasters must above all be weather resistant. According to their composition and associated properties, rendering mixes are divided into the following groups:

- *Lime plasters* are obtained from a mixture of slaked lime and sand. They harden with absorption of carbon dioxide and release of water. This reaction takes a very long time and is still not complete even after several years. Coatings applied to lime plasters must therefore demonstrate good water vapour permeability and be resistant to saponification.
- *Cement plasters* are made from a mixture of cement and sand and harden with water. Their highly alkaline properties only gradually disappear by means of carbonatisation, which means that a coating applied to these plasters must likewise be resistant to saponification. The moisture content must have fallen to below 3 % before a coating can be applied.
- *Gypsum plasters* harden exclusively by absorption of water. After they have dried out to about 1 % moisture content, they form porous and absorbent coating substrates.
- *Loam renderings*, especially the older type, are a critical substrate for coatings because they can vary so much in composition. Accordingly, their properties can also vary greatly. The constituent raw materials are loam, sand, straw, cow dung, and "quark" soft cheese in any proportions. Industrially manufactured loam renderings of consistent quality, and hence predictable properties, are now available.

1) The diffusion resistance value μ indicates how much higher the diffusion resistance of a material is than that of an equally thick, stationary layer of air.

External insulation and finish systems are composed of at least three layers. A thermal insulation layer con-sisting of various thicknesses of insulation materials, a reinforcing layer consisting of bonding coat and rein-forcing mesh, and a topcoat for texturing the surface and based on a mineral or synthetic resin render. The coat-ing properties are essentially determined by the topcoat.

Woodchip wallpaper is the most common type of wall covering in Germany. It consists of wood fibres which provide texture and are held firmly in place between several layers of paper. When bonded properly, they are a non-problematic substrate, and can be coated with all types of indoor emulsion paints, indoor silicate paints and latex paints.

5.6.2 Pretreatment of mineral substrates

Assessing the substrate

Before a substrate is coated, its nature and condition are assessed. The assessment must employ tools found on a typical building site. German BFS leaflet No. 20 "Assessing the substrate for coating and wallpapering work measures for repairing damage" describes suitable test methods.

Once the type of substrate has been established (e.g. plaster, concrete, metal, wood) and whether it is new, old, coated or uncoated, its condition is determined by the following tests:

- Scratch test with a knife or adhesive tape test (to determine its strength, load-bearing capacity)
- Wetting test with water (absorbency)
- Moisture test by inspection or moisture-measuring device
- Wipe by hand (chalking)
- pH measurement with indicator paper or phenolphthalein (alkalinity)
- Appearance (soiling, cracking, corrosion, mould and algae, salt efflorescence)

Cleaning and priming

For mineral substrates as for other substrates too, the prerequisite for a permanently adhering coating is firstly a thorough *cleaning*, which includes removal of all loosely adhering particles.

The nature of the primer depends on the nature and composition of the substrate and of the proposed coating system.

Most mineral substrates have a heterogeneous structure. They usually also have a certain capillary volume, particularly if a liquid, generally water, escapes during setting or drying. The smaller the proportion of these liquids at the manufacturing stage, the smaller the subsequent proportion of the capillary volume to the total volume. When water in particular is incorporated during the setting process, e.g. in the setting of concrete, the pore volume can be restricted to a minimum. These pores should not be imagined as a system of interconnected tubes of varying diameters, however. They are more like cavities irregularly distributed through the solid. Other substrates have little or no absorbency, e.g. fairfaced concrete or surfaces with old coatings. Overall, prim-ers play a number of roles depending on their composition:

- Consolidation of the substrate
- Adhesion promotion
- Equalising the absorbency
- Water repellency

Transparent primers mostly have a strengthening effect and regulate the substrate's absorbency. In order to work, they have to penetrate into the substrate and so are only used for porous, absorbent surfaces. They are often called impregnating agents.

Pigmented coatings promote adhesion and equalise the absorbency and differences in hue within the substrate. They are generally used on non-absorbent to weakly absorbent substrates. Where the substrate is too absorbent, pigmented coating materials may suffer from binder migration. As a result, pigments can be filtered off, as it were. The PVC of the coating rises and the surface may become chalky.

Impregnation refers here to the enrichment of a capillary-absorbent substrate with a highly liquid agent, which generally does not form a coherent film on the surface. Where an unpigmented impregnating agent does lead to a coherent film, the adhesive strength of the subsequent prime coat may be reduced .

Substrate absorbency

If the absorbency of the substrate is too high, binder migration out of the coating material may occur. As a result pigments may be virtually removed by filtration. The PVC of the coating increases and a chalky surface may be formed. If, however, the use of an unpigmented impregnating agent does cause a closed film to form, the adhesion of the subsequent primer may be reduced. The use of pigmented impregnating agents is therefore recommended primarily for only slightly absorbent mineral substrates.

The absorbency of mineral construction materials is conventionally determined from their capillary water absorption. This indicates the quantity of water soaked up in one hour by a surface area of 1 cm^2 (see Table 5.4). When the same type of construction material is used, absorbency increases as the density decreases (= increasing pore volume). Information of this type naturally offers only an approximation of the behaviour of coatings. Together with pore volume, the other parameters influencing capillary liquid absorption are, according to the Washburn equation (see page 259), pore radius, surface tension of both construction material and liquid, and also density and viscosity of both substances; this latter parameter may change during absorption. Manufacturing methods and chemical processes during the hardening of mineral construction materials may cause variations in the usual absorbency of the material.

On lime mortar plasters carbonate-rich sintered coatings may form that have a much lower absorbency than the layers of plaster below them. This becomes particularly striking if the surface of such a plaster is cracked. The differences in absorbency can then be seen from the emergence of dark cracks if the plaster is wetted with water or from swelling if it is painted.

The absorbency of a gypsum plaster is largely determined by its composition. Gypsum plasters mixed by hand often display a relatively high absorbency after setting, whereas mechanically mixed plasters have an impermeable, scarcely absorbent surface.

The absorbency of visible concrete surfaces is likewise dependent on a whole range of factors. They are very similar to highly absorbent gypsum surfaces in this respect.

Sand lime brick, brick, cement-bonded fibreboard and especially gas concrete are likewise all classed as highly absorbent, even if the absorbency of gas concrete is surprisingly low given its high pore volume.

Impregnating processes

The impregnating agent must be selected according to the physical and chemical properties of the substrate. The nature of the anticipated subsequent coating and the waiting time between impregnating and priming must also be taken into consideration. To ensure optimum adhesion of the primer to the substrate, it is preferable to use the same type of binder for impregnating agent and primer.

Before application of a coating, brittle setting coats or old coatings that have begun to chalk must be compacted in order that they can form a stable substrate for the next coating. Special impregnating agents with high penetrating power are now commercially available for this purpose, although unquestioning use of such agents is not to be recommended. In order to achieve good penetrating power, the non-volatile content of such agents is less than 20 %. Unrealistically large quantities of impregnating agent would therefore be needed to completely impregnate a brittle substrate. If only the uppermost layer were compacted, however, it in turn would form only an inadequate bond to the lower layers which would still exhibit brittle zones. To avoid a situation of this sort, the adhesion of an impregnated coating should be tested by means of a tear-off test using a piece of adhesive tape.

If there is no risk of damp penetrating back to the substrate, capillary-narrowing impregnation is a suitable means of insulation preventing calcium hydroxide or other salts from migrating out from the plaster. Fluosilicate treatment (see below) generally does not form an adequate barrier.

Primer-sealers

Walls and ceilings are frequently exposed to substances that stain, e.g. oils and greases in kitchens, nicotine in houses and restaurants, and soot or water stains following fires. These types of substance can show through a new coating. Primer-sealers are designed to prevent substances present in the substrate from acting on the coat-ing and vice versa. The choice of primer-sealer depends on the nature of the substance to be treated. Pigmented and unpigmented agents, both solventborne and in emulsion form, are used.

Cementitious or lime-containing substrates can be impregnated with hexafluorosilicate solution (**fluosilicate sealing**) to convert the readily soluble calcium salts into sparingly soluble hexafluorosilicates. This lowers the alkalinity of the substrate and prevents the water stains from showing through. However, if some of the soluble calcium salts in the substrate escape this process, efflorescence of calcium salts may still occur at the surface upon subsequent penetration of moisture. Such a complete reaction cannot be achieved in practice, however. In principle, fluosilicate-treated substrates should therefore always undergo additional impregnation.

Sources and references for Chapter 5

[1] A. Brasholz: Handbuch der Anstrich und Beschichtungstechnik, Bauverlag GmbH, Wiesbaden and Berlin 1989

[2] Brockhaus: Naturwissenschaften und Technik, F. A. Brockhaus, Mannheim 1989

[3] W. Endlich: Kleben und Dichten - aber wie?, DVS-Verlag GmbH, Düsseldorf 1996

[4] A. Goldschmidt, H.-J. Streitberger, BASF Handbook: Basics of Coating Technology, 2nd edition, Vincentz Network, Hanover 2007

[5] B. Hantschke: Glasurit-Handbuch der Bautenlacke, Laumanns Druck und Verlagsgesellschaft mbH, Lippstadt 1990

[6] C. H. Hare, Protective Coatings: Technology Publishing Company, Pittsburg 1994

[7] "I-Lack-Schwerpunktthemen", Vol. 5 (Vorbehandlungstechnik), Curt R. Vincentz Verlag, Hanover

[8] H. Kittel (Ed.): Lehrbuch der Lacke und Beschichtungen, Verlag W. A. Colomb, Berlin 1976

[9] H. Kollek: Reinigen und Vorbehandeln, Curt R. Vincentz Verlag, Hanover 1996

[10] D. Ondratschek (Ed.). Taschenbuch für Lackierbetriebe, 1994 to 1997, Curt R. Vincentz Verlag, Hanover

[11] O. Paprzycki, H. Pecina: Lack auf Holz, Curt R, Vincentz Verlag, Hanover 1995

[12] J. Ruf. Organischer Metallschutz, Curt R. Vincentz Verlag, Hanover 1993 C ZD

[13] F. Sadowski et al.: Lackierungen in der Metallindustrie, Expert Verlag, Ehringen 1981

[14] H. Saechtling: Kunststoff-Taschenbuch, 24th edition, Hanser, 1989

[15] K. Sponsel, W. O. Wallenfang, I. Waldau: Lexikon der Anstrichtechnik 1 - Grundlagen, 2 - Anwendung, 9th edn., Verlag Georg D. W. Callwey, Munich 1992

[16] D. Stoye (Ed.): Paints, Coatings and Solvents, VCH Verlagsgesellschaft mbH, Weinheim 1993

[17] Ullmanns Encyklopädie der technischen Chemie, 4th edition, Vol. 15, Lacke, Verlag Chemie GmbH, Weinheim 1978

[18] Ullmann's Encyclopedia of Industrial Chemistry, 5th edition, Vol. A 20, Paints, VCH Verlagsgesellschaft mbH, Weinheim 1992

[19] Umweltbundesamt: Holzschutz-Tips und Informationen zum richtigen Umgang mit Holzschutzmitteln. (Leaflet, approx. 1992)

[20] G. Vollmer, M. Franz: Chemie in Hobby und Beruf, Verlag Georg Thieme, Stuttgart and DTV, Munich 1981

[21] Verband der Lackindustrie: Betriebliche Fachkunde, 4th edn., 1996

[22] Z. W. Wicks, Jr., F. N. Jones, S. P. Pappas: Organic Coatings, Science and Technology, J. Wiley & Sons, Inc., New York 1992

[23] K. Winnacker, L. Küchler: Chemische Technologie, Vol. 4, Metalle, 4th edn., Carl Hanser Verlag, Munich 1986

[24] U. Zorll (Ed.): Römpp Lexikon Lacke und Druckfarben, Georg Thieme Verlag, Stuttgart 1998

[25] Company literature published by, among others, Paul Auer GmbH (Strahltechnik), Chemetall GmbH (Bondertechnik 23, 1987), Eisenmann KG (Leitfaden für den Lackierbetrieb, 24th edn., 1989)
Information supplied verbally by Desowag GmbH

6 Application and drying

6.1 Methods of application and criteria for use

Just as diverse as the various substrates that can be coated (as described in Chapter 5) is the choice of methods by which the coating can be applied to these usually pretreated surfaces. They include the manual methods, which are as important as ever and require great skill and experience, as well as coating processes performed in large-scale mechanised plants and, more recently, those using process-controlled automated equipment and robots, which can produce high-grade coatings extremely efficiently. Table 6.1 provides an overview of the main processes in use today and lists some of the technical and economic criteria for their use. It is increasingly the latter, along with the ecologically important aspect of solvent emission, that now govern the viability of the individual methods of application as well as determining the choice of coating material.

6.2 Manual application by brushing, rolling, trowelling, wiping

Brushing

Application of a coating with a brush is as indispensable as ever for small parts, for retouching sections of surfaces and in the DIY sector.

Advantages of brushing:
+ can be used for any shape of object
+ good wetting and covering of surface defects
+ low coating losses

Disadvantages and limitations of brushing:
- labour-intensive
- flow is critical (risk of brush marks), particularly in the case of the often pseudoplastic waterborne coatings
- uniform, precise film thicknesses are difficult to achieve

The material and shape of the brush used should be selected according to the coating material and the object. Thus, flat brushes are suitable for low-viscosity strippers and paints while round brushes are better for more viscous, e.g. pigmented, coatings. Different brush materials need to be used for waterborne or solventborne coatings. While animal bristles are suitable for use with solventborne coatings, they can split in waterborne types. Conversely, nylon brushes can swell up in solvents. Polyester brushes may be used for both types.

Paints applied by brush must be moderately viscous while also exhibiting good levelling (to counter pronounced brush marks). This calls for thixotropic behaviour which leads to slow internal build-up of structure after application. The solvent must also evaporate slowly to prevent thickening on the brush.

Rolling

As with brushing, in rolling too the nature of the coating material and the desired result must determine the choice of tool, in this case the roller: from long-pile lamb's fleece through plush-like fabric to foam, e.g. Moltopren®. Rolling is much less labour-intensive than brushing, and the film thickness is more uniform. Since no brush marks are left, a very smooth surface can be obtained.

Brock, Groteklaes, Mischke: European Coatings Handbook

ISBN 978-3-86630-849-7

Table 6.1: Overview of methods of application and criteria for their use

Method of application	Surface quality	Restrictions			Operating speeds	Solvent emissions	Application efficiency
		dimensions	geometry	other			
brushing	moderate to good	small areas	–	–	very low	low	very good
manual roller application	good		accessibility	–	moderate	low	very good
trowelling	–	small areas	–	–	low	low	very good
wiping	poor	large parts (pipes)	–		low	low	very good
curtain coating	very good	restricted working width	practically plane surfaces	–	high	low	very good
roller coating/ coil coating	moderate to very good	restricted working width	plane surfaces	–	very high	low	very good
conventional dipping	moderate	restricted object volume	not suitable for scoop-shaped parts	edge effects	high	low	very good
barrel coating	poor	small parts	pourable	–	high	low	very good
centrifuging	poor	small parts	pourable	–	high	low	very good
flooding	moderate	restricted object volume	not suitable for scoop-shaped parts	edge effects	high	low	very good
flow coating	moderate	restricted working width	not suitable for scoop-shaped parts	edge effects	high	low	very good
electro-deposition	low	restricted object volume	not suitable for scoop-shaped parts	–	high	low	very good
air atomisation, low pressure	good	–	–	–	low	high	poor
air atomisation, high pressure	excellent	–	–	–	low to moderate	very high	very poor
air atomisation, HVLP	very good	–	–	–	low	high	poor
airless atomisation	moderate	–	–	–	high	moderate	good
airmix atomisation	good	–	–	–	moderate	high	moderate
electrostatic-assisted air atomisation	very good	–	no Faraday cages	conductive substrate	moderate	high	good
rapid-rotation atomisation	very good	–	no Faraday cages	conductive substrate	moderate	high	good
powder spraying	good	–	–	conductive substrate	moderate	virtually none	moderate
powder sintering	moderate	–	–	high-build films	moderate	low	very good

The disadvantages with rolling are that the substrate surface has to be even, and substrate wetting is not as good as with brushing. Rapid rolling can lead to spattering, but the spattering tendency can be largely suppressed by adjusting the flow behaviour accordingly (→ 8.1.1).

Trowelling

Stoppers (highly filled, highly viscous to paste-like coatings) are applied either by mechanical roller coating (→ 6.4) or, particularly in the case of small areas where coatings are applied by hand for example, by means of a trowel. After trowelling, the layer of stopper first has to be smoothed out and then, after it has subsequently dried and hardened, it usually needs to be rubbed down vigorously.

The process can often be accelerated by using the somewhat more free-flowing sprayable stoppers.

Wiping

For rapid, even coating of large components, wiping with a special glove can be advantageous, particularly for pipes of all types, even those already fixed in place. The process has its limitations in regard to poorly accessible or elaborately shaped components, however. Moreover, the gloves generally cannot be cleaned and they are also extremely expensive.

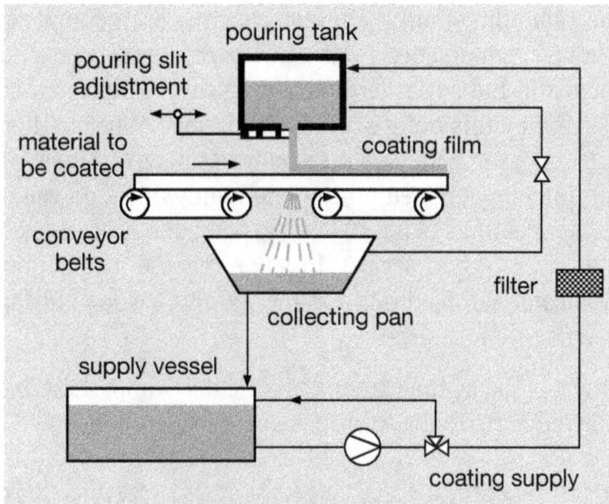

Figure 6.1: Curtain coating, according to [10]

6.3 Curtain coating

Curtain coating is used principally in the furniture industry to coat plane surfaces such as cupboard walls and door panels, etc. The process is also suitable for coating sheet metal and for paper and card. The coating is generally applied by transporting the workpieces on a conveyor belt through a curtain of coating material flowing from a narrow, adjustable slit (see Figure 6.1).

The coating material is continuously recirculated and replenished, and also filtered and cooled as required. The thickness of the coating film can be adjusted by altering the width of the slit and the throughput rate of the workpieces. Film thicknesses are usually between 20 and 500 µm. With narrower film thicknesses the surface smoothness may not be adequate. In such cases vacuum feed heads have proved useful: reduced pressure in the feed head expands the slit and allows a more uniform coating to be obtained.

Curtain coating is generally used with UV or electron beam curing coatings, but it is also suitable for two-component coatings (base and hardener are applied with different feed heads).

Curtain coating can produce a very good surface quality with almost 100 % yield, but it is only suitable for plane or very slightly convex components, or for profiles of a similar type. Curtain coating has mostly been replaced by roller coating (see next section).

6.4 Roller coating

Flat objects, such as sheet metal and metal strip, but also wooden parts, films and boxes, for example, can be coated very quickly and economically using one of the various roller coating techniques. As a preparation for coating, flat wooden parts are often rolled with an initial layer of filler before also being roller coated in some cases. A chromium-plated feed roll first transfers the coating material onto the rubber-covered coating applicator roll (see Figure 6.2), the amount applied being regulated by the gap between the two rolls. In many plants the feed roll is also supplied with coating from a separate paint trough.

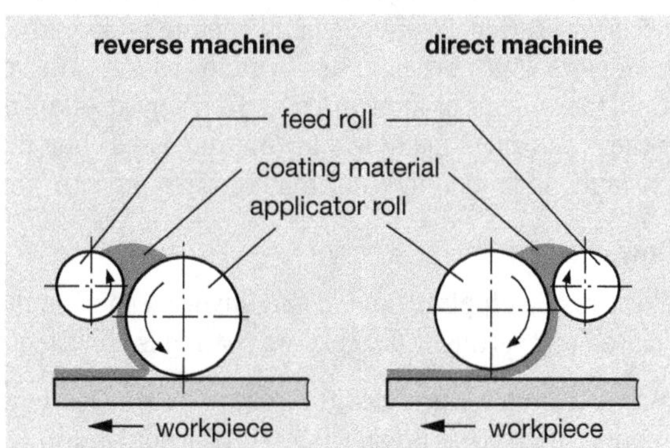

Figure 6.2: Roller coating, according to [10]

The applicator roll then transfers the coating onto the substrate. The earlier *direct process* (roll and workpiece rotating in the same direction) - characterised by an incomplete and uneven transfer onto the substrate - has been superseded by the newer *reverse process* and similar other, in which the applicator roll rotates in the opposite direction to the workpiece. The efficient and accurate transfer of coating assured by this process enables coatings to be obtained in a wide range of thicknesses (from 3 to 100 μm), even with very different coating viscosities. High-gloss, texture-free and uniform surfaces can thus be obtained on the one hand whilst, on the other, the application of ultra-thin adhesive primer coatings is possible. Two-component coatings can also be processed in this way (with two rolls operating in series). In some reverse installations, although a distributor roll (between the applicator and feed roll) rotates counter to the applicator roll, the latter rotates in the same direction as the workpiece.

This method of application is particularly important in *coil coating* (→ 7.6) and in the coating of flat wooden parts in the furniture and parquet industry.

6.5 Dipping, flow coating and related processes

Dipping, flow coating and also barrel coating and centrifuging are very simple coating methods used for mass production articles and small parts, as well as in assembly line production of agricultural machinery, for example.

Dipping processes

In conventional *dipping*, the workpieces, e.g. window frames but also electrical coils, etc., are immersed in the coating and removed again; the object is thus practically already coated. Care must be taken to prevent air bubbles from forming due to the presence of voids, however, and the coating must not be allowed to run out. Since impurities would accumulate in the bath, the objects must be absolutely clean. To ensure a consistent bath quality, extracted coating material and evaporated solvent or - in aqueous systems - water must be replenished, sedimentation in low-viscosity baths must be prevented by agitation and the temperature kept constant. Both solvent-based and waterborne dipping coatings can be used. So-called *Trilene* (dipping) coatings - with trichloroethylene as solvent - are no longer used for environmental reasons although they offered a number of technical advantages.

A benefit of all dipping processes is that workpieces are completely coated with only low coating losses, and the processes are very easily automated. At the same time, however, variations in coating thickness and tearing cannot always readily be avoided, so requirements in respect of surface quality should not be too exacting.

Electrodeposition coating is covered separately in the next section.

Barrel processes

Barrel coating and *centrifuging* are processes designed for small mass production articles. Here the special the low-viscosity specialty coating is poured or sprayed over a relatively large number of small parts which are held in a drum or basket. After the parts have been removed and allowed to drain, they are air or force dried or even stoved - depending on the binder - by spinning the drum, or they are centrifuged for a few minutes and then dried. During drying, the parts must be kept constantly in motion in order that flawless, smooth surfaces may be obtained.

Flow coating

In contrast to dipping, *flow coating* involves pouring the coating over the workpieces, nowadays in a closed compartment. The excess coating is collected and reused.

In a variant of the process, the coating is sprayed onto the workpieces through directional spray nozzles. The process is especially suitable for large and bulky objects such as radiators and machine parts.

Vacuum coating

Vacuum coating is a special technique in which liquid coating is applied in a closed installation under reduced pressure by means of a powerful siphon of air entering the compartment from below. This causes the coating to be entrained upwards and to atomise. It is then deposited forcibly on the workpiece, which passes through the compartment through very narrow apertures. The process is suitable for bars, wooden profiles, panels, metal profiles etc.

6.6 Electrodeposition coating

When a dipping process involves the use of an electric current to coagulate the binder and hence to deposit the coating film on the substrate, it is known as electrodeposition coating. In modern terminology the variants of electrodeposition coating are distinguished not so much by the intended purpose or by coating chemistry but rather by the electrical polarity of the direct current flow.

In *anodic electrodeposition coating*, the older process (in use since the 1960s), the workpiece is made the anode (positive pole). For high-grade primers, e.g. in automotive assembly line coating, and for single-coat topcoats (the two main areas of application for electrodeposition coating), it has largely been superseded, however, by the more recently developed method of *cathodic electrodeposition coating* (since about 1975) for the following reasons:

- outstanding corrosion protection
- uniform coating thickness distribution
- better throwing power (coating also deposited in voids, despite "Faraday cage")
- good coating over corners
- lower electricity consumption

In cathodic electrodeposition coating, the workpiece is correspondingly made the cathode (negative pole).

Workpieces are generally pretreated by alkaline phosphatising, also known as iron phosphatising, (usually in the case of anodic electrodeposition) or by zinc phosphatising (usually in the case of cathodic electrodeposition).

6.6.1 Principles of electrochemistry

The binders used in electrodeposition coating can vary enormously depending on their intended application. Polyacrylates, polyesters, epoxy resin esters and also maleinate oils and polybutadienes

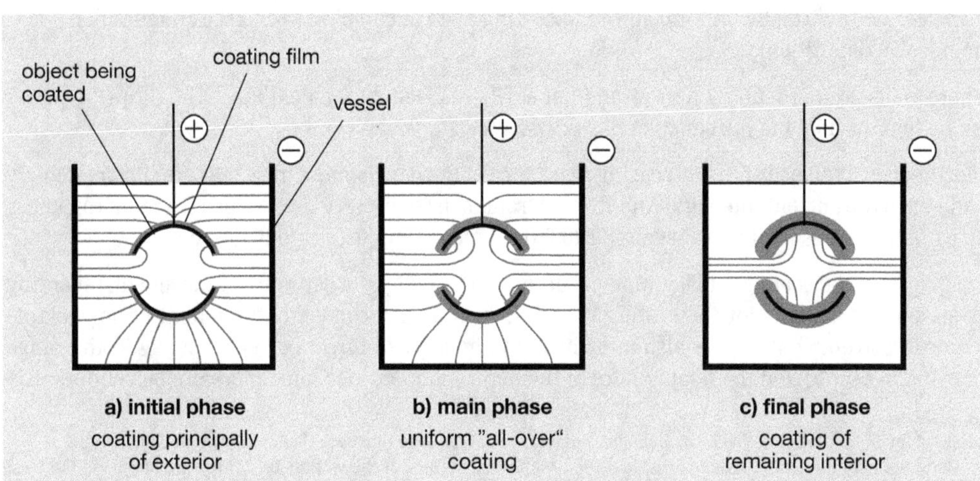

Figure 6.3: Electrodeposition coating in voids by throwing power: progressive deposition (here: anodic eloctrodeposition coating)

are used in anodic electrodeposition, epoxy amine adducts and polyacrylates in cathodic electrodeposition. The nature of the binder modification, through which water solubility and its reverse, deposition, are made possible, is of primary importance for the processes involved in electrodeposition coating, however.

Reaction sequence (anodic electrodeposition)

If the binder contains carboxy-functional groups (*acid-functional*), it can be made water-soluble (solution reaction) by neutralisation with amines (any sort of amines or ammonia, depending on the intended use):

$$\text{/\\/\\/\\-COOH} + NR_3 \rightarrow \text{/\\/\\/\\-COO}^- + H\text{-}NR_3^+ \qquad \text{/\\/\\/\\} = \text{polymer group}$$

insoluble — water-soluble

Colloidally dissolved binder or coating particles (bonded with fillers, etc.) of this type are also referred to below as *polymer ions*, since they are charged particles with polymeric substituents.

The negatively charged coating particles (colloid particles) are thus suitable for migration - or preferably diffusion and migration (see Figure 6.5) - to the anode (positive pole), i.e. for the *anodic electrodeposition* process.

They are not discharged immediately at the anode, however, but are first neutralised (deposition reaction) by the H^+ ions accumulating as a result of water electrolysis (drop in pH of up to around 2 !):

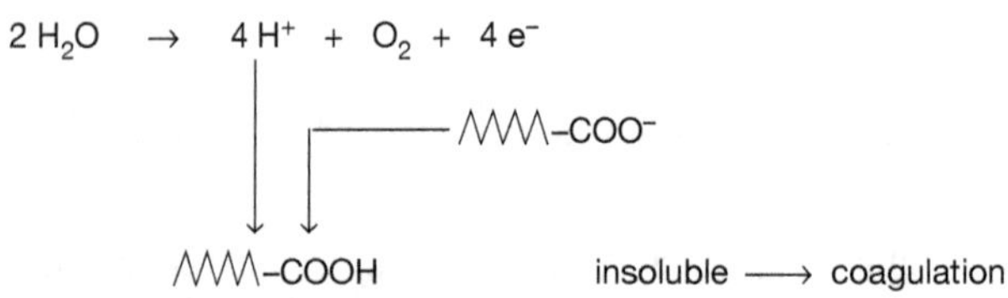

The coagulate, which is still not a closed coating film, still displays residual negative charges and so is deposited at the anode, where it thickens a little more (dewatering due to electro-osmosis, see page 281).

Deposition process

The coagulated film has a defined electrical resistance which, as the coating process continues, leads to a shift in the field lines[1] even in voids and Faraday cages, and so gives rise to good *throwing power*[2] with virtually uniform all-round coating even of complex workpieces (Figure 6.3: a to c; progressive deposition).

Where voids are particularly narrow and not easily reached by the field lines, the throwing power can be improved by the introduction of *secondary electrodes.*

A further consequence of the increasing resistance of the developing film is that the current density (and hence the current intensity) rapidly falls from its initially very high values. The film thickness too usually reaches its final value after 1 to 2 or at most 3 minutes (see Figure 6.4).

The quality of the coating - in addition to being influenced by the deposition time and disregarding the parameter settings for the coating material, such as pH, conductivity, content of non-volatile matter or solvent content - is also influenced by the applied direct current (usually in the range from 100 to 500 V) and the temperature of the bath. All quantities must therefore be continuously

1) "Field lines" are used here as an explanatory aid and are intended to illustrate the field strength distribution.
2) Along the field lines the coating particles move towards the surface of the workpiece. "Throwing power" is the term applied to the entirely desirable effect whereby the field lines also extend to less accessible parts of the object (voids, folds, etc.), as a result of which coating material is deposited there too.

monitored and kept constant in order to prevent surface defects. Otherwise the coating film would display sponge-like hollows, poor flow, pores, edge effects, defective corrosion protection, etc.

Deposition equivalent

The *deposition equivalent* (DE) is an important coating value in terms of the amount of energy required for electrodeposition coating. It is defined as the charge that has to be applied for 1 g of deposited film and is calculated from the current intensity I (as a function of time t) with m as the mass of coating deposited, according to the following equation:

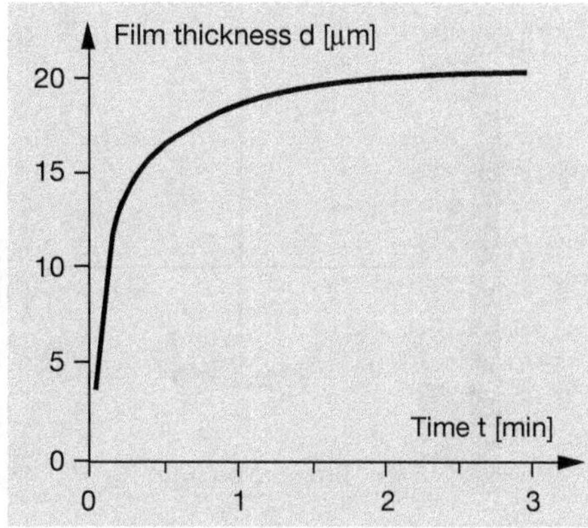

Figure 6.4: Film thickness as a function of deposition time

$$DE = \frac{1}{m} \int_0^t I \cdot dt \quad \text{[C/g (Coulomb/gram)]}$$

Typical values in practice are in the range from 40 to 150 C/g. (1 C = 1 A · s).

After the workpiece has been removed from the dipping bath and any uncoagulated coating material has been rinsed off (in a multi-stage rinsing process), the coating film is ultimately stoved (at 150 to 190 °C) to crosslink and to obtain its ultimate properties.
There is a growing trend towards the use of low-bake materials.

Reaction sequence (cathodic electrodeposition)

Binders used in the *cathodic electrodeposition process* contain basic groups (amino groups) which, when neutralised with acids, ensure the necessary water solubility by means of salt formation:

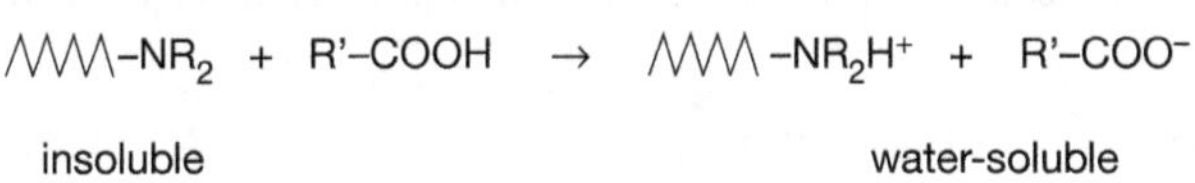

The principal acids used here are formic or acetic acid and occasionally also propionic or lactic acid.

Once the charged coating particles have diffused to the cathode (negative pole) they are then neutralised (this is naturally the reverse of the anodic electrodeposition process). The OH^- ions which have increasingly formed at the cathode by water electrolysis remove H^+ from the ammonium group (deprotonation); the pH rises sharply in the vicinity of the cathode. The following deposition reaction occurs:

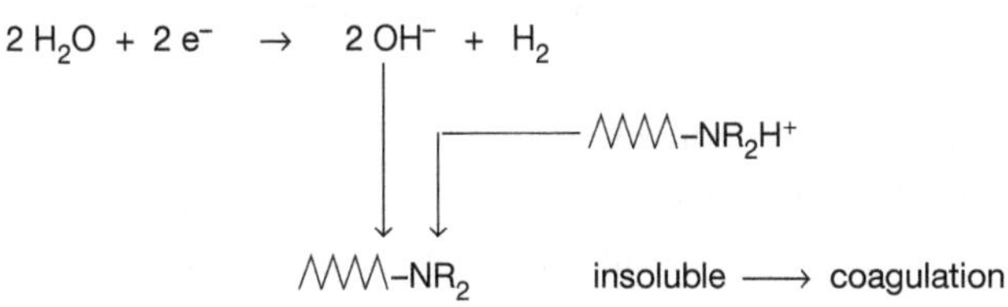

The further stages leading to the final coating film correspond in principle to those described above in relation to anodic electrodeposition.

Electrodeposition bath operation

To prevent pigments and fillers from settling out in the highly liquid electrodeposition bath, it must be continuously stirred or agitated, even outside the deposition periods. Amongst other things, this means that strictly speaking the commonly used terms "*electrophoresis*" and also "*anaphoresis*"

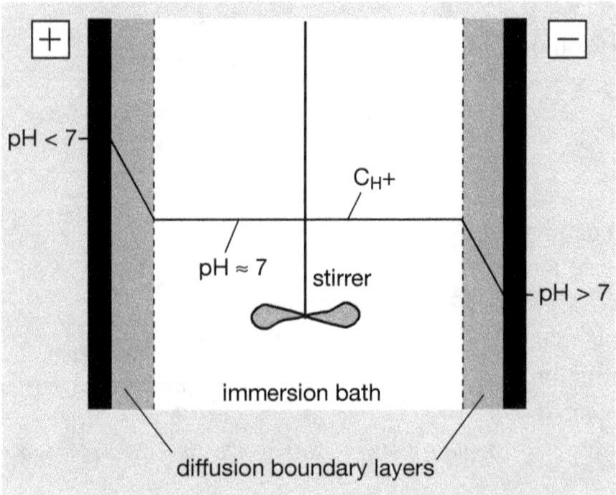

Figure 6.5: Diffusion boundary layers and H^+ concentrations in the electrodeposition bath during current flow (schematic flow)

or "*cataphoresis*" do not apply since no significant migration[1] of particles occurs in the electrical field. Instead, the coating particles that are deposited are principally those that are close to the surface of the workpiece, as a consequence of stirring and therefore more or less by chance.

The changes in concentration of the H^+ ions (= C_{H+}) or the OH^- ions, as well as those of the corresponding acids and bases, are restricted to the diffusion boundary layer at the electrodes, which is no more than 100 μm thick. Figure 6.5 provides a schematic view of these concentration gradients in the boundary layers. The migration of charged particles in the large, central part of the bath, made turbulent by stirring, is negligibly slow in comparison to the stirring speed (convection).

Since the polymer ions in an anodic electrodeposition bath undergo coagulation in the range above pH ≈ 4, this process will therefore begin somewhere in the region of the boundary layer. The corresponding process occurs in a cathodic electrodeposition bath when the pH value rises above 10 or 11.

Film precipitate and dewatering

The initially still loose film of coagulated coating particles is pushed towards the electrode (workpiece) by coating particles that are still electrically charged and hence migrating. This highly aqueous coagulate is fixed at the electrode and compressed to a greater or lesser degree by material moving in behind it as a result of the same process; the film thus already displays remarkably good mechanical stability.

The electrostatically induced removal of ions (together with their individual hydrate shells) from the vicinity of the electrode also leads to the dewatering and solidification of the film due to *electroosmosis*. Within a short time this causes the non-volatile content of the film to rise to over 90 %, although that of the bath itself is only around 10 to 20 %. Thus the coating attains sufficient strength for its subsequent rinsing (see below) and transportation to the stoving oven.

Secondary reactions

Secondary reactions also often occur alongside the electrochemical deposition of polymer ions, an understanding of which is important for the prevention of defects and hence for quality assurance.

The most important secondary reaction is the anodic dissolution of *iron* according to

$$Fe \longrightarrow Fe^{2+} + 2\,e^-$$

whereby under oxidation the iron ions then lead to iron(III) oxide slurry or - as the reverse process of dissolution - can be deposited as metal on the cathode.

The anodic electrodeposition process thus offers the possibility of dissolution onto the object to be coated. A metallic dipping tank can in principle serve as cathode, although separate cathodes are normally used. In the cathodic electrodeposition process, by contrast, special anodes made from stainless steels (previously carbon anodes) *must* be used to prevent anodic oxidation. The deposition of iron or even simply an accumulation of Fe ions at the cathode object would cause defects due to discoloration effects, for example. The metallic tank should be protected by application of an insulating coating.

1) As is normally implied by "phoresis".

There are also other radical organic secondary reactions initiated by electrolysis, but these are generally of lesser importance in regard to the coating result.

6.6.2 *Plant engineering and bath control (EDC)*

An electrodeposition plant essentially consists of the following components:

- dipping tank with electrodes
- circuits to maintain bath stability and quality
- supply and replenishment of coating material
- rinsing/cleaning system
- transportation system
- power supply

Figure 6.6 provides a schematic view of these basic elements of an electrodeposition plant. Their fundamental features are discussed below.

Dipping tank

The dipping tank must be adapted to the size and shape of the workpiece and to the other conditions; it is generally protected by means of an insulating coating. To prevent sedimentation it must be agitated vigorously, usually via an external circuit with circulating pumps and injection nozzles on the floor of the tank. The contents of the bath are also filtered during this process to remove entrained dirt.

The temperature of the bath (approx. 28 to 32 °C for cathodic electrodeposition) must be held within strict limits (approx. ± 1 °C) to ensure good coating quality. The *cooling circuit* prevents the temperature from rising as a consequence of the electrical current flow or of the workpieces having been heated in the pretreatment process.

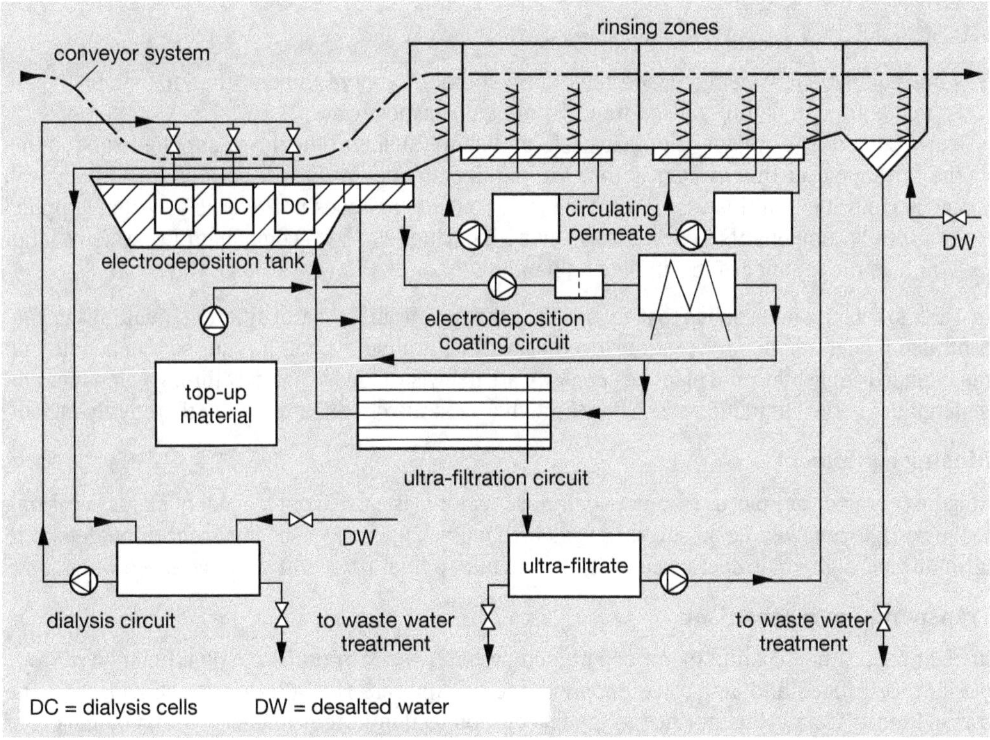

Figure 6.6: Electrodeposition plant (source: Herberts GmbH)

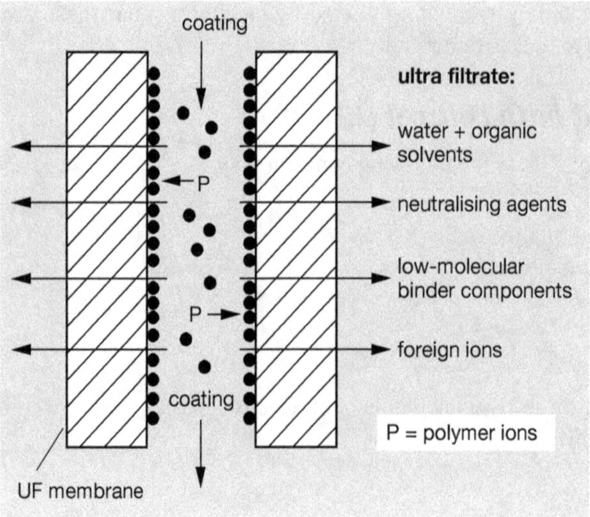

Figure 6.7: Ultra-filtration (UF)

Ultra-filtration

The principal objective of ultra-filtration is to provide sufficient (ultra-) filtrate to reduce the entrainment of adherent, uncoagulated coating during the rinsing stages. The coating is therefore pumped through filter modules in which the membrane separates out water and other low-molecular substances, as shown in Figure 6.7.

Ultra-filtration can also be used to remove the low-molecular impurities that accumulate in the bath liquid during operation. All or part of the filtrate must then be discarded. The concentrate and (after use) the undiscarded filtrate are returned to the bath.

In contrast to anodic electrodeposition, an additional **dialysis circuit** is needed in the case of cathodic electrodeposition to remove the quantities of acid that accumulate at the anode. (The acid is not deposited with the coating and would cause the pH in the tank to fall.) For this "*anolyte circuit*" the anodes are positioned in dialysis cells that are separated from the bath by membranes and in which the anolyte (circulating liquid) removes the acid that has accumulated there.

Supply systems

Coating supply and replenishment systems ensure that the contents of the bath are continually kept constant and ready for use and that the coating applied to the workpiece (particularly the solid, coagulated matter) is replaced.

The electrodeposition bath is supplied either as

- a one-component system (concentrated form containing all of the constituents; simply has to be made up with demineralised water before use) or mostly as
- a two-component system (pigment paste and semi-finished binder are supplied separately); the advantage of this system is that the pigment/binder ratio can be more accurately and easily controlled. In this case, therefore, "two-component" does not refer to a system supplied in multiple components to prevent subsequent curing (e.g. "two-component PU systems") but simply to the manner of its supply and handling.

In order to replenish the solids that have been consumed from the bath, different quantities of concentrated binder, pigment or coating, together with neutralising agent if required - depending on the intended application and plant technology - are likewise added separately through pipelines and metered in via the circulation system or through a separate circuit, after being thoroughly mixed.

Rinsing equipment

Rinsing of coated workpieces to remove entrained material is carried out in modern plants according to the *cascade principle*, i.e. in several stages with increasingly clean rinsing agent through to pure ultra-filtrate. High-grade coatings usually also require a final rinse with demineralised water.

Transportation mechanism

In electrodeposition coating the transportation system (→ 6.9.5) fulfils the dual function of *workpiece transportation* and also *power supply* to the workpieces. The latter can be provided via the transporter or via a cable attached to the object. Connection to the conductor rail is by means of sliding contacts.

Contamination of the bath by dirt and grinding dust falling from the components of the mechanical transportation system - which is positioned above the tank - is a particular hazard in electrodeposition coating plants and must in all circumstances be avoided by appropriate design of the conveying and conductor rails.

6.6.3 Developmental trends and fields of application

Besides the reduction in stoving temperature (low-bake electrodeposition coating) that we mentioned earlier, current research is principally centred on replacement of the lead compounds (anti-corrosive pigments) that are still used in some electrodeposition coatings and on the design of thick-film materials. The latter would allow the use of fillers to be reduced or even dispensed with altogether in view of the multi-layer construction of a coating system, leading to a considerable reduction in costs. There is also a trend towards further reducing the proportion of organic solvents in the dipping bath and minimising the release of decomposition products during the stoving process.

The main areas of application for electrodeposition coatings are in

- the automotive industry (cars and commercial vehicles), including accessories and spare parts (almost exclusively cathodic electrodeposition)
- agricultural machinery
- steel furniture, domestic, garden and leisure equipment
- building components such as radiators and profiles, components and housings for the electronics industry.

6.7 Spray application processes

The first air-atomising coating apparatus was in operation in the furniture industry as early as 1909. From 1923 onwards, spray guns - in conjunction with the newly emerging nitrocellulose coatings, which required or permitted rapid processing - also began to become established in the automotive sector, bringing about a revolutionary improvement in coating quality and productivity.

Depending on the object to be coated and the technological requirements, a wide range of atomisation methods is available today; they can be roughly categorised as follows:

- pneumatic atomisation (using compressed air)
- hydraulic atomisation (pressure release of liquid coating)
- hydraulic atomisation with air assistance
- atomisation by rotation (centrifugal forces)
- electrostatically-assisted atomisation
- purely electrostatic atomisation.

Although the use of electrostatic processes is on the increase, the first three processes - with no electrostatic charging of the coating - have also retained their importance, since

- their requirements in terms of investment, plant technology, substrates and coating materials are less demanding and
- they present fewer problems in terms of coating workpieces with voids or sharp edges.

Their fundamental disadvantage lies in the relatively high coating mist losses. Numerous developments have therefore been and are still being made with the aim of minimising these losses, even in conventional atomisation technology.

6.7.1 Atomisation methods without electrostatic charging

Atomisation of coating materials without electrostatic charging is achieved by mechanical forces alone. This involves using the effects of the speed

Table 6.2: Overview of spray coating applications without electrostatic charging

Pneumatic atomisation		**Hydraulic atomisation**			**Air-assisted hydraulic atomisation**
high-pressure spraying	low-pressure spraying ("HVLP")	with separate ultra-high pressure delivery pump	with integrated delivery pump (electric)	spraying nozzles with blowing gas	trade names: Air-Coat, Air-Combi, Airmix, Airless-Plus, Feinsprühen, etc.
atomising air approx. 2 to 7 bar 0.1 to 0.7 m^3/min	atomising air approx. 0.2 to 1.5 bar, 0.5 to 4 m^3/min	coating material pressure approx. 80 to 500 bar	coating material pressure up to max. 200 bar	blowing gas pressure approx. 2 to 4 bar	atomising coating material pressure approx.. 20 to 80 bar air pressure 0.5 to 2 bar

- of air jets (compressed air atomisation in low-pressure and high-pressure processes, pneumatic atomisation),
- of the flow of coating material itself (hydraulic or airless atomisation),
- of a combination of rapid coating flow and air jets (air-assisted airless atomisation, airmix) or
- arising from centrifugal forces (atomisation by rotation of a bell or disc, in combination with electrostatic charging).

Table 6.2 provides an overview of these process variants.

6.7.1.1 *Pneumatic atomisation*

Compressed air guns (an example is shown in Figure 6.8) are one of the most important applicational tools and are widely used because of the following advantages:

+ very good (fine) atomisation and (hence) generally good surface quality
+ uniform coating thickness and almost texture-free surface
+ good wetting of deep-pored substrates and elaborately shaped parts.

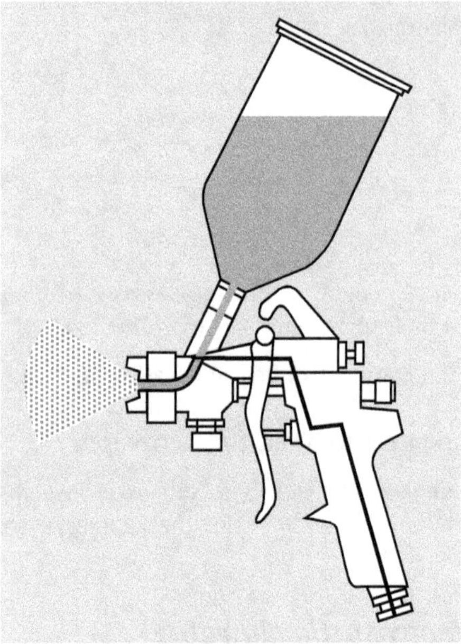

Figure 6.8: Cross-sectional view of a (gravity pot) compressed air spray gun. Coating supply by siphon from the gravity pot, air supply through the handle.

Several disadvantages are also associated with this good atomisation, however:

- high material losses (overspray), particularly with filigree parts or small surfaces
- high emission of volatile coating components due to the intensive air contact and because of the amount of overspray
- risk of surface defects due to spray mist
- experience of spraying is essential (difficult for semi-skilled labour).

Method of operation

To illustrate the principle, Figure 6.9 shows a longitudinal section of a pneumatic atomiser. This standard design can be modified by the addition of further control passages between the air circuit and the horn passage to improve the shaping of the spray jet.

Material supply is assured by the coating nozzle and is controlled by the injector (operated by means of a trigger guard). The *compressed air* required for atomisation – which can be controlled separately – is likewise released by the trigger guard; it emerges from the air circuit and only then entrains the coating from the coating nozzle (same principle as the water jet pump). This is followed by atomisation by the extremely rapid air jet, based on an impulse exchange[1)] between air and coating as two free jets.

In guns other than the completely round-section type, part of the compressed air is diverted by an additional air control valve and passed to the "*horns*" and possibly also to further control passages, from where it shapes the round-section jet into an oval; this oval should ideally be evenly flattened (see Figure 6.10).

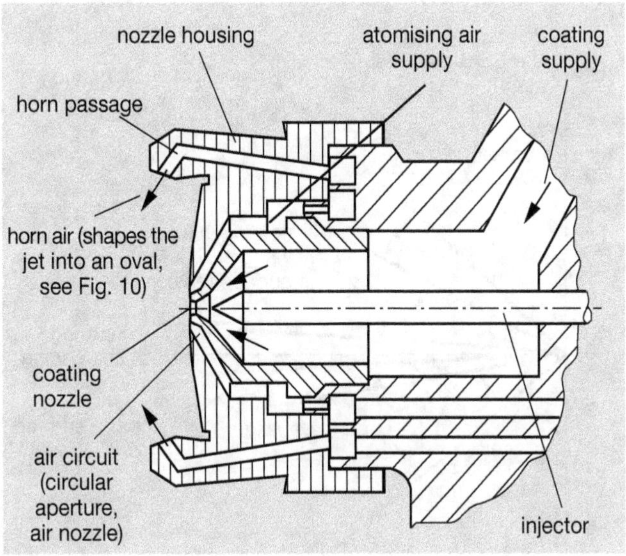

Figure 6.9: Sectional view of a pneumatic atomiser, according to [6]

This type of "spraying pattern" enables much more uniform coatings to be obtained in terms of coating thickness and surface appearance, provided that the oval is still surrounded by a soft "border", allowing fluid transitions.

Nozzle dimensions

Typical diameters for the coating nozzle are as follows:

- 1.2 mm for highly-liquid materials such as wood stains
- 1.3 to 1.5 mm for high-grade topcoats (e.g. for cars)
- > 1.5 mm for intermediates such as fillers
- a maximum of approx. 2.5 mm for highly viscous materials such as textured coatings.

A whole arsenal of special nozzle designs exists for special types of objects or of coating materials, including round-section nozzles, slot nozzles, disc-shaped internal mixing nozzles, crosshead extensions and others.

Coating supply

(see also section 6.9.3, Supply systems)

Coating can be supplied to the nozzle by various different means. The following designs have all proven effective:

- top-mounted *gravity pot* (coating delivered partly by the action of air siphon, partly by the inherent weight of the coating),
- *siphon pot* suspended below the gun (greater siphon action, i.e. higher air compression required),
- *pressure pot,* likewise suspended from the gun (compressed air from the injected air pushes the coating towards the nozzle)
- separate *autoclave* for handling larger quantities of coating.

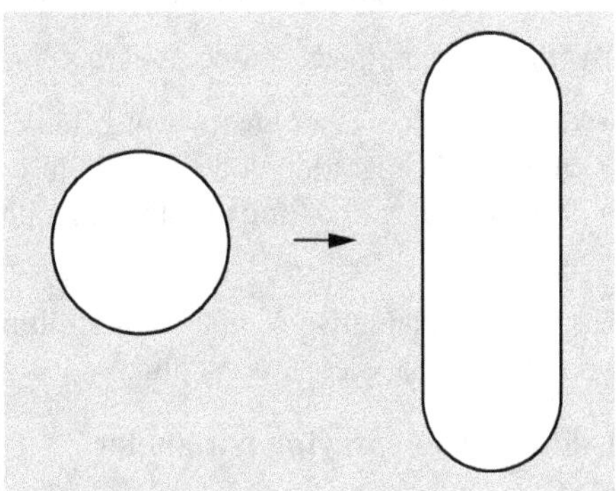

Figure 6.10: Spraying pattern for a round-section jet (left) and an oval-shaped spray jet (right)

1) Impulse = mass x velocity.

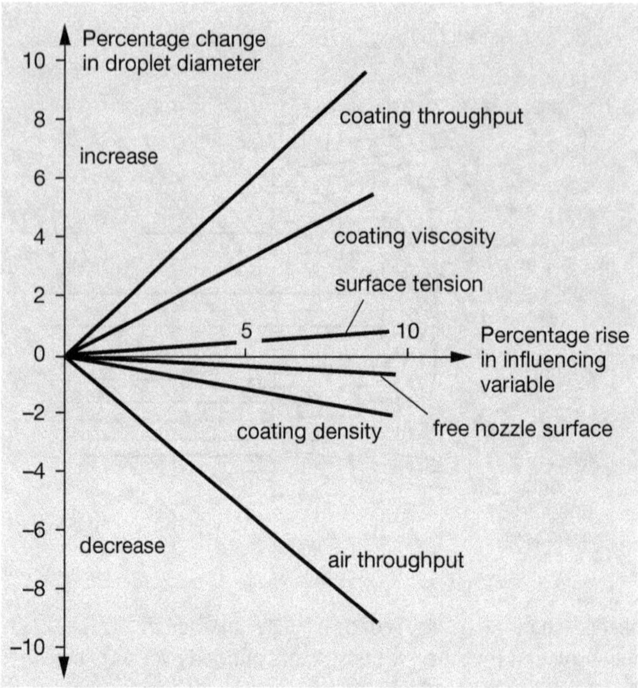

Figure 6.11: Influences on droplet size distribution during pneumatic atomisation, according to [5]

Generation of the coating mist

Droplet size

The average droplet diameter and hence the quality of atomisation and of the sprayed paint depend principally on the throughput of coating and of air, as suggested by the term "impulse exchange".

Figure 6.11 illustrates the increas-ingly coarse atomisation as the coating throughput increases and conversely as the use of injection air decreases. Too high a viscosity also has a distinctly unfavourable influence on droplet size.

Parameters such as surface tension and density, on the other hand, have very little influence on pneumatic atomisation.

Compressed air consumption

The obvious solution of maximising the quality of atomisation and hence of the surface by using extreme quantities of air (i.e. high air pressure) with reduced coating throughput is blocked by cost factors as well as by safety and environmental issues:

- excessive coating mist losses ("overspray") due to the small particle sizes and large airflows
- consequently, excessive losses of volatile components (solvents!) too
- increasing quantities of coating sludge or other coating residues
- larger proportion of respirable droplets in the coating mist aerosol
- greater consumption of compressed air
- associated excessive rise in energy consumption and
- unacceptably slow operating speeds (*surface throughput, surface efficiency*).

In practice, conventional high-performance guns have an optimum air pressure ("internal nozzle pressure") of 3 to 6 bar, with an average droplet diameter of approx. 25 to 60 μm.

The following methods can be used to *determine the droplet size distribution* in the spray mist:

- diffraction in a laser beam, evaluation of the interference sample
- graduated separation of droplets in a "cascade impactor" (see [3])
- freezing out the coating droplets in liquid nitrogen, followed by separation by low-temperature screening.

For a rapid qualitative comparison, an initial visual impression can also be obtained by quickly running the spray jet across a film.

Evaluation of spraying parameters

Let us briefly summarise once more the influence of the various coating and spraying parameters on the coating result:

nozzle diameter	large	small
average particle size	large	small
coating viscosity	high	low
spraying air pressure	low	high

These changes to spraying parameters have the following influence on the coating result:

overspray proportion	small	large
surface quality	poor	excellent
coating speed	high	low

A large nozzle diameter or too high a coating viscosity, for example, lead to a poor surface quality in the film but at the same time allow rapid film construction and a high operating speed.

Low-mist pneumatic application (HVLP, etc.)

Low-mist "*low-pressure guns*" have been available for some time now as a means of reducing solvent emission, but in their original form - due to their unsatisfactory atomisation quality - they failed to become generally established. The development of this technology was then accelerated from the 1980s onward by US environmental legislation, particularly "Rule 1151" (for automotive refinishing coatings) and "Rule 1136" (for wood coatings). Amongst other things this legislation required a maximum internal nozzle pressure (not gun supply pressure!) of 0.7 bar, which is now also the norm for most low-mist guns.

Method of operation

The term "HVLP" (**h**igh **v**olume **l**ow **p**ressure) suggests that although a low internal nozzle pressure is used, a large volume of air is also used for atomisation. Strictly speaking, however, the latter scarcely applies to this extent any longer, and the consumption of air is now practically the same as that in conventional guns.

The reduced internal nozzle pressure is achieved by means of an "air converter" built into the handle of the gun, which reduces the supply pressure of 5 or 6 bar in the gun accordingly. Additional or modified nozzle passages in HVLP guns ensure that the atomisation is almost equal to that in conventional high-performance guns. The only distinction lies in a - generally acceptable - deterioration in sensitive special-effect colours.

Transfer efficiency

The application efficiency or solids productivity (of HVLP guns in this case) is defined as follows:

$$AE\,[\%] = \frac{m_{nvm}}{m_{total}} \cdot 100$$

where m_{nvm} = mass of non-volatile matter in the coating deposited on the test piece, m_{total} = total mass of coating material required for the application aieless atomisation.

Depending on the size and geometry of the object, this may be around 5 to 30 % greater than in conventional high-performance guns. The extreme values that are occasionally seen in brochures and publications should always be checked closely to discover whether the application conditions, particularly details of the object being coated, are stated and whether they correspond to practice.

6.7.1.2 Hydraulic (airless) atomisation

In this process, which proceeds without air atomisation, the coating material is forced at a high pressure of generally from 200 to 400 bar, occasionally up to 600 bar, through a fine nozzle (of approx.

0.5 to max. 1 mm in diameter). The high speed and extreme turbulences cause the jet of coating material to be torn apart and atomised immediately after leaving the nozzle.

The advantages of hydraulic atomisation are:

+ low spray mist formation (less overspray, high material yield)
+ high spraying viscosity (i.e. high non-volatile content and thick coating films obtained with each spraying operation; materials can be used straight from the containers in which they were supplied)
+ high material throughput (2 to 10 l/min) and hence good surface efficiency (operating speed)
+ less spraying experience needed than in the case of compressed air spraying.

The following disadvantages must also be taken into consideration, however:

- inferior atomisation with correspondingly erratic surface appearance
- relatively sharp boundary to the spray jet (no "soft" transitions)
- quantities cannot be regulated during application
- high level of nozzle wear due to the enormous pressure, particularly in the case of abrasive materials
- limited processability of some water-borne coatings (dispersion damage by high shear in the nozzle)
- extremely noisy
- expensive plant required

The airless process is thus suitable for the efficient coating of larger surfaces e.g. in building conservation and physical protection, particularly where requirements in terms of surface smoothness and flow are not particularly exacting.

Hydraulic atomisation with air assistance (airmix)

This process is also known as "fine spraying", "aircoat" or "air-assisted airless" and represents the commonly chosen middle ground between hydraulic and pneumatic atomisation. Its advantage in comparison with the former methods lies in the greater droplet uniformity or - with the same particle size - a lower coating pressure requirement (approx. 60 bar). Its less sharply defined spray jet is a further benefit.

Hot spray application

By heating the coating material before atomisation to approx. 60 to 80 °C (further heating is pointless and only leads to disadvantages)

- the "low-temperature viscosity" can be increased from its usual level of approx. 20 s (4 mm DIN beaker) to 40 to 70 s: this allows a reduction in solvents of up to 6 %.
- a thicker coating can be sprayed onto vertical surfaces since a high viscosity is soon re-established when the coating meets the object; this also has a positive influence on non-sag properties (no tearing). (Flow deteriorates at the same time, however.)
- the amount of injected air can be reduced significantly (with a constant spraying viscosity) without impairing the spraying pattern.

In this process the coating material is usually circulated by a pump (with integrated flow heater). Occasionally the injected air is also preheated to slow the cooling of the coating and hence improve the flow on the substrate.

6.7.1.3 Recent process variants

Spraying with supercritical CO_2 ("Unicarb" process)

When spraying with supercritical carbon dioxide, part of the coating solvents is replaced by the "solvent" CO_2 in its supercritical state (reached at 31 °C and 75 bar). CO_2 acts as a thinner here rather than as a true solvent, which means that especially good solvents need to be chosen to make up the remaining part. This principle is thus unsuitable for conventional high solid coatings, particularly extremely high solid variants, and the formulation must therefore be modified.

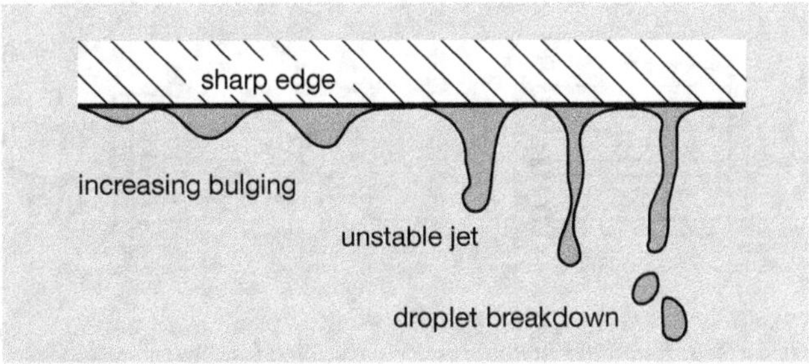

Figure 6.12: Phases of jet formation and spraying in purely electrostatic atomisation

To this end the process (developed by Union Carbide) allows the non-volatile matter (solids) to be increased from e.g. 30 to 70 % (relative to organic solvents) and dramatically reduces the amount of *overspray* (giving material economies) and *solvent emission* in airless application with or without electrostatic support. The sudden evaporation of CO_2 produces a good atomisation quality and also allows thick coatings to be applied without the risk of tearing.

Aside from the increased cost of training processors, a more widespread application of this process is naturally also restricted by the complex and expensive plant technology, and it is essentially only cost-effective when used with high production capacities.

The use of plants of this type is concentrated in the USA at present, principally in furniture and automotive component and agricultural machinery manufacture, although there are also a few examples in Europe.

Vapour injection cure process (VIC)

In contrast to the Unicarb process, the main purpose of the VIC process is to accelerate the curing reaction in two-component polyol/isocyanate systems without negatively influencing the pot life.

Earlier attempts to accelerate curing by vaporising relatively large quantities of catalysing amines in the spray booth ("vapour cure") have given way to a new technique:

Shortly before the atomisation process, a certain amount of a volatile amine is added to the compressed air in an amine generator; on contact with the two-component coating material the amine then deblocks (activates) a bismuth catalyst contained in the coating material. The catalyst in turn can then sharply accelerate the OH/NCO reaction (with no prior reduction in pot life).

In larger plants with a high throughput, this can significantly increase operating speeds by reducing the drying time and lowering the curing temperature. The method has failed to find widespread acceptance, however.

6.7.2 Electrostatic atomisation

The energy for surface expansion generated during air atomisation by compressed air has its origin in the case of purely electrostatic atomisation in the electrostatic repulsion between like charges. The coating material is thus first electrically charged in an electric field and the combination of

- the forces of attraction to the opposite pole (along the field lines) and
- the internal repulsion due to the accumulated like charge

increasingly causes the coating material to spread (see Figure 6.12).

Operating principles

The force acting on the drops in the direction of the counterelectrode is proportional to the electrical charge of the drops and the field strength E; this is obtained from the following relation:

$$\text{field strength } E = \frac{\text{voltage } U}{\text{distance between electrodes } d}$$

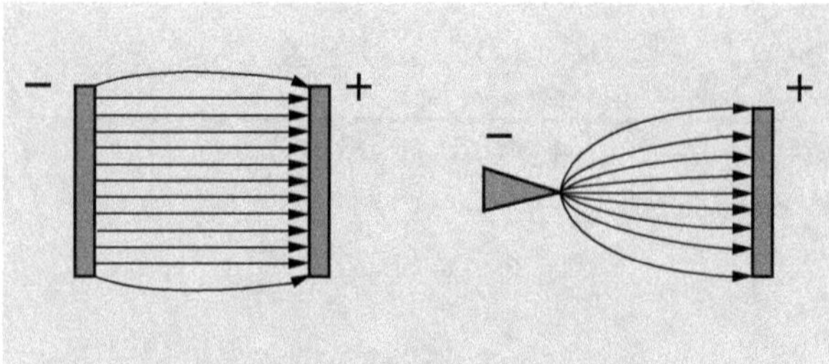

Figure 6.13: Field lines in uniform and curved field

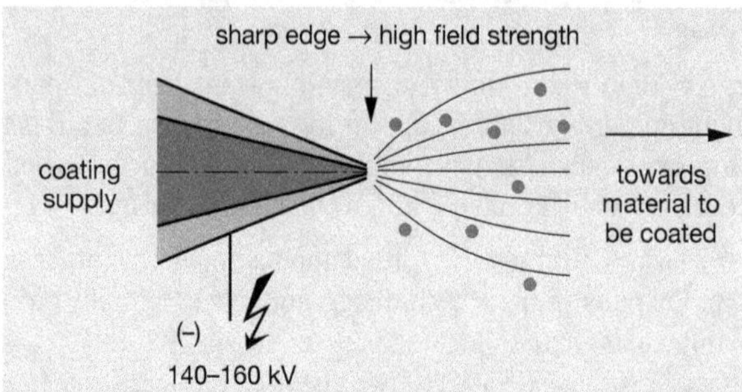

Figure 6.14: AEG slot

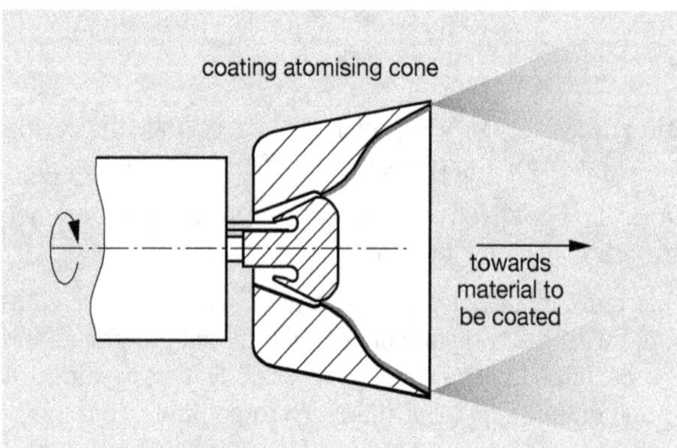

Figure 6.15: Spray bell

Characteristic of the progression of the field lines is their *concentration* preferably at tips and edges (Figure 6.13). The field strength here can be calculated approximately by $E = U/r$ (r = radius of curvature). The consequence of this increase in the density of the field lines in the vicinity of very sharp tips and edges with small radii is as follows:

At (negatively charged) tips of the charging electrode the breakthrough field strength required for *air ionisation* is exceeded, leading to the *"corona" charging* of air particles (molecules) which in turn then release their charge to coating droplets.

If the (positively charged or earthed) workpiece displays tips and edges, coating material is preferentially deposited there first in the form of *coating build-up*, which sometimes then spreads but can also lead to permanent excess coating.

In extreme cases, negatively charged space charge clouds can form with ionised air at very sharp tips and edges; these repel incoming droplets, which are likewise negatively charged, and can thus prevent a coating from being formed at all (*"backspray effect"*).

The backspray effect can also be caused by inadequate earthing of the object to be coated, since this means that the entire object is charged.

Electrostatic charging equipment:

Figure 6.14 illustrates the constructional principle of the now rarely used *spray slot* (*AEG slot*); permanently filled with coating material, the slot allows coating to be released when the field strength is increased. If there is no earthed object close by and hence no strong electrical field, the flow of coating dries up.

Slow-running *spray discs* and *spray bells* (Figure 6.15) are occasionally used to increase the flow of coating (these still operate on the principle of purely electrostatic rather than centrifugal atomisation, however).

Suitability features

Of prime importance in electrostatic atomisation is the *electrical conductivity of the coating material*, which can be adjusted by the choice of solvents and additives.

Application with purely elec-trostatic atomisation offers the following advantages:

1) **V**olatile **o**rganic **c**ompounds, i.e. principally organic solvents.

+ very good coating yields, almost 100 % application efficiency, i.e.:
+ reduced VOC[1)] emissions, coating sludge and contamination of the booth
+ very good throwing power
+ no wear on spraying edge due to reduced mechanical stress

The limits or disadvantages to the process are as follows:

- low flow rate (operating speed)
- inadequate coating of voids (Faraday cages)
- risk of excess coating at edges
- vulnerability of edge of spraying system to mechanical damage

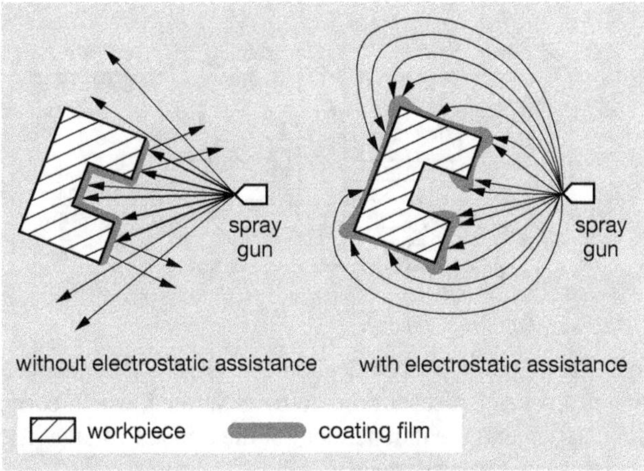

Figure 6.16: Coating droplet deposition without/with electrostatic coating charging, according to [9]

The unsatisfactory coating speeds and weaknesses in void coating in particular have resulted in the development of a number of variants, the purpose of all of them being to provide the coating droplets with additional *kinetic energy* along the way. Since these are the conventional methods that have already been discussed in this section, we can conversely now also speak of electrostatically assisted spraying processes.

Electrostatic assistance of conventional spraying processes

Electrostatically assisted air-atomising spraying systems are essentially designed in the same way as conventional systems. The coating channel also contains a *high-voltage electrode*, which extends up to 10 mm beyond the nozzle. Application of a high voltage causes a corona discharge to occur at the pointed end of the high-voltage electrode, as a result of which the pneumatically atomised coating receives a further charge in addition to the internal charge it (conduction charge) received *inside* the gun. The charged coating droplets are conveyed to the workpiece in the electrical field.

The advantages of this process over purely pneuma-tic atomisation (see Figure 6.16) are:

- improved coating yield, particularly with filigree workpieces
- a coating is also formed on the back of the part, due to throwing power.

In other gun designs the coating is charged in the gun alone by being passed in front of a live electrode (*"internal charging"*).

The effect of corona discharge mentioned above is also utilised in *hydraulic-electrostatic spraying*. It is generated in this case by one or more high-voltage electrodes attached to the core of the nozzle.

Electrostatic processing of water-borne coatings

If water-borne coatings are to be sprayed electrostatically, then special measures have to be taken because of their high conductivity (as mentioned earlier) and the associated risk of short-circuiting between the electrodes in the spraying system and the ground. These include:

- corona charging alone rather than internal charging (see above) or
- insulating the entire spraying plant, including the coating supply system (a cumbersome process), or
- insulating the spray gun only and supplying restricted quantities of coating by means of "shuttles" or similar vessels which are to be insulated separately ("canister system")

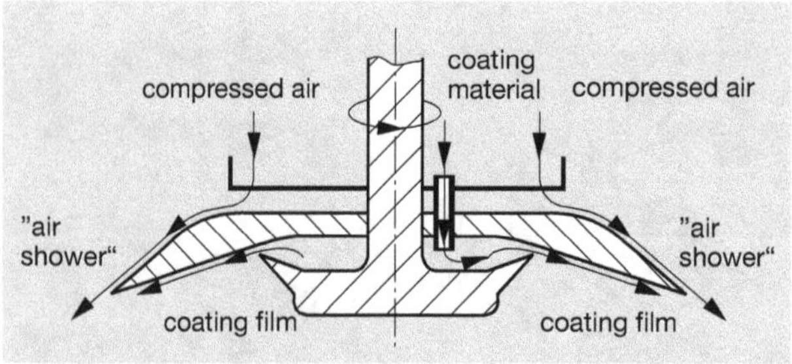

Figure 6.17: Rapid-rotation disk

6.7.3 Rapid-rotation atomisation

- **Types of atomiser**

In contrast to the slowly rotating spray bells and discs referred to above, which operate with electrostatic atomisation alone, rapid-rotation bells - although likewise under high voltage - atomise the coating material by purely mechanical means. High rotational speeds - of up to 60,000 rpm - cause very fine droplet size distributions to be generated, which correspond to those produced in pneumatic atomisation and are independent of the applied voltage. The voltage serves only to direct the spray mist towards the workpiece.

Figure 6.17 provides a schematic view of the "air shower" that in a rapid-rotation disc (which operates in a similar way to the bell) provides additional control of the spray jet and prevents contamination of the housing by backspray.

There is a pronounced difference between the two systems. Whereas *rapid-rotation bells* allow application directly onto the object, particularly in the principal area of application of automotive primary finishing, *rapid-rotation discs* spray the coating indiscriminately, although typically in one plane.

Figure 6.18 provides a schematic illustration, viewed from above, of the principle behind the so-called *omega loop*, in which the workpieces, usually flat parts, are conveyed on an omega-shaped track around the spray disc, which executes pendulum-like stroke movements in a vertical plane. For two-sided coating, the articles for painting can be additionally rotated about their own axes.

If we consider the parameters that influence the droplet spectrum, as we did in Figure 6.11, the following differences stand out in comparison to pneumatic atomisation:

- The *surface tension* is extremely important because of the special atomisation mechanism. The droplet size is proportional to the surface tension.
- As the *rotational velocity* of the disc or bell increases, the particle size falls sharply, as expected.
- The *coating density* and *diameter of the atomiser* are moderately important: higher values in both cases result in a finer atomisation.
- The *coating viscosity* is of very little importance here.

Charging methods

As we have already described in reference to electrostatic/pneumatic guns, in this case too there are two possibilities for charging the coating material:

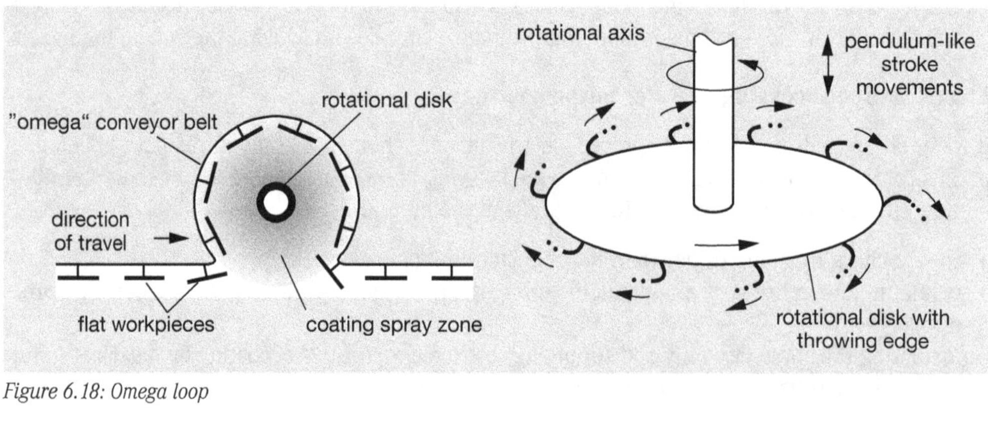

Figure 6.18: Omega loop

Table 6.3: Methods and trends in electrostatic coating of plastics

Method (●: indicates commonly used method)		**Notes**
incorporation of electrically conductive additives into the plastic	●	carbon pigment or fibres: black; addition of mica: expensive
application of an antistatic additive solution by spraying, immersion, etc.	●	*solvent*-containing antistatic liquids (NR_4^+), in some cases also aqueous
application of a temporarily conductive liquid film directly before coating; the coating film absorbs the moist conductive liquid	●	e.g. in solvent-based coating of bumpers; ultra-fine film!
application of a conductive coating ("conductive primer"), in some cases also on the back of the component (if possibly needed for other reasons too)	●	e.g. bumpers, bonnets, mud guards (goal: in-line coating as part of process)
optimising the earthing dynamics: spraying progressively out from the electrode		utilise conductivity of coating film while still wet
partial pre-spraying (1 to 10% of film thickness) with the coating to be used		
non-conductive hanger	●	no deflection of field lines by surroundings
treatment with bipolar ionised air from the front or (simultaneously!) from behind		charge equalisation by opposite charges
plasma, corona, flame pre-treatment	●	relatively low benefit; mainly used to improve wetting and adhesion
heating of plastic components		relatively low benefit

- internal charging - suitable for solvent-based coatings with relatively low electrical conductivity,
- corona charging - used primarily for water-borne coatings because of the risk of short circuiting in internal charging due to their high conductivity. The corona is generated by a ring of electrode pins arranged around the bell (Figure 6.19). This ensures that the charge is sufficiently high to achieve the necessary throughput in automotive primary coating, for example.

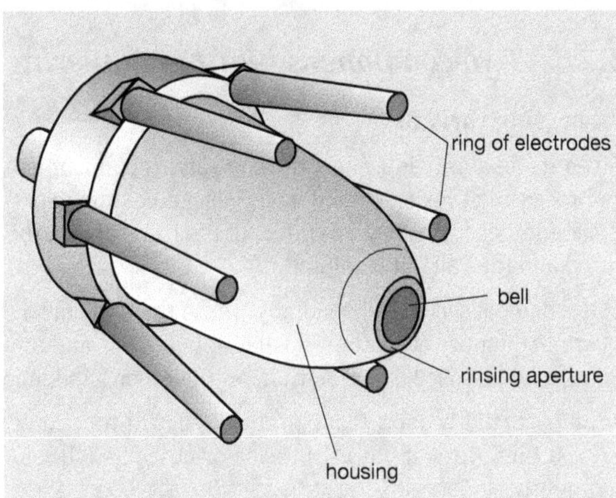

Figure 6.19: Rapid-rotation bell

Electrostatic coating of plastic components and wood

Plastics are electrical nonconductors, which means that they cannot be connected to an earth potential and hence cannot form the electrical counterpole to the gun or to the charged coating droplets. A number of measures can or must be taken, however, in order to be able to take advantage of the benefits of electrostatic coating for plastics too (Table 6.3 shows a selection of the main processes).

The solution is somewhat simpler in the case of *wooden workpieces* because of their capacity to absorb water at the right atmospheric humidity. Wood displaying a water content of only around 13 % or above, which is easily achieved by increasing the relative humidity and produces an adequate electrical conductivity, generally makes a good substrate for electrostatic coating.

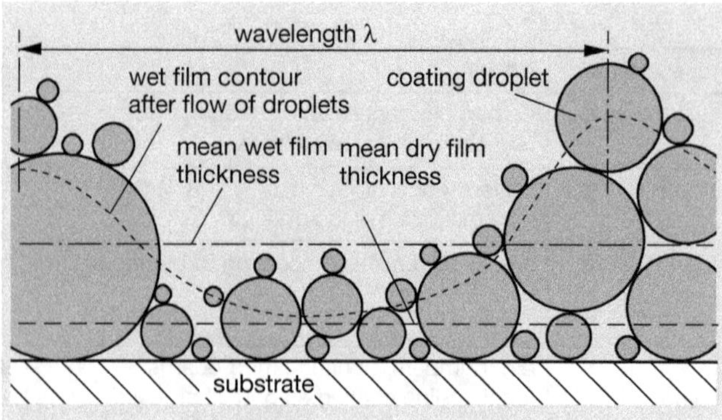

Figure 6.20: Surface structure ("droplet mountain range") after application by spraying, according to [5]

6.7.4 *Film-forming after spray application*

After application the coating film is formed from a multitude of droplets of differing sizes by the process of flow. Starting from a kind of "droplet mountain range" (see Figure 6.20), the drops of coating first flow into one another and produce a wavy wet film contour with an average "wavelength" λ. In a second phase this surface becomes smooth. The extent to which this happens decreases exponentially over time, mathematically speaking, although a high surface tension and a large wet film thickness encourage flow. The reduction in waviness is delayed by too high a viscosity and by wavelengths l that are too long, i.e. by too coarse an atomisation.

$$h = h_0 \cdot e^{-t/\tau} \text{ where } \tau = k \cdot \frac{\eta}{\sigma} \cdot \frac{\lambda^4}{s^3}$$

If we remember that the surface tension, wet film thickness and viscosity change dramatically (as the solvents are released) during the drying period, the race between flow on the one hand and drying and where applicable curing on the other becomes clear. *The "faster" a system is - at the surface at least - during chemical and physical film-forming, the less complete the formation of a perfect, smooth film will be.*

6.7.5 *Two-component plant engineering for spray application*

Processing variants

Given its flexibility in terms of storage and replenishment, spray application is particularly suited for processing two-component materials, i.e. coating materials with a pot life generally ranging from a few minutes to one day. After its pot life has elapsed, the material is unworkable, often even fully cured and thus all but insoluble.

In the manual sector and generally where there is only a small throughput of coatings or surfaces, a certain quantity of coating is usually "made up" and then used. Any leftover coating is discarded or must be disposed of correctly. If, on the other hand, importance is placed

- on particularly short reaction times (short processing times),
- on a high throughput of surfaces or coatings without the need to replenish supplies too frequently,
- where good reliability and accuracy in adding and mixing the reaction components are required in order to guarantee a consistent coating film quality,
- and on the avoidance of wastage losses with the relatively expensive two-component materials

then the prerequisites are in place for two-component plant engineering.

Plant design

Two-component plants are installations designed to meter and convey the individual components of reaction systems. The flow chart in Figure 6.21 shows a simplified model of a two-component application plant.

Components A and B are pumped from the metering pumps to the mixing chamber. The thoroughly mixed coating material is then conveyed to the spray nozzle. If necessary (after an extended shutdown or change of colour) the spray element, mixer, lines, etc., can be rinsed using the rinsing agent pump.

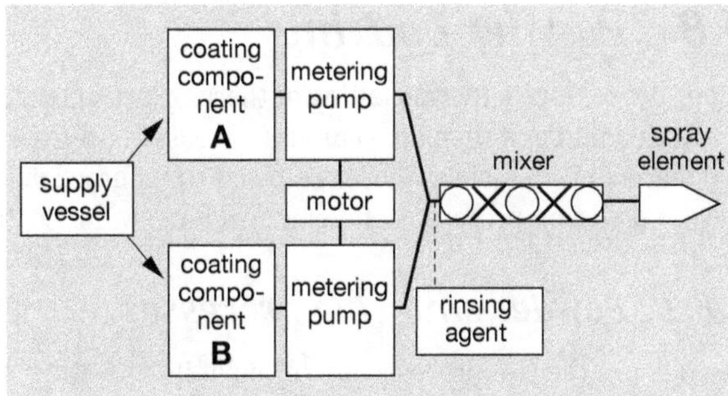

Figure 6.21: Flow chart for a two-component metering plant

Metering and mixing systems

The most important criterion in two-component plants is naturally the accuracy or consistency of metering and mixing. Metering variations should generally lie below ± 5 %. Metering in two-component plants is all the more accurate, the closer the viscosities and the mixing ratios of the components are to one another (i.e. comparable to the criteria for mixing by hand!).

Pumping is carried out by means of reciprocating and gear pumps. Built-in static mixing tubes ("*Kenics* helices"), in which numerous current alternations occur, are generally used for reliable and rapid mixing. Plants controlled by computers or microprocessors have been in use now for over 20 years. They allow a multitude of base coat shades and the addition of hardeners, thinners and rinsing liquids to be set at the touch of a button or (in the automotive sector) by means of codes printed on the bodywork. A colour change therefore takes only a few seconds.

Precautionary measures

In view of the intensive and extended contact of the materials with the plant components and possibly also with air, care must be exercised in two respects:

- Risk of *contamination by non-ferrous metals*, e.g. at joints and welds. Metal ions of this type can damage aqueous binder dispersions or as catalysts can accelerate curing - an undesirable consequence.
- Systems containing isocyanates form polyureas (note risk of flocculation) on *contact with atmospheric moisture* or with condensation and their curing quality is reduced.

Other aspects such as coating supply and closed circuits are covered in section 6.9.3.

6.7.6 Range of applications

In conclusion, Table 6.4 lists a number of examples of typical areas of application for each of the spraying processes described in this section, according to requirements in terms of surface quality, operating speed, degree of automation, etc.

Table 6.4: Typical areas of application for spray application processes

pneumatic atomisation, optionally low-mist pneumatic application (HVLP, etc.)	high-grade coatings, e.g. in plastics, automotive assembly line and automotive refinishing coating, in some cases also in furniture coating
hydraulic atomisation (airless)	for low optical surface requirements, e.g. shipbuilding, plant and railway engineering, underbody protection, etc.
air-assisted hydraulic atomisation (air-mix)	for moderate optical requirements, e.g. commercial vehicle painting (superstructures and chassis); also in carpentry and joinery
hot spraying	comparable to airless application
purely electrical spraying	for painting railings and fences (good throwing power!)
electrostatic assistance for conventional atomisation	as without electrostatic assistance (see above), principally in the small industrial sector, not in automotive refinishing (damage to electronic components)
rapid-rotation disk	high-grade industrial coatings (principally flat components)
rapid-rotation bell	high-grade coatings, particularly automotive primary coating

6.8 Powder coating

In powder coating, a thermosetting or thermoplastic powder material is applied to a workpiece, fused to a closed film and then also cured if necessary (in the case of thermosetting powders), a process that starts during fusing. There are two distinct groups of processes used for the application itself: powder sintering processes and electrostatic powder coating.

6.8.1 Powder sintering processes

The older of the two processes in terms of its operating principle, here the workpieces are heated before being coated, to temperatures above the melting temperature of the powder. The most widely used method is the *fluidised bed process*: when the hot workpiece is dipped into fluidised powder, the (electrically uncharged) powder sinters to it, fuses and cures, potentially forming a smooth polymer surface, for example on metal workpieces such as fittings, dishwasher racks, park furniture, etc. The process is also used for coating glass components and bottles (shatter-proof / low weight), ceramic components and heat-resistant plastics and textiles. Other powder sintering techniques include, briefly:

- the Skintraflux and rotary sintering process for pipe coating
- the loose powder sintering process for large pipes and containers
- the flame spraying process for very large components (e.g. internal coating of tanks).

6.8.2 Electrostatic processes

Advantages in the use of powder

The reasons for the continued rise in the use of electrostatic powder coating are partly economic in nature:

- high non-volatile content (almost 100 %, excluding decomposition products and volatile additives) and high application efficiency (90 to 98 % with recirculation and recovery)
- thick coatings can generally be produced in a single operation with no risk of running (up to approx. 150 µm thick, depending on the presence of decomposition products, e.g. hardener capping agents)
- cleanliness in coating stations

and partly also environmental:

- very low emissions (solvent legislation)
- very low coating losses
- energy conservation in drier (relatively high temperatures but low air throughput)
- non-toxic, inert dusts (but respiratory protection still required, as with all dusts!)
- not classed as hazardous in regard to transportation

Principle behind electrostatic powder spraying (EPS)

In electrostatic powder coating, the powder particles, similarly to the wet electrostatic coating process described earlier, are charged and conveyed in the electric field to the workpiece, where they adhere on the basis of Coulomb forces. Fusing and curing do not occur until the subsequent stoving process of the coating.

By far the most important process is *electrostatic powder spraying* (EPS). Here the powder is fluidised with air in the storage vessel (so that it can be handled and transported like a liquid) and delivered to the electrostatic powder gun.

There is also the *electrostatic fluidised bath process* designed especially for small parts, which are passed into or over a basin containing fluidised and charged powder.

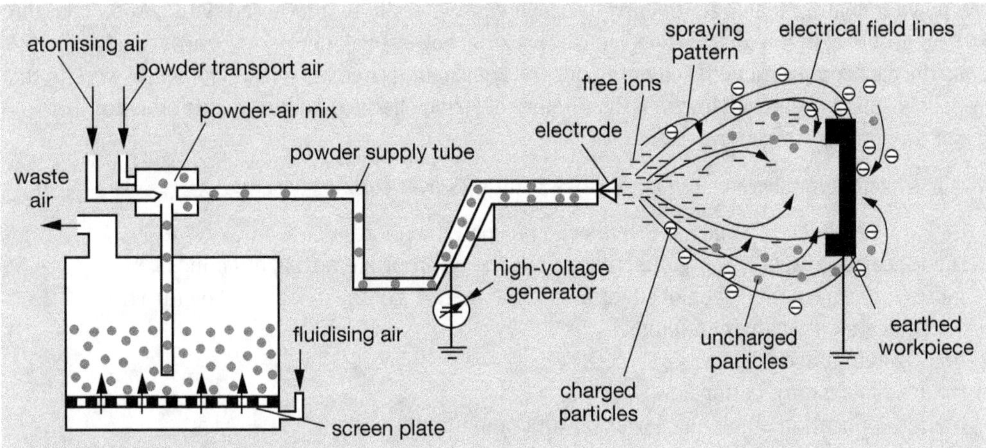

Figure 6.22: Principle behind corona powder spraying, according to [12]

In the EPS process the powder is made free-flowing in the fluidising vessel (see also Figure 6.22) and then, with the "powder gun" (the spraying element), is

- electrostatically charged and
- shaped into a defined spray cloud with additional air support.

The correct and most effective method of charging the powder particles is the central theme around which almost every problem and refinement in powder coating revolves.

Charging methods

Two different charging methods are used:

Corona process

In corona spraying systems the powder particles are (negatively) charged by the attachment of free electrons and air ions, which are generated in an applied high-voltage field with one or more electrodes, inside the gun or at the powder outlet (Figure 6.22). The high voltage is generally up to 90 kV, occasionally up to 120 kV, the level necessary for effective charging.

Corona charging is principally suited to smooth objects with simple geometric shapes. If the distance from the nozzle to the object is too short or if the object has sharp edges, the risk of the deposited powder coating being too highly charged by the flow of air ions is critical; this can lead to *backspray* and hence to surface defects. Moreover, accumulated clouds of air ions in corners, niches etc. can lead to repulsion of incoming powder particles. For this reason "*low-ion*" corona systems are increasingly being used, in which the flow of air ions to the workpiece is partly diverted by means of external, earthed counterelectrodes and hence is reduced at the workpiece itself.

Triboelectric process

In triboelectric powder spraying, the particles transported in the fluid are not charged by an external electrical field but exclusively by triboelectric, i.e. friction-electrical effects (Greek: τριβειν = to rub) in the turbulent, rapid flow through a plastic channel in the gun (Figure 6.23). The charge - at least in the case of the material used nowadays, PTFE - is exclusively positive.

A high (negative) charge rapidly forms on the inside of the plastic channel; this charge would subsequently interfere with further frictional charging and must be diverted via an earthed counterelectrode (also known as an electrostatic ioniser). Since there is no externally applied high-voltage field nor charged clouds of air ions, the coating of edges and also of voids (Faraday cages) is facilitated. The induced mirror charges, the formation of which is shown schematically in the diagram, have their origin in attached charged powder particles and ensure very stable adhesion to the substrate surface. Similar considerations apply to the corona method.

The powder coating materials in current usage can generally be processed with corona charging with no problem; in the case of triboelectric charging, however, all powders are not equally suitable, since the capacity for frictional charging depends on coating chemistry. Developmental work in this respect is still in progress. However, the number of "tribo-chargeable" powders is growing, as is the use of this process in general.

A tribo system demonstrates the following advantages over corona charging:

+ no Faraday effect: recesses and voids can be coated with greater accuracy
+ the jet can be directed more effectively with the appropriate nozzle dynamics
+ more uniform coating, even of edges and at the margins of the object (no "picture frame effect" via preferred attachment to edges)
+ readily automated
+ reduced backspray, better flow
+ higher application efficiency, lower powder consumption.

The main disadvantages of tribo-charging are as follows:

- more strongly influenced by air flows due to absence of external field
- in some cases poorer and also slower charging ability of powders, particularly of the fines
- higher investment costs (more guns) for the same output
- heavier wear (abrasion) in the guns.

Instrumentation and plant engineering

For optimum powder feed and application, both gun systems also require (clean and dry) air for conveying, accelerating and shaping the spray cloud.

Spraying systems

In addition to the appropriate charging method, the selection of the correct *spraying nozzle* for the workpiece is also important. The choice includes fan nozzles with increased flow rate, round-section nozzles with a soft, voluminous spray cloud, and multiple head nozzles which improve the uniformity of the coating system and the operating speed in tribo guns in particular

Powder spray bells, which rotate at approx. 5000 rpm, generate a very wide, well charged spray cloud and are particularly suitable for large workpieces. Similarly to "rapid-rotation" in wet coating systems, there are also *round-section spray systems* for powders, with rotary disc and omega loop.

As well as the traditional steel plate spray booths, glass or plastic *booths* are increasingly being used for powder coating, since they are easier to clean and tend to repel rather than to collect powder because of their inherent charge.

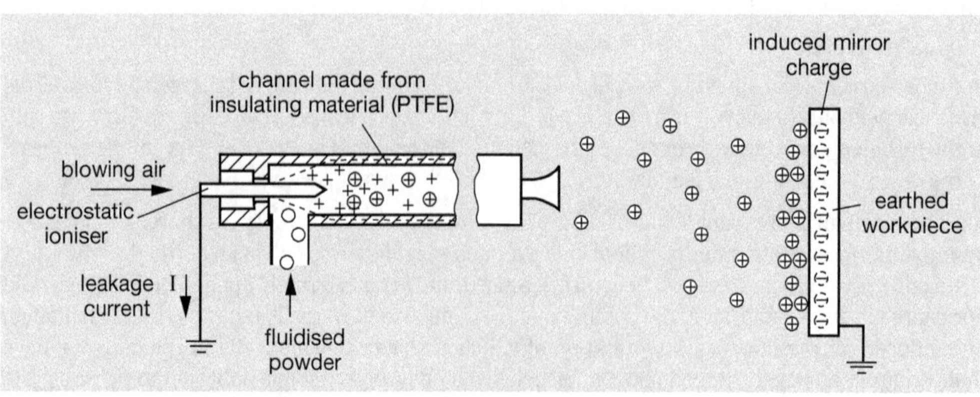

Figure 6.23: Principle behind tribo charging, according to [10]

Application efficiency

In most cases the first application efficiency, depending on the object, is well below 50 %. In order to achieve a total application efficiency of over 95 %, the powder must therefore be conveyed and applied as a *circulation* process (in-line recovery). However, since the particle distribution of the powder overspray usually differs from that of the applied powder (enrichment of fines because the fine particles are more difficult to charge), (thoroughly mixed) new powder must be added continuously and a high first application efficiency rate achieved to ensure a uniform coating result. Various types of filters and also cyclones are used to separate (and recover) powder from the air. The latter have their limitations in regard to retention of fines, however.

Suitability aspects

Powder coatings - particularly the latest reactive systems with lower stoving temperatures - have a limited *shelf life*. They must therefore be stored in a cool, dry place to prevent caking. Some systems must even be transported and stored under refrigerated conditions (e.g. clear automotive powder coatings).

The *principal areas of application* for electrostatic powder coatings are domestic appliances, office equipment, building components (primarily cladding panels), the general metal-working industry ("white goods" and car accessories, spare parts), garden furniture, etc., and - with a very promising future - the automotive industry (fillers and topcoats, especially clear coatings).

Several problems or limitations arise from the special properties of the powder coating:

For good coverage and good levelling, a relatively high *coating thickness* is usually needed. Exception: specialty thin-layer powders still undergoing development.

The *optical properties*, such as gloss, development of special effects and levelling mostly fail to attain the standard of liquid paints. The range of hues is limited, especially for metallic and pearlescent effects. Good effect development relies on paint shrinkage combined with solvent evaporation.

Hue tolerances cannot be maintained as closely from batch to batch as is the case for liquid paints, e.g. automotive finishes. The reason is that the coloristic or application properties of powder coatings cannot be corrected during production or in the coating line. Here, too, though, there is light on the horizon in the form of tinting systems.

Colour-change times are relatively long compared with those for liquid paints. Various paint systems have now become available that offer improved cleaning possibilities, including powder-booth modules that can be swapped and moved around.

The high *stoving temperatures* of at least 120 to 140 °C, and mostly >160 to 200 °C, limit the choice of both pigment (as regards heat resistance) and substrate. For plastics and wood, infrared melting of the powder coating can be combined with downstream UV curing. An alternative is to use low-temperature powder coatings that can be stoved at <120 °C.

6.9 Spray booth technique

6.9.1 Booth ventilation techniques

The spray atomisation of liquid coating materials abruptly releases large quantities of often combustible or toxic solvents and sometimes also other coating constituents; in powder coating the coating material has an extremely large (combustible) surface area. Waterborne aerosols, too, usually contain droplets that can reach the lungs and so must not be inhaled. Even grinding work can generate fine dust that should be avoided. In the interests of safety and also to avoid introducing dirt into the coating system, spraying is generally carried out in closed booths or at least at/in spraying stations equipped with the appropriate extraction equipment.

Operating parameters

The preconditions for an accurate and high-quality coating operation are:

- irrotational descending air flow or air feed, uniform in time and space
- constant pressure ratios between the booth interior and its surroundings
- constant temperature control
- constant relative humidity (particularly important for water-borne coatings, especially water-borne undercoats)
- reliable removal of overspray
- separation of coating mist from outgoing air by wet or dry separation
- integration of the spraying equipment required for coating with any conveying or coating supply equipment
- good illumination of the objects to be coated
- compliance with safety regulations.

Figure 6.24 shows a cross-section through a spray booth designed for the automotive industry with integrated air supply.

The *descending air speed* of the incoming air, purified if necessary by means of preliminary and fine filters (depending on surface quality requirements), should be between around 0.2 and 0.5 m/s. Where application is by automated systems or robots the descending air speed may be considerably lower than in manual spray application , because a higher solvent burden in the air is then acceptable. Horizontally ventilated booths are still found occasionally, but they display disadvantages in terms of flow ratios and are also prone to dust raising. The booth is usually set at a slight *overpressure* of approx. 5 Pa to minimise the ingress of contaminated air from the environment, preparation zone, etc.

Air-exchange can be greatly reduced in powder coating, too, even to the extent of recirculation in a hall (provided filters are installed) since virtually no volatiles are released.

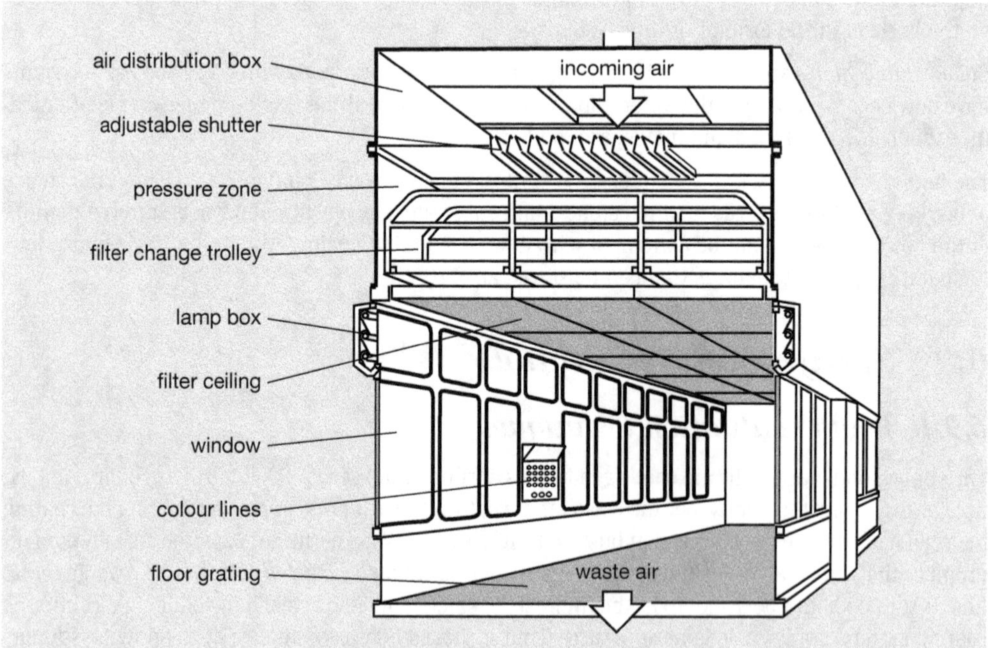

Figure 6.24: Tunnel-type spray booth in the automotive industry (source: Dürr AG)

Safety aspects

Given the ready *combustibility* of many solvents and - at a certain concentration - the *risk of explosion* of solvent/air mixtures, professional associations have produced standard calculation templates for plant operators, published in the appendix to **VBG 23** and **24**[1]. These can be used to calculate the *minimum quantities for fresh air supply* to ensure safe operation, depending on the amount of solvents used and on other criteria, for both spray booths and tunnel ovens and for drying ovens with batchwise operation (including laboratory drying ovens). Nevertheless, explosion-proof equipment is usually also necessary, of course.

6.9.2 Waste air purification

Types of plant

Coating mist separating systems are based on either dry or wet separation. The *dry separators* (perforated baffle plates, paper filters or felt filter mats) are considerably cheaper but - because of their relatively low storage capacity for separated coating particles - they permit continuous operation only in smaller booths and spraying stations or those subject to only low stresses. Booths in smaller industrial plants or automotive refinishing shops are also included here. The contaminated filter substrates generally have to be disposed of as special waste.

Wet rinsing systems offer superior separating performance and, above all, constant operating conditions over extended periods, even when booths are in constant use. Here the air to be rinsed is brought into intimate contact with the washing liquid, e.g. by means of nozzle, cascade, vortex or venturi systems, and the coating droplets are separated from the air, freed from adhesive residues and bonded with coagulating agents to form coating sludge. Once as much water as possible has been removed, the sludge is disposed of. (→ 7.11).

Criteria for use

In addition to coating mist separation, *solvent removal* from the waste air is now also an essential requirement for the operation of medium to large coating plants that are covered by the provisions of TA Luft [German clean air act]. The following basic measures are those most commonly taken to reduce VOC emissions (**v**olatile **o**rganic **c**ompounds) in the manufacture and processing of coating materials: first of all use of

- high solid materials
- water-thinnable materials
- powder coatings or
- improvements in the application and handling of coating materials (improvement in application efficiency, introduction of a coating circulation system during the application process, reduction in the dry film coating thickness, etc.).

Alternative purification methods

The use of such methods is sometimes restricted, for reasons of quality for instance. The following additional principles must therefore be applied to remove the solvent content from the waste air flows.

• *Condensation*
Solvents can be condensed by cooling or devolatilisation, but it is an energy-intensive and hence costly process.

1) VGB = Verhütungsvorschriften der gewerblichen Berufsgenossenschaften - regulations published by the industry professional bodies in Germany in respect of accident prevention.

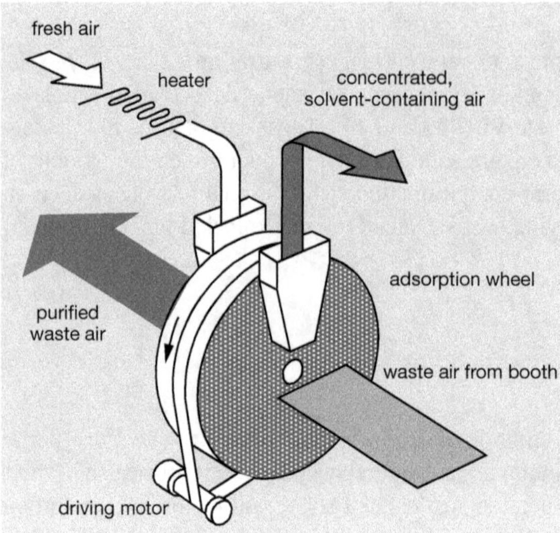

Figure 6.25: Adsorption wheel with adsorption and desorption zones (by means of hot air or steam)

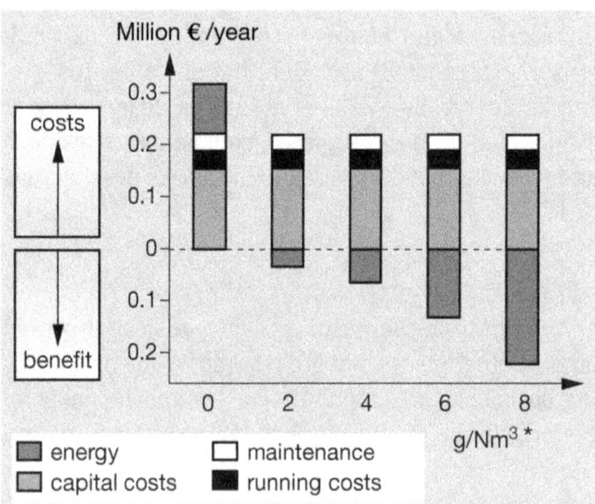

*Figure 6.26: Thermal afterburning: cost/benefit ratio as a function of solvent contamination; * properly: m_N^3*

- *Physical or chemical absorption*

This is usually achieved by rinsing with a suitable liquid, in which the substance to be separated accumulates. It can then be recovered by regenerating the liquid, by distillation for example. The process is restricted to special applications, however, being very suitable for the removal of amines for example, but otherwise useful only for small quantities of solvent in the waste air flows.

- *Adsorption on porous surfaces*

This is a widely used, very cost-effective method for large waste-air flows with low solvent contamination in particular. Suitable substan-ces include activated carbon or molecular sieves. The cycle includes desorption of the solvent, as illustrated schem-atically in Figure 6.25 by means of the commonly used "adsorption wheel"; this step allows the solvent to be recovered or increases efficiency in an afterburning stage (see below).

- *Oxidation processes*

Such processes include *thermal, regenerative* and *catalytic afterburning*. Depending on the level of waste air contamination, the solvent vapours can be selectively burnt at varying temperatures, sometimes with the use of catalysts, the cost-effectiveness of the process being very much dependent on the solvent content. As the content increases, the positive effect of energy recovery from solvent combustion ultimately predominates. Figure 6.26 illustrates the shift in the cost/benefit ratio with reference to thermal after-burning. This high-temperature combustion is generally only practi-cable if sensible use can also be made of the heat generated by the process.

- *Membrane filtration*

This relatively new process, to be found in various pilot plants, is based on the selective permeability for solvents of a semipermeable polymer membrane. The membrane is attached to a stable, porous carrier to increase its mechanical resistance. The solvent-enriched air flow can be used again either for recovery or combustion of the solvents.

Biotechnological processes

The three variants - biomass, biofilter and bioscrubber applications - are becoming increasingly important, particularly among coating manufacturers or processors whose waste-air flows are not constantly and heavily contaminated with solvents, but vary in regard to their volume, kind and degree of solvent contamination, with alternating peak loads and quiet periods for VOC emissions.

- *Biomass* is a natural filter substrate such as peat or bark mulch, heavily populated with microorganisms that display a high adaptability to solvents as nutrients. The solvents must be biodegradable and at least slightly water-soluble, however; the waste-air temperature must not exceed 40 to 50 °C and must have a sufficiently high relative humidity.
- *Biofilters* or *biofilter mats* function simultaneously as adsorbers, buffers or storage units, and nutrients. They are still at a developmental stage.
- *Bioscrubbers* contain microbes either in suspension (activated sludge) or in a steady state settled on "bioturf".

Scope for applications

Most practical applications are based on a *combination of various methods* to improve their cost-efficiency and effectiveness, e.g. an adsorption/desorption stage for solvent vapour accumulation, followed by one or more further stages to make use of the previously attained elevated solvent concentrations. Two typical applications are shown below:

adsorption + afterburning or

adsorption + membrane filtration + condensation

In addition to such processes, it is sometimes helpful to operate a plant under elevated solvent concentration by reducing the proportion of fresh air and increasing the air circulation.

The *optimum purification process* in each case has to be determined by a careful analysis of the quantity and nature of solvent contaminants and of the economic and technical conditions on site. Customised plant engineering is often required for special applications.

6.9.3 Supply systems

Storage vessels such as gravity, suction or pressure pots or autoclaves are unsuitable for handling larger quantities of coating (> 100 to 200 kg/day). Instead a paint or coating supply installation is used, such as the one illustrated schematically in Figure 6.27.

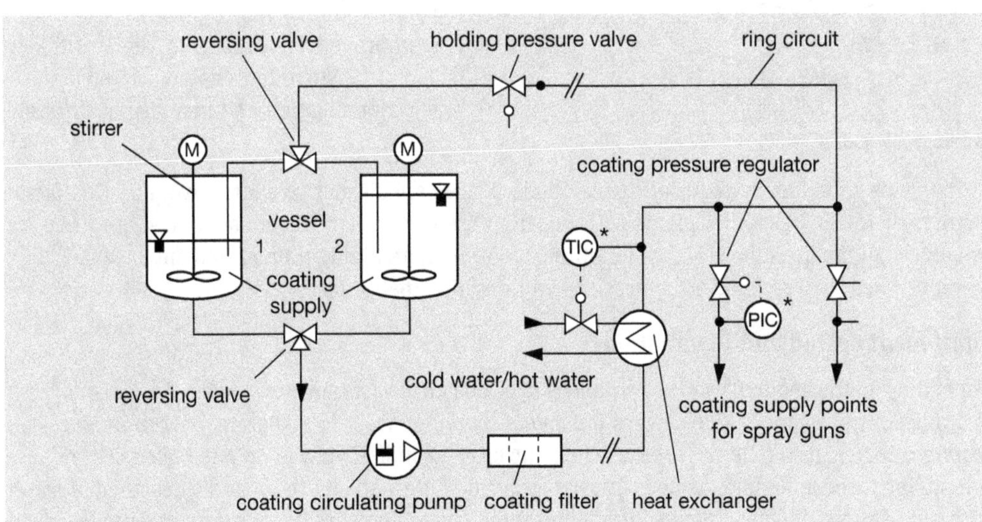

*Figure 6.27: Example of the construction of a coating supply installation; * pressure or temperature adjustment*

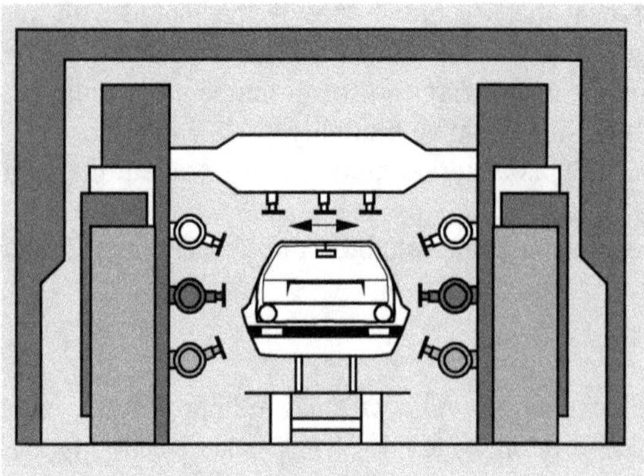

Figure 6.28: Painting of sides and roof with automatic devices (source: Behr)

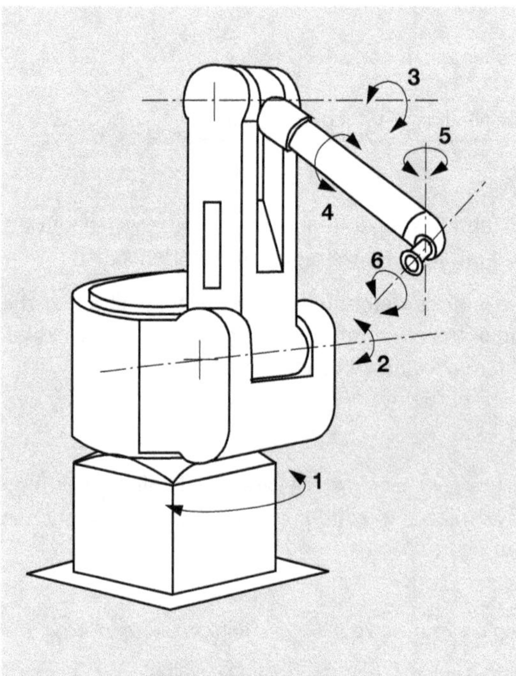

Figure 6.29: Coating robot with 6 programmable axes of movement, according to [10]

To prevent signs of sedimentation, the material is circulated, i.e. excess material is returned to the storage vessel, where it is continuously stirred if necessary. Any entrained dirt is removed by a filter. The heat exchanger ensures that the temperature and hence indirectly the viscosity remain constant (the latter being important for the coating result).

Coatings that are intended to be processed by means of a closed circuit of this type must be checked for the necessary *ring circuit stability* (→ 8.4.4).

6.9.4 Automated coating processes

Requirements in terms of the functional and aesthetic quality of coatings are becoming ever more exacting, as are industrial safety and environmental directives. Against this background there is a whole range of both economic and ecological arguments (the two having become more or less inseparable now) in favour of an increasing automation of the coating process:

- improvement in coating quality, quality consistency and process safety
- higher "first run OK" rate
- coating material economies (less overspray, reduced losses)
- reduction in environmental pollution, less waste for disposal
- lower energy requirements
- improved conditions in the workplace and its surroundings
- rationalisation and increase in throughput.

At the same time automated coating installations are capital- and cost-intensive, and this factor restricts their use. In smaller plants with varying workloads, such as manual paint shops, they are usually out of the question. They generally also require more highly trained staff and in some cases new, flexible organi-sational structures and working time patterns.

Equipment design and installation

Depending on the geometry of the workpiece to be coated, a vast range of *moving coating equipment* is available, from simple side-coaters *automatic coating devices* to elaborate, user-programmable *coating robots*. Figure 6.28 provides a schematic cross-sectional view of an automated coating line comprising three automatic devices fixed at each side (alternatively they can be designed to travel with the workpiece along a "travel axis") and three spray elements positioned horizontally, which perform characteristic oscillating movements to produce a more uniform coating.

The term *robot* usually describes an articulated automatic device with at least 6 user-programmable axes (see Figure 6.29). Robots are preferably used for coating more elaborately shaped objects, e.g. for painting vehicle or for coating bumpers in the automotive industry, and generally also throughout the whole sector of industrial coating. In the past the use of robots seemed to have reached its limits, however, because - so far at least - they were often regarded as being less flexible and "trainable" than their human "colleagues", in addition to which their programming can be extremely complex. Coatings personnel are now a rarity in modern paint lines.

6.9.5 Conveyor systems

Cost-effectiveness, flexibility, ease of automation: these are the three criteria by which every aspect of the design and operation of a conveyor system must be evaluated. The following individual factors are involved here:

- shape, size, thermal endurance, etc. of the object
- suspended or upright transportation
- throughput, number and sequence of coatings involved in the construction of the coating system
- flexibility of coating production, sorting, distribution operations, etc.
- spatial and design constraints on site

Of the broad range of technical possibilities, let us briefly consider the most widely used systems.

Circular conveyors
The most commonly used conveyors are *overhead* systems. They are usually circular conveyors (with an endless chain guided on a rail) which continuously transport workpieces suspended from carrying rollers. Circular conveyors are robust, uncomplicated and inexpensive conveyors designed to carry identical workpieces through an unchanging process at constant speeds. Smaller objects such as chair legs and other turned wooden parts or shock absorbers are transported upright on circular table conveyors.

Power + free conveyors
The use of circular conveyors is on the decline, whereas the far more flexible power and free conveyors are growing in popularity. In P+F conveyors the load hangers are not fixed to the endless power chain but instead run on their own guide track ("free section") below the endless conveyor track ("power section"). As required and as directed by the control system, pusher dogs on the conveyor chain can engage with corresponding cams on the trolleys (or vice versa) and then carry the latter along.

Using a combination of various P+F sections, speeds, stoppages, buffer zones, distribution functions and loading and unloading zones can be controlled as required.

Automatic feed units
These are programmable lifting and carrying devices, used particularly in the pretreatment and coating of metals. The units run forwards and backwards above lines of troughs and dip the workpieces (usually suspended from hangers) vertically into the troughs below in a cyclical operation. The principle is similar to that employed in container loading. They are usually combined with other conveyor systems.

Floor-level drag conveyors
Designed on the principle of the drag chain conveyor, these systems are driven by chains laid below or above the floor surface; bogies matched to the size of the workpiece are latched into the chain pushers. They are used for conveying very heavy objects or objects that must be kept free from falling dirt at all costs (e.g. automotive topcoating) and for wood finishing (chairs, etc.).

Skid conveyors
These are special floor-level conveyors with individual conveying platforms (which can also be turned crosswise or vertically). Highly flexible, they can be adapted to match the individual coating process.

Conveyor belts
Conveyor belts made from various materials are used for the coating of flat wood components in the furniture industry, for example, particularly in conjunction with curtain coaters. *Roller conveyors* carry the components away from the coating area.

In modern industrial coating shops, the various types of conveyors described above are linked by automated feeder stations, optionally with automatic position detection of the workpieces on the floor conveyors or hangers.

6.10 Drying installations

As we have already described in detail in other contexts, the transition of the still liquid or powdered coating material into the desired mechanically solid state can be based on either purely physical drying (physical film-forming) or chemical crosslinking (curing) or on a combination of the two. In industrial coating facilities and to a certain extent also in manual operations, the entire process (which can rarely be broken down precisely into individual processing or temporal steps) is generally accelerated by the effects of energy.
Depending on the substrate and coating material, a distinction is made between

- drying at room temperature (ambient temperature), also known as "air drying",
- forced (accelerated) drying (up to approx. 80 °C)
- stoving (100 to 200 °C, occasionally even higher)
- radiation curing (UV and electron beam curing)

6.10.1 Stoving conditions

The quality of the thermally or radiation cured coating film is guaranteed only by a sufficiently long and sufficiently great energetic effect. This means that fixed minimum and also maximum *stoving conditions* must be complied with.

"*Understoved*" coatings may demonstrate inferior mechanical resistances (reduced hardness) as well as diminished chemical, particularly weather, resistances. "*Overstoving*" can lead to yellowing or embrittlement of the coating film. If the still wet coating is heated too quickly, solvents can also become trapped below the film forming on the surface (→ page 392), leading to blistering, popping or later to wrinkling.

In order to be absolutely clear about the stoving conditions, however, it is important to state not only the stoving temperature and stoving time but also to clarify whether the former refers to the *oven* or the *object temperature*. The geometry and mass of the object have a huge influence on the degree to which the actual surface temperature of the object lags behind the temperature of its surroundings. In many cases it never reaches the oven temperature at all, particularly in the lower parts of the object, e.g. the door sills in car bodies.

The applied coating system must therefore be able to withstand a varyingly wide temperature range without exhibiting any notable difference in quality in individual zones. Materials that cure at ambient temperature can compensate for any "deficit" remaining from the coating process by subsequent postcuring (for up to several weeks). Coatings designed specifically for stoving, on the other hand, in which a crosslinker is only unblocked at elevated temperatures for example, can rarely make up for any deficit in curing arising from understoving.

6.10.2 Overview of drying processes

Table 6.5 (page 320) provides an overview of the main drying processes and their applications. The individual drying processes are discussed below.

6.10.3 Circulating air (convection) drying processes

Heat input into the object to be coated with the wet or powder film on top of it proceeds by means of convective heat transfer. This is described by the following equation:

$$\dot{Q} = \alpha A (\vartheta_L - \vartheta_F) = \alpha A \Delta\vartheta = \alpha A \Delta T$$

$\dot{Q}$: heat flow across surface A
α: heat transfer coefficient
ϑ_L: Temperature of oven air in °C
ϑ_F: Temperature of the film surface
T: absolute temperature
Δ: Difference

It can be seen that the heat flow into the object is proportional to the temperature difference and to α. α increases with decrease in thickness of the fluid boundary layer between the oven air and the surface, i.e. the faster the air flows along the surface. At the same time, the rapid inflow also boosts diffusion of solvent or water from the surface into the air and rapid removal of the vapour. For rapid setting and heating of the film or object, a strong air flow is advantageous, but harbours the risk of over-rapid surface drying. The permissible upper limit of the flow velocity must be determined experimentally, if need be. When air is blown onto the object (in jet driers) or directed by air jets to certain areas of the object, the flow velocity can reach 15 m/s.

Any known energy source can be used to heat the air in convection driers. Mineral oil, natural gas, electricity, fuel oil, hot water and steam are all used in practice. The choice is governed by local conditions: economic and plant-related factors as well as by the temperature required.

Plant design

There are two possible systems for use with gas or oil heating:

In *indirectly heated driers* the energy supply comes from a heat exchanger. In *directly heated* driers the ambient air itself is heated; heating gases produced by gas or oil combustion are thus mixed with ambient air. Direct heating is more energy-efficient but involves the risk of the coated surface becoming contaminated with soot particles due to incomplete combustion (yellowing or black specks).

Depending on the level of usage of the spray and drying booths, a further option in addition to compartment driers for batch processes is offered by *combi-booths*, particularly in manual operations such as automotive refinishing paint shops. These *combined coating and drying installations* are used first for spray coating at elevated descending air speeds of approx. 1 to 5 m/s and then, at reduced circulating air speeds, for heating the coated object. Given their higher energy consumption and smaller capacity, combi-booths are naturally only suitable for operations with limited booth requirements.

In industrial coating, *compartment and booth driers* of the type described above are joined by the widely used tunnel ovens, which are equipped with air locks (hot-air curtains) to reduce heat loss or - even better - are constructed in an A-shape called *A-driers* (see Figure 6.30). In the latter design, the heat is mostly trapped in the upper part of the oven.

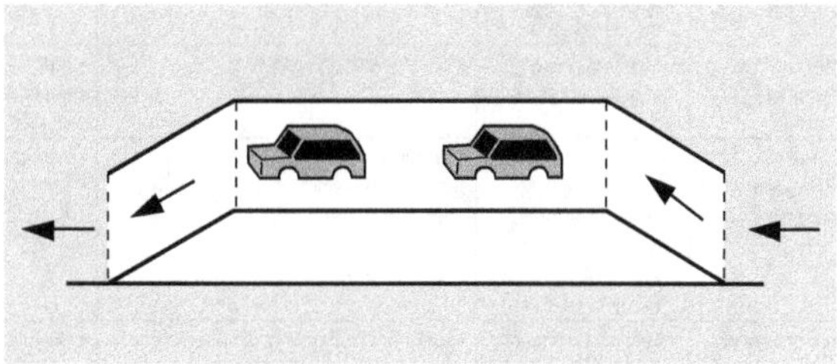

Figure 6.30: A-shaped tunnel oven (A-drier)

Table 6.5: Overview of methods for accelerating drying (adapted from [7])

Drying method	Energy input	Practical information		Installations
		Coating material/characteristics	Workpiece	
Drying by means of heat carriers				
air (circulating air) **(convection drier)**	• circulating hot air flow 2 to 15 m/s, in special cases up to 25 m/s.	• heat-curing systems, solventborne or solvent-free • no restrictions in terms of film thickness or pigmentation • drying time: 3 to 40 min. • influence on coating quality: temperature and time, hence risk of entrapped dirt in coating film while still open	• all non-heat sensitive materials • limited use: plastics and wood • all workpiece geometries can be dried, since good transfer of heat	• long drying times, therefore large footprint • relatively high energy consumption, especially due to insulation and aperture losses and discharge of heat; can be minimised by structural measures (A-shape, hot-air locks, heat exchangers, etc.) • good capacity for automation, highly flexible
dehumidified air ("Low X")	• cold to weakly heated, dried circulating air ($X < 6$ g/kg)	• water-dilutable coatings only physical setting and pre-drying, e.g. in wet-in-wet mode	• all materials and workpiece geometries especially advantageous for plastics and wood	• dry zones/booths with upstream dehumidifiers (different physical principles)
oil, steam	oil: immersion	special applications (generally low grade)		
Drying by radiation				
IR (infrared radiation drying)	• electromagnetic waves: • λ 1 µm to 1mm • IR absorption in coating film dependent on pigmentation and wavelength • drying from inside out	• same as for circulating air drying • solvent composition must be matched to radiation energy • faster heating, short-term excess temperature, hence shorter drying times • good curing of powder coatings • no restrictions in terms of film thickness or pigmentation • improved surface quality can be obtained after optimisation	• same as for circulating air drying • limited use: transparent materials such as glass and plastic, also materials with a tendency to yellow, e.g. plastics and wood • shadowing at inaccessible cavities and niches • coating material can withstand higher temperatures than substrate	• IR zones or tunnel, perhaps contoured; mobile static equipment • electrically or gas-heated radiators of different designs for short-wave, medium-wave and long-wave IR radiation. • mostly, pyrometers for temperature detection and intensity control (programmable controls) needed
UV (ultraviolet radiation) curing	• Absorption of UV(A) radiation (0.32 to 0.4 µm). • photochemical reaction: free-radical or cationic polymerisation • duration: a few seconds at most	• unsaturated polyesters/acrylates for free-radical curing; epoxy resins/vinyl ethers for cationic curing • (cationic: postcuring for up to 24 hours) • clearcoats and glazes can be applied as high-build films pigmented systems: • as thin films (coatings) or printing inks good surface quality given adequate levelling time	• drying only of areas directly accessible to the radiation (principally flat parts and sheeting) • wood, plastics: possibility of yellowing and embrittlement • widespread use with wood and timber derived materials, paper, cardboard, plastics, sheet steel • changes to process and coating chemistry enable three-dimensional workpieces to be UV as well	• high Hg pressure vapour lamps with electrodes or microwave excitation low-pressure lamps for pre-gelling • curing installation and process • have low footprint and energy consumption • low investment costs • environmentally friendly, since virtually no VOC emissions • ozone extraction and perhaps inert gas feed needed
pulse radiation drying (IST*)	• UV pulses with λ=0.1974 µm, (= resonance frequency C=C), very rapid polymerisation	see UV	see UV • also suitable for complex geometries (e.g. chairs)	see UV
electron beam curing (EBC)	• electron beam from 150-400 kV incandescent cathode, very rapid free-radical polymerisation	• unsaturated polyacrylates; rarely: cationic curing systems • no restrictions in terms of film thickness and pigmentation • good coating quality • no photo-initiators required, unlike UV radiation	• all substrates; risk of embrittlement in some plastics • widespread for paper, film, • wood coating at very high throughput speeds	• very low footprint • high investment costs (radiation protection, inert gas supply, ventilation) • low energy consumption • good capacity for automation • only for very high surface throughputs, perhaps beneficial
laser beams	• high energy CO_2 laser	• all heat-curing systems	• very limited drying possible (printing platen etc.)	• special installations
Drying by means of electrical processes				
resistance drying	• heating by passing a current through the workpiece	• all heat-curing systems	• example: wire enamelling	• special
incuctive drying	• heating by means of induced vortexes	• all heat-curing systems	• ferromagnetic subtsrates • especially suitable for thick workpieces	• special
high-frequency drying	• heating by high-frequency, electromagnetic waves: (primarily 27.12 MHz) in the capacitor field	• coatings with highly polar solvent and high dielectric loss factor (e") (primarily water-borne coatings)	• non-metallic substrates only; particularly suitable: flat wood sections	• special
microwave drying	• heating by means of electromagnetic waves	• coatings containing dipole-type solvents (predominantly water-borne coatings)	• only non-metallic substrates	• special

Often, tunnel driers are divided into several zones of different temperatures and air supply/removal means. This makes the heating or oven curve easier to adjust and saves on energy.

Aspects of use

As well as being used for drying/curing paints and coatings, etc., drying installations also find application as:

- *contact moisture driers* (after pretreatment and rinsing zones, to remove residual water)
- *glue driers* (to permit further processing after joint bonding, e.g. coating)
- *gelling driers* (for solidifying surface drying of e.g. plastisol-based underbody sealants)
- *surface and preheating driers* (to combat cold surfaces or entrapped air, e.g. in wood finishing).

The primary advantage of circulating air driers - and hence the reason for their widespread usage - lies in their ability to be used universally for components of any shape or size.

Moreover, the uniform heating process means that the risk of *overstoving*, i.e. damage to the coating film or substrate surface caused by local overheating, is very low, even if the conveyor mechanism is temporarily stopped.

6.10.4 Infrared drying

Definitions

In infrared (IR) drying, the coated workpiece is heated by absorption of IR radiation; the process thus involves no specifically radiation-induced curing.

IR radiation - like UV radiation, visible light and microwaves - is electromagnetic radiation. The energy or intensity of IR radiation depends on the wavelength range and hence on the temperature of the emitting elements. Classification of IR-drying (roughly):

IR radiation	Radiator type	Wavelength [μm]	Temperature [°C]	Radiator heating-up period [s]
long-wave	ceramic	3.0 to 6.0	200 to 700	120
medium-wave	quartz tube	1.5 to 2.5	750 to 1500	30 [1)]
short-wave	quartz tube	1.0 to 1.5	1500 to 2700	1
near-infrared (NIR)	high-energy halogen lamp	0.7 to 1.2	3000 to 3500	<1

A distinction is occasionally made on the basis of radiation wall temperature. *"Dark radiators"* correspond to the lower end of the long-wave range In the short-wave range, electrical radiators are generally used, which permit a more accurate and faster temperature control (*"bright radiators"*).

During heat input by means of IR radiation, both the intensity of the IR source and the absorption behaviour of the film and the substrate material must be considered. The emission intensity is given by the *Stefan-Boltzmann* law:

$$P = \varepsilon C_S \left(\frac{T}{100}\right)^4$$

P: radiation power per unit area
ε: emissivity of the radiator surface (<1)
C_S: radiation characteristic of the black body (5.76 W m^{-2} K^{-4})
T: absolute temperature

1) Medium-wave IR radiators with heating-up times of around 1 s are now also available.

Owing to the fourth power of T, the radiation power achieved technically increases from approx. 10 to 1000 kW/m^2 from dark radiators to NIR radiators.

Operating features

The extent to which IR radiation is absorbed, which is also crucial to the speed and final value of the temperature rise, depends on

- the chemical composition of the coating (especially the pigmentation)
- the absorption/reflection behaviour of the substrate (where the coating is IR permeable) and
- the surface quality of the coating

and the interplay with the wavelength distribution of the radiator.

Most coatings absorb very little in the wavelength range of the dark radiators. For this reason, the coating heats up mostly from the substrate up and solvent evaporation occurs from the depths, without surface setting, which is particularly advantageous for thick wet films. Short-wave radiators, by contrast, heat the film directly, which primarily has an accelerating effect on water evaporation.

Since solvents are released by radiant heat in just the same way as in circulating air drying, they must likewise be removed by air. This means that a suitable *fresh air supply* must always be provided.

Processing advantages

The advantages of IR radiation drying are as follows:

- more rapid heating up of the coated workpieces, since more energy can be transferred in a shorter time than is the case with circulating air drying
- more rapid recooling, since the heat does not penetrate deep into the substrate during short exposure periods
- driers made ready for use more quickly
- higher throughput
- lower energy costs (heat is only generated where it is required)
- lower thermal stress on the workpiece
- space-saving due to shorter physical length of drier
- relatively low investment costs and small amount of equipment required
- good controllability of the object temperature by means of the power/intensity of the IR radiator
- plant technology can be adjusted to dry small surfaces too
- the coating dries from the inside out (due to absorption within the coating).

Aspects of use

The following aspects should be taken into account when using IR radiation drying processes:

- The shorter the wavelength of the IR radiation,
  - the less permeable the film is to the radiation (absorption and reflection increase)
  - the higher is the radiator temperature,
  - the greater is the proportion of directly effective radiation energy and the lower is the less effective convective heat through the oven air also heated,
  - and the greater is the heat-transfer performance.
- The darker the shade, the greater the degree of radiation absorption, i.e. the coating film heats up more intenisvely. In metallic and light shades this can lead to correspondingly increased losses in reflection.
- It is important to comply with the prescribed radiant powers or distances between object and radiator to avoid overcuring.
- Drying takes longer in voids and recesses that are not directly accessible to the radiation.

Possibilities for use

Other than for the industrial coating of simply shaped workpieces, the principal application for IR drying of liquid coatings is in the automotive industry, often in combination with circulating air drying. Thus, for example, all water-borne automotive assembly line coatings are dried by IR and circulating air. IR tunnel driers or IR cartridge radiators are also used for retouching work in car manufacture and particularly for sectional coating in manual automotive refinishing operations.

IR drying seems to be particularly suited - and will certainly grow in importance in the future - for use with powder coatings. When the powder coating on coated workpieces melts very quickly under the action of IR radiation and then reaches the final curing temperature, i.e. the heating-up time is kept very short, benefits can be seen in regard to flow, adhesion, lower thermal stress on the workpiece (of particular importance for wood) and in terms of cost-effectiveness. Protection against overcuring is especially important here, however, due to the risk of changes in hue. In this case, protection against overstoving is crucial owing to the risk of changes in hue. The rapidity of IR heating has also led to the introduction of IR drying into coil powder coating.

NIR radiation drying leads to extremely high power input into the film and brings about intensive drying and especially hardening of waterborne and powder coatings. In powder coatings, the entire process of film formation is completed in about 10 seconds. The speed and extensive selective heating of the film renders the method specifically suitable for wood as well. Predrying with cold, dry air is beneficial in the case of waterborne coatings (see Table 6.5). Owing to the possibility as well as the need to focus the NIR radiation onto a particular layer in the coating or the coating structure, the method is limited to flat objects, e.g. strip metal or uniformly cylindrical bodies.

6.10.5 Radiation curing

In *UV* and *electron beam curing* (EBC), the purpose of the radiation - unlike IR - is not to heat the coated object, but to trigger a free-radical or cationic polymerisation as a chemical crosslinking reaction in solvent-free, waterborne or powder coating systems formulated specifically for this process. Binders suitable for free-radical curing are unsaturated polyesters and especially acrylates while special epoxy resins and vinyl ethers are suitable for cationic curing. The higher-energy EBC is significantly faster; unlike UV curing it requires no photoinitiators, but involves considerable investment in plant technology, including industrial safety precautions.

Use with pigmented systems

UV curing of pigmented systems presents certain difficulties in the case of thicker coatings due to the absorption of UV by many pigments. In EBC, with its far greater depths of penetration, the problem of pigments does not arise. Porous substrates can likewise be coated more reliably since the electrons also penetrate into the substrate surface.

Both curing methods are of limited importance in metal coating, being instead more commonly used in the coating of paper, cardboard packaging, plastics and wood-based materials as well as in the manufacture of printed circuit boards and printing plates. They are ideally suited for the rapid curing of large surface areas or large numbers of items but have their limitations in regard to elaborately shaped workpieces with areas inaccessible to the radiation. Deficient curing due to shadowing cannot be compensated for by heat or subsequent storage, unless a chemically curing coating is also present (*dual-cure system*).

Due to the high levels of investment required (see above), EBC is restricted to drying in processes involving particularly large numbers of items - e.g. for flat furniture components - but has recently gained some ground.

Radiation curing of powder coatings

A relatively new area offering interesting possibilities for the use of radiation is that of powder coating of heat-sensitive substrates such as wood, chipboard and plastics. The powder is sprayed on and then melted by IR radiators (without heating the substrate too greatly) before finally being cured instantaneously by UV radiation.

Comparison of processing features

The advantages of chemical radiation curing can be summarised as follows:

+ extremely rapid and complete cure of the coating
+ very high throughput rate
+ usually little or no accumulation of volatile solvents, virtually no evaporation periods required (except in the case of aqueous UV-curing coatings)
+ low energy consumption
+ "cold" curing, so also suitable for heat-sensitive substrates
+ drying installation takes up little space
+ quickly made ready for service, flexibility in use
+ high-quality surfaces

The disadvantagesor troublesome as regards the process, however, are:

- somewhat higher coating raw material costs in comparison with other coating systems
- relatively high expenditure in terms of industrial safety (due to the radiation and the curable materials)
- occasional adhesion problems with UV-curing coatings, particularly on metal
- UV materials have limited pigmenting and flatting capability
- shadowing occurs with elaborately shaped workpieces
- elaborately shaped workpieces suffer from shadowing or are not irradiated strongly enough

Sources and references for Chapter 6

[1] A.Goldschmidt, B.Hantschke, E.Knappe and G.-F.Vock: Glasurit Handbuch Lacke und Farben. 11th edition, Curt R. Vincentz Verlag, Hanover 1984
[2] A. Goldschmidt, H.-J. Streitberger: BASF Handbook: Basics of Coating Technology, 2nd edition, Vincentz Network, Hanover, 2007
[3] H. Kittel: Lehrbuch der Lacke und Beschichtungen. 2nd edition, Vol. 9, S. Hirzel Verlag, Stuttgart 2004
[4] R. Laible et al.: „Umweltfreundliche Lackiersysteme für die industrielle Praxis", Expert Verlag, Ehningen bei Böblingen 1989
[5] P. Mischke: „Filmbildung in modernen Lacksystemen", Vincentz Network, Hanover 2007
[6] B. Meuthen, A.-S. Jandel: Coil Coating. 2nd edition, Vieweg & Sohn Verlag, Wiesbaden 2008
[7] D. Ondratschek (ed.): besser Lackieren! Jahrbuch 2008, Vincentz Network, Hanover 2007
[8] Z. W. Wicks, F. N. Jones, S. P. Pappas, D. A. Wicks: Organic Coatings. 3rd edition, J. Wiley & Sons, Hoboken/New Jersey 2007
[9] J. Pietschmann, Industrielle Pulverbeschichtung, 2. Aufl., Vieweg Verlag, Wiesbaden 2003
[10] RömppOnline Lexikon, mit updates, s. www.roempp.com
[11] H.-J. Streitberger und K.-F. Dössel (Eds.), Automotive Paints and Coatings, 2nd ed., Wiley-VCH Weinheim 2008
[12] J. Prieto und J. Kiene, Holzbeschichtung, Vincentz Network, Hannover 2007
[13] www.aktuelle-wochenschau.de/2007/index07.htm und (darauf basierend): HighChem hautnah - Aktuelles zur Chemie der Lacke und Farben, Hrsg. GDCh, Frankfurt, 2008
Zusätzlich sei auf die Tagungsbände, Lehrgänge, Fortschrittsberichte und Lackier-Treffs der DFO und des IPA zu allen wesentlichen Themen der Lackier- und Anlagentechnik verwiesen:
• Deutsche Forschungsgesellschaft für Oberflächenbehandlung e.V. (DFO www.dfo-online.de
• Fraunhofer Institut für Produktionstechnik und Automatisierung (IPA) http://www.ipa.fraunhofer.de/Arbeitsgebiete/lackiertechnik/index.php
EGL und des IPA zu allen wesentlichen Themen der Lackier- und Anlagentechnik verwiesen:
• Deutsche Forschungsgesellschaft für Oberflächenbehandlung e.V. (DFO) Adersstr. 94, 40215 Düsseldorf, Tel. (0211) 370457
• Fraunhofer Institut für Produktionstechnik und Automatisierung (IPA) Nobelstr. 12, 70569 Stuttgart, Tel. (0711) 970-1759
• Europäische Gesellschaft für Lackiertechnik e.V. (EGL): Regelmäßige EGL/DFO-Lackiertreffs in Berlin, Dresden, Krefeld, Magdeburg, Osnabrück, Paderborn, Stuttgart. Auskunft über Termine und Themen beider EGL, Adersstr. 94, 40215 Düsseldorf, Tel. (0211)370457
[14] Empfehlenswerte Fachzeitschriften zu den Themen Lackier- und Anlagentechnik sind:
Besser Lackieren!, Curt R. Vincentz Verlag
Journal Oberflächentechnik (JOT), Vieweg Verlag, München
Metalloberfläche (mo), Carl Hanser Verlag, München
Fahrzeug- und Metallackierer, Audin Verlag, München
[15] www.powdernet.de/manual.htm

7 Painting and coating processes

7.1 Paints and coatings: market and fields of application

Any coating process is first of all expected to satisfy as far as possible the three criteria of

- cost-effectiveness
- quality (in regard to performance and appearance) and
- environmental compatibility;

all three influence the choice of a given method of application.

The only thing to do is to optimise: to find the right combination and strike the best compromise between the often contradictory requirements.

In view of the extent to which the requirements for the final coating differ according to the individual object, however, it is virtually impossible and would in any case be unreasonable to seek always to maximise all three criteria.

Table 7.1 illustrates the variety and in some cases the contradictory nature of the requirements. It shows the factors other than coating material that determine the method of application. (For the range of possible methods of application see also Table 6.1.)

These multiple requirements can be dealt with somewhat more systematically by categorising them by area of application. Figure 7.1 (page 326) provides an overview of these areas, each of which is characterised by its typical processes, some of them confined exclusively to particular sectors. A breakdown according to turnover would look rather different, since pricing structures in the various sectors differ quite considerably.

Table 7.1: Criteria determining the range of applications for coating processes

	Requirements	
	high/large	**lower/smaller**
technological requirements	steel superstructure with long-term corrosion protection	furniture and equipment for interior use
optical requirements for surface	cars	(agricultural) machinery
number of objects to be coated	giveaways, car accessories	automotive refinishing
size of workpiece	ships, cooling towers	handles, chips, CDs
geometry of substrate	hollow components: cars, pipes	cladding panels, furniture doors
special requirements, e.g. for electrical conductivity of coating	conductive coatings on plastic housings; insulating wire coatings	other consumer coatings
environmental requirements through to statutory regulations	extreme differences in trends and legislation, specific to regions and sectors	

Brock, Groteklaes, Mischke: European Coatings Handbook

ISBN 978-3-86630-849-7

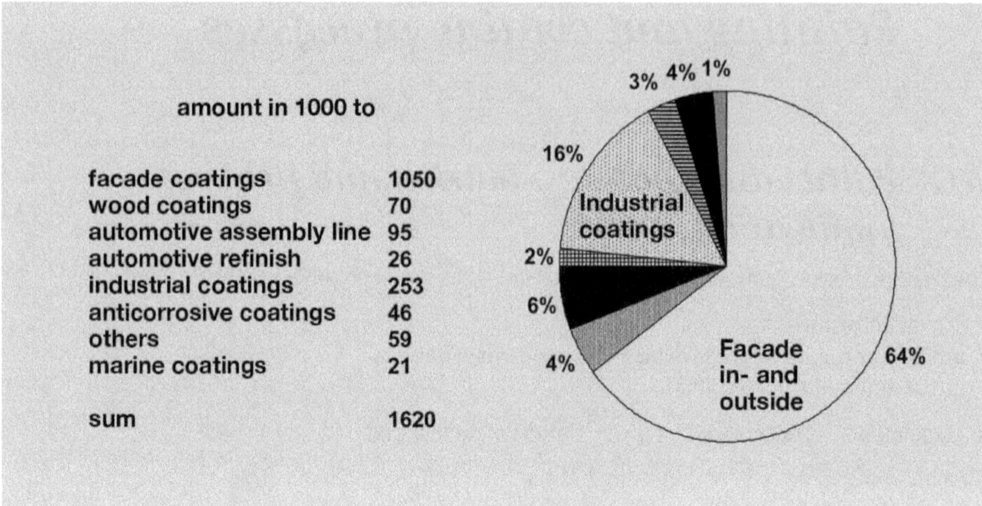

Figure 7.1: Breakdown of usage of coating materials (by volume) in Germany (1997) (source: Deutsches Lackinstitut)

7.2 Automotive assembly line coating (automotive primary finishing, OEM[1)])

Although with around 5.9 % of the market (see Figure 7.1) automotive assembly line coating is not a dominant force, it is described here in some detail since it is representative of many other metal coating processes. For given the extreme *demands placed on the coating process and on the final coating and in view of its complexity* and high *degree of automation*, many of the issues and difficulties that are a feature of this process are encountered to a lesser or limited extent in other coating processes too.

List of requirements

Some of the principal requirements of the final coating are listed below:

- outstanding long-term protection against corrosion, weather, chem-ical influences (ranging from bird droppings to acid rain), stone chipping, demands made by car wash installations, etc.
- perfect surface appearance: gloss, build, freedom from haze, together with evenness and consistency of shade and special effects
- extremely high demands regarding the flawlessness of the coating (no trapped dust, etc.)

Such a multitude of requirements clearly cannot be satisfied by the one or two coating films which are all that can be justified from a cost perspective in other applications. Here there is good cause to use at least three, but usually and increasingly four, five or more "high-chem" coatings applied in layers and matched to one another, each of which is responsible for fulfilling the various requirements.

Process steps

By way of illustration, the principal steps in a flow diagram of the processes involved in modern automotive coating practices are described in more detail below with reference to the relatively simple example of a three-coat system (plain-colour coating system, see also Figure 7.2). The rinsing, cleaning and drying stages have been omitted from the sequence.

1) **O**riginal **e**quipment **m**anufacturing.

degreasing of the shell
↓
phosphatising
↓
electrodeposition coating
↓
weld seam sealing / underbody sealing
↓
application of filler
↓
topcoat (one or more coats)
↓
quality control, secondary finishing if necessary

Further information about the individual process steps is given below:

Pretreatment

The shells nowadays consist of steel, galvanised steel, aluminium or Al/Mg alloys. They are first *degreased* (at 50 to 60 °C, with aqueous alkaline cleaning agents, for approx. 2 minutes) and then rinsed thoroughly (for approx. 1 minute). This step is followed by *zinc phosphatising* as a basis for corrosion protection and adhesive strength in the coating composition. A 1.5 μm thick phosphate film is formed within approx. 2 minutes at 50 to 60 °C (→ 5.3.4, 5.3.5).

After being rinsed again, the phosphate film is then usually, but not in every case, additionally "*passivated*" or "*compacted*", e.g. by post-treatment with chromate for 2 minutes at 50 to 60 °C ; nowadays, mostly chrome-free alternatives are used..

Afterwards the shell must be thoroughly rinsed again, finishing with demineralised water. Before the cathodic electrodeposition coating stage, any contact water is removed by drying, e.g. for 10 minutes at 140 °C with plenty of air, and the shell is then cooled down again.

Electrodeposition coating (cathodic)

Apart from a few exceptions, only cathodic electrodeposition is now used for automotive coating for the reasons set out in section 6.6. Typical deposition parameters are:

- applied voltage: 350 to 500 V
- coating thickness: 18 to 25 (in some cases only 12 μm or up to 40 μm) μm.
- deposition time: 2 to 3 min
- pH of the bath: 5.6 to 6.6
- bath temperature: 26 to 32 °C
- bath capacity: 200 to 500 m^3.

A throughput of around 50 units per hour and line can be achieved by using continuous overhead conveyors. After being rinsed with ultra-filtrate and finally with demineralised water, the shells are stoved in the primer drier at 180 to 190 °C, often undergoing a programmed temperature cycle to obtain better surface results. Nowadays grinding of the primer is no longer necessary other than to remove minor defects.

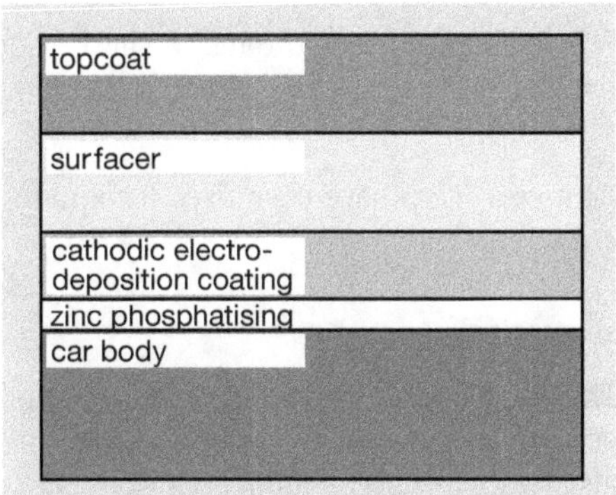

Figure 7.2: Construction of a three-coat automotive coating

Weld seam sealing and underbody sealing

Before the surfacer is applied, joints and seams must be *sealed* and the holes that were needed for the electrodeposition process (to prevent sagging or air escape!) must be closed up (to prevent dripping). This is performed using special guns and various PVC materials, both by hand and by robots. Next to be applied is the *underbody sealant* (100 to 1000 μm), which is likewise usually based on PVC plastisol or sometimes also polyurethane see further. The sealant is generally applied by robots by means of an airless process. In some cases an additional coating to protect against *stone chipping* (up to 300 μm) is also applied to especially vulnerable areas, again using a very restricted airless spray jet. Some manufacturers delay the application of the above underbody and stone chipping sealants and even of the weld seam and cavity sealants until a later stage of production, sometimes even waiting until after completion of the topcoat.

When the PVC plastisol are then heated in a *gelling drier* at around 130 °C, the PVC particles first dissolve into the surrounding medium (plasticiser) and then gel after cooling to form a mechanically solid film. Sometimes this step is instead combined with the subsequent filler drying stage.

Before the filler is applied, the shell is cleaned with dust bonding cloths, by blowing off with air or increasingly also by means of rollers covered with emu feathers (see also the start of the section on topcoating). From this stage of the coating process onwards, the slightest irregularity or fragment of dirt will be apparent in the finished coating.

Application of the surfacer

The surfacer coat (between 30 and 40 μm in thickness) serves a wide variety of functions, the nature of which has altered considerably over the years. The filler serves to

- fill in minor substrate irregularities and prepare the surface for application of the topcoat
- provide adhesive strength and corrosion protection, especially on bright metal, i.e. where the primer coat has been damaged or ground through
- provide protection against stone chipping thanks to special mechanical properties.

At one time an important criterion for a surfacer was its ability to be quickly and easily ground. This is far less relevant now since grinding is used in automotive primary finishing only for repairs or at the very most for "topping" (i.e. light grinding). In the manual sector, e.g. in automotive refinishing, surfacers have a number of additional functions too (see appropriate section).

With the exception of conventional (stoving) surfacers, the majority of fillers used today are water-thinnable products. They are cured with melamine or urea resins or increasingly with blocked isocyanates. The products are almost exclusively applied electrostatically by means of rapid-rotation atomisation, with inaccessible areas being "filled" by hand, either before or after the automated application stage.

The filled shells pass through a short evaporation zone to release solvent and allow trapped air bubbles to escape before reaching the surfacer oven, where they are dried at temperatures of up to 175 °C.

Powdered fillers have been used for a number of years now in the USA with a good deal of success.

Topcoating

The visual quality of the uppermost coating film has an important part to play in engaging the interest of the buyer and thus its first, short-term function is to present an attractive appearance. It must also retain this over a number of years, however, with ever greater expectations of the length of its service life, while at the same time the conditions to which it is exposed are becoming progressively more taxing. It must be able to withstand

- increasing chemical influences on the coating system from acid rain, industrial pollution, chemicals, changing fuel grades, etc.
- regional variations in increasingly strong, aggressive UV radiation and
- greater use of gritting products in winter (grit rather than salt).

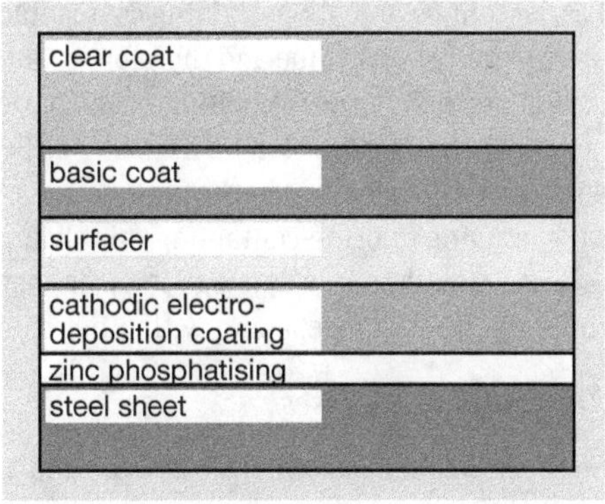

Figure 7.3: Construction of a three-coat automotive coating with two-coat topcoat

Types of topcoat

The challenge presented by these greater demands is principally being met by the use of modified binders and new or modified pigments. The trend is moving away from the traditional one-component alkyd, polyester or acrylate stoving systems towards even more highly crosslinking one- or two-component coatings, many of them isocyanate-curing, and in both water-thinnable and conventional solvent-based variants.

Only extremely light and weather resistant pigment types can be used. Here too progress is constantly being made, especially in the area of red, yellow and special effect pigments.

Cleaning

Before the topcoat is applied, a meticulous cleaning cycle is absolutely essential. The process is always completed by blowing off with ionised air (to combat electrostatic attraction of dust) and by the "emu zone", in which rotating rollers covered with emu feathers (which have favourable electrostatic properties over very large surface areas) collect even the smallest particles of dust.

Application

The topcoat material is usually applied with automatic electrostatic-assisted rapid-rotation atomisers, which also permit rapid colour changes within a matter of seconds. Robots have already become established in most areas (→ 6.9.4). Even pre- and post-spraying of areas which are difficult to access and the coating of chambers, previously done by hand (pneumatically), is now done by robots - on new lines, at any rate.

The required dry film thickness (for one-coat topcoats) of 35 to 50 µm - depending on hiding power - is applied in one or two spray cycles, sometimes with a short intermediate flash-off stage. It is then stoved in the topcoat drier at 130 to 140 °C (depending on the binder system).

Multi-coat topcoating

The one-coat topcoat process described above is now virtually only used for few solid shades, and here too two-coat or multi-coat systems are increasingly taking over because of their greater durability and enhanced appearance. It consists in the application of a colour-giving *base coat* which is then covered by a *clear coat* (Figure 7.3). This process makes even plain colours more attractive, with greater visual "depth" and a substantial improvement in durability.

One-coat metallic coatings, in particular, are now almost never used in automotive assembly line, specifically because of their disadvantages in terms of weather resistance and repairability.

Base coat materials

Nowadays, thanks to the benefits described above, over 85 % of *all* automotive assembly line coatings are produced with aqueous or conventional base coats. The base coat is applied in a thickness of only 13 to 20 µm, although this leads to problems with the *hiding power* of some red and yellow shades. Tinted surfacers are often used to overcome this drawback. A range of from 3 to a maximum of 7 differently coloured surfacers can cover most requirements.

Pearlescent base coats containing coated mica or other special effect pigments as essential pigments are generally not opaque and therefore require an additional base coat film as substrate. A white base coat is normally used, although a dark base coat is sometimes chosen to enhance the interference effect ("three-coat topcoat"). This task has been taken on to some extent by appropriately pigmented surfacers. Other new effect techniques, such as colouring the base coat or clear coat with soluble dyes or using fluorine-containing (highly dirt-repellent) clear coats are mentioned here only briefly, as is the use of nanotechnology in clear coats. Primary goal of these and other efforts is to increase the scratch resistance, along with other chemical and mechanical properties.

Application processes

Surfacers plain base coats, clear coats and also the first spray cycle for special effect base coats are applied in the same manner as one-coat topcoats using automatic electrostatic-assisted rapid-rotation devices, sometimes after pre-spraying of the interior by hand or by robot. In contrast, the second (final) spray cycle in the case of metallic and pearlescent colours is applied almost exclusively by purely pneumatic methods, since rapid-rotation application would produce rather different, darker results that in many cases could not be reproduced in any subsequent refinishing work (which is carried out pneumatically). Figure 7.4 provides a detailed illustration of an example of the application and drying of a two-coat topcoat.

Both aqueous and conventional base coats are not stoved at first but receive their clear coat (30 to 50 µm thick) after only a brief physical (surface) drying period, in a process known as "*wet in wet*" coating. A brief convection drying cycle is sufficient for conventional base coats; aqueous base coats are usually dried in a combined IR/hot air zone (object temperature approx. 50 °C), sometimes only with heat supply.

Clear coat materials

After the introduction of aqueous surfacers and base coats, the conventional clear coat is the last significant source of solvent emission involved in automotive coating. It is also the most difficult to convert to an aqueous base, however. At the same time a pronounced improvement in resistance characteristics has also been required to allow a switch from solvent to water, two essentially apparently contradictory concepts. Nevertheless, the goal is now within reach thanks to highly modified new binders, newly formulated hardener systems, predominantly polyisocyanates, and new two-component mixing techniques, which can be used for the reliable and homogeneous mixing even of reaction components with very different viscosities.

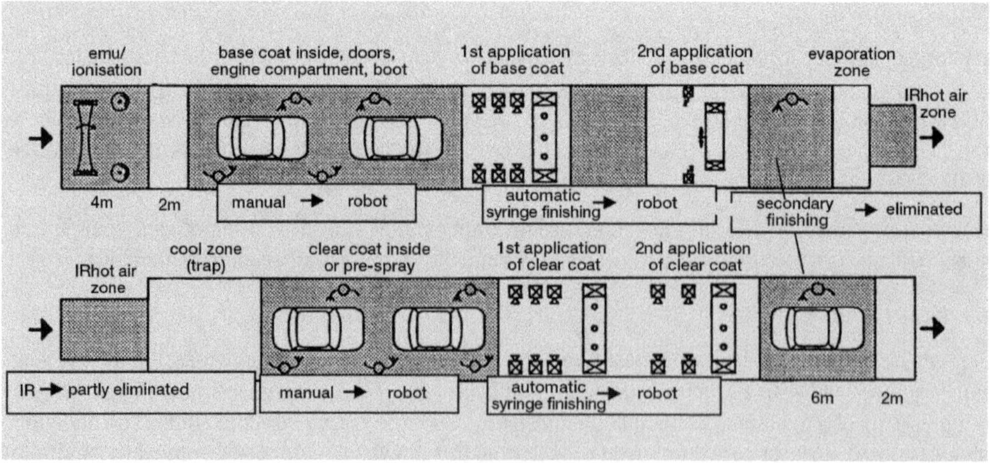

Figure 7.4: Examples of changes (see arrows) in 2-coat topcoating with water-thinnable base coats and clear coats.

The clear coats employed are conventional one-component and two-component types, usually based on polyurethane. Both are increasingly available as aqueous systems, with water-thinnable 2-component coatings on the verge of being launched. Several plants employ powder clear coats which offer impressive ecological benefits and promise to enliven competition in the future. Powder clear coats which for some time have been dispersed in water are increasingly evolving along the lines of traditional waterborne paints.

Secondary finishing

Despite all of the sophisticated measures employed to avoid dust, defects (trapped dust or mechanical damage) are unavoidable in a few per cent of finished automotive coatings. Any necessary regrinding is followed either by a complete respray of the car or at least by partial retouching, depending on the size of the defect, in the latter case using specially catalysed assembly line coatings or, particularly for small marks, refinishing grades. Local polishing may be sufficient to remedy very minor and superficial defects.

Development trends

Even now, and increasingly so in the future, the aforementioned traditional coating system and its technical application are undergoing numerous changes. Cost pressures are impacting the materials employed, energy consumption and process times, while legal and health specifications as well as new quality requirements in respect of effects, functionality and performance are coming to the fore.

Solvents

One specification, for example, is the use of lower quantities of solvent. High solid systems are particularly in demand in the USA, where special binders of narrow molecular weight distribution find application; depending on the application area, two-component systems may be used. In Europe and to some extent in Japan, too, it is waterborne and powder coatings which are popular. Experience shows that switching to waterborne coatings reduces solvent emissions from production lines by two-thirds to three-quarters - even if the clear coat contains solvent and only the base coat, filler and electrodeposition paint are waterborne.

Energy

New drying and curing methods are also gaining hold. For "low-bake" methods, the industry is looking for ways of being able to stove or cure at lower temperatures. Drying methods of the kind already being used in wood coatings - employing dry, moderately warm air - could possible be transferred to waterborne automotive finishes. This technology makes most sense for new lines, with the result that only a few manufacturers are using it at the moment.

In the medium term, the automotive industry will dispense with surfacers in coating systems - indeed, some coating lines have already done so. The role of surfacers will be played by electrodeposition coatings and basecoats, and thereby help to save 30% of energy and 25% of costs (see Figure 7.5 on page 332).

In the long-term, the body will not even need the electrodeposition coating. Its place will be taken by other metal pre-treatments, such as the sol-gel processes or self-organising mono-layer coatings.

Property profile

For vehicles in the premium price range, UV curing will soon be used because the resultant coatings can be further processed quickly and feature high scratch resistance. Since UV is not good at reaching cavities, it will be complemented by two-component isocyanate curing. Such dual-cure or twin-cure clear coats contain one or more base binders and are cured by means of two successive reaction mechanisms. Example: UV curing followed by additional polyol/isocyanate crosslinking.

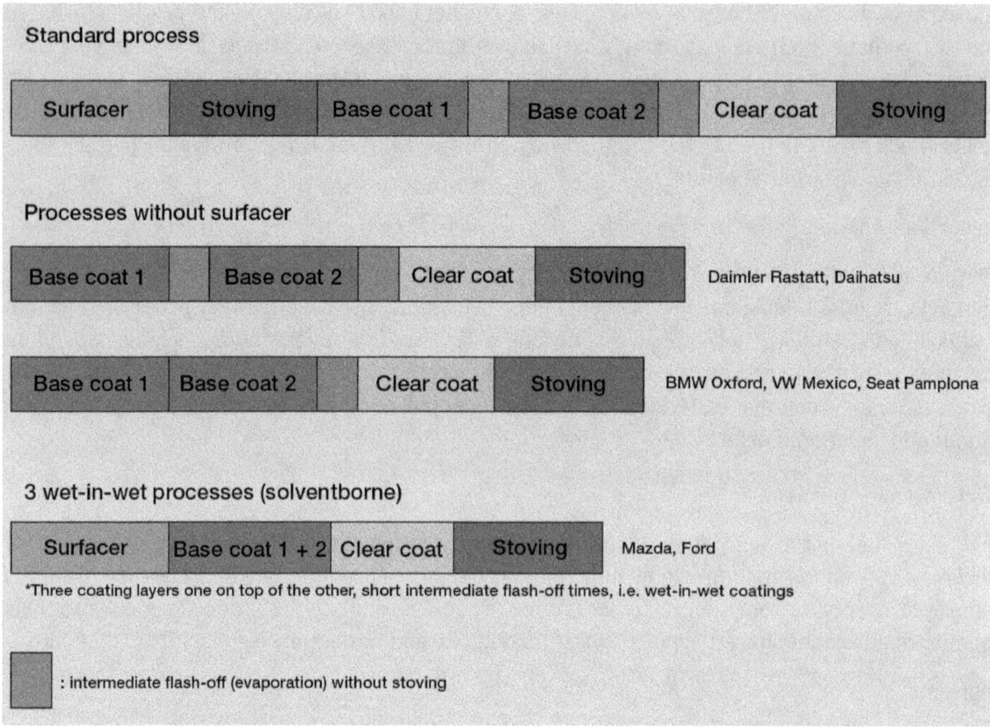

Figure 7.5: Trends and outlook at automobile coating process

Nanotechnology, too, will change the world of automotive finishes. Nanoparticles, for example, provide for better crosslinking of the paint layer. Applications tests in accelerated weathering installations and with abrasive materials have shown, for instance, that coatings containing nano-linked polymers retain their gloss longer than conventional clear coats.

7.3 Automotive refinishing coatings

Automotive refinishing coatings, unlike assembly line coatings, are not stoved but instead dry or cure at temperatures above around only 10 °C. They are usually force-dried, however, at 50 to 60 °C for example, or a maximum of 80 °C. Smaller coated surfaces can also be heated by means of portable or swivelling IR radiators.

Nevertheless, the coating result is both visually and technologically comparable with the stoved primary coat - this is in any case a continual requirement.

Other characteristic differences can be seen from the following list:

	Assembly line coating	Refinishing coating
Application technique	automated process technology	individual manual process
Method of application	electrostatic, rapid-rotation, etc.	pneumatic atomisation only
Climatic influences on coating result	climatic control	climate-independent
Curing process	full cure in oven	postcuring, sometimes even after several weeks
Variety of products	small product range	wide assortment

This means that the coating systems which can be used are completely different (particularly in regard to binders and solvents) and yet they must be closely matched to the primary coating - especially in regard to colour.

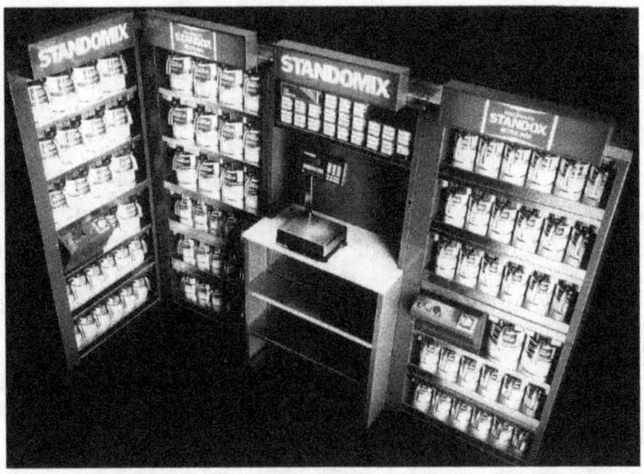

Figure 7.6: Mixing station with mixing coatings, scales and screen

Refinishing materials

The range of products is extensive, partly because of the large number of available colours, partly also because of the variety of areas of application - cars *and* commercial vehicles - and because of the diverse processing methods and regional requirements prevailing in different parts of the world. Some of the areas of application are listed below (note that automotive refinishing paints are also used for primary coating!):

- complete, partial or component respraying in the event of damage, i.e. reproduction as new (insurance requirement!)
- used car coating ("secondary coating") to ensure a good appearance, at least visually, for resale, through to a "quick" low cost cosmetic respray
- special finishes of new cars, applied either at the original factory or later by the dealer
- primary coatings for HGVs and buses, especially
- multi-coloured liveries for buses and
- box body and fuel tank coatings for utility vehicle, in some cases using special chemical-resistant materials

The vast number of shades of car paints that need to be matched (e.g. around 15,000 base coat colours) clearly cannot be manufactured, stored and sold as finished coatings. A mixing system comprising around 50 to 80 individual mixing coatings is used instead, on which very high demands are placed in respect of the necessary accuracy of shade and shelf life, and which involves extremely high standardisation costs. Any mixing coating of this type contains in each case a coloured or special effect pigment and must be stirred regularly to prevent colour defects (due to settling out of the pigment). Figure 7.6 illustrates a complete mixing station with scales and screen.

In addition to the products of the rapid and high-performance two-component polyurethane and two-component epoxy coating chemistry, alkyd resin, nitrocellulose combination coatings and other one-component coatings are also widely used, being tailored to local equipment and to requirements and coatings practices.

Coating process

Refinishing differs from primary coating in that the condition, age and requirements of the substrate are different in every case. Bright metal parts must be primed, damaged parts levelled, ground and filled, for example. Old coatings must either be protected by insulation against excessive dissolution or removed altogether if a dissolution test (by rubbing with solvent) indicates a too readily swellable substrate. Plastics too must be pretreated by appropriate means e.g. with a primer.

Coatings chronology

The nature of the final coating composition depends very much on the substrate but also on the optical and technological requirements. For the best possible coating result in terms of resistance characteristics and surface quality, a three-layer composition is normally recommended:

- primer, adapted to the type of substrate
- filler
- topcoat (one-, two- and three-layer topcoat compositions, see above).

Depending on cost and quality aspects, variations can of course be made to this ideal composition. For example, separate priming and filling stages (with a grinding cycle) can be avoided by using a *primer surfacer* followed by a wet-in-wet application of the topcoat - admittedly with clear losses in gloss.

Commercial vehicle coating

The coating of commercial vehicles, both primary coating and refinishing, involves a number of other aspects in addition to those already mentioned in regard to car coating:

- Following a thorough surface grinding with vibrating grinders, corners and recesses must also be reground.
- Weld seams must be meticulously deflashed and thoroughly degreased.
- Aluminium tailboards and superstructures, stainless steel parts and hot galvanised surfaces, particularly in the form of composites with other metals, represent particularly difficult substrates. In such cases the surface is either "washed" with phosphoric acid (etched, pickled) or at least primed with a wash primer containing phosphoric acid.
- Old thermoplastic coatings must be insulated with highly crosslinkable two-component fillers.

Further information about the coating of plastic add-on components can be found in section 7.4.

Coating equipment

The layout of a specialist refinishing shop is naturally much less technologically complex than that of a primary coating installation. A number of basic rules should be mentioned, however; thus, for example, the following areas must be provided:

Preparation zone

This must be very spacious, being used for cleaning, degreasing, dust removal, levelling, priming, filling, grinding, glue removal, etc. Modern preparation zones are often equipped with floor-level extraction systems via grates or at least with static extractors fitted to the machines (grinders, etc.) to reduce contamination, of the coating area in particular.

Coating shop

The actual coating station may consist, for example, of separate coating and drying booths, arranged either in series, connected by longitudinal feed through a gate, or side by side with cross feed. A combined coating and drying booth ("combi-booth") may also be provided, either additionally or as the only booth. The advantage of this arrangement lies in its compactness, the disadvantages being its high energy consumption and low throughput. The latest technology provides a down draught through the appropriate booth. There may also be separate spraying booths for coating small parts and accessories or (a recent trend) for job coating work outside the automotive sector (lamps, furniture, etc.).

Finishing zone

This area - for polishing, assembly, etc. - should be separated from the preparation zones. A separate coatings store (also known in Germany as a "VbF[1] store") is required for safety reasons.

Common terms

Some of the key specialist terms used in refinishing work as well as in other manual coating sectors and in industrial installations are briefly described below:

Stoppers are highly-filled coating materials used to level out substrate irregularities that are too large to be corrected by primers or fillers. They can be applied by wiping, brushing or spraying.

1) Verordnung über brennbare Flüssigkeiten [German regulations on combustible liquids], see also section 9.4.

Primers are applied in relatively thin coats. Their principal purpose is to ensure adequate corrosion protection and adhesion of the coating composition to the substrate.

Surfacers are designed as intermediate coating materials to level out irregularities in the substrate (e.g. to *fill* grinding grooves), to support adhesion and corrosion protection and to ensure that the coating composition has good resistance to stone chipping. They usually also having priming properties nowadays, particularly as **primer surfacers**, the use of which can eliminate the need for additional priming.

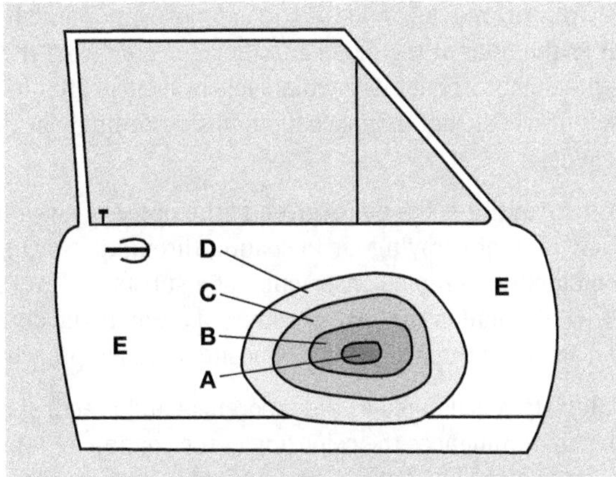

Figure 7.7: Schematic view of touching up, with blended transitions (A →D) into the surrounding area E

Insulating surfacer are especially highly crosslinking fillers used to insulate the substrate against excessive dissolution by solvents from the topcoat. *High-build surfacers* allow the use of stoppers to be dispensed with in the case of more irregular surfaces.

Topcoats cover the coating composition from the top and protect it from weathering and other influences. The most commonly used variants are one-coat and two-coat topcoats (base coat + clear coat).

Base coats are generally applied on top of the filler or priming filler and in multi-coat topcoat systems they are responsible for the colour and/or effect. Consequently they may contain coloured and special effect pigments. They must be covered with a clear coat for protection. Since about 1990, water-thinnable base coats have increasingly been used in the automotive refinishing sector too. They have been mandatory in Europe since 2007 due to EU solvent legislation.

Clear coats are transparent, colourless coating materials that usually form the uppermost layer of a coating system, in which case they can be defined as topcoats with no (optical) hiding power. They protect the colour-imparting layer or also the substrate itself against mechanical, atmospheric and chemical influences.

Spot repairs are made by applying a complete coating system to the smallest possible area of the spot to be repaired (up to around the size of a two-pound coin) and spraying out over the surrounding regions to match colour and gloss. The transitional zones must then be polished. Given the complexity of this operation, touching up (see below) is usually more practical.

Touching up is used to "invisibly" repair small and medium-sized defects such as scratches or dents, even if they are located in the middle of intact original coating. The process involves spraying pigmented coating, usually base coat, from the centre of the repair area outwards, gradually increasing the dilution of the coating if necessary, until the new coating blends into the original finish. Figure 7.7 shows these transitions (A>>D). The entire component E or all relevant parts are then coated with clear coat.

Dissolution ("swelling") of the coating substrate by solvent from the coating system can - to an extent - be a desirable means of promoting subsequent adhesion and water resistance. Too great a degree of dissolution, however, can lead to loss of adhesion and to lifting[1)] of the old coatings, as well as to general softening and wrinkling of the coating and ultimately - in metallic coatings - to clouding of the effect.

1) *Lifting* can occur as a defect during the overcoating of precoated substrates if previous coatings, surfacer, etc., are not adequately cured. Swelling followed by an increase in volume leads to creasing, irregularities, blisters, etc. - a common source of defects in the area of general industrial coatings and wood coating too.

Peripheral marks and **sinking** can pose a problem if an old coating has been ground through as far as the filler or the metal and then solvents from the refinishing coating penetrate into partially exposed, absorbent intermediate layers such as filler or base coat. The solvents then evaporate after days or weeks and the topcoat slumps accordingly at these points. Insulating the substrate can be beneficial.

Wet spraying is the name given to the process variants in spray coating that are characterised by a low ratio of spraying air to coating throughput or in which the spraying distance is very short, combined with rapid construction of a substantial wet film thickness. The effects described above under dissolution can then become apparent, to the detriment of the coating, and there is also a risk of tearing ("curtaining"). The opposite of this approach is known as **dry spraying**.

Coloristics is defined as the theoretical and practical knowledge of everything involved with coatings that influences the colouring of the coating system: from pigment testing and standardising to coating batch approval, via mixing, formulating and correcting colours through to quality assurance. The *tinter* is responsible for mixing and in particular for approving or correcting colours and mixing formulae. The painter in the workshop must also be capable of compensating for colour variations in the coating system (changes due to weathering and chemical influences, factory colour variations) using the mixing system (see above) or of correcting them by means of tinting.

7.4 Industrial plastics coating systems

List of requirements

Having already covered the various types of plastics as coating substrates and the methods of pretreatment they require (section 5.4), in this section we will deal with specific practical aspects of plastics coating technology. Of particular importance in this respect are the following material requirements for plastic objects:

- Suitable, constant surface tension
- Wall thicknesses as uniform as possible (otherwise temperature differences occur due to the poor heat conductivity of the material)
- Closed, uniform surface
- No surface defects (pores, fibres) or stresses/orientations (from injection moulding)
- Minimal influence from internal or external release agents; they should at least be easily removable

A consistent quality of plastic and/or an appropriate incoming inspection is therefore even more important here than it is in metal coating. Due to the different mechanical/thermal behaviour of thermoplastics in particular, great care must be exercised when removing flashes, weld seams, etc., to ensure that the heat generated by grinding does not cause the material to reach its *softening point*, leading to smearing or to fissured surfaces. The tendency of plastics to become electrostatically charged must also be offset, principally by wiping with anti-static cleaning agents such as water/alcohol mixtures, or (in automated installations in particular) by blowing off with ionised air shortly before application of the coating.

Coating materials

In order to be suitable for plastics coating, a coating material must, above all

- contain suitable solvents that, in particular, are not too aggressive ("sharp" solvents can lead to swelling and to embrittlement and cracking in the plastic as well as to loss of adhesion) and
- be more flexible than the substrate to enable it to resist any type of deformation.

For optimum coating results a two-, three- or four-coat construction is recommended, again depending on technological requirements:

- Primer ("primer for plastics"), the main purpose of which is to ensure an adequate bond; an adhesion promoter may need to be added for particularly difficult plastics such as polypropylene.
- An additional insulating or barrier primer often has to be used for plasticised PVC or flexible PU foam to prevent migration of the plasticiser.
- Elasticised filler to compensate for surface roughness.
- Topcoat, likewise elasticised, often a two-coat variant.

With a view to further cost reductions, demands are increasingly being made for topcoats (traditional and also waterborne base coats) that can be sprayed onto the plastic direct.

Method of application

With the exception of electrodeposition coating and electrostatic application (see Table 6.3, page 305), all of the methods known from wood or metal coating can also be used for plastics coating. In drying and curing processes the sensitivity of the substrate to heat must also be taken into consideration, of course: drying is generally performed at a maximum of 80 °C. UV and electron beam curing can cause yellowing in some plastics.

7.5 Painting of rail vehicles

The history of the painting of locomotives and wagons has developed in a similar way to that of car painting: the early phase of manual application of black oil paints (black was especially practical because of the heavy soot deposits), was followed from around 1930 by the introduction of spray application of nitrocellulose paints.

Since the late 1970s, due to the particularly high resistance characteristics required for rail vehicles (see below), two-component PU coatings have been used almost exclusively:

- in three-coat compositions (primer/intermediate coating/topcoat) for primary coating and
- in two-coat compositions (priming filler/topcoat) for maintenance work.

Water-thinnable systems are now increasingly being used, although they generally display a lower resistance to alkaline and acid cleaning agents. A gradual basic soiling is usually tolerated. Solvent-based coatings are used only for prestige trains such as intercity and high-speed trains because of their enhanced gloss. On ordinary goods wagons, where requirements in respect of appearance are not particularly high, one-coat coatings based on aqueous acrylate dispersions, for example, are also used (directly after descaling, i.e. after removal of the Fe_3O_4 mill scale).

Since, as in other coating applications, about 80 % of the costs involved in coating are labour costs and only a maximum of about 20 % are materials costs, savings are sought by allowing long intervals between coating cycles, e.g. about 8 years when conventional coatings are used.

A particular requirement of rail vehicle coatings is a resistance to highly acid cleaning agents, e.g. to mixtures of phosphoric and nitric acid, that have to be used to remove flying spark particles from current collectors (iron oxide, copper oxide, etc.).

7.6 Coil coating

The continuous organic coating of strips of - usually galvanised - steel or aluminium by means of rollers (see also section 6.4) is one of the most rapidly growing areas of coating technology. It provides a very cost-effective and low-emission coated coil strip that often saves processors from having to set up their own coating shop with pretreatment area, coating store, test laboratory, waste disposal systems, etc. The coating application process provides an almost 100 % yield; waste water from the pretreatment stage is collected in the plant and treated if necessary and the circulating air supplied to the thermal afterburning process.

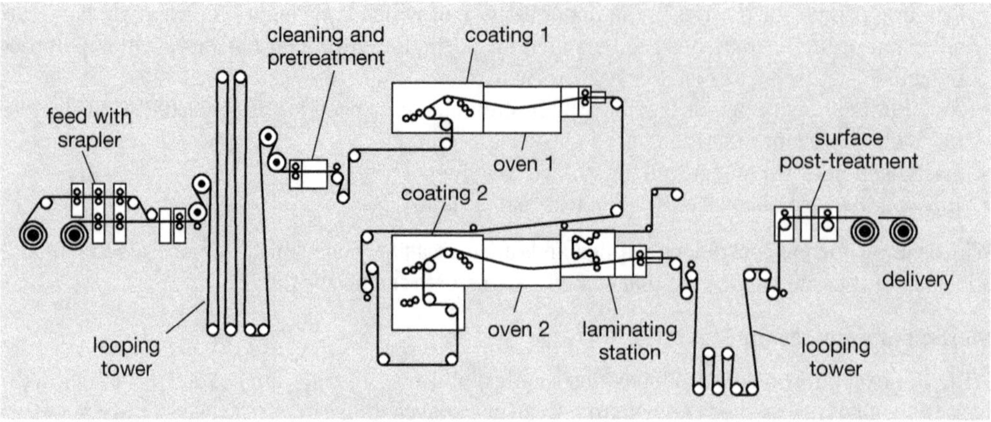

Figure 7.8: Diagram of a coil coating plant (source: Chemetall GmbH). The pretreatment zones and ovens are shown in very abbreviated form

Process sequence

The organic coil coating process, as shown in Figure 7.8, essentially involves the following stages:

- Cleaning and chemical pretreatment (usually phosphatising) of the metal surface.
- Single or double-sided roller application of the coating material (the diagram shows two double-sided applications).
- Drying and curing in the stoving oven; high stoving temperatures of 160 to 290 °C are typical here.
- Laminating with plastic film, if required; also: decorative embossing of PVC paste coatings or application of removable protective film. The latter can also facilitate subsequent forming operations and thus prevent the surface of the coating from being damaged.

The bright metal sheet is supplied as a coil, which can if required be welded or stapled to the previous unwound sheet to avoid the need to stop the plant. Looping towers are used to store the material during such processes and also at the end of the process; the distance between their upper and lower rows of rollers can be altered. The running speed of the strip is up to around 3 m/s in the case of thin Al foil 6 m/s. The coated strip is either coiled up again or - depending on the fabricator's requirements - cut straight into sheets.

Areas of use

The use of coil-coated metal sheet has proven itself in a number of different sectors, in which coating processes can thus be eliminated, including

- the construction industry first and foremost: for cladding panels and structures, trapezoidal sheeting, garage doors, entry doors and partitions; secondly
- the automotive industry, particularly for caravans and commercial vehicle superstructures (it is also being tested as an alternative to cathodic electrodeposition); and finally
- white goods, i.e. larger domestic appliances, together with housings for audio and video equipment and similar products.

Trends

For several years now, electrostatic "powder cloud" coating has been employed on a number of coil coating installations as it is technically less complicated for smaller coil batches and can be less expensive. Moreover, several coil coaters are already employing near-infrared radiation (NIR; see also section 6.10) for drying and curing instead of energy-intensive hot air circulation. They are exploiting the fact that near-infrared radiation can be focused on to a defined, constant distance.

7.7 *Electrical insulation systems and electronic assembly systems*

These purely functional coating materials insulate and strengthen electrically conductive carrier materials such as wires, electronic components (blank), motors, transformer components and machine components, for which as a rule there are no requirements in terms of optical properties. The following system components - matched to one another as closely as possible - are used here:

Wire enamels

Used to give bright wire an insulating surface, they are applied continuously in several layers (up to 8) and stoved at over 400 °C. The process does not affect the ability of the wire to be tin-plated (or soldered). The principal binders used for wire enamels are phenolic resins, polyesters and esterimides, and in combination with polyurethane systems provide an especially good basis for tinning requirements.

Impregnating agents

These materials are then used to reinforce the layers of wire and the coils by filling the voids. Impregnating agents also serve to carry away lost electrical heat during operation. They are applied by flow coating, dripping or dipping and stoved at around 120 °C, in some cases by the passage of an electrical current. In addition to conventional stoving systems, unsaturated polyesters with appropriate hardeners are also principally used here.

Embedding compounds (casting compounds)

These materials are used to provide components with additional protection, specifically by means of all-over casting (embedding, usually in a block-like, compact compound). They are formulated without solvent (with the exception of any reactive thinners that may be present, e.g. styrene) because of the thick film dimensions, and they include fillers.

Electrical insulation systems are specifically required to display mechanical hardness, together with heat resistance, chemical resistance and tracking resistance (i.e. freedom from pores), as well as good stress resistance and dielectric strength. Modern electrical insulation systems satisfy these requirements with only minimal film thicknesses (a few μm); this has permitted the reduction in size and miniaturisation of many electrical and electronic components and devices.

Electronic coatings

These usually highly-crosslinked coatings are intended to protect circuit boards and other electronic modules against climatic influences, especially moisture (spray and condensation). At the same time, they must resist high temperatures (-40 to 140 °C), chemical influences (salt, fuels, acid and alkalis), leakage currents and mechanical vibration. This requires the use of such systems as two-component polyurethanes, epoxy/amine systems, silicone resins, UV-curing materials, and - in some cases - dual-cure systems (see also section 7.2).

Here, too, the trend is away from traditional solventborne coatings and towards high solid, waterborne and 100 percent systems, such as UV coatings.

The growing use of these protective coatings in electronics parts in vehicles and all types of equipment also extends to applications in aerospace, military, shipbuilding, oil production and other production systems, electronics and other telecommunications sectors, vehicle manufacture, and all kinds of instrumentation and control equipment.

7.8 *Other metal coating systems*

In the area of general industrial coating, the coating materials, methods of application and requirements are far more diverse than is the case in automotive assembly line coating. Other than primers, most of the materials used here are one-coat topcoats. The following breakdown of some of the principal areas of application for coatings illustrates the possibilities for variation that exist.

Domestic appliances such as refrigerators, freezers, washing machines, etc. (white goods) are phosphatised and in some cases primed by electrodeposition before being coated, generally by electrostatically-assisted spraying, and then stoved. The use of water-borne coatings and particularly of powder coatings is increasing in this sector. Coil-coated metal sheet is also proving increasingly popular with appliance manufacturers who have no in-house coating facilities.

The same range of coating processes is available for metal components and furniture (fittings, shelving, etc.) as well as for **radiators** and **electrical elements**. The wider the desired choice of colours, however, the less suitable is powder coating, which performs less well in this respect. Stoving coatings are then often applied manually by means of pneumatic spraying.

Car wheel rims are increasingly being coated with special-effect powder coatings, which are little used in other applications. The process now imparts a considerable (metallic) effect, albeit with less sparkle than is typically produced by liquid coatings.

Bicycle frames, grids and other filigree workpieces are ideal for electrostatic-assisted applications and/or powder coating. These methods offer the only cost-effective means of maintaining (overall) application efficiency.

In the **packaging coating sector** (which includes can coating), the **container coating system** takes on a stabilising protective function against often aggressive contents, a precondition for which is freedom from pores. Fruit acids (e.g. from tomatoes), solvents and chemicals would otherwise quickly cause the aluminium plate or tinplate to corrode or the contents themselves would deteriorate on contact with the corrosion products (rust). Epoxy/phenolic resin coatings (which because of their colour are also known as "gold coatings") are therefore often used for internal coating, in some cases with an additional primer. Internal can coatings must be sterilisable, non-toxic and mechanically resistant (to scratches and dents).

In **can coating** (packaging coating), generally smaller cans are normally made by taking a sheet of plane metal (often precoated in white), dyeing and then coating the exterior with a clear printing ink, coating the interior from the other side and finally forming the can. The weld or solder seam produced must be carefully sealed with "seam sealant" to ensure that it is equally resistant. The coating is applied by rolling or spraying and after stoving it must above all display good deformability. Larger cans and drums are often formed first before being coated by rolling or spraying on the outside and spray coated on the inside. Hot spraying is also used here. Depending on the size and degree of automation of the installation, up to 20 cans per second can be produced and coated.

Punchable finishes are used in particular for coating the outside of lead packaging. They are generally pigmented and, in addition to satisfying the usual requirements for exterior coatings, they must also be stackable, often need good surface slip (for filling machines) and must be capable of being printed with UV-curing printing inks. Alkyd, epoxy, polyester and acrylate stoving systems can all be used as film formers for punchable finishes.

The range of objects, coating materials and processes is particularly diverse in the **mechanical engineering** sector. The choice of topcoat materials includes gloss and matt coatings as well as textured and hammer finish paints. Here too the coating system (usually applied to metal) is matched to the intended area of application and to the mechanical and optical requirements. A number of illustrative examples is provided below:

Although *agricultural machinery* is at great risk of corrosion, manufacturers are under pressure to keep costs as low as possible. In most cases, therefore, two-coat systems (primer and forced or oven drying topcoat) are applied by dipping powder coating or spraying, wherever one-coat anodic or cathodic electrodeposition coating cannot be used.

Relatively simple *building tools*, for the DIY market for example, can be subject to extreme mechanical stresses, as a result of which even the most resistant coatings can rarely keep pace with the service life of the tool. For that reason low-cost but easily repairable two-coat or even one-coat coatings ("cosmetic coatings") are chosen. In the professional sector (industrial, crafts, agricultural), by contrast, more and more emphasis is being placed on durability and greater longevity.

Machine tools and metal working machines are exposed to aggressive oils, coolants and other liquids and therefore require an especially resistant coating system, such as PU filler and topcoat.

In the case of *wood working machines* on the other hand, synthetic resin topcoats offer adequate protection.

Corrosion protection is of enormous importance in **marine coatings**. Substrates are subjected to a thorough blast cleaning treatment followed immediately by application of the primer (based on polyvinylbutyral or epoxy resin, pigmented with zinc dust if required). The subsequent coats can be based on PU, epoxy and vinyl resins and on liquid polymers. To prevent the growth of algae, accumulation of barnacles, etc., an additional *antifouling* coat is also applied, based principally on organometal (tin) compounds that in water are gradually released to the surface. This fact is also the reason for numerous efforts in recent years to find an alternative to tin compounds, which are toxic to microorganisms. The coatings are predominantly applied by airless spraying because of the large surface areas and low optical requirements. Coatings by roller is also possible.

Aviation coatings are subject to a different set of special requirements. In this case the focus is on resistance to de-icing agents, aviation kerosine and oils (e.g. "Skydrol"), to extreme changes in temperature and to UV radiation. Coating systems and application methods are similar to those used in automotive refinishing with two-component PU materials (wash primer, filler and topcoat). Before maintenance coating work can be carried out, however, the entire coating system generally has to be removed, with great care being taken to ensure that the metal substrate is not damaged (no hydrogen embrittlement of the metal due to the action of coating strippers).

7.9 Coating of wood and wood-based materials

Wood and wood-based materials have already been covered in detail in section 5.5. The key facts are reviewed and enlarged upon below - primarily in order to clarify the interrelations between the coating and the wood.

Of all of the different types of substrates that can be coated, wood is certainly the least uniform, and its properties and coatability are extremely dependent on its origin, age and storage conditions. The main wood substrates and hence the most important areas for wood coatings are as follows:

- solid wood, for furniture, window frames, etc.
- wood core plywood (composite of glued solid wood chips), for hand-built furniture
- chipboard, for furniture and panels
- fibreboard for doors, rear panels for furniture and drawer bottoms
- plywood, for light-weight structures.

Features of wood of relevance for coatings

The most important characteristic of wood that must be taken into consideration before it is coated is its *water content*, which must be appropriate and not too high. Wood is hygroscopic, i.e. it adjusts its moisture content to that of the surrounding air and in so doing undergoes considerable variations in volume (contraction or swelling). Outside, wooden products retain an average moisture content of between around 12 and 25 %; inside it is only between around 6 and 18 %. Green (new) wood exhibits still higher moisture contents and must therefore be selectively dried (nowadays usually in seasoning ovens) before being used. During fabrication too, a consistent moisture content must be assured by means of appropriate climatic control in workshops, warehouses and production areas, for example.

Table 7.2: Densities of wood and changes in volume with maximum water absorption, according to [9]

Type of wood	Density [kg/m³]	Maximum contraction/swelling [%]		
		radial	tangential	longitudinal
pine	300 to 860	4	8	0.4
beech	490 to 880	6	12	0.3
oak	390 to 930	4.6	10	0.4
fibreboard	400 to 1000	5 to 20		

Table 7.3: Dependence of apparent density of wood on pore volume (according to [9]*)*

apparent density [kg/m³]	Pore volume in total (including submicroscopic pores) [%]	Pore volume of macro- to microscopic pores [%]
200	86.8	71.2
400	73.5	44.2
600	60.3	19.5
800	47.3	7.1
1000	33.8	2.3

By reference to various types of wood, Table 7.2 illustrates the wide variation in the density of wood as a material and also the degree to which it expands when the moisture content is altered.

The three different *growth directions* of wood, known as radial, tangential and longitudinal, were illustrated in Figure 5.7. The various *apparent densities* of the wood correlate to the extremely divergent *pore volumes* (Table 7.3), which when the coating is applied have a huge influence on the absorbency of the material as well as on the resistance that has to be overcome by the spreading coating droplet.

One consequence of the continuous alternation between swelling and contraction, assisted by the porosity of the substrate, is the well-known phenomenon of "warping". This effect is intensified by thermal stresses and by external mechanical stresses: stresses and cracking can occur in the longitudinal direction, perpendicular to the longitudinal direction or in no preferred direction, depending on the influence. Apart from the loss of protection on account of the coating, adhesive bonding can dissolve and glazing start to leak. The escape of resin from resinous woods (such as pine) represents a further, temperature-related, defect.

The protection of wood by coating must always be complemented by constructive wood protection in order that damp, microbial attack etc. may be reduced from the outset. Most of all, this necessitates protection against precipitation and ground damp, along with good ventilation of the wooden construction as well as the covering or slanting of horizontal surfaces.

Additional chemical wood protection through the use of preservatives should be mentioned, too.

The porous wood structure not only absorbs water; other solvents too are admitted and cause short-term swelling during coating or paint stripping, for example, although their swelling effect is less pronounced than that of water.

The following sequence of various short-term swelling behaviours applies to swelling in ash veneers in the tangential direction. The order of magnitude of the swelling when stored in liquid for 30 minutes is 0 to 2 %.

water > ethanol > methylene chloride > nitrocellulose dilution, turpentine

As a result of the absorption of liquid during swelling (up to the "fibre saturation limit"), the fibres at the surface stand up like bristles, the effect being even more pronounced in the case of water or water-based coatings than with solvent-based liquids, for the reasons described above. Wood-based materials therefore generally need to be sanded before and often also several times during the construction of the coating system.

Coating materials / processes for wood and wood-based materials

Staining, glazing, bleaching

Staining is used to compensate for or to match colour differences in solid wood and veneers for interior. The use of stains for colour matching is also known as "patinating". Staining can also serve to emphasise the structure (grain) of the wood and to introduce an individual character to the wood in keeping with the design of the furniture. Both solvent-based and water-based stains are used, enhanced with dyes or with pigments ground until transparent. Water-based stains are characterised by their ease of use but require longer (circulating air) drying times and in some cases IR drying, and, as mentioned above, they produce a rougher surface as a result of swelling and protruding fibres.

The "enhancement" of a grain (optical strengthening of the grain contrast) can often differ markedly in its intensity and direction of action for water-based and solvent-based stains. The conversion of a wood-finishing process to low-VOC products may sometimes involve considerable difficulties.

Glazes for outdoor contain pigments ground until transparent in addition to binders. Thin-film glazes (impregnating glazes) have an impregnating action, whereas high-build glazes form a protective film on the surface. The surface of wood is occasionally also *bleached*, i.e. lightened, by means of hydrogen peroxide.

Grain printing

In order to create the visual impression of high-grade veneers, plain or uneven veneers are first coated with a printing substrate. Once it has dried, the actual wood pattern (in other words the veneer texture) is printed by means of a rotary printing unit. Sometimes this can involve complex printing in multiple colours using several printing units. Neither of the two coatings completely hides the base veneer, so the character of the wood is retained. Chipboard and fibreboard can also be enhanced by grain printing, in which case they must of course first be thoroughly levelled.

Priming processes

Before wood chipboard and fibreboard is coated, the surface must first be sealed, i.e. insulated. This can be achieved by means of film lamination, or by *levelling*, whereby irregularities on the surface are smoothed out at the same time. The stopper is usually applied with rollers and then sanded. Nowadays ultra-quick-drying water-based UV or electron beam curing stoppers are increasingly being used.

For exterior applications, wood must generally be protected against wood-damaging fungi by means of *dipping primers* based on water or synthetic resins ("blue stain protection"). For high-grade opaque coatings, a *spraying filler* based on acrylate, for example, is often applied as an additional coating to cover and completely smooth the surface of the wood.

Topcoat application methods

The two main types of wooden workpieces - flat and carcass components - determine to a great extent the nature of the topcoat application process and hence the design of the coating installation. Flat components, e.g. boards, furniture components, doors, panels, etc., are coated by *roller coating* or *curtain coating* (decolaging), whereby they are transported horizontally on conveyor belts or platform conveyors. Spraying devices are occasionally also used for automated edge coating, for example. Elaborately shaped objects such as furniture carcasses or chairs, as well as small batches of flat objects are usually coated by *spraying* and transported by overhead conveyors or apron conveyors.

Depending on visual requirements, the criteria mentioned in Chapter 6 for the use of pneumatic and airless spraying, optionally with electrostatic assistance, and of rapid-rotation atomisation, apply here too.

Flow coating and *dipping* are both suitable methods of application for window frames, with flow coating being especially advantageous for wooden components that are capable of floating. *Vacuum*

Table 7.4: Binder systems and their properties in relation to wood coating

System	Use	Advantages	Disadvantages	Market trend
nitrocellulose coatings, physically drying	indoor furniture	glossy, quick-drying, scratch-resistant, alcohol-resistant, thin films possible	poor light and solvent resistance, low non-volatile matter, high solvent content	declining
two-component PU ("DD coatings")	(kitchen) furniture, exterior applications, marine paints	high chemical and mechanical resistance, flexible	relatively high price, limited pot life	increasing
acid-curing coatings, urea/formaldehyde resin + alkyd resin	similar to two-component PU		formaldehyde emission	declining
oil-based coatings (alkyd resin)	exterior applications	good gloss and flow, weather and chemical resistant	slow cure, embrittlement common	declining
unsaturated polyester coatings (UP coatings)	high-grade, glossy surfaces, e.g. pianos, stoppers	scratch-resistant, high-gloss, low-solvent, high-build films possible	poor light resistance, short pot life (two-component), poor storage stability	still used for special applications
water-borne coatings: acrylate and PU dispersions	furniture for the home and office	low-solvent, lightfast,resistant	slow drying, extremely rough surface	increasing
UV curing coatings, also: electron beam curing	furniture, especially for schools, parquet floors	extremely rapid curing, capable of withstanding high stresses, very low solvent content	UV: not readily pigmentable	growing rapidly
coating powders (e.g. IR melting + UV curing)	(expected to be versatile)	solvent-free, rapid curing	heating of workpiece	good prospects (future)

coating is the preferred method for wooden profiles and beams, as well as plank-like profiles of wood, metal and plastic.

An interesting, albeit somewhat disputed current development is the powder coating of MDF fibreboard; even on wood, the results are unsatisfactory. See also section 6.8.2.

The variety of substrates and specifications has resulted in a broad range of binder systems for wood coating materials. Table 7.4 lists these systems together with their principal advantages and disadvantages.

Drying processes

The choice of a suitable *drying* or *curing process* depends to a great extent on the binder system used, but the temperature should not exceed 60 to 80 °C over extended drying periods. Higher temperatures of 120 to 140 °C should be sustained for no more than 1 to 2 minutes to prevent the heat from penetrating deep into the wood. In addition to circulating air and forced convection drying, the action of IR radiators is also increasingly being utilised, particularly for water-borne coatings. The wooden component is often preheated to between 40 and 80 °C before being coated, to reduce the damaging effect of air trapped in the pores.

7.10 Building protection/coating of mineral substrates

In terms of price and manufacturing costs, exterior paints certainly do not rank on a level with other paints and coatings. Nevertheless, the fact that around one-half to two-thirds - depending on the definition - of the production of coating materials (by volume) is made up by exterior paints (see also Figure 7.1) is frequently overlooked.

Substrate materials

The principal mineral and silicate substrates and their special features were covered in detail in section 5.6. We will review them here briefly in order to clarify their interactions with coating materials.

Concrete, exposed aggregate concrete, gas concrete, wood cement and asbestos cement all undergo a pronounced alkaline reaction; binders that are sensitive to saponification or that are not alkali resistant for other reasons should therefore be avoided.

Brick, klinker brick, natural stone can be absorbent to varying degrees but tend to be chemically inert.

Sand lime brick is likewise alkaline and highly absorbent.

Mortars and *plasters* (lime and gypsum plasters, cement rendering) must be inspected before being coated to ensure that the structure is intact and not loose. Plasters that are not too brittle can be treated by means of plaster strengtheners in the form of highly penetrative, low-viscosity binder solutions. For the professional decorator an understanding of the chemical characteristics of the various plasters and mortars is vital. Lime mortars and plasters are extremely alkaline.

Gypsum surfaces are not water resistant (being somewhat water soluble) and are often highly, but unevenly absorbent, particularly in respect of aqueous media.

Mineral-based facing plasters and *synthetic resin plasters* are generally left uncoated when they are new, being coated only in connection with maintenance work.

Coating factors

The following properties of mineral substrates should generally be taken into particular account when selecting coating materials and processes:

- High degree of roughness.
- Elevated porosity and strong absorption capacity, particularly in the case of highly (mineral) filled substrates.
- Swelling and contraction with subsequent cracking, depending on moisture absorption and temperature, particularly in exterior applications. The coating should be able to bridge the resulting cracks to the greatest extent possible (i.e. it must be flexible).
- Water vapour permeability of the substrate; depending on requirements, the permeability of the coating must be matched to that of the substrate to control the overall moisture balance of the component, e.g. the wall of a house.
- Water-soluble inorganic salts can often crystallise out of the substrate ("*efflorescence*"). They must always be removed using a dry process.

All mineral substrates must be adequately clean and dry before being coated.

Coating materials

Exterior and interior wall paints are now largely based on aqueous polymer dispersions, ranging from polyvinyl acetates through to saponification-resistant copolymers, and polyacrylates in particular. Purely organic silicate systems are also used.

Modern masonry coatings consist of combinations of silicone resins and polymer dispersions. They are usually applied in several coats for exterior applications. The (more expensive) silicone resin emulsions produce enhanced water vapour permeability and weather resistance.

Where elevated resistances to chemicals or to cleaning are required, two-component epoxy or alkyd resin emulsions or chlorinated rubber materials are used.

The low-viscosity primers - which are absolutely essential for exterior coatings on porous substrates - are increasingly being switched to silicone microemulsions.

The coating technique is governed entirely by the object in question: depending on the surface area and on the equipment available, coatings can be applied by brushing or rolling through to large-scale airless spraying.

New materials

Recent years have seen numerous innovations in masonry paints and coatings as well, mostly in the form of novel properties obtained by the use of nanotechnology.

These include the Lotus effect, self-cleaning surfaces, beading, antimicrobial effects and photocatalytic properties.

7.11 Separating, preparing and recycling paint and coating residues

The coating of an object is only ever a part - albeit the most "eye-catching" part - of a production process. In the same way as in the manufacture of coatings, all manner of coating residues or wastes are inevitably generated during the course of coating processes, particularly in the case of application by spraying:

- overspray (mis-sprayed paint mist)
- non-recyclable residues of coating batches (especially in the case of two-component coatings)
- residues or coating batches that have exceeded their shelf life, defective batches
- diluted coating residues from washing and rinsing processes.

Separating and washing systems

Apart from the removal of volatile solvents from the waste air (see section 6.9.2), the main issue in industrial spray coating is the collection of coating mist losses. Two methods of dealing with this problem have become established: dry separation (in the manual sector) and in particular the "wet washing" process with water.

Dry and wet separation

Although the investment costs for dry separation are lower, the cleaning costs are higher and the risk of fire is greater. Dry separation is therefore generally only used when the overspray quantities are small or where coating work is carried out only occasionally or with only low coating throughput. Wet separation makes up around 80 % of the total in metal coating and almost 100 % in plastics coating. The process operates on the basis of an initially very intensive intermixing of air, water and coating, with a subsequent separating stage (Figure 7.9).

The mixing together of the air laden with coating particles and the water from the booth is a technically complex aspect of wet washing. A number of different systems are available, some with and some without the use of wall trickling. The most effective system is based on the *venturi principle,* in which an especially rapid and finely divided vortexing occurs at a die-shaped constriction as a result of rapid acceleration and turbulence, followed by a release zone where the components separate.

Coagulation

The purpose of the subsequent coagulation stage is to "unstick" the coating sludge and to separate it from the surrounding liquid phase, whereby the coagulant and any defoaming agent used must be precisely matched to the coating system. There are basically three known methods of coagulation, the first of which (the physical method) is increasingly gaining in importance for technological and ecological reasons:

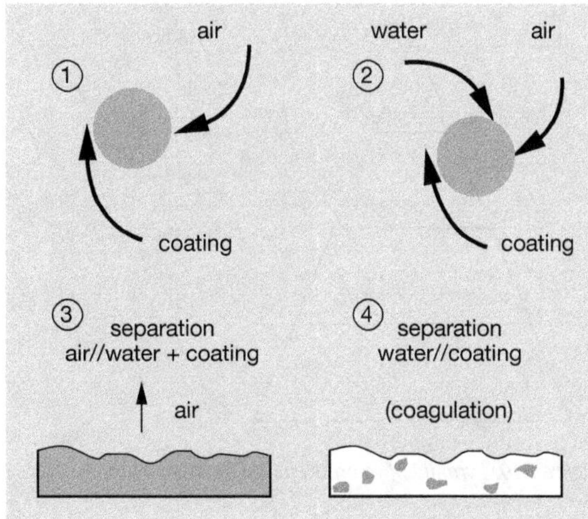

Figure 7.9: Principle behind wet coating separation with coagulation
1 = removal of coating mist by directed air flow,
2 = wetting and bonding of coating droplets by intensive contact with water,
3 = collection of booth water,
4 = coagulation and unsticking, followed by coating sludge removal and dewatering

- Physical coagulation by adsorbents and fillers, e.g. phyllosilicates. Waste water is pH neutral and low in ions.
- Chemical coagulation by addition of acids or bases (denaturation of the binder in particular) or of polyvalent metal salts, e.g. iron(III) or aluminium salts ("breaking" of dispersions, especially emulsions).
- Electrocoagulation in a coagulating tube, by means of electrostatic influences without the use of chemicals; its action is highly controversial, however.

Waste water

Before reusing the purified waste water or returning it to the supply system, it is essential to check that the pH value, the COD[1] value, the solvent and heavy metals content, etc., comply with local *discharge guidelines*. The local water supply company or water authority should be consulted.

Recycling of coating sludge

Depending on the coagulation behaviour of the coatings used, the coagulate may float on the surface or may undergo varying degrees of sedimentation, with differing levels of bonding in the process. In all cases it is useful for the coagulate to be promptly (and automatically) removed and skimmed from the sedimentation vessel by means of

- centrifuges / decanters,
- surface / bottom scrapers or
- flotation vessels with overflow.

Finally the coating sludge produced as a result is dewatered as far as possible by means of centrifuges or filter presses in order for it to be disposed of or recycled as compactly and cheaply as possible. The associated difficulties are illustrated by means of the following points.

Quantities

The manufacture of coatings generates 1 to 2 % of residual coat-ings and coating waste relative to production quantities. The amount of coating sludge (or coagulate) generated at the fabricator's premises (industrial assembly line coating) is around 20 % (coagulate solids) relative to the total quantity of coatings used[2]! The total amount of coating sludge waste (solids) generated across all sectors in Germany in 2007 was around 200,000 t.[2]

1) Chemical oxygen demand
2) Source: Chemische Werke Kluthe GmbH

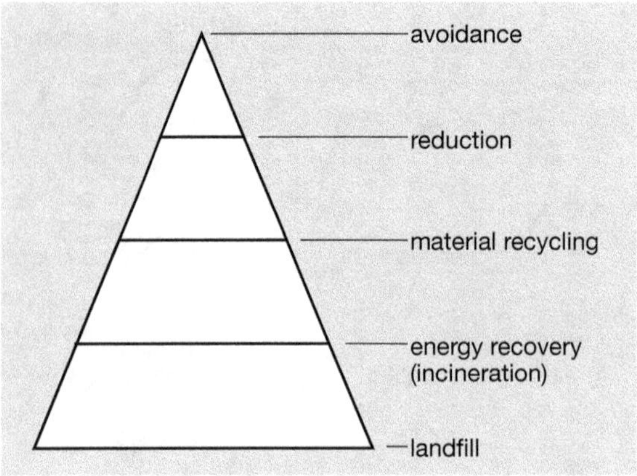

Figure 7.10: Pyramid of priorities for waste treatment

In order to

- conserve resources and relieve the pressures on the environment and to
- save costs in production *and* in disposal (the current cost of disposal of coagulate special waste in Germany is around 0.5 euro/kg, in some cases even higher), the need to reduce the amount of residual coatings and coating sludges generated, following the hierarchical principle illustrated in Figure 7.10 at all times, is already important today and will become even more urgent in future.

Pyramid of priorities

The highest priority in development and use has always been the avoidance or at least the reduction of waste, e.g. by the use of methods of application that generate fewer residues, by optimising processes (greater application efficiency) or by means of more environmentally friendly coating systems. Only when these measures do not succeed in their intended aims (or are only partially successful) should material recycling - preferably still as a coating material - be pursued. See also section 9.3 for information about the treatment and disposal of waste.

Recycling methods

In the industrial processing of one-component materials, *recirculation* of the coating is sometimes possible, particularly where materials are segregated by type and by colour. This can be managed internally by feeding the overspray straight back to the stock of fresh coating material or by recovering the collected material from the sedimentation pool and dewatering it, occasionally also by means of ultra-filtration. By contrast, only a few attempts - external to the plant - have proved successful so far, e.g. the recycling of filler coagulate in automotive assembly line coating.

In principle, coating sludges can alternatively be used outside the coating cycle in new areas of application, e.g. as sealants and insulating materials, etc. For technical reasons this route has so far gained only a limited importance, however. Chemical recycling processes (hydrogenation or oxidative cracking to produce mineral oil-type products) are still at a developmental stage.

Incineration

The (least valuable) method of utilisation by incineration should be chosen only as a last resort. Moreover, coating sludges must not be landfilled because of their high content of organic substances. (Exceptions with regional differents include small quantities from manual or other operations, contaminated filter mats, etc.). See also Chapter 9 for more information about the problem of waste. In the process of "energy recovery" from coating sludge, e.g. as a fuel additive or substitute fuel, at least the chemical energy content is thus utilised. Its calorific value is equal to and in some cases even higher than that of brown coal. Tried and tested methods have been practised sometimes in Germany for some 40 years, e.g. use as an additive in cement production.

Against the background of the Kreislaufwirtschafts- und Abfallgesetz (KrW-/AbfG) [life-cycle management and waste act], which came into effect in Germany in October 1996 and which ranks energy recovery on a level with material recycling under certain conditions, the former is once more the subject of fierce debate. Waste legislation and the waste economy are discussed in section 9.3.

1) Source: Deutsches Lackinstitut, "Sinnvolle Verwertung von Lackrückständen - heute und morgen", 1993

7.12 Removal of coatings

In coating plants there are essentially two categories of components that require removal of coatings:

- curtains, goods conveyors, grids and other accessories that are coated incidentally as part of the application process and
- coated objects that were incorrectly coated or have to be recoated, including large test and demonstration items to enable them to be reused.

Having invested a great deal of effort both in development and in particular during the coating process to obtain good adhesion and resistance in a given coating, it is often all the more difficult to completely expose the substrate again without subjecting it to severe mechanical or chemical damage.

The processes concerned can initially be roughly categorised into:

- chemical coating removal methods and
- physical methods (mechanical and thermal).

The choice of equipment for chemical coating removal ranges from small static units through to fully automated tank installations in the case of toll strippers, with treatment and rinsing baths and waste water treatment.

Chemical coating removal

The three principal stripping media are listed in Table 7.5, which clearly shows how the substrate on account of its individual sensitivities influences the choice of stripping agent. These agents are

Table 7.5: Chemical coating removal processes (with reference to [8])

Coating binder basis	Alkaline, aqueous hot stripping media	Organic coating solvent (anhydrous)	Acid, aqueous stripping media (oxidising, concentrated acids)
coating powders (polyester, PU, acrylate)			
on steel	+++	+	++
light alloy	-	+++	+
galvanised steel	-	++	-
epoxy coating			
on steel	+	+	++
light alloy	-	+	+
cured liquid coatings			
on steel	+++	++	++
light alloy	-	+++	+
galvanised steel	-	++	-
plastic	+	+	+
bath temperature	40 to 90 °C	20 to140°C	20 to 55°C ; in some cases over 100°C
notes	very economical, long shelf life, low emissions	very effective, relatively expensive	low running costs but high plant costs, long shelf life
+++ = very suitable + = suitable with limitations	++ = suitable - = unsuitable		

now also required to display good biodegradability and to be free from halogens, which means that low-temperature strippers such as methylene chloride, which at one time was used because of its excellent dissolving power, are now of very little importance.

Stripping chemicals act in two ways:

- By swelling followed by detachment, usually assisted by mechanical means. The coating detaches in flakes or strips.
- By dissolution (oxidative or saponifying), particularly in the case of non-crosslinked coatings. Once the coating has been completely dissolved, a very fine (pigment) sludge remains on the substrate.

Stripping times can be reduced by operating at elevated temperatures or with ultrasonic assistance.

Physical coating removal

Physical coating removal processes can be subdivided into mechanical and thermal methods. In mechanical processes a coating is removed by abrasives or by grinding; in thermal processes it is

Table 7.6: Physical coating removal processes (according to [8])

	substrate	Area of use	Notes
Mechanical			
grinding	all materials	smaller surfaces repairs	subject to geometry of component relatively expensive
blast cleaning	all materials, but with some restrictions for plastics	for thin coating films after-cleaning required!	subject to geometry of component high wear/expensive
high-pressure water jet	see above	for heavy workpieces	subject to geometry of component high plant investment costs
thermal			
distillationchamber pyrolysis*	all metalls, no plastics	universally applicable	parts must be heat-resistant to approx. 450 °C low waste generation! time consuming
fluidised bed stripping*	see above	for regular, small to medium batches	as above no scoop-shaped parts (removal of sand!) to approx. 500 °C
low-temperature stripping: embrittlement with liquid nitrogen at −196 °C and mechanical post-treatment	see above	especially for high-build coating films on flat parts	high cost of consumables risk of embrittlement of material
"CO_2 stripping"; blasting with dry ice (CO_2)	see above	large surfaces: exterior applications thin coating films	for special applications (e.g. aircraft)
CO_2 laser (short term heating of coating film)	all materials	for sensitive materials layers can be removed one by one	see above

either decomposed by heat or made brittle at extremely low temperature and then loosened mechanically from the substrate. A summary of the various methods is provided in Table 7.6.

For heat-insensitive workpieces, good results have been obtained with oxidising, nitrate-containing fused salts that can clean very thoroughly at 300 to 500 °C. However, these are fairly costly.

7.13 Quality management, process safety and quality assurance

For some 25 years now the focus of commercial activity has no longer been simply the production of "quality" (= optimum fitness for purpose) but primarily also the control and assurance of quality by means of appropriate quality assurance systems.

Quality management or **QM** essentially refers to all organised measures aimed at improving products, processes and performance of any kind. QM is a key management responsibility. Quality management systems are mandated in such areas as aerospace, medical, health-services, pharmaceuticals and food production.

Objectives

With such a system - introduced, certified and inspected by specialist service providers -

- the risk of providing defective products or services not in accordance with requirements can be reduced,
- complaints or demands for replacements can be handled more effectively,
- effectiveness and efficiency can be increased and internal processes made more transparent,
- competitiveness can be assured,
- product or service inspections on the part of the customer can in some cases be avoided or reduced
- customers can often regard the suitability of suppliers as a given and
- contact between customer and supplier can concentrate on the product and service.

The prerequisite for achieving quality is a clear statement of the conditions of supply, i.e. the required product properties or the services to be provided, between customer and supplier, including the tolerances accepted by both parties.

Management tasks

The following key steps must be complied with in the introduction of a quality management system (QMS):

- nomination by the company management of a quality manager with the appropriate skills and qualifications. His or her principal task is the
- production of a QM manual that regulates all of the responsibilities and procedures involved in the operational process, from order processing through production and release to the handling of complaints and the
- motivation and training of employees, and finally the
- enshrining of "quality management" not only in the IT system but in the consciousness of the company management and of all employees.

Certification

The development of a certified quality assurance system involves organisational procedures and continuous internal and third-party inspection across all areas of the company. The **EN ISO 9000** series is a collection of standards that enable the principles of quality management measures to be documented. Together they form a coherent set of standards for quality management systems aimed at facilitating mutual understanding at national and international level.

All products have to meet different specific requirements and therefore can only be generated by applying specific quality *assurance* measures. In contrast, quality *management* systems are not product-specific and can therefore specify a similar structure across all sectors and for all products.

Organisations need to be managed and run systematically and transparently if they are to thrive. One path to success can lie in introducing and maintaining a management system which focuses on continuous improvement and takes account of all stakeholders' requirements. Aside from other management skills, managing and running an organisation requires quality management.

The EN ISO 9000:2000 series of standards are process-oriented. The predecessor standards defined 20 elements of quality management which corresponded to standard processes within the manufacturing industry, from development to production and assembly to customer service, such that the structure of the ISO 9000:1994 ff. hampered its transferability, for example, to service companies.The method by which these organisational actions are to be established and proven is set out in national and international standards, e.g. in **ISO 9000** to **9004**. ISO 9000 contains fundamental information relating to the use of the series of standards, the "instructions for use". Depending on the nature of the company's business and the type of processes under consideration (= quality elements), the system can then be certified to ISO 9001, 9002 or 9003.

Following the internal audit (inspection of all processes of relevance to the quality management system), the external audit results in certification provided that all of the requirements of quality management are complied with. Industrial safety and environmental management can be integrated into the quality management system, in the same way as human resource planning, since timely and appropriate (further) training of employees can have just as much influence on quality assurance as the reliability of a raw material or the production design, for example.

Literature is available that provides a more detailed introduction to the subject of quality assurance [4, 13].

Sources and references for Chapter 7

[1] A. Brasholz: Handbuch der Anstrich- und Beschichtungstechnik, Bauverlag 1989

[2] K. Dören, W. Freitag, D. Stoye, Wasserlacke: Umweltfreundliche Alternative für Beschichtungen, TÜV Rheinland 1992

[3] G. Fettis, Automotive Paints and Coatings, VCH Weinheim 1995

[4] A. Goldschmidt, H.-J. Streitberger:, BASF-Handbook: Basics of Coatings Technology 2nd edition, Vincentz-Network, 2007

[5] Company brochure published by Herberts/Standox: "Entsorgungsleitfaden für Lackierbetrieb", Wuppertal 1992

[6] H. Kittel, Lehrbuch der Lacke und Beschichtungen, Verlag W.A. Colomb, Berlin-Oberschwandorf 1975-1979

[7] Lexikon der Anstrichtechnik, Vols. 1 and 2, Callwey Verlag Munich 1994 and 1995

[8] D. Ondratschek (Ed.): Jahrbuch für Lackierbetriebe 1998, 55th edition, Vincentz Verlag Hanover 1997 and earlier editions (up to 1997: Taschenbuch für Lackierbetriebe)

[9] H. Pecina and O. Paprzycki: Lack auf Holz, Curt R. Vincentz Verlag Hanover 1995

[10] VdL-Leitlinien "Umwelt, Gesundheitsschutz und Sicherheit", Frankfurt 1995

[11] U. Zorll (Ed.): Römpp-Lexikon "Lacke und Druckfarben", Thieme Verlag, Stuttgart 1998

[12] Further recommendations: see [13] and [14] in Chapter 6

[13] H.-J. Streitberger, K.-F. Drössel, Automotive Paints and Coatings, Wiley-VCH, 14v edition, 2008

8 Test methods and measuring techniques

The quality of a coating material - and subsequently that of the finished coating system - lies in its serviceability and/or suitability for the customer's requirements, namely its

- reproducibility from one batch to the next
- with consistent (storage) stability.

Consistently good processability takes precedence here over the obvious requirement of practically identical material composition, which can be very complex. Even if an identical composition is assumed, such a batch can still exhibit fundamental differences from the previous delivery in terms of its processing or resistance.

As a consequence, traditional physical or chemical analytical methods are not particularly informative when it comes to quality assurance or to the suitability testing of coatings. It is much more important to test the material under the application and drying conditions, environmental influences, mechanical stresses, etc., that correspond to the subsequent conditions in reality. Naturally, one must try to accelerate the testing processes or to perform them as a simulation in cases where a strict recreation of the actual situation would simply involve too long a wait for the results or where the particular influences cannot fully be reproduced on a laboratory scale.

8.1 Rheology and rheometry

8.1.1 Rheological principles

Of the many rawmaterial- an coating properties that influence the final coating result, one of the most important (even in the case of powder coatings) is of course the flow behaviour of the system, which to a large extent governs the behaviour of the coating and the effort required for its processing and application. Unfortunately, flow behaviour cannot as a rule be described by a single characteristic, such as viscosity for example, since the viscosity of pigmented systems - and also of clear coats in some cases - generally alters with the level of (mechanical) stress and/or with the length of time for which the stress is applied.

As the property of a liquid substance, viscosity expresses the manner in which or the resistance against which it can flow under the influence of force.

Flow curve diagrams

When ideal (Newtonian) behaviour prevails, flow behaviour is defined solely by the *dynamic viscosity* η, which is obtained from the following relation, the term "viscosity" and its determining quantities τ and γ being clarified in Figure 8.1[1)]:

$$\eta = \frac{\text{shear stress } \tau}{\text{shear rate } \dot{\gamma}} \; [\text{Pa} \cdot \text{s}]$$

If the shear rate $\dot{\gamma}$ (which for historical reasons is also represented by the symbol D) is plotted against the shear stress τ required for the flow process or vice versa (as is even more commonly the case), then for *ideal (Newtonian)* liquids a straight line passing through the origin is obtained as

Brock, Groteklaes, Mischke: European Coatings Handbook

ISBN 978-3-86630-849-7

1) is the derivative of shear deformation γ as a function of time

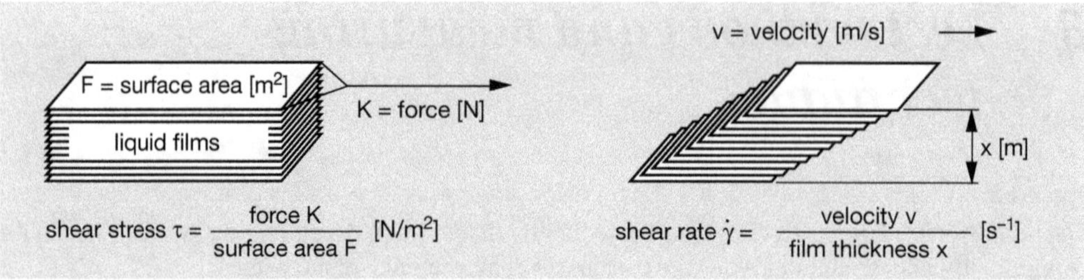

Figure 8.1: Clarification of the term "viscosity"

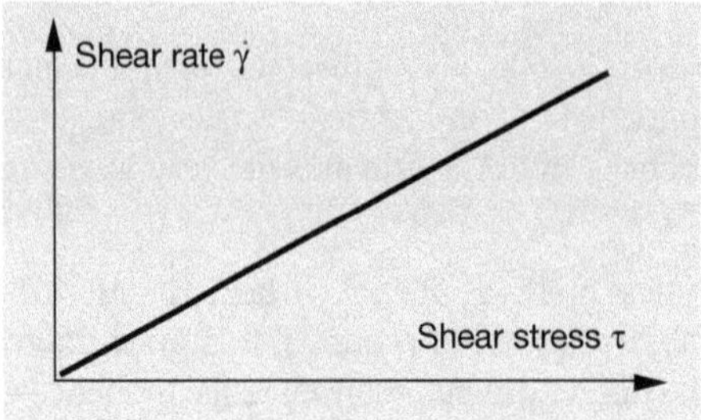

Figure 8.2: Flow curve for a Newtonian liquid

the *flow curve*, as shown in Figure 8.2. The viscosity is therefore a constant in this case and in the graph of shear stress against shear rate it is equal to the gradient of the straight line.

By contrast, non-linear flow curves are frequently obtained for pigmented systems and sometimes also for clear coatings (particularly in the case of dispersions/emulsions), which may but do not necessarily have to pass through the origin. In such cases the shear rate is therefore not proportional to the shear stress, and the viscosity changes as the shear stress or the shear rate increases. The flow curves for *pseudoplastic* (or *shear thinning*) systems are illustrated in Figure 8.3.

It is clear from this that - as is often the case with coatings - in order to be able to describe and have a better understanding of the flow behaviour the yield point needs to be introduced as an additional quantity (see below).

Figure 8.4 shows an alternative means of illustrating flow behaviour: the viscosity curve, i.e. the change in *viscosity* with rising shear rate or rising shear stress (e.g. increase in stirring speed or stress during spraying, etc.).

The reduction in viscosity in a pseudoplastic coating can therefore be explained in that

- on the one hand the flowing particles (pigment particles with binder shells, for example) are oriented, deflocculated and/or deformed under the shear stress, causing their active cross-section and hence their flow resistance to be reduced,
- on the other hand skeleton structures (*"gel"*) between pigment particles and/or binder molecules, depending on energy supply and the strength of the skeleton, are gradually broken down (to a *"sol"*).

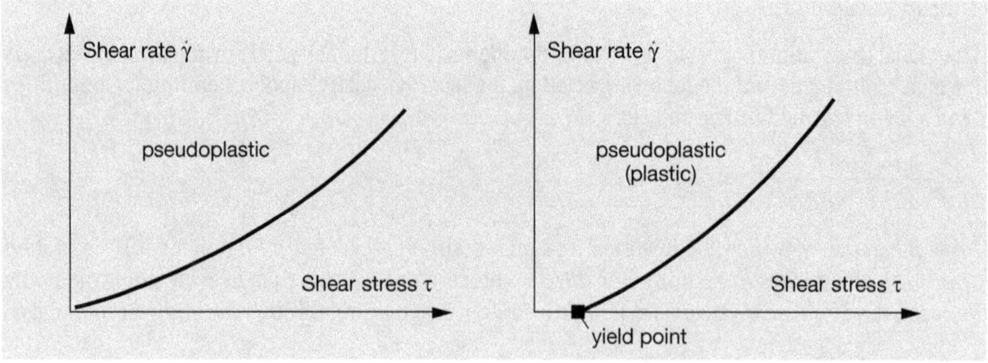

Figure 8.3: Flow curves for a pseudoplastic system without and with a yield point (plastic behaviour is characterised by a yield point)

The *yield point* can be understood as the *precise* force required to induce the system to flow in the first place (see Figure 8.3). This condition occurs when the strength of the skeleton or gel is very great, for example in coatings with an elevated pigment concentration, high flocculation and/or pigment particles with a pronounced needle shape. If the shear is too slight, the skeleton remains stable; the system only begins to flow (plastic behaviour) when a certain force threshold (= yield point) is passed. As the energy supply increases, the viscosity continues to fall, corresponding to the progressive breakdown of the skeleton, until the limiting value - the final viscosity - is reached.

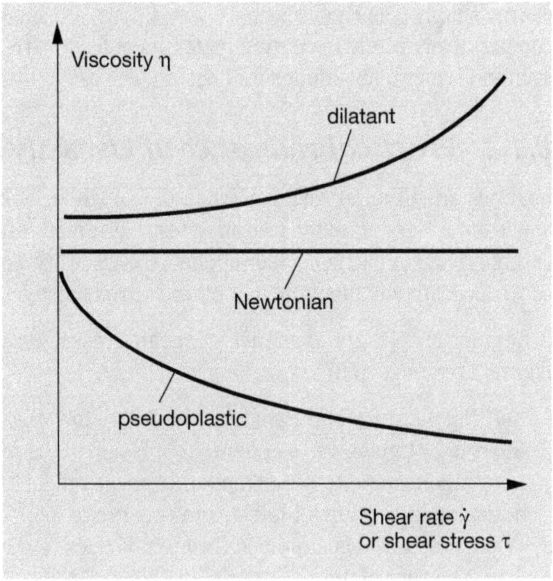

Figure 8.4: Viscosity behaviour as a function of stress

Examples of liquids in the coatings sector exhibiting pseudoplasticity and a yield point include

- pigment concentrates
- non-drip decorating paints ("solid colour")
- coatings before or during the action of shear stress in the atomiser nozzle.

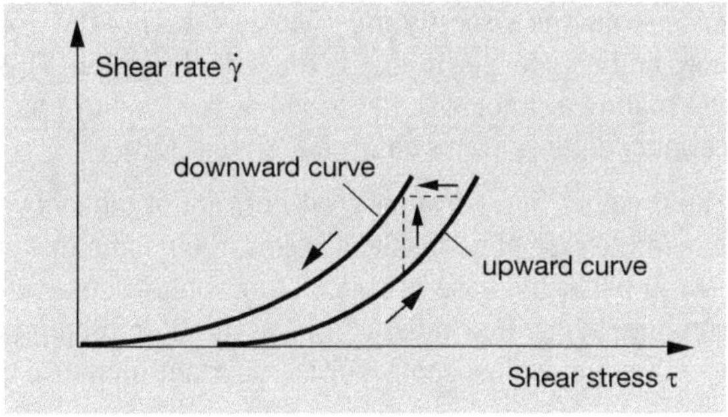

Figure 8.5: Thixotropic flow behaviour

Another feature of flow, *thixotropy*, must be clearly distinguished from pseudoplasticity and plastic behaviour. Whereas pseudoplasticity refers to the reduction in viscosity with increasing mechanical stress, thixotropy signifies the reduction in viscosity with increasing *duration* of stress under a constant shear rate (see ← arrow in Figure 8.5) or under a constant level of stress (↑).

If in order to determine the flow behaviour of a thixotropic substance the shear stress is gradually increased ("upward curve"), by increasing the speed of the rotary element in a *rotational viscometer* for example (see below), then the flow curve of a plastic substance is initially obtained. If the shear stress is then maintained for a certain time, a reduction is observed in the shear stress that is needed to maintain a given shear rate, i.e. the viscosity continually falls (horizontal dotted line in Figure 8.5). A gradual reduction in the shear stress then produces a flow curve that does not coincide with that obtained under *rising* shear stress, labelled "downward curve" in Figure 8.5. In the same way, when a given shear stress is held constant, the resulting shear rate can be observed gradually to increase (vertical dotted line in Figure 8.5).

The area enclosed by the upward and downward curves and by (one of) the dotted lines can be regarded as a measure of the *thixotropic effect*. Thixotropic systems are also pseudo-plastic as a rule and generally display a yield point.

There are also other special features of flow behaviour, such as *rheopexy*. This is the opposite of thixotropy: as the stress duration increases - under a constant stress level - the viscosity rises. Another is *dilatancy*, which is the reverse of pseudoplasticity: the viscosity rises with increasing energy supply (see Figure 8.4); it is also known as "shear thickening". Rheopexy and dilatancy occur comparatively rarely in coating materials at least, although dilatancy can occasionally be observed in high-pigment, low-binder mill bases.

8.1.2 Practical relevance of viscosity behaviour

Coatings are either at rest, e.g. when in storage, or being moved, e.g. when being stirred or during application. The state of the returning rest, e.g. directly after application, must also be taken into consideration. If Newtonian behaviour applied, the viscosity would be the same in all three phases. However, as we have already mentioned, it often depends instead on the shear stress at any given time.

A number of relevant examples of shear stresses on coating materials are given below (see also Figure 2.59, page 173):

1. Settling out, flow and running under very low shear rates (0.001 to 1 s^{-1}).
2. Stirring by hand gives a general impression of the material, e.g. "thick, thin, viscous", etc. These perceptions correspond to moderate shear rates (1 to 1000 s^{-1}). Brushing, flow coating, dipping and curtain coating all fall within this range too.
3. When a material is atomised through nozzles (spraying), the shear rates are then increased by 1 to 2 powers of ten.

A fundamental requirement of paints and coatings is that they must flow well, but not sag. This means that the viscosity must increase again after application; not too quickly, in the interests of flow, and not too slowly, due to the risk of sagging. This represents a real optimisation problem for the coating technologist and has to be resolved by appropriate pigmentation and/or the addition of certain additives (see section 2.4.3, page 173).

The dividing line between good flow and a tendency to run (curtaining, sagging) is particularly fine in the case of high solid systems; their sometimes extremely high binder and pigment content - with slight increase in solids after application - requires a strong thixotropic effect to prevent running, whilst flow must still be assured. In dipping coatings and in paints applied by brushing too, thixotropic flow behaviour is important (non-drip decorating paints).

A further emphasis of rheological measures lies in the prevention or slowing down of sedimentation (settling out), especially of inorganic high-density pigments in low-viscosity coating systems (e.g. white coatings). A distinct pseudoplasticity is important for the formation of the special effect in metallic coatings, especially water-borne variants, preventing turbulence in the still wet coating film.

In addition to pure flow behaviour, elastic features (normally typical of solids) are also frequently found to a varying extent in liquid coatings. Together these are also known as "viscoelastic behaviour" (→ 8.1.4). In section 2.1.1 the nature of viscoelasticity and its effects were discussed in detail, predominantly in relation to solid polymeric materials. However, viscoelasticity is also important for good flow, in water-borne coatings for example, and generally also for a good formation of metallic effects.

8.1.3 Measuring flow behaviour

Given the dependence of flow characteristics on shear stress, as described above, the fundamental principle in selecting the appropriate measuring technique (*viscometer*) is that the measurements should be conducted under the same (shear) conditions as those to which the coating will subsequently also be exposed in the critical phase (application, flow). A coating for application by flooding or a decorating paint need not be tested under the same intense shear as a coating intended for spray application, for example.

Nevertheless, it is often possible or even necessary to dispense with complex and expensive viscometers in practice. In the manufacture of materials, for example, whose flow behaviour is well known, and also in the manual coating sector, partial information and single-point measurements are often sufficient to enable the application viscosity to be adjusted. Simpler measuring systems such as flow cups (see page 368) can then be used.

Rotational viscometers

Types of instruments

In the category of high-performance instruments, one of the best and most versatile viscometers is the rotational viscometer which - depending on the type of coating, viscosity range and measuring requirements - is avail-able in the following variants:

- plate-plate systems (Figure 8.6)
- cone-plate systems (Figure 8.7) - with constant shear rate v/x across the entire test medium range
- cylindrical coaxial systems (Figure 8.8).

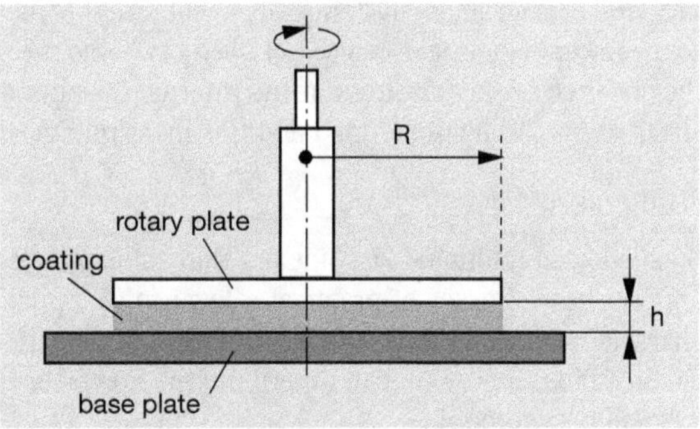

Figure 8.6: Plate-plate rotational viscometer

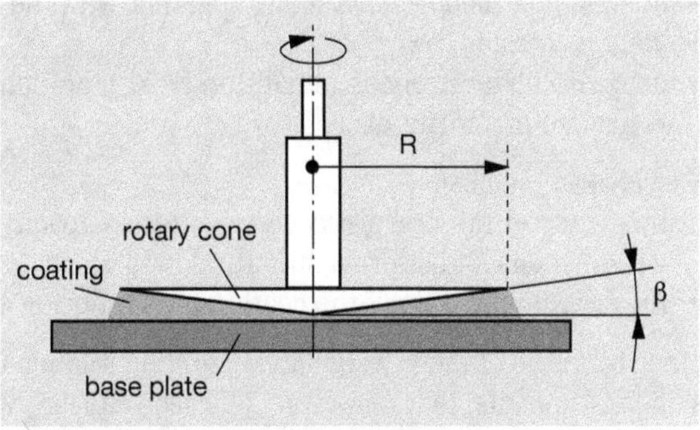

Figure 8.7: Plate-cone rotational viscometer

Instruments of this type can be used to measure absolute values for viscosity [mPa·s] rather than merely relative values (see below), as is the case with simpler instruments.

The following relations apply in respect of the unit in which dynamic viscosity is expressed:

$$1\ \text{Pa·s} = 1000\ \text{mPa·s} = 1\frac{\text{N}}{\text{m}^2}\cdot\text{s}$$

i.e. for the definition of a "Pascal second" (as it is usually known in layman's terms). Incidentally, 1 mPa·s corresponds to the earlier centipoise unit (1 cP), which is useful since it allows numerical data for viscosity values - from old sets of tables, for example - to be adopted unchanged.

The conditions for absolute viscosity measurements as referred to above are:

- laminar flow in a defined shear aperture (corresponding to Figure 8.1)
- stationary, constant flow state
- homogeneous liquid remaining unchanged during the measurement
- no significant elastic component in the sample.

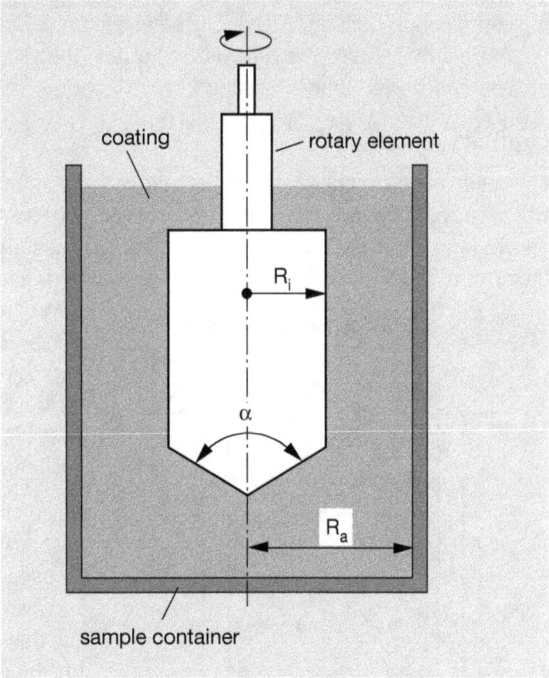

Figure 8.8: Cylindrical coaxial measuring system to DIN 53019

The requirements in respect of equipment are best met by the cylindrical and plate-cone rotational viscometers (Figures. 8.7 to 8.8; these diagrams also indicate some of the principal apparatus constants). A further distinction is made between the following equipment variants:

In *Searle* measurement systems, the shear stress is measured at the rotor shaft; the measuring cup or the lower plate is stationary. In *Couette* systems the measuring cup or the lower plate rotates and the shear stress is measured at the internal cylinder or at the cone/upper plate. Alternatively the shear stress can be predefined and the shear rate measured in both cases.

Measuring technique

Sophisticated rotational viscometers that subject the sample progressively to all of the shear rates and/or shear stresses of practical relevance ("simulation") are especially versatile. A meaningful characterisation can thus be made of virtually any situation that can arise during the application of coatings. In this way we can determine the flow behaviour of

- ideal (Newtonian)
- pseudoplastic and thixotropic and also
- dilatant and rheopectic systems, together with the
- yield points and,
- using oscillating probes (small torsional vibrations), also the elastic component of the flow behaviour of a material.

We can also establish

- by means of the creep/recovery test: the capacity of the (liquid) coating to compensate for deformations and internal structural breakdown following mechanical stresses,
- the crosslinking rate of the coating binder during and at the end of its pot life.

Since by virtue of their design and operation rotational viscometers are sophisticated and usually static instruments, they can rarely be used "on site", which means that frequently they cannot be utilised during the preparation for application. *Immersion* measuring instruments (hydro-meters) - likewise based on the rotational principle - offer a compromise for measurements taken during the course of production. Although they have no defined shear aperture and strictly speaking can therefore only be used for Newtonian liquids, they are perfectly adequate for acceptance tests or process controls on known products.

Capillary viscometers

The other main practice-related alternative for on-site measurements is the capillary viscometer, which is available in various designs according to the degree of precision required. The most well-known (and simplest in design) is the *flow cup*, usually according to ISO 2431 (Figure 8.9), e.g. in the form of the Ford or ISO standard cup, although here the "capillary" is reduced to a hole in the lower end of the cup. Given its relatively wide opening by comparison and the flow ratios within the cup, it can only be used to measure relative values for the viscosity rather than absolute values, i.e. it is possible only to state the "drain time" (in comparison with other samples).

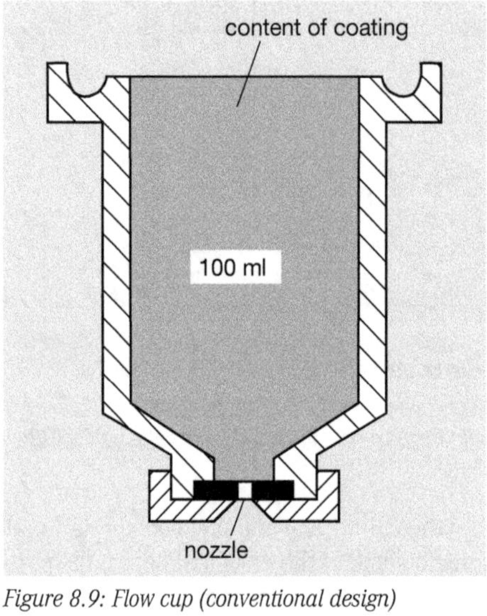

Figure 8.9: Flow cup (conventional design)

Since the shear stress falls as the material drains out, strictly speaking such cups can likewise only be used to measure Newtonian liquids. In daily use, however, for adjusting spraying viscosity for example, the flow cup delivers perfectly useable results even for slightly pseudoplastic water-borne coatings or for conventional base materials.

However, attention is drawn to the scope of validity mentioned in the corresponding standards. Overly long or short flow-out times are misleading and cannot be used!

The design of the *Ubbelohde viscometer* (Figure 8.10) is the reverse of that of the simple flow cup described above. The high ratio of capillary length to capillary diameter gives rise to a prevailing laminar flow, and very accurate viscosity values can be obtained for low-viscosity, Newtonian liquids (diluted binder solutions). Since the density of the liquid plays a part in this method, it can be used to determine the *kinematic viscosity*, according to the following relation:

$$\text{kinematic viscosity } \nu = \frac{\text{absolute viscosity } \eta}{\text{density } \rho} \quad [\text{mm}^2/\text{s}]$$

The kinematic viscosity can be measured by the time taken for a certain amount of liquid to flow through the capillary or, more precisely, the time taken by the meniscus of the liquid in the reservoir to drop from mark M_1 to M_2. In addition to the Ubbelohde viscometer, other similar capillary viscometers based on the action of gravity or force are available for special applications.

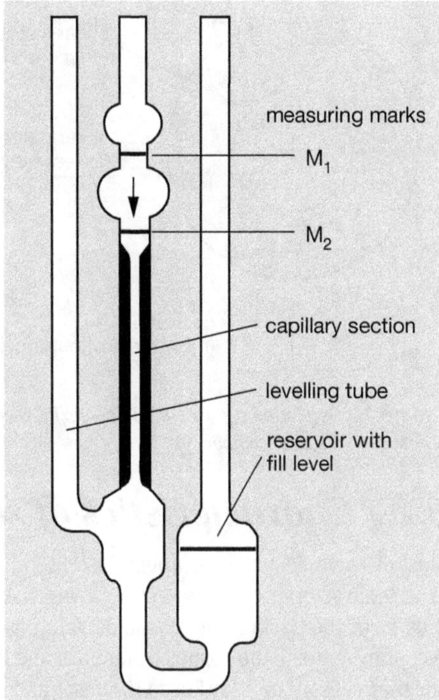

Figure 8.10: Ubbelohde viscometer, DIN 53177

Other viscometers

A number of other types of viscometer are described briefly below:

The *Höppler viscometer* measures the time taken by a ball to travel a given distance in an inclined tube filled with the viscous liquid. This method is particularly suitable for more highly viscous, transparent liquids (binder solutions, printing ink varnishes).

The *ball draw viscometer* operates on a similar principle. In this case, however, the ball is pulled upwards by a rod with a given force and can therefore also be used for non-transparent liquids.

In the *Laray viscometer* (which is frequently also used for printing inks) a rod coated with a (highly viscous) liquid has to slide through a narrow aperture. Here too the time taken for it to travel a given distance is measured.

8.1.4 Viscoelasticity

In section 2.1.1.10 it was explained that a coating film should not only be regarded (strictly speaking) as a material with elastic behaviour but that it also displays features of viscous liquids, in other words it should actually be classed as a viscoelastic liquid. The same characterisation can or must be made in the case of liquid coating materials too. If the coating displays a relatively high elastic component in its flow behaviour, however, although unexpected difficulties may occur during application this behaviour can sometimes also be entirely desirable. Too high an elastic component may lead to problems as the material leaves the spray nozzle or can give rise to a textured surface, whilst if it is too low the consequence may be an unsatisfactory effect formation in base coats, for example.

In order to determine the viscoelastic properties of a liquid, and hence also of a coating, the material is subjected to a sinusoidal oscillating deformation (see Figure 8.11), for example in the coaxial cylinder system described above and known as a "rheometer", i.e. a special design of rotational viscometer. The shear deformation γ depends here on the deflection and the geometry of the aperture.

The shear stress signal τ resulting from the shear γ is likewise sinusoidal but may also exhibit a phase shift δ in comparison to the standard signal. The possibilities are as follows:

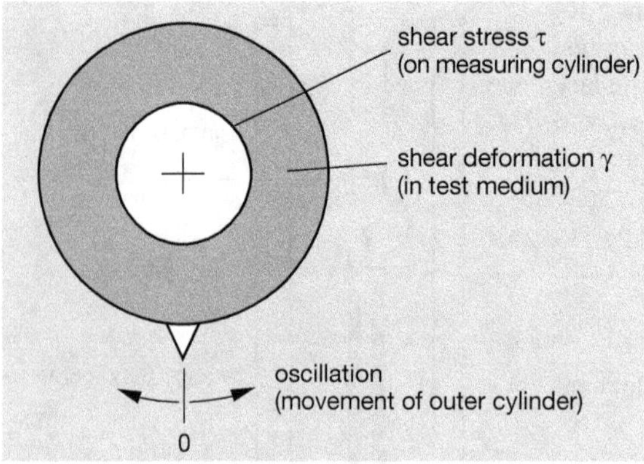

Figure 8.11: Oscillating deformation in a coaxial rheometer to determine the elastic component

$\delta = 0°$
→ perfectly elastic liquid

$\delta = 90°$
→ purely viscous liquid

$0 < d < 90°$
→ viscoelastic liquid

The angle of phase shift d is hence a measure of viscoelasticity.
The literature should be consulted for further information about the theory of viscoelastic behaviour and the methods used to determine it, including the creep/recovery method, for example (e.g. [3, 13]).

8.2 Characteristics of solvents and liquid products

Most of the measured variables described here - which in the industry tend simply to be referred to as "characteristics" - are used in everyday production to identify a particular technical grade and thus form the basis for a yes/no decision as to whether an incoming product or one awaiting final acceptance meets the required standard and lies within the (sometimes broad) acceptance tolerances. For that reason, in addition to the tests and characteristics that have already been described (most of which are also laid down in standards), there are also numerous variants and alternatives - originating in technological requirements and the "traditions" of the manufacturer or customer - which lack of space prevents us from covering at this point. In many cases they are perfectly suitable for general usage too and could consequently be included in subsequent standards.

For the reason mentioned above, naturally only the most important of these tests and characteristics can be described below.

8.2.1 Composition and purity of liquids

Gas chromatographic analysis (GC)

This standard method (see also draft DIN 53 174 and draft DIN 55 682) for testing the composition and identity of volatile substances, especially of solvents, is mentioned here only briefly since it really falls within the field of general analysis, instructive literature about which can be found elsewhere [5, 14].

Distillation analysis to DIN 51 751 or ISO 4626

100 ml of a volatile substance are distilled (rate 20 drops/min) and the temperature/boiling point curve plotted at 5, 10, 95 and 100 ml. The boiling curve (change in temperature as a function of the volume distilled) for a solvent has implications for its evaporation behaviour following application of the coating. Boiling ranges are defined as follows (see also section 2.2.3):

low boilers	< 100 °C
medium boilers	between 100 and 150 °C
high boilers	> 150 °C.

Evaporation index (EI) to DIN 53 170

This is another means of assessing the volatility of solvents, in this case by evaporation at 23 ± 2 °C. 0.3 ml of test substance is dropped onto filter paper and the time t for it to completely evaporate is recorded, using diethyl ether as a reference, i.e.:

$$EI = \frac{t_{sample}}{t_{ether}}$$

The classification of solvents is similar to that for boiling ranges:

highly volatile	EI <10
moderately volatile	EI 10 to 35
difficultly volatile	EI 35 to 50
scarcely volatile	EI >50.

The relative evaporation rates from a *coating film* in the process of being formed can fluctuate enormously, however, due to varying solvent behaviour and also to retentions in the binder. The EI thus gives only a first indication of the composition of the solvent blend. To ensure an effective film formation, the EI of the good solvent must always be higher than that of the poor, "false" solvent (→ also 2.2.4).

Refractive index to DIN 51 423

Reference is made to the cited standard and to other relevant literature for details of how to determine the refractive index. Its advantage lies in the fact that in transparent liquids it can be determined very quickly and it is remarkably informative about the purity and concentration of solvents or solutions. Furthermore, the refractive index of pigments (which is determined in a particular way) plays a key role in the hiding power of coatings and in cloudiness phenomena (→ also section 2.3.2.2).

Density to DIN 53 217

Definitions

The density of a coating is defined as

$$\text{density } \rho = \frac{\text{mass m}}{\text{volume V}}, \text{ e.g. in g/ml}$$

The relative density is also used:

$$\text{relative density d} = \frac{\text{density of coating}}{\text{density of water}} \text{ (dimensionless)}$$

Measuring techniques

The simplest method of determining density uses a *pycnometer* (DIN 53 757 and 53 217 or ISO 787 and ISO 2811). The pycnometer vessel is accurately weighed with and without a content of coating. The density measurement is thus based on the evaluation of a difference in weight. The buoyancy of the air must also be taken into account, and this evidently characterises the high degree of measuring accuracy that can be obtained.

Measurement by immersion of an *araeometer* - equivalent to a battery tester - is relatively inaccurate in comparison, but very quick. Measurement with *gamma balls* (or immersion balls) is recommended as an accurate laboratory method of a similar type; here the buoyancy of a ball in the test medium is measured with a balance (Archimedes' principle).

By far the most accurate method of density measurement uses a *torsion pendulum*, which operates on the principle that the frequency of a glass tube made to vibrate (perpendicular to the bend of the tube) (see Figure 8.12) is dependent on the density of the filling medium. The method can also be used to determine binder concentrations accurately (even with only a slight difference in density between binder and

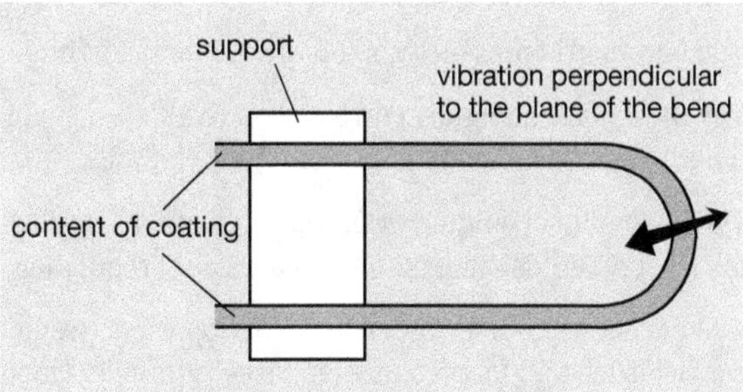

Figure 8.12: U-shaped glass tube as torsion pendulum

solvent). In coating production this means that, amongst other factors, temperature-related density differences (due to seasonal influences, for example) can be determined and taken into consideration in volumetric metering. The *Mohr balance*, which measures the buoyancy of a sinking body of a known volume (suspended from a balance), is regarded as outdated in comparison.

Significance

The densities of conventional coating materials vary in the range from 1.0 to 2.5 g/ml, depending on the content of binder (in the case of clear coatings) and in particular on the content and nature of any pigments and fillers they may contain. It should always be borne in mind that the density can exert a variety of different influences, on sedimentation behaviour (of heavier pigments and fillers), on particle fineness in rotary atomisation and especially also on the yield of a coating for example, and is thus an argument for selling by weight or mass.

Colour value

To evaluate and classify the colour of clear liquids such as clear coatings, drying oils, resin solutions, etc., a visual or colorimetric comparison is made with various liquid or glass standards.

The three most important colour values and the comparative series of standards in each case are as follows:

- Gardner colour value to ISO 4630, with 18 glass or liquid standards; the liquid standards consist of transition-metal chloride solu-tions of various concentrations.
- Hazen colour value (APHA colour value) to ISO 6271, with a colour value range from 0 to 500 based on a concentration series of cobalt chloride/hexachloroplati-nate.
- The iodine colour value to DIN 6162 and 53 403, with a range from 1 to 1100 (corresponding to mg iodine in 100 ml potassium iodide solution).

An approximate conversion can be made with the help of comparison tables.

8.2.2 Safety data

Flash point

The flash point is the lowest temperature at which a test liquid in a closed cup (at 1013 mbar = 760 torr) gives off sufficient vapour that a vapour-air mixture is formed in the cup that is capable of being ignited by an external source of ignition (small flame). It must not be allowed to continue to burn after ignition!

There are two main measuring techniques, depending on the level of the flash point:

- according to *Abel-Pensky* (DIN 51 755 and 53 213) in the temperature range from 5 to 65 °C (heated in a water bath).
- according to *Pensky-Martens* (DIN 51 758) in the temperature range from 65 to 200 °C (heated in an air bath).

The barometric pressure should always be taken into consideration!

The flash point generally correlates to the boiling point of a liquid, but even slight contamination of a liquid can bring about considerable variations.

The flashpoint of a liquid or preparation affects its classification according to the hazardous materials law on liquids subject to German safety regulations (*BetrSichV*):

The *"Verordnung über brennbare Flüssigkeiten"* (regulation on flammable liquids; VbF) has been replaced by the *"Betriebssicherheitsverordnung"* (regulation on industrial safety and health; *BetrSichV*)

The new *BetrSichV* imposes formal changes on installations subject to *VbF* (e.g. stores for combustible liquids). Since the VbF passed out of law on 1 January 2003, such installations are now subject to the provisions of the *BetrSichV*. The previous hazard classes for categorising the combustible liquids, namely A I, A II, A III and B, have been replaced by the terms from the hazardous substances legislation (flammable, highly flammable, extremely flammable). According to the regulations of the BetrSichV, VbF installations previously assigned to class A III are no longer classified as "installations requiring monitoring". Furthermore, some provisions regarding VbF installations have been relaxed: for example, the obligation of disclosure has been replaced by a licence requirement. Furthermore, the threshold quantities have been exempted from the licence requirement.

Ignition temperature

In contrast to the flash point, the ignition temperature is the temperature at which a liquid ignites spontaneously, i.e. without an external flame. The *ignition groups* G I to G V cover the range from 450 to 100 °C.

Explosive limit

Explosive limits refer to the upper and lower *limiting concentration* of (in this case) a vapour, of solvents for example, or also dust (pigments, powder coatings) in a mixture with air, such that in the range between these limits the mixture can be caused to explode by being heated or by the action of sparks.

8.2.3 Application-related data

Surface tension

Surface tension expresses the tendency of a substance (in this case a liquid, hence also a coating) to occupy the smallest possible surface area (for a given volume). It is explained by coherence due to oriented forces inside the liquid. Strictly speaking, the surface tension is the interfacial tension with the air.

Significance for coating technology

Surface tension is of relevance in the coating sector for a number of reasons:

- A low surface tension facilitates the atomisation of a spraying coating into fine droplets (see page 298).
- As a precondition for good substrate wetting by the coating and also for good subsequent adhesion, the surface tension of the coating must be lower than that of the substrate. Otherwise cratering, pinholes, buckling and loss of adhesion may occur (see also section 2.4.2.2).
- Wetting of the pigment surfaces - as the first step in the process of pigment dispersion - is likewise encouraged by a low surface tension in the dispersion medium.
- A high surface tension on the other hand encourages flow due to breakdown of surface waviness.

Additives that can be used to influence surface tension are covered in section 2.4.2.2.

Methods of determination

The surface tension can be determined by means of the *ring detachment method*, for example (a straight wire or a quartz plate are also commonly used) according to DIN 53 914 or 53 593 (Figure 8.13, page 364). The power of resistance of the surface is measured up to the point at which the cylindrical lamella of liquid formed as the ring is withdrawn is detached. The method may not work with pigmented systems since the heterogeneity of the material disrupts the coherence of the developing lamella.

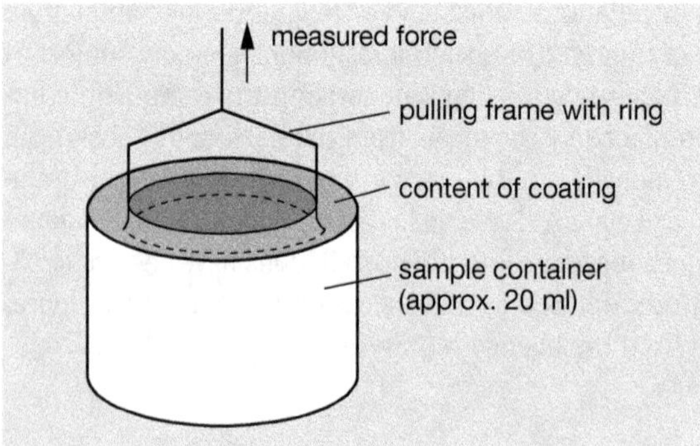

Figure 8.13: Measurement of surface tension by means of the ring detachment method

Another method that can be used to measure the wettability of the substrate in particular, as already described in detail in section 2.4.2, is based on observation of the (equilibrium) *angle of contact* Θ, which is formed at the edge of the droplet when liquid is dropped onto a substrate.

Wetting of the substrate surface, which can just as easily refer to that of a pigment particle as to the surface of a workpiece, is encouraged when the angle of contact Θ is < 90°. Spontaneous wetting (spreading) occurs at 0°.

Fully automated angle of contact measuring instruments have now been developed in which the settling droplets are monitored by video and their dome shape can then be evaluated by image analysis.

One well-defined industrial tech-nique involves the use of *testing inks* of a defined composition and hence also a defined surface tension (σ_{ink}), and is performed as follows. The testing inks are brushed onto the surface to be tested ($\sigma_{substrate}$) and after 3 seconds assessed to determine whether the ink is spreading (if $\sigma_{ink} < \sigma_{substrate}$) or contracting (if $\sigma_{ink} > \sigma_{substrate}$). By the use of various testing inks the likely coatability or (in films) printability of the substra-te, e.g. a plastic surface, can be narrowed down.

In such cases it is only right to make a distinction between

- the disperse (non-polar) component of the surface tension based on the only moderately effective intermolecular dispersion forces with low range,
- the polar component of the surface tension, which stems from hydrogen bonds, electrostatic forces and dipole interactions; which in non-polar plastics in particular is very small but can be significantly increased by means of appropriate pretreatment methods, such as flame treatment (→ 5.4.2) (and as its name suggests is based on the forces due to dipolar effects) and
- the overall surface tension, a characteristic with a marked practical significance, although it is usually used as an indication of coatability or of the wetting power of the coating.

Range of values

In practice the values for the (overall) surface tensions of paints and coatings lie in the range from 25 to 40 mN/m, depending on solvent and binder basis. Conventional coating systems tend to be found at the lower end of the range and hence are "better-natured" in regard to wetting and adhesion, whereas water-borne coatings mainly display a higher surface tension and hence are also likely to give rise to problems with wetting.

The surface tension values for most metals - at least after contact with air, i.e. once displaying an oxide film - as well as those for (clean) glass, for example, are considerably higher than the above range and thus permit good wetting. Zinc and/or zinc-coated surfaces behave more like coating materials, particularly the more non-polar plastics, in terms of surface tension - an indication that wetting or adhesion problems are likely to be encountered unless additional measures are taken. See also section 5.2 in regard to the surface tension of substrates.

Non-volatile matter (nvm), solids content, residue on evaporation
to DIN 53 216 or ISO 1515 and 3251

The non-volatile matter of a coating material or binder remains as a residue after drying (→ 3.2). Temperature and drying period have a significant role to play here, since high-boiling solvents, monomer components, reactive thinners and decomposition products - depending on their retention behaviour - are very slow to leave the film as it forms. It is therefore important to agree and to adhere to defined drying conditions that resemble actual practice as closely as possible. (Examples are listed in the DIN standard).

In order to determine the non-volatile matter, a defined, not too large amount of sample material (0.5 to 2 g) is dispersed in a small metal lid, weighed and then reweighed after drying or stoving. The result of the determination is expressed as:

$$\text{non-volatile matter} = \frac{\text{reweighed amount}}{\text{weighed amount}} \cdot 100\ \%$$

Electrical conductivity

The electrical conductivity of a coating and also of coating raw materials is important

1. for the ease of electrostatic application
2. as a measure of the condition and purity (stability!) of water-borne coatings and
3. for the suitability and stability of electrodeposition coatings.

In regard to 1.: the conductivity must lie within a certain range in order to avoid short circuiting on the one hand and defective chargeability on the other.

In regard to 2: too high a content of electrolytes and hence excessive conductivity can damage the dispersion binder and lead to pigment flocculation. Excessively low conductivity values indicate a deficit of neutralising agent (amine or acid).

In regard to 3.: in electrodeposition coating the rate of deposition and hence the surface quality of the coating are highly dependent on the conductivity. The comments in respect of point 2 also apply here.

Conductivity κ is defined as

$$\kappa = k \cdot \frac{I}{U}$$

where I = current strength, U = voltage, k = form factor or cell constant; it is determined according to DIN 53 779 by means of alternating voltage measurement and expressed in μS/cm. More recent equivalents of the old *Ransburg* measuring cells (defined electrodes, distances, etc.) are used as measuring vessels.

pH value and **MEQ value**

These two quantities are used as special criteria for water-borne coatings. They have very little relevance in conventional coating systems (in terms of their stability and application, at least).

pH value

The **pH value** (= negative decimal logarithm of the concentration of H^+ ions, $pH = -\log(C_{H+})$) is closely linked to electrical conductivity but is not the only quantity that influences it.

As a measure of the concentration of H^+ ions, i.e. for the acid/base behaviour of aqueous solutions in the ranges

- pH < 7: acid behaviour
- pH = 7: neutral
- pH > 7: alkaline, basic behaviour,

the pH value is particularly important for the stability of the dispersion or emulsion binders (see also section 2.1.4.6), but also for that of dissolved film formers, for which the following principles must be taken into consideration:

- If binders or coatings/paints are acid-neutralised, i.e. acid-stabilised, their resistance is usually increased (to a certain limit) due to an additional acid excess, i.e. a lower pH value; a rise in the pH value due to the addition of alkali, on the other hand, leads to rapid coagulation.
- Conversely, alkali-neutralised binders and coatings (neutralised with amines or alkali solutions) can readily tolerate a somewhat higher pH value but definitely not an addition of acid.

The colour intensity and stability of pigments as well as the subsequent weather-proofing and corrosion protection action of the coating system also depend on the correct pH value.

Determining the pH

The pH value is measured in some cases using special pH indicator papers, provided that the coating has no inherent colour to interfere with the process. Usually, however, the more accurate and safe measurement using pH electrodes to DIN 19 261 (and others, e.g. ISO 787) is preferred. In this case, though, it is important always to check the electrode to ensure that the diaphragm (the cell membrane) is clean and to clean it if necessary, and above all to calibrate it regularly with the accompanying standardised buffer solutions. This is because measurement errors in pH determination - leading to adjustment errors at the acceptance stage and to defective batches, to changes in coating properties and ultimately to misunderstandings in the chain linking raw material manufacturer, coating manufacturer and user - are unfortunately a common occurrence and one of the most frequent sources of error in the development, production and application of water-borne coatings.

Even the universal chemical demineralised water (deionised water, distilled water) is occasionally used in a contaminated state - containing hydrochloric acid esters, for example (due to column regeneration) - and can cause a great deal of damage if the contamination is not detected. Checking every batch of demineralised water by means of pH determination can thus likewise avoid considerable loss and expense.

The latest instruments, which are less sensitive to mechanical and chemical influences, use an ion-sensitive field effect transistor to detect the H^+ ions and hence to display the pH value.

MEQ values

MEQ values provide even more reliable and meaningful information about the presence of neutralising agents in the coating or raw material by indicating the actual quantity of neutralising agents rather than merely the pH value arising from their presence which, due to the effect of the buffer, is relatively inaccurate.

The MEQ value is the content of neutralising agents (acid or base) in a water-borne raw material or coating, expressed as the quantity of milliequivalents of the neutralising agent, relative to 100 g of coating solids.

Given the purpose of the neutralising agents as described earlier (see pH value), the MEQ value of a water-borne coating or binder is also a measure of the solubility of the water-thinnable resin component. If it falls below a certain minimum value the consequence will be the irreversible coagulation (precipitation) of coatingcontents. In this context

1 milliequivalent = 1 ml 1-molar acid or base[1].

The abbreviation MEQ_{amine} is commonly used to distinguish amine-neutralised coatings and MEQ_{acid} for acid-neutralised types. Accordingly the MEQ value is calculated as follows:

$$MEQ_{amine,\ acid} = \frac{C_{acid,\ base} \cdot 1000}{m_{coating} \cdot nvm}$$

where $C_{acid,base}$ = consumption of 0.1 m acid or base[1] during titration to the equivalence point in ml, $m_{coating}$ = supply of coating (usually 10 g) and nvm = non-volatile matter in % ("coating solids").

1) More properly expressed as: 1 ml of a 1-molar *monobasic* acid or base (previously: 1n)

The MEQ value can be determined to VDA[1] 621190 (relative to electrodeposition coating) by titration of the electrical conductivity, which changes with the amount of neutralising agents (in other words by conductometry), or from the pH value, i.e. by potentiometry. When this type of measurement is undertaken in practice, an emulsifier must be added to stabilise the binder , as otherwise it coagulates due to a lack of electrostatic stabilisation as the neutralisation point is approached.

8.3 *Analytical values for solids*

Acid value (AV) to DIN 53 402 or ISO 2114

The acid value indicates the amount of KOH in mg that is needed to neutralise 1 g of test substance. It is used in the production of resin to characterise the resin and to monitor the conversion and involves titrating with KOH solution and phenolphthalein as indicator. Anhydrides are first converted to acid with water under the action of heat.

Hydroxyl value (OH value) to DIN 53 240 or ISO 4629

The hydroxyl value refers to the amount of KOH in mg that is equivalent to the amount of acetic acid bonded during the acetylation of 1 g substance (→ 2.1.4.9). After acetylation of the substance with acetic anhydride/pyridine, the remaining excess of acetic acid is back titrated with KOH/phenolphthalein. The OH value is an important characteristic for describing the reactivity of a coating (film former) component with implications for the subsequent coating film properties after crosslinking.

Other values

There are several other values of similar importance that are principally used in the characterisation of coating resins. Due to space constraints they are simply listed below together with the relevant standards, in which the method of determination is described in detail in each case.

Isocyanate value to DIN 53 185
Saponification value to DIN 53 401 or ISO 3681
Iodine value according to various methods and/or standards, e.g. to DIN 55 936.

Dry film density

The density ρ_{dry} of the cured coating film, like that of powder coatings or pigments, is either

- determined according to DIN 53 193 or ISO 3507 by means of a pycnometer (cf. page 361), in which the solid (scraped off if necessary) is made up with petroleum ether, or
- calculated as follows from the composition of the coating, which must of course either be known or determined:

$$\rho_{dry} = \frac{\rho_{coating} \cdot \rho_{solvent} \cdot nvm}{100 \cdot \rho_{solvent} - (100 - nvm) \cdot \rho_{coating}}$$

A certain degree of inaccuracy must be taken into consideration here, since strictly speaking the densities of the components of a mixture are not additive.

In principle the dry film density can also be calculated from the solids volume and increase in mass of a substrate when it is coated; this method is very imprecise and laborious, however.

8.4 *Testing of liquid paints and coatings*

8.4.1 *Optical properties*

Hiding power to DIN 55 984, 55 987 and 53 162 or ISO 2814, 3905, 3906 and 6504

2) VDA = Verband der Automobilindustrie [German automotive industry association] (here: collection of industry standards and specifications). An overview of VDA and other industry standards can be found in [9].

Definitions

The hiding power is the capacity of a coating material to hide the colour or differences in colour of a substrate. It is determined by comparative measurement of the reflection of a coating material on a black and on a white substrate. Black and white glass plates or contrasting cards, for example, can be used as such a substrate.

The hiding power value indicates the surface area (m^2) of such a contrasting substrate that can be covered with a litre of the dry coating material such that a fixed hiding criterion is satisfied (e.g. a brightness of 98 % over black, relative to the brightness over white).

Source

Hiding power as a consequence of scattering power, the fundamental details of the source of which were provided in section 2.3.2.3, depends on the *refractive indexes* n of the pigments and binders contained in the coating film or, more precisely, on the ratio of the various refractive indexes:

$$n_{rel} = \frac{n_{pigment}}{n_{binder}}$$

The higher the value for $n_{rel.}$, the greater the scattering power and hence the greater the hiding power of the pigment. Accordingly, TiO_2 (rutile) with its high refractive index of n=2.75 has the greatest hiding and whitening power of all white pigments. The refractive index of an organic binder (dry) is around n=1.5. A filler such as $BaSO_4$ (blanc fixe) with n=1.64 consequently exhibits virtually no scattering power.

Influencing variables

As well as the refractive index, the particle size of the pigments (degree of abrasion/flocculation!) also has a role to play, the optimum particle size (with the greatest scattering power) being highly dependent on the individual pigment - and specifically also on its refractive index. Up to a certain limiting value of pigment concentration, the level of which depends on the given binder, the number of scattering particles also influences the hiding power, of course.

The hiding power of coloured and black pigments is strongly influenced as a rule by light absorption (*selective* light absorption in the case of coloured pigments). It is further proportional to the concentration of the pigment and its degree of abrasion, at least up to a certain limit, where the hiding power reaches its maximum. A lower hiding power in the coating resulting from the (partial) substitution of the expensive TiO_2 white pigment with a less expensive filler exhibiting a lower refractive index, for example, leads both in theory and in practice to a higher consumption, i.e. the hiding power value and the yield (see below) fall.

Determination

Light absorption and/or light reflection and deflection are therefore critical for the hiding power of a coating. Expressed in physical terms, the absorption coefficient K and the scattering coefficient S determine the loss of light dI as it passes through a film of thickness dx:

$$-dI = (K + S) \cdot I_0 \cdot dx$$

K and S together can be a maximum of 1; I_0 is the intensity of the incident light.

The relation between absorption and scattering, i.e. the quotient K/S, can be determined using the *Kubelka-Munk* equation:

$$\frac{K}{S} = \frac{(1 - R_\infty)^2}{2 R_\infty}$$

where R_∞ = reflection at an infinitely large film thickness (i.e. larger film thicknesses no longer produce a change in the degree of reflection).

The *Kubelka-Munk theory*, from which this relation derives, and which assumes perfectly diffuse radiation distribution on the illuminated side and within the coating, is today the most important and successful basis for the treatment of absorption and scattering processes in the area of pigments and dyes.

The hiding power is usually measured simply and pragmatically by applying a coating of defined thickness onto a high-contrast substrate (usually black and white), called a contrast card. After the coating has dried or cured, the colour difference between uncoated and coated area is measured. One method for measuring the contrast ratio of interior emulsion paints is described in ISO 13300

The thickness of a coating capable of hiding the substrate is determined by preparing coatings of different thickness. The coating thickness above which no further change occurs in coloristic properties (or only to a defined extent) is the hiding thickness. These limits are codified in DIN 55987 and in ASTM D 2805-70, but not in agreement with each other. Thus, in DIN 55987, an overall colour difference of 1 max. is defined as hiding. This applies to coating system which has been applied over a black (reflection value R = 0.05) and white substrate (R = 0.8). According to ASTM D 2805-70, the ratio of the Y values over black and white must be greater than or equal to 0.98. However, the use of just one tristimulus value instead of the overall colour difference fails to adequately account for differences between regions of different hue.

Yield

Yield is defined as the surface area that can be coated by a given amount (mass or volume) of coating or paint in the minimum opaque film thickness d_{min} or in the adequate film thickness for the particular function. Hence

$$\text{yield} = \frac{\text{surface area}}{\text{mass}_{\text{coating}}} = \frac{\text{nvm}}{\rho_{dry} \cdot d_{min} \cdot 100}$$

being calculated from the value for d_{min}, the non-volatile matter nvm (solids) and the dry film density ρ_{dry} (this data must be determined by experiment or taken from the coating formulation). See also p. 186 in regard to the dependence of yield on the formulation.

Transparency

According to DIN 55 988, transparency T indicates the thickness of coating d that can be applied until the overall colour difference value Δ^*_{ab} reaches a value of 1 relative to the substrate; the following relation thus applies:

$$T = \frac{d}{\Delta^*_{ab}(d)}$$

Colour strength

Definition

Colour strength is a measure of the tinting strength of a colorant. A distinction is drawn between absolute and relative colour strength (see also DIN EN ISO 787-24). The colour strength of a pigment is its most important property.

The *absolute colour strength* is the quantity of coloured pigment needed to adjust a defined quantity of white pigment in a specific application medium to a standard depth of shade. This means that the colour strength is not a fixed parameter, but rather essentially depends on the application medium (e.g. paint, plastic, printing ink...) in which the pigment is tested and on the dispersion conditions. It also means that the colour cards issued by manufacturers cannot be compared as they employ different test systems. It is normal practice to quote the absolute colour strength on colour cards and technical datasheets; all the products within a product class will have been dispersed in the same application medium under defined dispersing conditions.

The *relative colour strength* is usually used in the production of pigments, and is measured against a standard. For this, each test batch is dispersed and coloured under the same conditions as the standard. The ratio (samples/standard) of the K/S values (according to Kubelka-Munk), which represents a measure of the absorption of the overall system, is called the relative colour strength and is expressed as a percentage. A figure of 100 percent means that the sample has the same colour strength as the standard. Higher values indicate higher colour strength, a fact which means that less pigment is needed for producing a certain hue. The relative colour strength is quoted for all direct comparisons between two products, e.g. when testing alternative products or for quality control purposes.

Method of determination

In order to establish their colour strength, coloured coatings are therefore mixed with a standardised white coating (see also "white blend"), and white coatings correspondingly with a standardised black coating.

Transparent coatings such as glazes, stains, etc., are often assessed as a transparent film (on a glass plate or card) or, if the material is to be used for the manufacture of special effect colours, mixed with pure silver special coatings, e.g. metallic coatings.

In physical terms the colour strength corresponds to the selective light absorption in the specific wavelength range in which the selective absorption of light by the given pigment is at its strongest. Expressed another way: in the range of the maximum of the spectral absorption curve for the pigment.

Hence:

In colorimetry, only those coatings or samples which have approximately the same colour location, or in layman's terms the same shade or tint, should ever be compared in terms of colour strength. Otherwise the results obtained can completely contradict one's own perception. The eye as a measuring system is superior to any colour measuring device in this respect; it can more easily make an approximate comparison between the colour intensities of different hues (although it cannot supply numerical values, of course).

Colour strength must be differentiated from the constants used in colorimetry (see section 8.6.2), i.e. from the numerical recording and description of a colour location in a colour coordinates system. To determine the colour strength, the white blend is usually spread or sprayed onto a test panel or card. It can be especially cost-effective to perform the first evaluation stage as a wet measurement; for a final evaluation or acceptance, however, a test piece coated and dried according to the conditions of its ultimate application is usually necessary.

Significance

Colour strength is not only important for supply agreements and acceptance inspections. It is also a good means of determining optimum dispersion times and other dispersion parameters, since it increases sharply during the dispersion process (covered in detail in section 4.9) as a result of the growing number of particles, until it reaches a limiting value which thereafter remains unchanged (colour strength development curve or dispersion curve according to DIN 53 238, see Figure 8.14). As a consequence of overdispersion and possible subsequent flocculation, the colour strength may even decrease again over a continued dispersion period.

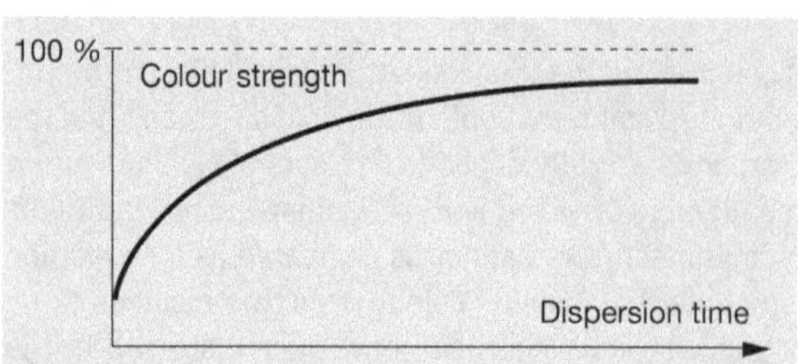

Figure 8.14: Dispersion curve

The slope and course of the dispersion curve are very much dependent on the coherence and wetting ability of the pigment particles and on the primary particle size and shape of the pigments

(→ 2.3.2). Finely divided, organic pigments are usually significantly more difficult to disperse than inorganic pigments containing coarser particles (not to be confused with Mohs' hardness and the associated risk of abrasion). Accordingly, micronised, transparent oxide pigments (iron oxides, titanium dioxide) are likewise more difficult to disperse than conventional, opaque oxide pigments (see section 2.3 in general for further information).

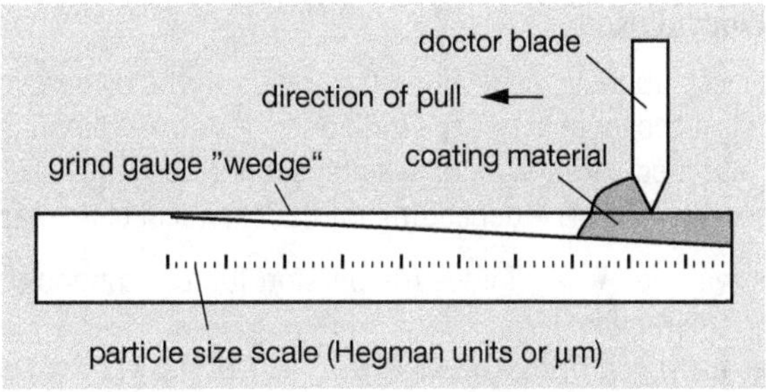

Figure 8.15: Schematic cross-section of a grind gauge with doctor blade

Granularity and particle sizes

The degree of pigment grinding can be simply and rapidly determined by measuring its fineness using a grind gauge according to *Hegman* (Figure 8.15), as described in DIN 53 203 or DIN EN 21 524 and ISO 1524. Strictly speaking it is the granularity, i.e. the maximum agglomerate size, that is measured here and not the average particle size or even the particle size distribution.

The test is performed by distributing a sample of coating or mill base with the doctor blade along the wedge-shaped indentation in the grind gauge. Then the film thickness at which the coating film appears rough due to the emergence of streaky or dotted tracks (where coarser particles push through the surface) is determined by eye and using the lateral scale.

A device known as a "*Garmsen wedge*" is used in an alternative method for measuring particle size. Here, two glass plates held apart by platelets at one narrow end enclose a flat, wedge-shaped cavity. Again, this method also allows a rough assessment of the hiding power to be made (in addition to the granularity), but it is not always successful.

In the case of powder coatings, use of the traditional *screening method* and the *air-jet screen* is still recommended for determining particle sizes.

Electronically assisted analysis of particle size distribution

The modern methods of determining particle size distributions (evaluation according to DIN 66 142) are now indispensable in the development, production and quality control of pigments, binder dispersions and even of coating materials. The most important methods are briefly described below:

- The "*Coulter counter*" measures the number and size of dispersed particles indirectly, by means of the current flow through a capillary. This is typically interrupted when a particle passes through the capillary.
- In *photon correlation spectroscopy* (PCS) - also known as quasi-elastic light scattering (QELS) - the laser light scattered by particles fluctuates due to Brownian motion. The method is suitable for particles in the range from 1 nm to 1 µm, i.e. not for coarser pigment agglomerates or flocculates.
- *Laser light diffraction* at dispersed particles permits the particle size distribution to be determined from the diffraction sample detected. The measuring range in this case extends from around 0.05 µm to 800 µm and hence follows on from the PCS method.

  These three methods can only be used with extremely diluted dispersions, in which the particle sizes can quite easily deviate somewhat from their original state.
- The traditional assessment using a light microscope (possibly with an image analysis system) and *transmission electron microscopy* (*TEM*) are still important, particularly as they provide valuable information about the form and speed of flocculation processes whilst at the same time enabling the various pigments present to be identified.

Flocculation

Flocculation or reagglomeration refers to the congregation of already dispersed pigment particles. It is a consequence of the high energy state of the pigment particles with their large surface areas, combined with inadequate stabilisation. Destabilising forces of attraction (as discussed in detail in section 2.3.2.4) - depending on the chemical character of the pigment surface - include:

- van der Waals forces (dispersion forces, London forces, dipolar interactions, induced dipolar interactions)
- hydrogen bridge bonds
- ionic interactions (Coulomb forces).

Flocculation is usually an undesirable state since the colour of flocculated pigments is less intense than that of their finely divided counterparts and they can occasionally lead to a loss of gloss, display an altered colour location (flocculated iron oxide red has a bluer tint than its finely divided equivalent), or cause flooding. There are some perspectives, however, from which a certain - or selective - flocculation is actually desirable, to prevent hard sediment for example, e.g. in filler-rich preparations.

Assessment of the pigment state

Rub-out test

To reveal flocculation or flooding (visible accumulation of *one* pigment at the surface of the coating, important in pigment blends) in coating films, a rub-out test is performed by rubbing the film with a finger while it is still damp but already starting to dry. In doing so the flocculated or flooded, i.e. separated, pigments are rubbed together and the colour of the homogeneous blend is reinstated by means of a kind of "redispersion". The difference in colour strength from the unrubbed film indicates the extent of the problem. Thus the *rub-out value* RV is obtained by measuring the colour intensities:

$$RV = \left(\frac{CI_{rub}}{CI_{unrub}} - 1 \right) \cdot 100$$

where $CI_{rub.}$ = colour strength of the rubbed surface and $CI_{unrub.}$ = colour strength of the unrubbed surface.

Occasionally the overall colour difference value ΔE (→ 8.6.2) is also quoted to characterise the rub-out effect. The effect can occur even in single pigment systems, e.g. iron oxide red, because important optical properties such as hue and/or scattering power are very much dependent on the current particle size, which is altered by flocculation. If a rub-out test on white/coloured pigmented systems shows the colour strength of the rubbed coated surface to be greater than that of the rest of the surface, we refer to a positive rub-out effect, and in the opposite case to a negative effect.

Other methods of determination

In addition to the rub-out test, which represents an equally practical as it is important method of characterising the dispersed particles, a number of other methods exist, as listed below:

- microscopic assessment (light microscope, scanning and transmission electron microscope)
- IR scattering: determination of the current particle size
- opacity measurements
- photoacoustic spectroscopy
- ATR-IR spectroscopy
- chromatography
- indirect methods: gloss and colour strength measurements.

Settling out (sedimentation) of pigments/fillers

Stokes' law gives the following equation for the sedimentation rate:

$$v = K \cdot \frac{r^2 \cdot (\rho_P - \rho_B)}{\eta}$$

where v is the settling speed, K is a constant, r is the particle radius, ρ_P is the density of the pigment particle, ρ_B is the density of the binder solution and η is the viscosity of the binder solution.

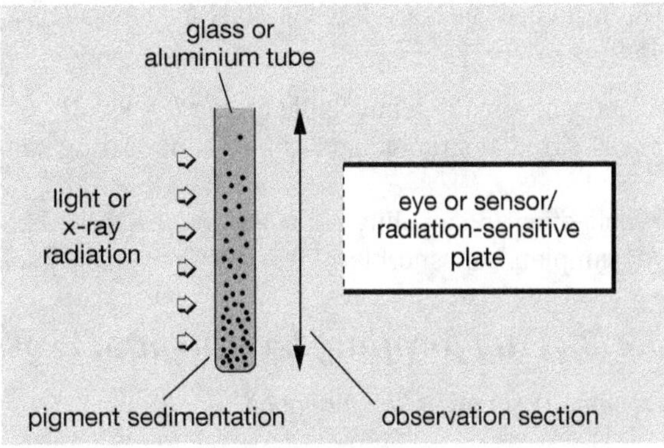

Figure 8.16: Sedimentation analysis under light or X-ray radiation

In this relation we can see:

- the influence of the pigment density: the higher it is, the more quickly the pigment settles out.
- the greater the viscosity of the system, the more slowly a sediment is formed.
- the particle radius is squared. Large, unground or flocculated particles thus sink much more quickly than small ones.

Sediment formation

Agglomerates or flocculates are very large particles and hence they settle out more quickly but form only a soft sediment. Fine particles sink more slowly to the bottom but form a tightly compacted, solid sediment that in extreme cases can be virtually free from binder.

Measurement and characterisation

- The extent and rate of sedimentation can be analysed visually or by means of X-ray radiation, as shown in Figure 8.16.
- The quantity and (indentation) resistance of the sediment can also be determined using a sedimentation balance (e.g. "dynometer").
- Sedimentation can be accelerated by centrifuging to facilitate a rapid separation and examination of sediment or of supernatant binder solution.

8.4.2 Emissions

Given the importance currently attached to solvent reduction and industrial safety, emission measurements and - prescriptive limits are important tools; they are handled very differently from one sector or country to the next, however. Values relating to the emission of solvents ("solv.") can be expressed in a variety of ways, such as:

g solv./m² surface area	g solv./l coating (ready for use)	VOC value calculation (USA) $\frac{mass_{solv.}}{volume_{coating} - volume^*_{water}}$
g solv./nvm$_{coating}$	g solv./kg coating	* for use with water-borne coatings

VOC = volatile organic compounds

As a measure for air pollution, emission values can also be quoted cumulatively (as the total for one day, for example), as average values calculated from many individual measurements, or as single values (maximum values, peak values). It is therefore important to clarify the nature of the measured value in each case.

The principal measuring techniques for emissions analysis are as follows see also ISO 11890, draft:

- determining the non-volatile matter (→ 8.2.3)
- gas chromatographic analysis (GC) of coating samples (applied to aluminium foil, for example)
- collection in sampling chambers (which have been defined as specifically as possible) with "samplers" or "snorklers" (e.g. "Dräger tubes"), followed by GC.

8.4.3 Film-forming, leveling and crosslinking

Preliminary stages of film formation

Leveling and running are a direct consequence of the flow characteristics of coatings.

Leveling is defined as the capacity of a still liquid coating to compensate independently for any irregularities introduced during the application process. *Running* (also sagging, curtaining, tearing, dripping, etc.) is a coating defect and occurs espe-cially on vertical surfaces and at edges and corners.

The slower the progress of (chemical and physical) film-forming after application, the more time is available to the coating for both of these leveling processes. In other words, an inert, unreactive crosslinking agent and a slow ("long") thinner improve leveling but involve the risk of the onset of sagging, i.e. running, as well as of increased dust occlusions.

Measuring techniques

A number of different designs of doctor blade (according to Stieg) are available for testing both of these features. A combined test blade as shown in Figure 8.17 is especially practical.

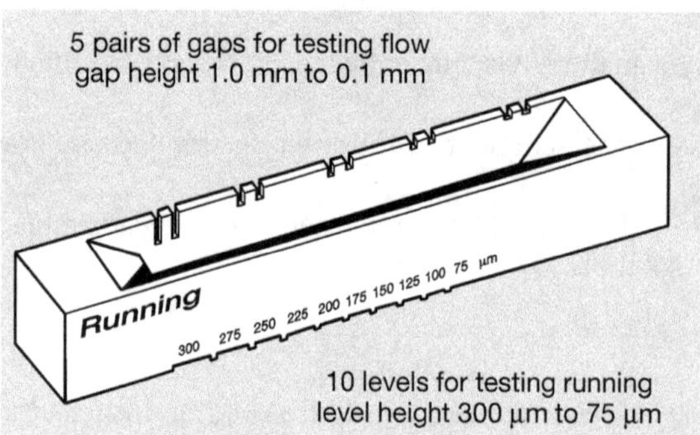

Figure 8.17: Test blade for measuring flow and running behaviour

Leveling is tested by applying the coating with the blade and then reading the gap height (wet film thickness in µm) relating to the pair of stripes between which the gap is still just visible. The closely related "wedge blade" method is used for exterior coatings.

An alternative method that can even be used for continuous testing of leveling during a production process, is the optical *interference process*, by means of which surface irregularities are optically detected, intensified and recorded.

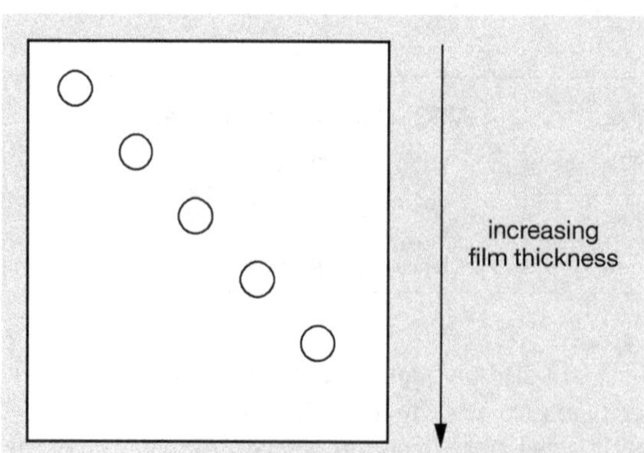

Figure 8.18: Perforated plate for testing running

Running is tested by using the other blade edge of the device illustrated in Figure 8.17 to apply 10 stripes of different thicknesses. The specimen is then immediately placed in a vertical position and the film thickness determined for the stripe of coating that has not yet run into the next lowest (thicker) stripe.

In practice, especially in the case of spray coatings, the tendency to run is often established using *perforated plates* (Figure 8.18), onto which various thicknesses of coating are applied (in the form of a "wedge"). As the film thickness (which is measured) increases towards the bottom of the plate, there is a limiting film thickness below which distinct bead-shaped flow marks are visible at the sides of the holes.

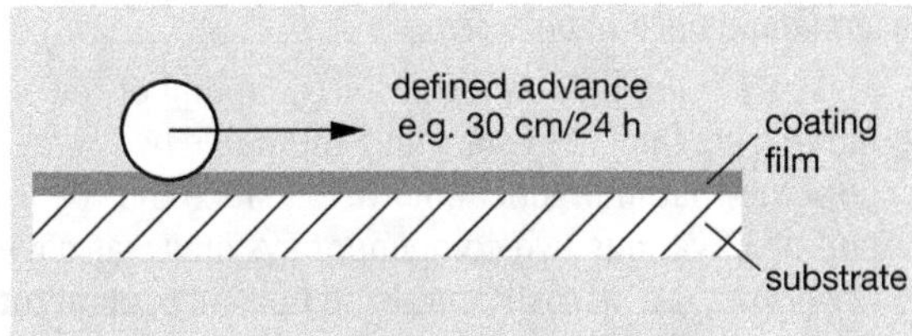

Figure 8.19: Coating drying test with "rolling ball" (or wheel) (according to Rossmann)

Monitoring the film-forming process

Definitions

Depending on the nature of the film-forming process, the coating system and the requirements in respect of speed of further processability, a whole range of tests is available for monitoring the film-forming process. In every case, however, a defined wet film of a known film thickness must first be applied, using a doctor blade for example.

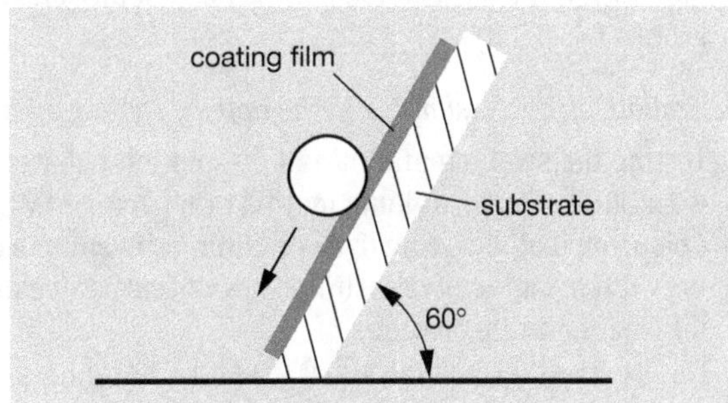

Figure 8.20: Monitoring viscosity with a "rolling ball"

This consolidating film can then be examined in the following ways.

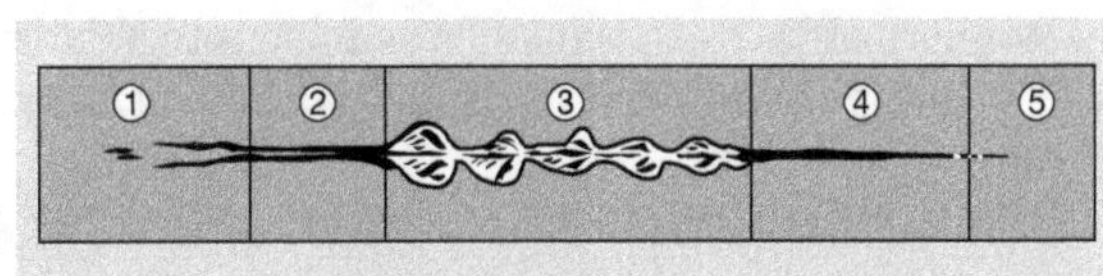

Figure 8.21: Scratch made by a drying time recorder, showing the different drying stages 1 = flow, 2 = base track, 3 = film tearing, 4 = surface track, 5 = dry

- Rolling ball principle according to *Rossmann* (Figure 8.19) to determine dust dryness.
- According to *Wolff and Zeidler* a similar principle can be used to understand the changes in viscosity after application (Figure 8.20): the speed and distance travelled by glass or steel balls moving through a coating film on an inclined substrate serve as a measure for the increase in viscosity.

Determining the drying time

The drying time of coatings can be measured using drying time measuring instruments ("recorders") by differentiating the successive drying stages. This is done by drawing a pointed instrument, i.e. a sharp spike, needle or similar device, continuously across the drying film, generating a characteristic injury pattern (Figure 8.21), the various stages of which are allocated to a defined time sequence.

Degree of dryness

The degree of dryness (to DIN 53 150) can be determined in a graduated test using scattered beach sand or glass beads or (from degree of dryness 2 onwards) by printing on paper using a test stamp.

In dryness tests it is especially important to provide the "previous history" of the drying coating film and the external conditions, since film thickness, temperature and relative humidity in particular all have an extremely marked influence on the drying results.

Crosslinking

The chemical degree of conversion of the functional groups in the binder (e.g. OH and NCO groups) that are used for crosslinking can only rarely be determined with sufficient accuracy by means of the drying tests described above (in the case of oxidatively curing systems, for example). Spectroscopic and other physical methods together with hardness and pendulum damping measurements are more useful here. A selection of interesting methods is mentioned below.

Determination before processing:

- Measuring the pot life (criterion: doubling of the setting viscosity see also DIN EN ISO 9514); monitoring the pot life of water-thinnable two-component systems with this aim in mind would be somewhat impractical, however, since the coating batch often changes little outwardly. The pot life of various two-component systems was covered in detail in section 2.1.4.
- Observing the viscosity increase in a cone/plate rotational viscometer (→ 8.1.3).
- IR and/or UV spectroscopy, depending on the nature of the functional groups under observation.
- Monitoring the surface tension, which usually falls due to the disappearance of reactive polar groups (→ 8.2.3).

Determination of crosslinking in the applied coating material:

- Testing the swelling and partial dissolution behaviour according to specific requirements, e.g. with butyl acetate, xylene, methyl ethyl ketone (MEK rubs), benzine or alcohol.
- Colouring method using dye-containing solvent: the coating film is coloured differently depending on the degree of crosslinking, and this can be determined by colorimetry. Preferably used for clear or white coatings.
- Differential thermoanalysis (DTA) and differential scanning calorimetry (DSC). These methods principally provide information about reactivity, especially in the case of powder coatings, for example, and less about the degree of conversion.
- Monitoring the glass transition temperature T_g by measuring hardness and elasticity.
- Measuring the dielectric constants (DC) or *relative permittivity* (which change during crosslinking).
- IR spectroscopy (see above).

8.4.4 Ring circuit stability

When coating material is supplied via a coating circulation system, as is generally the case in the industrial coating of cars, machines and other large objects, the coating material is sometimes subjected to considerable shear forces in the valves, pumps and feed mixers. This is associated with the risk of

- post-dispersion of the pigments (leading to changes in hue and sometimes also to flocculation)
- damage to special effect pigments (distortion of aluminium platelets and hence loss of effect, "greying")
- damage to dispersion film formers in water-borne coatings (gluing, "breaking" of dispersion droplets).

Furthermore, deposits can build up in the ring circuit. If they are then dissolved by new batches that possibly contain better solvents, problems may arise. These points must all be taken into consideration in the development and approval of coatings.

8.5 Specific tests for powder coatings

In addition to the usual test methods that are also used with conventional coatings, a number of methods specifically designed for powder coatings are described in DIN 55 990 (see also ISO DIS 8130).

We have already mentioned that the application of powder coatings involves various temperature zones that have to be adjusted to one another in sequence and hence need to be measured:

- softening temperature 70 to 80 °C
- extrusion temperature approx. 100 °C
- flow temperature (coating film-forming) 120 to 140 °C
- curing range (crosslinking) from approx. 160 to 220 °C.

The values are average values and deviations may be acceptable in special circumstances. In any event, the "safety gap" of around 20 °C between the individual zones should ensure that in each case, for example,

- (too rapid) bonding is prevented during storage and
- there is an adequate temperature gap before curing to permit flow.

Gel time is another criterion affecting powder coatings: it is the time in seconds taken by a coating powder melt to crosslink at the end of the melting process and at a given temperature. This is manifested by a sudden change in the melt (which must be well stirred) to a rubber-like state. In industrial powder coatings the gel time is measured at 180 °C.

A sufficiently long gel time permits the passage in the developing powder coating film of gases (air, water vapour, etc.) from the substrate, which may possibly be gaseous due to porosity.

The **stoving loss** provides information about the presence of any residual solvents, capping agents (for hardeners) or other decomposition products, and about any residual water content.

The **particle size distribution** of powder coatings, on which application and flow are extremely dependent, is measured either by

- screen analysis (individual screens connected in series or air-jet screen) or
- laser diffraction (more modern and faster, if not always more accurate).

The **free-flowing properties** of the material are evaluated by pouring out a defined cone of powder, followed by measurement of the cone angle. The steeper the cone, the less free-flowing is the powder. Other methods are based on measuring the time needed for a defined quantity of powder to flow out of a cup fitted with a fluidising device or from a hole made in the side of a fluidising cylinder. Exceeding the shelf life of coating power materials (exceeding the permissible storage time or stored overly warm and/or damp) can result in a significant deterioration in their free-flowing properties (and in their subsequent fluidising capacity and transportability). Modern, reactive powder coatings often have only a very limited shelf life at temperatures of above 20 to 30 °C, whilst those clear powder coatings designed for car coating generally have to be stored and transported at temperatures of below 15 °C. Storage and transport temperatures exceeding 35 to 40 °C must be avoided at all costs.

The specific **chargeability** of powder coatings is determined by measuring the leakage current (the current flow between gun and earthing) in proportion to the processed quantity of powder coating. The method only works perfectly for tribo charging, however, because corona charging also charges numerous air molecules, and this naturally leads to overly high values for current flow.

Non-sag properties or the risk of tears, drips, curtains (as sags are also called), can be estimated with the tablet method. This consists in compressing the powder coating into a tablet, placing it onto a pre-heated panel at a defined angle to the vertical and stoving it under the usual conditions.

8.6 Features of coatings after application

Precisely the same observation that we made in regard to the properties of a coating *before* application also applies to the finished coating:

The quality of the finished coating system lies in its serviceability and/or suitability for the customer's requirements, specifically in its

- reproducibility from one batch to the next
- and from one application to another.

This naturally assumes that the conditions and methods of application do not alter significantly or at least not beyond the agreed tolerances, unless none have been agreed or they cannot be precisely defined. Since the latter is the case in many manual applications, from decorative paints through exterior coatings and automotive refinishing paints to industrial coatings, they must often demonstrate a considerable breadth of variation in terms of their processability whilst still producing results that vary only slightly.

8.6.1 Film thickness measurement

One of the first tests performed immediately after application is the

Film thickness measurement (wet).

In continuous processes such as coil coating or automotive assembly line coating in particular, it is vital to check immediately after application and regularly thereafter that the correct film thickness is maintained. A good overview of the methods involved (and of the measurement of dry film thickness) is provided by DIN standards 50 981 and 50 982 as well as EN ISO 2808.

A simple example of a wet film thickness gauge is the (Rossmann) *comb gauge*, on each side of which are teeth of varying lengths located between two feet. It is drawn along the test piece perpendicular to its surface or also just placed into the still-wet coating. and the relevant wet film thickness is read from the first tooth that is wetted. The *rolling rim* (also "twin wheel" or "measuring wheel") to ASTM D 1212 is a useful auxiliary tool, particularly for convex surfaces. As it rolls, the second of its three wheels, being slightly eccentric, touches the liquid coating only when it reaches a certain rotational position, at which point the wet film thickness can be read from a scale at the side.

Two examples of radiometric film thickness measuring techniques, which are quick, involve no contact with the coating and are also suitable for continuous measurement, are:

- the beta backscattering method to DIN 50 983, in which the number of backscattered beta particles (electron beams) represents a measure of the film thickness. It requires an adequate difference in the atomic numbers of the elements comprising the material of the substrate and that of the film to be measured, i.e. in the chemical composition of the layers. This method is due to extensive occupational health and safety increasingly being replaced by the
- X-ray fluorescence method (DIN 50 987: in this case the intensity of fluorescent radiation is a measure of the mass per unit area and hence of the film thickness). It is much simpler and more versatile in its use than the beta backscattering method, and can be used to measure virtually any type of coating film.

Film thickness measurement (dry)

The methods for measuring the thickness of the dry film (clearly set out in ISO 2808) may be either *non-contact* (like the beta backscattering and X-ray fluorescence methods described above) or *with contact*. The latter group is subdivided into "nondestructive" and "destructive" (of the surface).

Nondestructive measurements

This category includes methods whose operating principles are based on the use of any or all of the following:

- *magnetic adhesion* (DIN 50 981), in which the magnet is a permanent magnet (the adhesion to iron depending on the thickness of the intermediate coating film),
- *magnetic induction flux* of an electromagnet, dependent on the film thickness, according to DIN 50 981, likewise for ferromagnetic substrates,
- measurement of the film-dependent *eddy current loss* to DIN 50 984 on non-ferrous metals (change in the inductive resistance of a coil through which a high-frequency alternating current is passed).

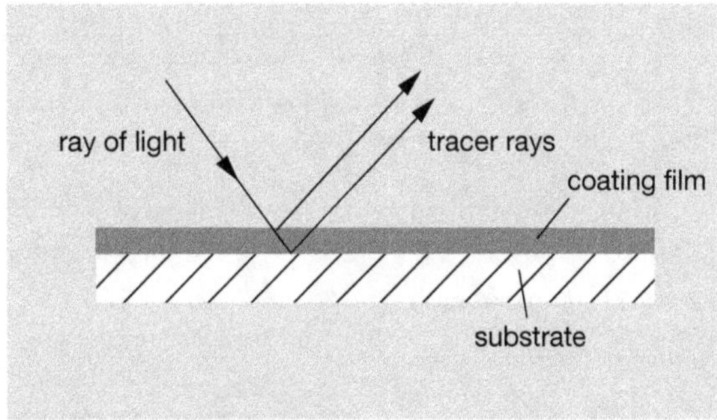

Figure 8.22: Principle behind the light section microscope

Combined instruments are now available that utilise both magnetic induction (on iron) and eddy current (on non-ferrous metals) and are able to store the measuring results or feed them into a computer.

Ultrasonic testing, in which time lags for reflection of ultrasonic waves at coating/coating and coating/substrate interfaces are evaluated, is increasingly being used for measuring film thicknesses on plastic substrates, wood and other non-metals.

Further, modern methods, which also lend themselves to online measurements, employ thermal waves (thermal waves following irradiation with modulated IR laser light, measurement of phase displacement) or the measurement of short-term heating by laser pulses (photo thermal method). The latter method even allows the thickness of powder coatings to be measured *prior* to stoving, and so allows a prediction to be made about the thickness of the finished coating.

Destructive measurements

The following methods are destructive but can consequently be used on multi-coat systems comprising different individual coating films: cross-sectional image measurement, wedge cut (PIG) and depth meter (IG meter), and also the light section method. All of these methods operate on the principle by which a V-shaped cut or other injury is made in the coating down to the substrate and its depth then gauged either optically or mechanically.

In **cross-sectional testing** to DIN 50 950 the sample must be cut perpendicular to the surface, a thin section prepared from it if necessary, and then measured under a light microscope.

In the **wedge cut method** (using a paint inspection gauge - PIG) to DIN 50 986, a V-shaped cut is first made in the coating down to the substrate and is observed with a cathetometer using low depths of focus. The method is very inaccurate for small thicknesses, especially below 10 µm.

The **depth meter** (or IG meter) is a tripod with a linear arrangement to its feet; it is set up such that the centre foot is used as a probe to measure the film thickness in terms of the difference in height at a point that has first been laid bare. If skilful preparation permits the individual layers of a multi-coat construction to be removed separately, then they can even be measured one at a time. The depth meter can also be used for wet film thickness measurement by immersing it in the wet coating film and adjusting the probe by hand from above until it touches the coating film.

Light section microscopes (EN 12373-3) allow the film thickness of a clear coating to be measured by comparing two rays of light ("tracer rays") using a micrometer eyepiece, these rays being produced by reflection of incident radiation at the upper and lower margins of the coating film (Figure 8.22). In the case of pigmented coating films, a strip of coating must first be removed to reveal the substrate.

8.6.2 Optical film properties, colour and colorimetry

Classification and definitions

The first and not uncommonly also the deciding impression when buying a car, piece of furniture, appliance, etc. is the visual impression, determined primarily by the colour or hue and the perfect gloss. One quarter of all car buyers - so we are told by members of the trade - are prepared spontaneously to change the make of their car purely on the basis of an especially attractive and glossy *hue*, e.g. a special effect hue.

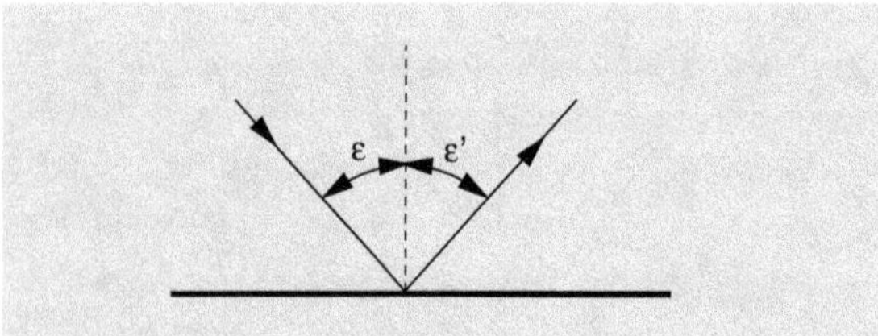

Figure 8.23: Mirror reflection

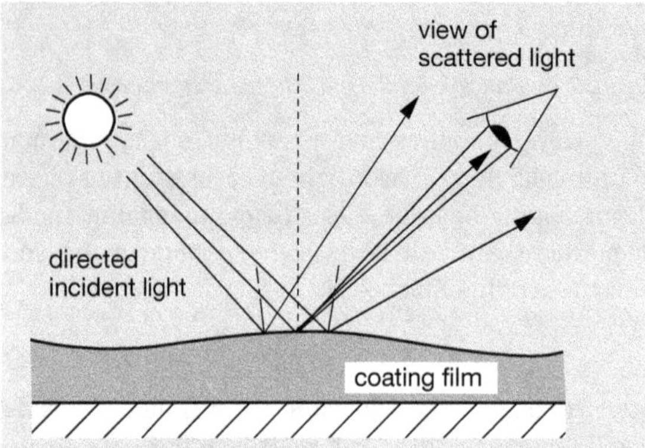

Figure 8.24: Scattering effect at wavy surface

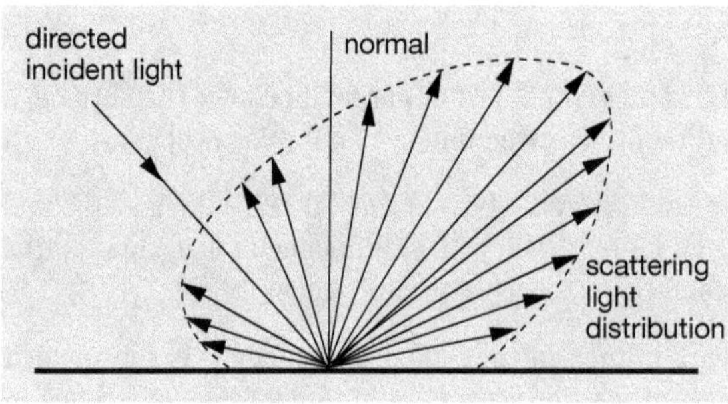

Figure 8.25: Diffuse reflection at matt but light surface

The physical definition of *gloss* on the other hand is strictly objective: according to DIN 67 530 or ISO 2813 gloss is the proportion of directed surface reflection; the ideal case (at a perfectly smooth surface) results in a mirror reflection (Figure 8.23) with the same angles of incidence and reflection **ε** and **ε'** respectively.

In many cases, however, a light scattering effect of varying intensity (e.g. orange peel effect, Figure 8.24) or even a diffuse reflection (Figure 8.25) occurs at the surface of the coating.

Wavy structures ranging from 0.1 to 10 mm in size are known as orange peel or *flow interference*, also "pebbling" (Figure 8.26, right).

A *haze* manifests itself by scattered light of low intensity close to the main reflection; here the mirror image is not distorted, as in the case of orange peel, but surrounded by a diffuse halo (Figure 8.26). The haze is measured close to the principal direction of (20°) reflection, at an angular variation of ± 0.9°. This principle is thus very different from that used for measuring flow interference, being based on the different intensity progression of angle-dependent reflection curves.

Gloss measurement

Gloss itself is measured using *reflectometers* to DIN 67 530 (ISO 2813), in which nowadays the angles of incidence (= angles of reflection, measuring angles, measured from the perpendicular, from the normal) can be selected according to the character of the gloss:

- 60°: for the first measuring step for universal measuring angles, especially suitable for moderate degrees of gloss (30 - 70 gloss units)

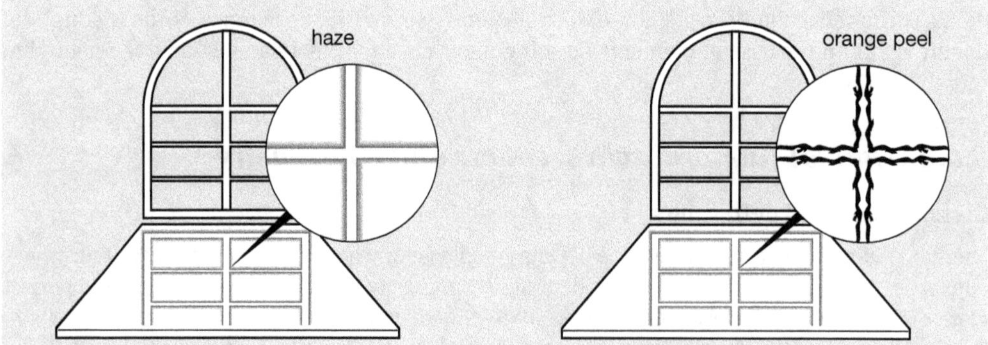

Figure 8.26: Gloss with haze (left) and gloss with flow interference (orange peel, right)

- 20°: especially for high-gloss surfaces (> 70 gloss units at 60° geometry)
- 85°: for matt surfaces (< 10–30 gloss units at 60° geometry)

The best correlations of measuring result and visual impression (due to the angle-dependent sensitivities in reflectometry) are obtained when these categories are observed.

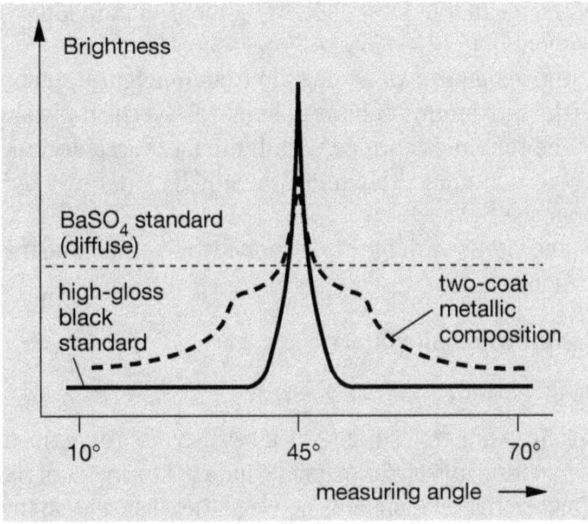

Figure 8.27: Goniophotometer curves for typical coating surfaces; angle of illumination 45 °

Waviness measurement

The waviness structure is determined by taking punctiform measurements of reflection from a scanned section of wave peaks and troughs. The optical profile of the surface is recorded; among other things, the profile can be used to distinguish between coarse and fine waviness. More recent devices (wave scanners) can determine separate characteristics for 3 to 5 different wavelength regions, plus a value for very fine structures in the haze region.

Effect measurements

Measurement of the metallic effect or mother-of-pearl effect in special effect coatings (also "two-tone effect") is a variant of gloss measurement. A *goniophotometer*, which can measure the brightness of the reflection at any given measuring angle (e.g. 0° to 75°), is used to obtain a characteristic brightness curve (Figure 8.27). Furthermore, the "two-tone factor" can be calculated according to DIN 1319 from the ratio of the reflections at 0° and 38° (see also page 144).

Distinctiveness of image

Distinctiveness of image, expressed as the DOI value, is one general measure of the condition of the topcoat, as a cumulative effect of gloss, haze and flow. The instruments used to measure distinctiveness of image are similar in design and usage to the reflectometers described above, but there are also a number of sys-tems that reproduce the subjective impression particularly accurately. In each case a series of samples (or letters) with textures of varying degrees of fineness is detected by reflection, i.e. the limit of sample recognition or readability is deter-mined to a certain extent by light/dark edges, either visually or by means of instruments.

Determining roughness
(covered here because of its effect on surface appearance)

Another means of evaluating the surface smoothness or roughness consists in scanning the surfaces by one of the following techniques:

- Electromechanical methods use a *contact stylus instrument* to DIN 4768 and 4762 (ISO 8503 "record player" system, with electronic conversion of the signals). The measuring process is relatively slow and sensitive to shock.
- Optical processes using reflected *laser beams* are quick and non-contact but in the case of clear coatings involve greater effort, such as metal coating of the surface for example.

Profilograms are obtained from which the instruments can calculated determine the desired characteristic variables, such as

- the arithmetic mean peak to trough height R_a (arithmetic mean of all distances in the roughness profile from its median line, the most frequently used roughness variable)

- the mean roughness height R_z (mean of individual roughness heights from a series of measurements), to DIN 4768
- the geometric mean peak to trough height R_s (geometric mean)
- the maximum roughness height R_t (vertical distance between the highest and lowest point of the roughness profile within the measured section)
- the maximum roughness height R_{max} (largest individual roughness height recorded in the measured section)
- and other variables determined by similar means, sometimes in response to specific questions.

Colour and colorimetry

Definitions

Colour is understood here as a sensory impression, as the perception of colour gained by looking at a coating object, for example, induced by incident light. The purpose of colorimetry is to produce dimensional relationships between the colours such that measurements of their order of magnitude and of their relation to one another correspond to our visual impression.

The light or type of light used to illuminate the specimen is also of influence here. Yellow-tinted incandescent light (standard illuminant A), from which the blue and UV component of daylight is largely missing, is not generally recommended. Of the various synthetic types of standard illuminants, *standard illuminant D65* (corresponding to a colour temperature of 6500 K) most closely resembles natural light. However, since direct sunlight is likewise most unsuitable, colour technologists still prefer a large north-facing window with diffuse light. See also ISO 3668 in regard to visual colour comparisons.

How then do we make the leap from the light reflection from a (coloured) test panel (coating) to a colorimetric system of coordinates that at the same time corresponds to our visual perception (in accordance with the basic principles set out in ISO 7724)?

Factors influencing colour impression

It requires a multitude of mathematical conversions and relations, which above all are based on or make use of the following physical conditions:

- The *sensitivity* of the eye to the various components of visible light (to blue, red or green depending on photopigment in the retinal cones), which can be described according to DIN 5033 by the *standard colour matching functions* x (λ), y (λ) and z (λ).
- The *chromatic stimulus* φ (λ) that strikes the eye, formed by a combination of the type of light and the object surface reflection.
- The *angle of observation* from which we view the object and which must be reproduced in the instrumentation.
- The measured *light intensity* in the wavelength ranges corresponding to blue, red and green vision.

These measured light intensities, allocated to the three ranges blue, red and green (with which the entire spectrum of visible light can be covered), are also known as *tristimulus values* X, Y and Z. Mathematically they can be regarded as surface integrals of the standard colour matching functions, multiplied by the chromatic stimulus function.

Since the tristimulus values are not particularly easy to understand, they are converted into standard colour coordinates x and y, which are used to produce the chromaticity diagram xy (Figure 8.28), also known as the "colour triangle" or "colour pyramid".

$$x = \frac{X}{X+Y+Z} \quad \text{and} \quad y = \frac{Y}{X+Y+Z}$$

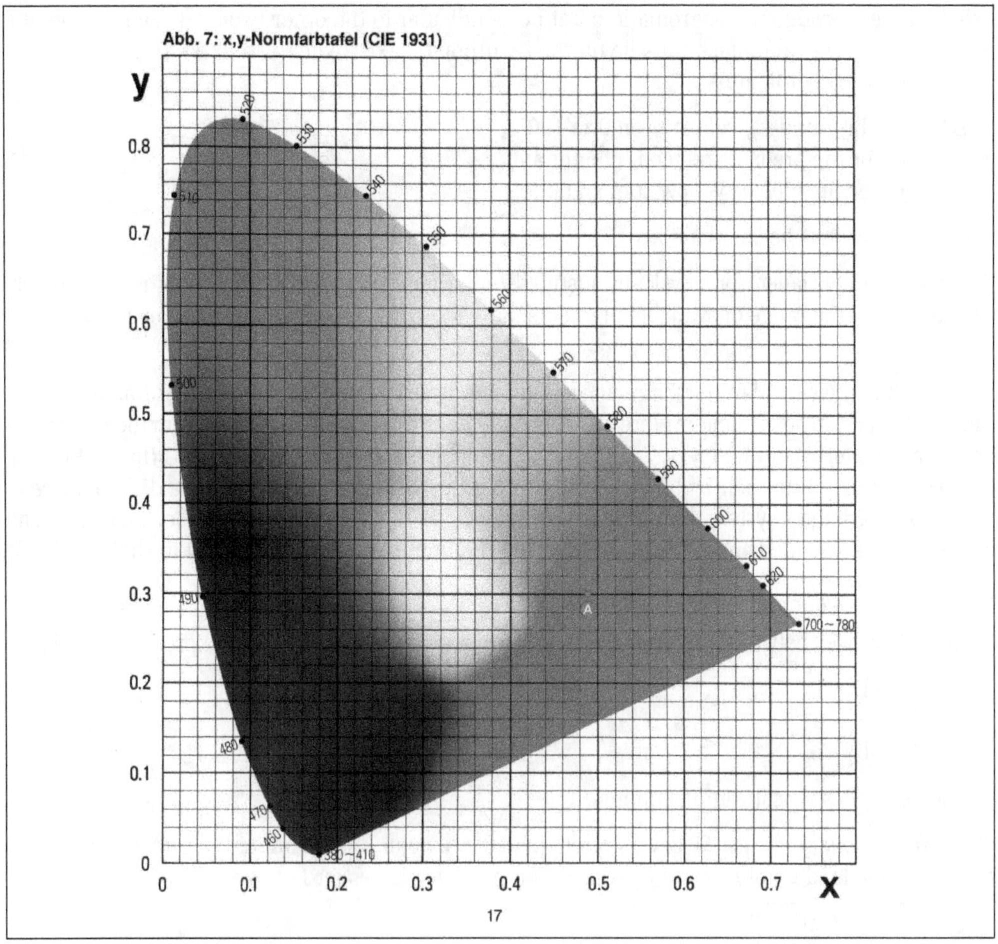

Figure 8.28: CIE chromaticity diagram (x, y)

The standard colour mixture curve y (λ) corresponds to the spectral sensitivity curve for the eye; hence the tristimulus value Y is also a measure of *brightness* (perpendicular to the plane of the chromaticity diagram at the achromatic point). One disadvantage of the chromaticity diagram is that distances within the triangle that mathematically are of equal magnitude are not perceived by the human eye as being so.

The best known and most widely used colour system, an improvement upon the previous model in this respect, is the CIE-L*a*b* system (Figure 8.29), which is often simply known as the CIELAB system. (The * symbols indicate that this is a successor to the earlier CIE-Lab system, but in the interests of simplicity they have been omitted in the following discussion). The a and b axes are perpendicular to one another and intersect at the achromatic point. The a axis represents the red-green and the b axis the yellow-blue dimension. The brightness

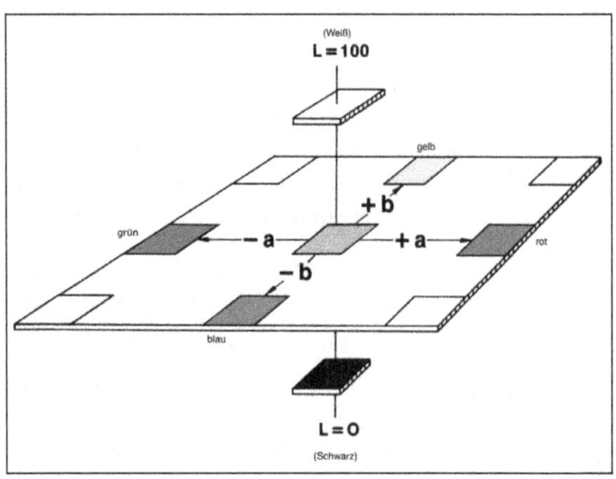

Figure 8.29: CIE lab uniform colour space

axis L passes through the achromatic point perpendicular to the other two axes. The achromatic colours white, grey and black thus lie on the “achromatic axis” where a and b = 0.
By stating the 3 coordinates

- L (for the brightness), values from 0 to 100
- -a to +a for the green or red com-ponent and
- -b to +b for the blue or yellow component

any colour can thus be characterised.

There are no permanent or physically justified limits for values of a and b. They are ultimately determined by the range of available pigments, i.e. they are “outwardly displaced” over the course of time.

In addition to the cartesian (rectangular) coordinates system described, a polar coordinates system can also be used to define colour. Here again, the greater the distance of a colour from the L-axis, the brighter and more saturated it is. This distance is therefore known as *saturation* C (from chroma). If two colours differ in their a and b coordinates but not in saturation, they have a different *hue* h. The polar coordinates system is accordingly also known as the **LCh** colour system (or CIELC, h). The method for calculating Lab (C, h) values from the tristimulus values is described in DIN 5033.

Conditions for colour difference

Colour differences between a reference R (standard) and the sample S can be expressed as follows:

$$\Delta L = L_S - L_R$$
$$\Delta a = a_S - a_R$$
$$\Delta b = b_S - b_R$$

The same applies for saturation and hue.

The *overall colour difference* ΔE is very often quoted as a quality standard for the coincidence of two hues. It is obtained graphically or by calculation from the values for L, a and b:

$$\Delta E_{ab} = \sqrt{(\Delta L)^2 + (\Delta a)^2 + (\Delta b)^2}$$

While quoting the likes of DE permits a rapid statement to be made about the overall difference between two or more hue patterns, it naturally fails to provide any information about whether yellowing (b value), a red tinge or other deviation is involved.

Depending on the optical requirements and color range for the coated object, colour tolerances may lie in the range of $\Delta E = 0.3$ to 1.0 or higher. They are often dependent on hue because of the differing sensitivity of the eye to different hues. In any case they have to be agreed between the coating manufacturer and processor and recorded for each batch if necessary.

Special cases in colorimetry

In the L*u*v* colour system defined in 1976, also known as the **CIELuv** colour system, an even better coincidence is obtained between the geometric and perceived colour differences. Similarly to the CIELAB system, it uses the coordinates u* and v* instead of a* and b*; calculation of the colour coordinates from the tristimulus values is quite different, however.

The phaenomenon that two hues with different reflection curves can appear identical to the observer or also be clearly distinguished, depending on the type of incident light, is known as **metamerism** (DIN 6173). Metamerism thus always relates to a *pair* of samples.

The standard illuminants A and D65, which differ considerably in their spectral distribution, can produce pronounced metamerism effects, especially for mixtures of different pigments. Samples or tints of hues must therefore always be tested for metamerism by comparing them against the reference or standard under different types of light.

Colorimetry methods

Measuring techniques

The two principal measuring techni-ques for colour or, more precisely, for the light reflected by the object are the

- three-colour process and the
- spectophotometric method.

The *three-colour process,* which can be performed with particularly compact and portable instruments, employs three sensors fitted with filters such that they correspond to the sensitivity curves $\bar{x}$ (λ), $\bar{y}$ (λ) and $\bar{z}$ (λ) for the eye. Although the instruments do not generate reflection curves, they are perfectly suitable for quality control and particularly also in the production and processing of coatings.

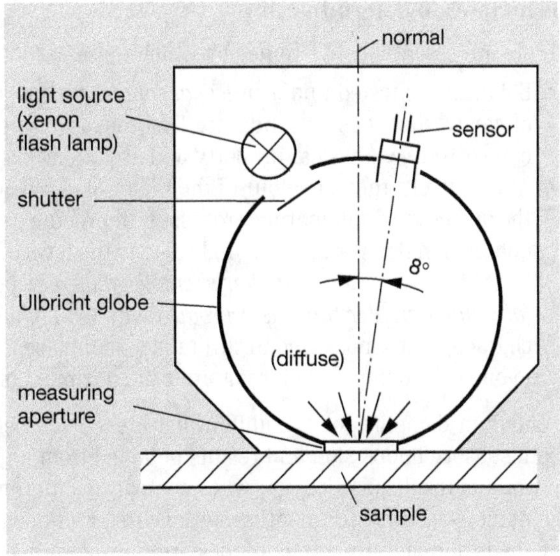

Figure 8.30: Measuring geometries d/8 ° in the spectral photometer

In the *spectophotometric method* a spectral sensor line (multiple sensor) measures the reflection across the wavelength range (reflection curve). The tristimulus values can be calculated by integrating and comparing the measured values with the stored colour mixture curves. Spectro-photometers offer a very high degree of precision for absolute colour data.

Given the advances made in performance and miniaturisation, nearly all manufacturers of colorimetric devices employ a spectral method.

Measuring geometries

The construction "diffuse d/8°" is an important measuring geometry in a spectrophotometer. This means that the measuring head (see Figure 8.30) has the following features:

- diffuse illumination of the sample
- the reflected light is measured at an angle of 8° to the surface normal (perpendicular)

The inner surface of the globe is coated with a diffusely nearly ideal scattering material, e.g. $BaSO_4$. The shutter prevents interfering reflections that might be formed at the sample without this screen.

In this way gloss and surface effects are largely excluded. Depending on the substrate, an additional *gloss trap* can produce a further improvement by eliminating interfering direct reflections (where angle of incidence = angle of reflection); their effect and their application are disputed, however. In any event, their use, like all other measuring details, must be recorded in the results. The measuring geometry 45/0 (angle of incidence 45°, observation perpendicular to the sample surface) or similarly 0/45, is also widely used.

For very rough surfaces, e.g. grained plastic surfaces for automotive interiors, illumination of circular-45° (a series of diodes in a circle about the measuring surface) with measurement at 0° is recommended.

The choice of measuring geometry is generally determined by the surface structure (which may possibly be oriented) and other specific requirements and conditions.

8.6.3 Mechanical engineering film properties

This group of coating properties includes *adhesive strength* or the bond formed with the substrate, and primarily the associated resistance to all conceivable mechanical influences: slow and rapid *deformation* and *external impact* (penetrating bodies of all kinds) and *abrasion stress.*

Principles of determination

There are essentially two routes by which these particular film properties can be investigated:

- Behaviour is tested under *practical conditions*, e.g. stone chip resistance, deformations, effects of car washes, desert sand, etc. This means that the results are complex, cannot be clearly categorised by physical property and are often "yes/no" or "good/bad" decisions.
- Alternatively, the behaviour of the material can be described by *physical variables* alone, with the objective of automating processes, improving quality records, etc. Characteristic values are obtained in this way, increasingly also in the form of digital values. The problem can then often be that *the precision of a characteristic value may be much greater than its practical equivalent (and often greater than its reproducibility too!)*. One should resist the temptation to interpret characteristics measured to several decimal places if the accuracy of the test method is limited to one or two figures before or after the decimal point.

Pendulum damping (or "pendulum hardness", see below) is a classic example of this problem: both sensory impression and results obtained from a practical test series can often contradict the pendulum results, particularly when comparing different binder systems. Nevertheless, the method is perfectly suitable for monitoring the cure of a known system, for example. In this case too, test methods and measuring techniques are thus often not entirely satisfactory - in other words there is still room for improvement.

Measuring adhesive strength

Adhesion is generally used to refer to coherence with the layer below, bond (of a filler, for example) to describe adhesion to the layer above (to the topcoat in this case) if one's perspective is reversed. In physical terms there is no difference between the two, however. Nevertheless, a distinction must be made between

- *adhesive fracture*, that can occur at the interface with the next layer (defective interlayer adhesion) and
- *cohesive fracture*, that occurs within a layer and indicates poor coherence within a system - due to inadequate film formation, for example.

Measuring techniques

Three key methods for determining adhesive strength should be mentioned:

The **cross-hatch adhesion test** to DIN 53 151 (and DIN 53 211, ISO 2409), conventionally performed with a multi-blade knife (an example of which is shown in Figure 8.31). The cutting area is brushed clean and inspected for flaked strips and areas. The standardised test specifications set out various cutting intervals depending on the requirements and film thicknesses. The test result values 0 (0 % detached) to 5 (> 65 % detached) can be established from the damage patterns, which are described in the DIN standard for example.

The test is often made more stringent by the attachment and detachment of a strip of adhesive tape ("sticky tape detachment"). The influence of any embrittlement of the film arising from the cross-hatching can cause problems: in very brittle material the notching causes the coating to flake off at the sides as a result of an unfavourable distribution of forces in a horizontal direction. For the same reason, it is essential for notching to be absolutely perpendicular. In practice a knife is often used to scratch the film, or a cross-hatched, barred or similar pattern is applied by hand and the film tested for adhesion by removal of a chip or by means of adhesive tape.

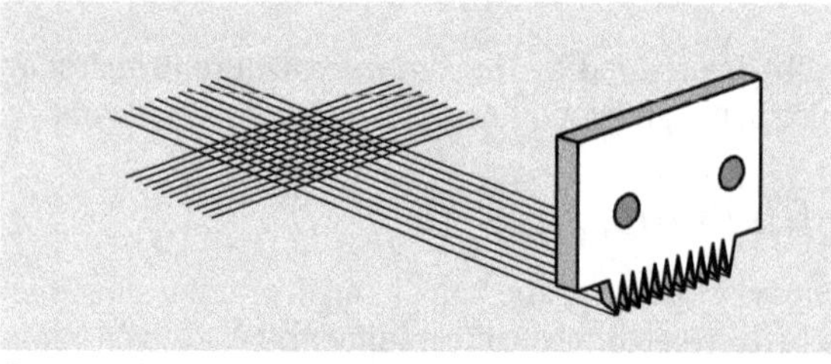

Figure 8.31: Principle behind cross-hatching

Adhesive strength can likewise be measured by the **pull-off test** using a disc bonded to the surface and a tensile testing machine to DIN 53 232 or ISO 2 4624 (Figure 8.32). The test must be performed with great care. The brittleness of the coating and the risk of the disc's tilting both influence the process considerably

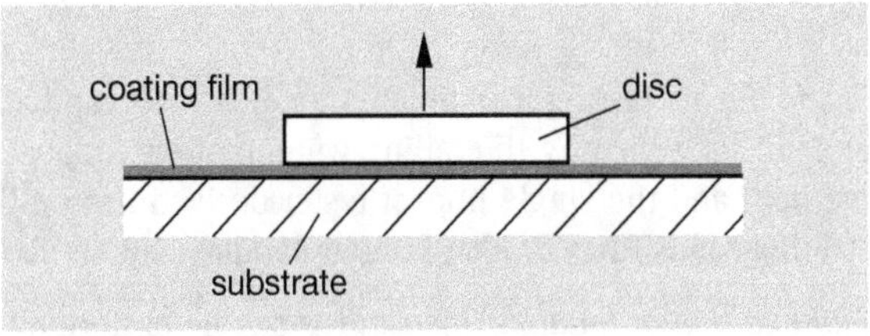

Figure 8.32: Pull-off test

More readily reproducible results can be obtained in certain circumstances using **torsion testers**, which function in a similar way to the pull-off testers. This method differs from the pull-off test in that the disc or pin attached to the surface is twisted off rather than pulled off. The adhesive strength is calculated from the torque required to do so.

Measuring deformability

Erichsen **cupping** (DIN 53 156, ISO 1520) serves to determine the deformability of coating films and makes use of a coated test panel. A round-ended indenter 20 mm in diameter (Figure 8.33) is pressed into the panel, usually from the back, to reveal both deformability and loss of adhesion during a given deformation. As soon as the first crack appears on the surface of the coating, the pressing action is stopped and the indentation measured. Since in this test the result is influenced by the entire coating composition and any weak points that may be present, the panels used must always be comparable with those used in practice.

The **mandrel flex test** (DIN 53 152, ISO 1519) is likewise used to test deformability or flexibility, or more accurately to determine flexural elasticity. Test strips are bent around bending mandrels (rods) of various diameters. The smaller the diameter, the more readily cracks and loss of adhesion occur in the coating film. The bending mandrel may also be conical in shape (ISO 6860). In the latter variation information can be gained from only a single bending operation since a whole range of bending radii is obtained.

Rapid impact deformations

All of the deformability tests mentioned so far have concentrated on very slow deformations, during which the coating composition therefore has a relatively long time to adjust. In practice, of course, in the case of minor impacts or particularly of flying stones, for example, no such time is available. A number of tests involving very rapid impact loading or deformations are therefore described below.

Application criteria

The most important resistance characteristics in respect of rapid impact, from flying stones for example, are

- good adhesion to the substrate,
- good coherence in the film,
- elasticity of all layers of the coating,
- short recovery times after extension (deformation),
- good distribution of impact energy in the environment of the contact point and on the surface of the substrate.

Impact loading can be exerted in principle either as

- a "single" impact loading, with defined mass, speed and angle of contact by the impact body, or as
- "multiple" impact loading, with random distribution of the above variables and a correspondingly diverse damage pattern, which can only be evaluated as a whole.

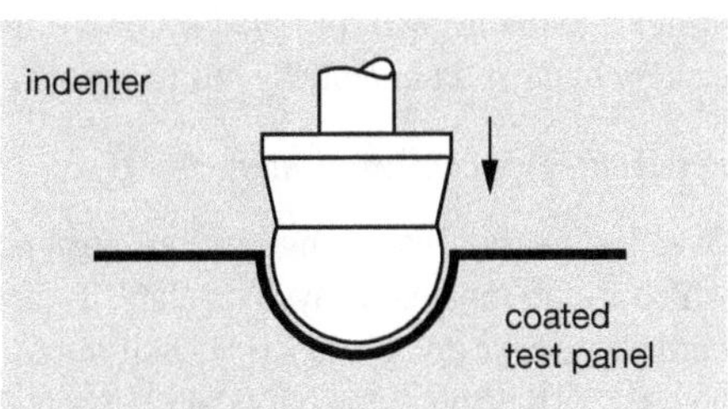

Figure 8.33: Erichsen cupping

Measuring systems

Single impact testing instruments include the drop ball impact tester with downpipe, height scale and slit for adjusting the falling weight, single impact testers to EN ISO 6272 ("pecking chicken" method) and the single impact tester derived from a BMW standard, in which a surface-mounted striking pin is fired by a ball under compressed air see also DIN 55 996.

Multiple shot or *multiple impact tests* can be performed as a shot blast test to DIN 53 154 for example, in which 1000 steel balls 10 mm in diameter drop onto a metal sheet in a defined pattern. As with all such tests, however, the correlation with real impact from flying stones is not particularly good, since the balls are too heavy, their surface too smooth and their fall too slow. Another method involves the use of "gravelometers" of various designs according to the requirements of the various car manufacturers, in which grit or shot is fired at the test piece under compressed air. A common feature of all of these test devices is that the coated test panel is always fixed at a defined angle at the end of a blasting channel along which a diverse range of impact materials is accelerated under compressed air. In some cases these tests are performed in conjunction with a corrosion test, which is normally carried out first.

There is a multitude of test procedures and specifications associated with these methods, but space does not allow them to be covered in any detail here.

Thermomechanical analysis (TMA), penetration method

With the growing proportion of coated plastic components used in motor vehicles, for example, attention must also be paid to the adhesive strength of the coating composition on very diverse plastic substrates; testing of this property is thus becoming increasingly important and widespread. Without special plastic surface treatment and appropriately formulated coating materials, coated plastic components often display inadequate impact resistance characteristics, especially in the low-temperature range down to around -30 °C, and the tests described above provide only an incomplete picture of these characteristics.

Dynamic TMA offers a suitable alternative, however, enabling viscoelastic properties (flow/elasticity behaviour) to be examined as a function of temperature, following which conclusions can be drawn about the impact resistance of the coating/plastic system. Specifically, dynamic TMA measures the reaction of the coated part to a periodically changing force.

Instrumented *puncture* is an **impact resistance test method** based on a VDA guideline and covered in DIN 53 443. A graph is plotted of force against time in the millisecond range, from which the amount of force needed to destroy the coated plastic test piece can be determined. In this way the viscoelastic behaviour of the coating film can be adjusted to that of the substrate - a precondition for good impact resistance. Ductile - through to brittle - coatings (and substrates) can thus be assessed reproducibly and also at various defined temperatures by means of their characteristic force/time curves.

Hardness measurements (general)

In spite of its name, hardness is one of the "woolliest" terms to describe, i.e. it is one of the most difficult to define with physical accuracy. It can be used to convey the *static resistance exerted by a coating against an indenter* - but in practice this definition remains unsatisfactory.

Possibilities for characterisation

Over the years a vast range of hardness concepts has been developed, the predication datas for which by no means always correlate. To prevent misunderstandings, one should avoid the use of hardness *per se* and always refer to it in tandem with the feature of the associated test method (see below). Furthermore, the boundary between *hardness* on the one hand and *elasticity and deformability* (flexibility) on the other is indistinct (see also Table 8.1).

Table 8.1: Methods for testing elasticity and hardness properties

Property	Test
elasticity/deformability	indentation test, mandrel flex test drop ball test, flying stone test thermomechanical analysis, puncture test
	damping hardness measured as pendulum hardness (pendulum damping) or rocker hardness: provides information about elastic behaviour, despite its name
hardness	Buchholz, Knoop, scratch, pencil hardness indentation hardness (micro-indentation hardness) abrasion tests [1)]

Earlier we discussed rapid and slow deformations. In principle the tests for "damping hardness" belong to the same group, but traditionally they are usually classed together with "hardnesses".

By low hardness - in this broad sense - we therefore mean a strong damping of moving bodies (e.g. pendulums) placed on the coating film, as well as a high susceptibility of the coating to mechanical damage from sharp objects, and finally also inadequate abrasion resistance of the coating system.

The elasticity of a coating system, on the other hand, refers to its tendency to reverse a deformation caused by an external influence and regain its original surface.

The best possible combination of elasticity and hardness - the two often being directly opposed - must therefore be found. The aim is not to achieve maximum elasticity alone nor extreme hardness alone (with its risk of brittleness) but to find a suitable compromise.

This background and customers' rising demands explain why so much is currently being done to launch self-healing effects featuring new physical and chemical methods.

General conditions

In any measurement of hardness, the substrate and the condition and film thickness of the coating film are particularly important: the substrate must be very level and hard (sheet metal or glass). The surface of the test film must likewise be level and free from flow defects and inclusions of any sort, and above all the film must be of an adequate thickness (at least 30 µm), since otherwise a hard substrate could give rise to misleadingly good hardness results. The film thicknesses of coating films under comparison should differ by no more than 10 %.

Since most coatings soften with increasing temperature and rising moisture content, tests should always be performed under a constant test atmosphere (if possible standard atmosphere, i.e. 23 °C/50 % r.h.; see also EN 23 270), at any rate if results are to be documented and exchanged with other laboratories.

Methods for determining damping hardnesses, e.g. the familiar property of pendulum "hardness", are based on a periodic loading. However, such variables are not so much an expression of hardness as a consequence of elastic behaviour, and this particular property is therefore more correctly known as "pendulum damping".

Periodic loading in any case charac-terises the fundamental tests for elasticity (and hence also tests for crosslinking, solvent retention, etc.).

Vibration methods for measuring hardness

König or *Persoz* pendulum damping (DIN 53 157, ISO 1522) is the best known of these methods. It determines the damping of the oscillations of a pendulum with two supporting balls lying on a coating film. The number of oscillations taken for the oscillation amplitude to fall below a given level (gener-

1) Abrasion resistance is generally classed with hardness but is naturally also responsible for surface slip, brittleness, etc.

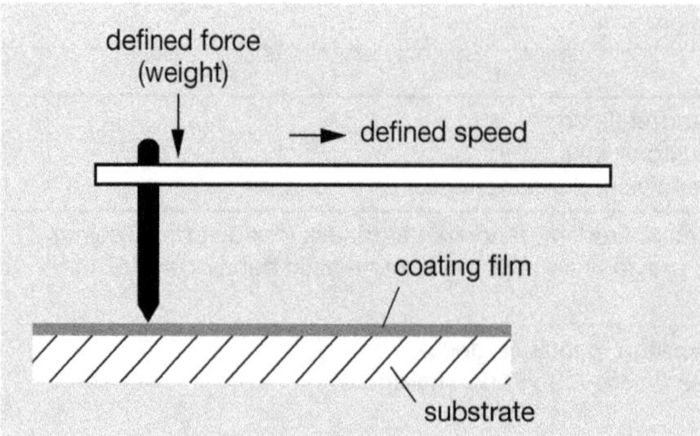

Figure 8.34: Principle behind scratch hardness measurement according to Clemen

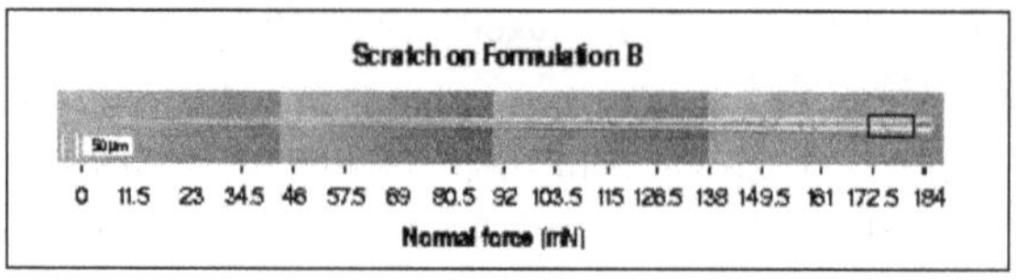

Figure 8.35: Scratch made by the nano-scratcher

ally half the initial value) is measured. The softer the film, the greater the damping of the pendulum oscillations. The two methods according to *König* and *Persoz* differ in terms of pendulum mass and diameter of the supporting balls, the pendulum excursion and the oscillation time. The pendulum damping is expressed as the number of oscillations or (more commonly) as the oscillation time in seconds, calculated from the number of oscillations. One oscillation lasts 1.4 s with the König arrangement and 1 s with the Persoz one.

The *Sward-Zeidler* rocking (damping) hardness tester is based on a similar measurement of oscillation. It is a kind of gymnastic wheel comprising two annular steel rockers connected by crosspieces. The wheel rocks backwards and forwards on the rockers until, as in the pendulum damping method, the amplitude falls below a certain level. The amplitude is indicated by ring marks on two bubble levels (comparable with those on a spirit level).

Both of these instruments, the one used to test pendulum damping as well as the rocking hardness tester, must be calibrated regularly, e.g. with glass or metal plates on which a certain oscillation period must be reached.

Measuring scratch hardness

Scratch hardness is tested under relatively normal processing conditions with a stylus (or cutter) (Figure 8.34) and increasing indentation force until a permanent line appears on the surface (ISO 1518). The weight is a measure of the scratch resistance. The highly disruptive influence of the surface structure must be taken into consideration, however.

Sclerometers produced by various manufacturers operate on the same principle, enabling a hardness test to be performed quickly even on curved surfaces. In the sclerometer a calibrated spring, whose tension can be adjusted, presses onto an engraving point extending from the instrument. The sclerometer is drawn vertically over the test piece at a given speed and the spring tension recorded at which a scratch mark is left.

In the **pencil test**, various pencils of different degrees of hardness are drawn across the test piece at a constant pressure and a defined angle of 30° to 45° in increasing order of hardness. When a visible mark is left by one of the pencils, the hardness value ("pencil hardness") is expressed as the grade of the pencil. The hardness scale for pencils increases in the following sequence: 6B-B, HB, F, H, 2H-9H see also EN 13523 (draft).

The **nano-scratcher** is undoubtedly the best method at the moment. It works on the same principle, but combines extreme sensitivity with high-precision metering of indenting force and speed. Depending on these specifications, the depth of scratching and the type of damage can be observed very precisely. (Figure 8.35). The technique is fairly complicated but can also be used to trace the recovery, especially in correlation with the depth of scratching.

Indentation hardness test

Indentation hardness is generally determined by measuring the depth of indentation into the coating film by indenters of a defined geometry within a given period and under a given load.

The **"Buchholz" method** (DIN 53 153, ISO 2815), in which an indenter with a conical point (cutting edge of a wheel) leaves a spindle-shaped indentation, is the best known indentation hardness test. The 500 g indenter stands on the test surface for 30 s, and 35 s later the indentation length l (in mm) is measured using a cathetometer. This measurement gives the "Buchholz hardness" in the form of the expression

$$\frac{100\ \text{mm}}{l}$$

The **Tukon** hardness test produces a "refined" Buchholz hardness, although in this case a test diamond with a rhombic cross-section and pyramidal end face is pressed down with relatively low force. Measured values are expressed as **"KHN" (Knoop hardness number)**.

In the **Vickers** method (DIN 50 133) a measuring pyramid is pushed down under a load and withdrawn again; the deformation pattern is then evaluated. This method also provides information about elasticity and cohesion.

The *Fischer* **micro-indentation hardness** test (to DIN 55 676: *universal hardness*) is even more informative. In this method a small, pyramid-shaped diamond with a square base is pushed down into the coating film under a measured force and then withdrawn. This very accurate measuring technique can be used for films from only 2 µm in thickness (provided that the maximum depth of indentation is no more than 10 % of the film thickness). Information is thus provided not only about the indentation resistance but at the same time about the *overall* viscoelastic behaviour (elastic and residual components of deformation, elastic modulus).

Examples for the use of this accurate measurement of hardness include the following:

- indentation resistance of binder films or pigmented coatings
- elastic component of deformation, possibly in connection with stone chipping and car wash resistance ("reflow effect")
- determination of the depth-dependent "spring component" (identification of solvent retentions within the coating film, for example); see also section 2.1.1.10 in regard to the mechanical behaviour of polymers
- monitoring of crosslink density in relation to time and space
- depth of penetration of chemicals during resistance testing of the coating film

Figure 8.35 illustrates the influence of the stoichiometric ratio of base coat to hardener on the indentation hardness of the dry film by reference to the *curing of a two-component clear coat.*

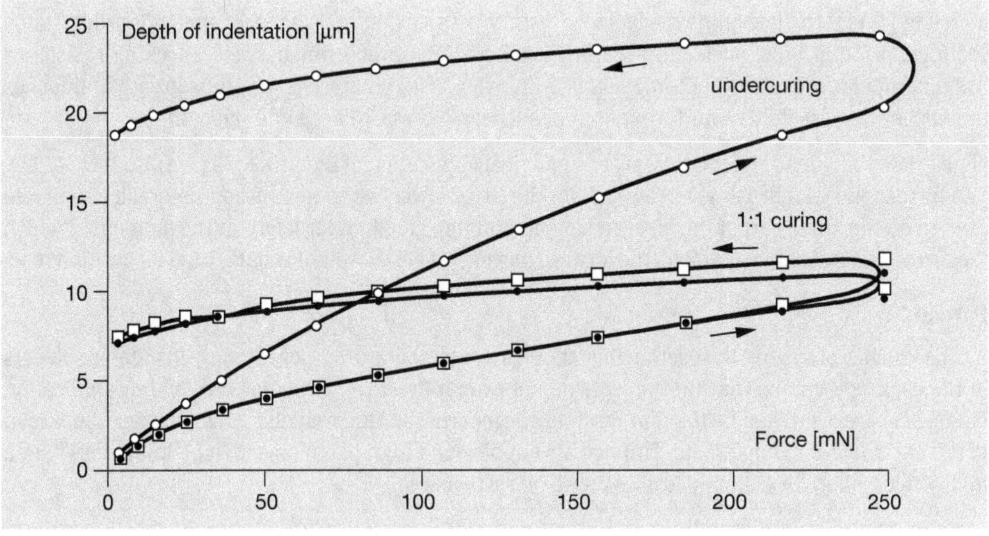

Figure 8.35: Influence of curing ratio on (micro-)indentation hardness

The greater depth of indentation - under the same force - exhibited by the undercured film (too high a ratio of base coat to hardener) in comparison to the correctly cured clear coat can be clearly seen. At the same time the distance between the outward and the return curve is a measure of the flow component of deformation. The smaller the flow component, the greater the elastic component. The latter is characteristic of a good cure.

Abrasion tests

The abrasion resistance of coatings is generally tested by means of abrading surfaces (covered with sandpaper) or with sand itself, the resulting loss in mass then being determined by gravimetry.

In the **sandblasting abrasion** test according to *Stock*, the abrasive falls freely onto the test piece, which is secured on an incline. The quantity of sand needed to abrade down to the substrate is given as a measure of the abrasion resistance.

The well known **Taber abraser** comprises two abrading wheels which turn on the rotating test piece, the type of abrasive used varying according to the level of abrasion required. The **Weinmann-Bosch** system is a similar test apparatus. Here an abrasive wheel covered with sandpaper and driven by an eccentric cam is pushed over a surface and turned slightly further on every second stroke so that a fresh abrasive surface is always available.

The *Gardner* **scrubbing tester** is used to determine *washing and scrubbing resistance* to DIN EN 13300 (formerly per DIN 53 778). On completion of a defined number of scrubbing cycles, the damage pattern is evaluated by eye, or the number of cycles required to achieve a particular damage pattern is taken as the characteristic value.

The **scratch resistance of automotive finishes** (car-wash resistance) is particularly critical for automotive topcoats. To replace the large number of "in-house methods" by a generally accepted test method that closely resembles practical conditions, as well as to reduce testing and development costs, DIN EN ISO 2066 and DIN 55668 simulate machine washing in a laboratory car-wash fitted with rotating brushes and a standardised dirt composed of quartz flour. The painted test panels are immobilised on a bench and moved back and forth at a defined speed and a defined distance from a rotating wash-brush. When the wash cycles are complete, the test panels are cleaned. The residual gloss, expressed in relation to the initial gloss, serves as a measure of the scratch resistance.

Determining blocking resistance

The **blocking resistance tester** (usually made in-house nowadays as they are no longer made commercially) is used to determine the *blocking resistance* (pressing resistance) of emulsion paints, latex coatings, printing inks, packaging coatings, coil coatings, wood finishes, etc. Blocking resistance has a particular impact on the *stackability* of (newly) coated materials and thus greatly influences the further processability and throughput of such coated workpieces and papers.

Figure 8.37 provides a schematic view of a practical application of the principle (see also ISO 3678[1)]). The further the weight is moved away from the pivot towards the test piece, the greater the force exerted on the lower part of the blocked sample (forcibly glued under load). At the moment at which the two sections separate (detach), the forward movement of the weight is halted and its position read.

Pore analysis

A consequence of stoving the coating film at excessive temperatures - or of using unsuitable solvents in the coating - can be that the solvents do not evaporate at the anticipated rate after application but instead are overtaken by the film-forming and/or crosslinking reaction. This can sometimes lead to *blisters, bubbles, pinholes*, etc. Trapped air bubbles and high pseudoplasticity ("lumpiness"), in a highly filled filler, for example, can likewise be responsible.

1) Test of stackability: ISO 4622

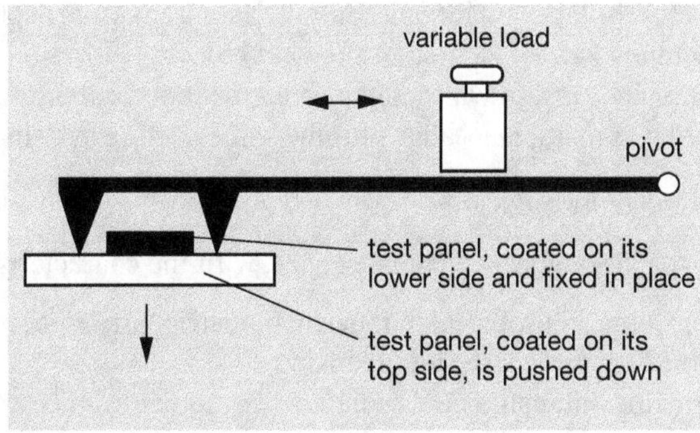

Figure 8.37: Principle behind blocking resistance testing

Pore detectors according to DIN 55 670 offer a means of testing a coating for susceptibility to these phenomena. In macroscopic terms a perfectly dry coating - even if it is pigmented - is a homogeneous dielectric with a high dielectric constant (provided that it is a non-conductive coating). At the points at which blisters or pinholes occur, however, the dielectric constant falls to around 1. If the coating film, applied to metal, is then used as the dielectric of a capacitor, the electrical *breakdown voltage* can be established. Since the electrical field strength under an applied constant capacitor voltage is inversely proportional to the dielec-tric constant, the defects lead to punctures in the presence of a sufficiently high voltage. The metal substrate (concrete can also be used) acts as an electrode in the capacitor. The other electrode is a wiper blade made from graphitised rubber or a fine metal brush. When a voltage is applied between the metal and the wiper, the sample can be scanned and inspected for punctures. A puncture can be recognised both by black specks on the surface (which should be as light as possible) and also by an optical and in some cases an audible signal in the instrument. Commercial pore detectors of this type are generally now automated devices with electronic pore meters. Depending on film thickness, the applied voltage can be varied between 200 V and 15 kV, or applied in pulses of up to 35 kV in the case of very thick coatings. Thanks to their integral current limiting device, they are nevertheless perfectly safe to handle (similarly to electrostatic coating).

Other instruments operate on the basis of the **sponge method** with *conductivity detectors*. In this case the difference in resistance between the insulating coating and the water-filled pores is measured. A moistened sponge serves as the sensor and the voltage source is a 9 V battery.

The earlier test according to DIN 53 161, in which pores are made visible by *precipitation of copper or cadmium* onto bright base metal, is scarcely used nowadays.

8.6.4 Light stability and weather resistance

In the majority of cases, the test methods and coating properties described so far have been concerned with purely physical stresses and resistances. In this section we will look at the types of attack and also the causes of damage or resistance that can be attributed to chemical processes, namely chemical changes in the surface and/or within the coating film. These changes manifest themselves in the form of:

- change in gloss and colour
- cracking and blistering
- swelling or shrinking or
- changes in technological properties, such as hardness, adhesive strength or corrosion resistance.

Outdoor weathering

In order to test all of the above resistance characteristics, a test piece must first be manufactured under practical conditions, i.e. by conventional manufacturing practices, and then aged, normally by artificial means, to ensure that post-curing and/or solvent evaporation likewise correspond to practical conditions.

For these resistance tests it is once again the case that *the more closely the test conditions correspond to conditions in practice, i.e. to exposure to sunlight and weathering, the more meaningful will be the results.*

Since the latter conditions naturally vary from place to place, there are countless outdoor weathering stations located all around the world for this purpose (see also DIN 53 166, ISO 2810; wood: draft DIN EN 927), to which samples are sent according to requirements and customer specifications, often to multiple locations for the same coating system.

Weathering zones

Typical stations and climate zones, with the characteristic weather details for each, include:

- Central Europe/inland: partly industrial atmosphere
- North Sea coast: salt-laden sea air
- Alps: intensified UV radiation
- Florida: humid, relatively uniformly warm tropical climate
- Arizona: dry but hot desert climate with widely fluctuating temperatures and high UV radiation
- Australia: tropical climate in the north, extremely high UV radiation in the south, with a dominant short-wave component.

Acceleration

The reduction factor for solar radiation in Florida is around 50 %, which means that one year's exposure in Florida is the equivalent of approx. 2 years in central Europe. This is an average figure, however, and actual results can of course differ substantially, according to meteorological conditions and the chemical compounding at the coating. Different authors sometimes use quite different correlation factors for their analyses, depending on their particular experience, and in some cases they are dispensed with altogether.

Type of stress

Test pieces are usually set out at various *angles of inclination* according to the degree of latitude (for maximum benefit from the radiation), and are either fixed in place or are capable of being adjusted to the position of the sun. Test pieces laid out at a shallow angle (e.g. 5 ° in Florida) must be washed down regularly to remove accumulated dust. Such measures, i.e. dusting or taking a gloss measurement without prior dusting, must be documented in every case.

The test pieces reach a higher temperature, leading in turn to faster degradation of the coating, if they are placed on top of a "*black box*" (black painted box). The boxes can even be heated (*"heated black box", "HBB"*).

An instrument known as "**EMMA**®" (**e**quatorial **m**ount with **m**irrors for **a**cceleration), which is available in Arizona, uses mirrors to intensify the solar radiation on the test panel. However, this arrangement too greatly diminishes the influence of humidity - which is just as necessary to the weathering process - and the rate of more hydrologic film decomposition is consequently too low. In a sister instrument known as "**EMMAQUA**®", an accompanying spray of demineralised water is therefore also provided.

Accelerated weathering

Light ageing devices

Outdoor weathering processes as described above are a fundamental requirement in the automotive industry in particular. They are naturally far too slow to be used as tests alongside the development of new raw materials or coating formulations, and must therefore be supplemented by accelerated weathering and exposure to artificial light sources (see also DIN 53 231 and 53 387 or ISO 2809). Such measures should ideally at least enable a trend to be established after a few weeks. It must be remembered, however, that *a reliable correlation with outdoor weathering cannot always be assumed.*

Suitable light sources include xenon, fluorescent, metal halide or in the past carbon arc lamps. They change all the time and must therefore be replaced on a regular basis. They must also be operated with optical filters since the primary radiation, especially below 300 nm, is too energy-intensive and induces coating damage that is atypical of results obtained in practice.

Weathering instruments

Some of the best known instruments for accelerated weathering are briefly described below (this list is by no means comprehensive). Examples worthy of mention here include the **Weather-o-Meter (WOM)** and the various **Xenontest** models (in accordance with ISO CD 11 341) manufactured by Atlas BV. The test pieces are attached to the inside of a drum within the instrument and rotate around the radiation source, with intermittent dark phases. In order to accelerate more hydrolytic degradation they are simultaneously sprayed with water in defined cycles that vary from one specification to the next.

One of the principal radiation and humidifying cycles in use today is the "**CAM 180**" test, developed in the USA (draft US standard SAE J-1960), which allows relatively good transferability of results to outdoor weathering conditions.

The **UV-CON** instrument (Atlas BV) and the **Q.U.V.** system (Q-Panel Lab Products) operate with UV radiation from fluorescent tubes and likewise incorporate humidification. When used with UV-B fluorescent tubes they are unsuitable for one-coat plain-colour coatings because of the pigment changes induced by this rather harsher UV radiation, but they allow rapid testing of clear coats for crazing in particular. They are sometimes operated with UV-A tubes, in which case they are less aggressive.

Lightfastness test

The lightfastness of paints for interior use and of other materials such as textiles is determined by means of a light source defined in ISO 2809, using the *wool scale* as reference standard (see also DIN 54 003 and EN ISO 105).

Chalking test

A weathering test is often followed by a chalking test (together with gloss and adhesion tests). The chalking test involves taking an "impression" with a tacky surface, to which pigment particles loosened by the degradation of the binder adhere and are made visible.

The swollen gelatine side of photographic paper is used in the version according to *Kempf* (draft DIN 53159), and a similar process is followed in the *adhesive tape method* (DIN 53 223, ISO 4628).

Thermal cycling resistance (of furniture surfaces)

Heat and moisture fluctuations generate large mechanical stresses in coated wood surfaces that can lead to cracking and lifting of the coating.

Special cold-check tests have proven useful for testing thermal cycling resistance. These consist in subjecting the samples to numerous cycles of abrupt temperature changes, e.g. from 60 °C to -20 °C. Cyclic loading with linear cooling and heating periods, accompanied by long holding periods at high and low temperatures, yield an optimised test method.

Corrosion tests

In the case of corrosion protection tests, the correlation with outdoor experiments is even more limited than it is in weathering tests. This is explained by a marked temporal distortion of the processes that occur in corrosion or that can prevent it. In *atmospheric* corrosion, many anti-corrosive coatings exhibit a "retard" effect, which means that the anti-corrosive pigments contained in the coating gradually release certain ions (zinc, phosphate, etc.) to the same degree to which they are consumed in outdoor testing too. Other, corrosion-promoting substances are in some cases eliminated by ion-exchange effects. If corrosion tests are conducted too quickly, these processes cannot keep pace with the ag-

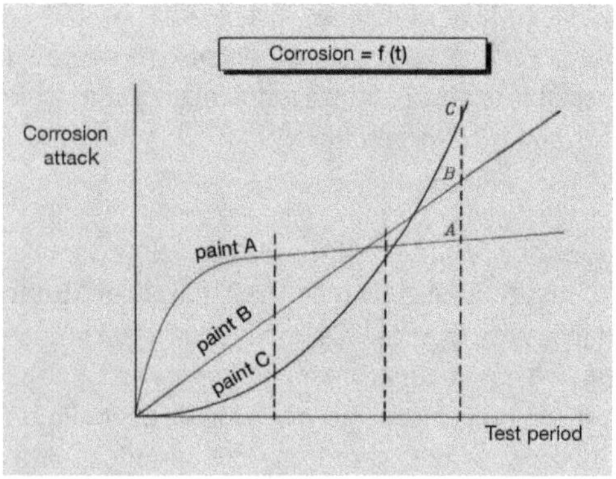

Figure 8.38: Corrosion by using different protection coatings

gressive advance of corrosion, and the results obtained are thus considerably poorer than those observed in practice. In spite of this, accelerated corrosion tests are necessary and are also sufficiently reliable to permit rapid comparisons to be made within a particular systemand even enables defects in the coating to be discovered.

Corrosion tests are normally performed to a very precise schedule, with the exposure time adjusted fairly arbitrarily to the quality of an organic coating. During subsequent comparison of the corrosion-protection properties, the time factor can cause problems with evaluation.

Figure 8.38 exaggerates the change in corrosive attack on three different protective coatings over time. The attack can be linear, exponential or logarithmic and involve different incubation periods. The middle value in the example shown indicates that all three coatings would have suffered the same attack at that point in time. At the earlier test time, the coatings would be ranked A, B, C. The exact opposite order would be obtained after the longest test period.

This is an important problem when correlations are attempted with other tests. Users are advised to perform interim assessments during testing with a view to acquiring more accurate information about the corrosion-protection behaviour.

The **salt spray test** to DIN 50 021 or ISO 7253 is the best known test of this type; it consists in the continuous spraying of a salt solution in a salt spray chamber at elevated temperature for a period of from 3 to 10 days. The surface of the test panels is usually abraded first. The test is used in accordance with ISO 4628 to evaluate rust formation (submigration) and blistering (categorised in the DIN standard by number and size of blisters formed).

In **saturated atmosphere testing** (also known as damp heat testing) to DIN 50 021 (similar to ISO 6270), high relative humidity at elevated temperature (usually 40 °C) leads to the continuous formation of condensation on the test piece. Depending on the test variant, a distinction can also be made between a constant atmosphere and an alternating heat/humidity atmosphere. Regeneration phases are generally also included at intervals to test or make use of the capacity of the coating to “recover”.

The **Kesternich test**, involving exposure to an alternating saturated atmosphere with presence of SO_2 to DIN 50 018 or draft DIN ISO 3231, is rarely used now because the results generally bear little resemblance to reality. It has brought many a good coating system into disrepute without justification. The test is still of some importance for the testing of resistance to industrial atmospheres, however.

The **alternating atmosphere test** to VDA 621-415 (“VDA test”) is a combination of the salt fog and saturated atmosphere test and is used predominantly in automotive coating, and in other industrial sectors in the meanwhile. The results it delivers are more realistic than those obtained from the salt spray test alone, especially on non-ferrous substrates such as zinc or galvanised steel.

Filiform corrosion occurs on steel, but primarily on aluminium alloys and takes the form of a fine network of thin threads where dehesion occurs under the coating; it has the appearance of the damage done by the bark beetle. In aircraft construction and, e.g., in coated window profiles, this special type of corrosion is feared because of the damage it wreaks in a very short time. For filiform corrosion to take place, there must be flaws in the coating, salt must be present as a trigger, and a definite climate is needed, including a relative atmospheric humidity of between 65 % and 95 %. It is tested in accordance with DIN EN ISO 4623 and DIN 65 472: The surface is scratched, seeded with salt or hydrochloric acid vapours and then stored in a conditioning cabinet at 40 °C and 80 % relative air humidity.

Water immersion is a simple test for simulating the influence of moisture (DIN EN ISO 2812). Immersion in demineralised water at a specific temperature can lead to surface changes, blistering and loss of adhesion. The blisters are mostly "osmotic blisters", i.e. blisters that can form by osmosis on the metal surface in the presence of salts and contaminants

Testing for chemical resistance

Loading of organic coatings by chemicals occupies a great deal of time in coatings testing. For example, containers for the chemical industry must have liners which resist aggressive media, such as acids and bases. Radiators, painted cupboards and laboratory furniture must also exhibit high resistance to all kinds of chemicals. Household appliances and furniture are tested for their resistance to diverse foodstuffs and cleaners. The demands of the automotive industry are even more varied. Coatings there must be resistant to mineral oil products, automotive chemicals, atmospheric contaminants, natural aggressive substances, industrial grime, tree resins and others. These tests are usually of the visual type.

To ensure good reproducibility and comparability, defined substances (purity, concentration, etc.) must always be used in testing for chemical resistance.

Test media

Some of the typical test liquids that are frequently used include:

- fuels, lubricants, brake fluids, antifreeze, oil black
- tar and tar removers
- insect fluids, bird droppings, tree resin, pancreatin
- detergents of all kinds, preservatives
- hydrochloric acid, sulphuric acid, caustic soda
- household chemicals, alcohol
- foodstuffs such as fruit juice, tomato juice, ketchup, fats and oils
- fungal and bacterial cultures for wood finishing

Alternative test methods

The effects of the above test media on the test piece can be examined either at room temperature or under progressively increasing temperatures (e.g. using a temperature *gradient bench*). Then, the effects already mentioned in connection with weather resistance testing are examined as a function of exposure time and temperature (see also ISO 2812).

8.7 Damage to coatings and coating systems

Damage to coatings and coating systems covers a whole variety of manifestations of damage, with the most diverse origins or causes. It includes:

- field damage due to biological, chemical (also industrial) and weathering influences
- mechanical damage caused by the effects of dirt, flying stones, scratches, etc. and
- coating defects, where the search for the cause - whether the coating manufacturer or user bears the responsibility for the damage - is often difficult and in some cases never leads to conclusive results.

Finding the cause of damage

A number of examples are set out in Table 8.2 to illustrate that damage to coatings can originate both in unsuitable coating materials and also in incorrect application.
In many cases analytical methods can help to move the search forward. Microscopic analysis with photographic documentation has proven itself in numerous cases. In addition to the traditional cross-section or microtome section methods (see also section 8.6.1), which also enable the coating thicknesses to be measured, special illumination techniques can also be beneficial in standard microscopy.

Table 8.2: Selection of types of damage to coating systems and their possible causes

Damage to coating system	Cause lies in	
	coating material	application
wetting defects	• surface tension too high, inpurities, inhomogeneity	• inadequately cleaned or pretreated surface (greasy residues or similar)
bubbles/blisters	• unsuitable solvents / mixture • "lumpy" flow behaviour	• wet film thickness too great • too rapid heating up in drier • IR radiator too close
grinding traces	• inadequate "body" in coating • unsuitable pigmentation	• abrasive paper too coarse or dirty • film too thin
lifting/wrinkling	• unsuitable solvents	• coating too wet • film too thick • unsuitable old coating
curtaining, sagging, running,	• yield point too low	• coating wet • film too thick
incorrect shade	• defective (mixing) coatings	• metering error • dirty spray gun • film too thin (not opaque)
entrapped dirt	• fish eyes • lint in coating • shelf life too short	• dust in booth • inadequate cleaning, spray mist

Novel damage analysis methods

FTIR microscopy ("IR microprobe") replaces the visible light in a microscope by IR radiation and - as in traditional IR spectroscopy - measures the absorption of IR light in a sample, though in this case in the microscopic range, on sample surface areas in the mm^2 range. In this way occlusions and other defects can be selectively triggered and analysed by IR spectroscopy in respect of *chemical groups, binder or pigment type*, etc.

The mechanism of the *laser microprobe* ("LAMMA") is much more complicated. Material is atomised and partially ionised by means of a laser beam, permitting the identification of most *chemical elements.*

Sources and references for Chapter 8

[1] A. Berger-Schunn, Praktische Farbmessung, Muster Schmidt-Verlag, Göttingen 1991

[2] A. Brasholz: Handbuch der Anstrich- und Beschichtungstechnik, Bauverlag 1989

[3] L. Gehm, Rheologie, Vincentz Verlag, Hanover 1998

[4] A. Goldschmidt, H.-J. Streitberger:, BASF-Handbook: Basics of Coatings Technology 2nd edition, Vincentz-Network, 2007

[5] L. A. Hütter, Wasser und Wasseruntersuchung, Verlag Salle+Sauerländer, 5th edn., Frankfurt/Main 1992

[6] P. von den Kerkhoff, H. Haagen, Lackschadenkatalog, Vogel Verlag, Würzburg 1995

[7] H. Kittel, Lehrbuch der Lacke und Beschichtungen, Verlag W.A. Colomb, Berlin-Oberschwandorf 1975-1979

[8] Lexikon der Anstrichtechnik, Vols. 1 and 2, Callwey Verlag, Munich 1994 and 1995

[9] O. Lückert (Ed.): Prüftechnik bei Lackherstellung und Lackverarbeitung, Curt R. Vincentz Verlag, Hanover 1991

[10] H. Pecina and O. Paprzycki: Lack auf Holz, Curt R. Vincentz Verlag, Hanover 1995

[11] W. Richter, Einführung in die Farbmetrik, 2nd edn., Walter de Gruyter, Berlin 1981

[12] J. Ruf, Organischer Metallschutz, Curt R. Vincentz Verlag, Hanover 1993

[13] G. Schramm: Einführung in Rheologie und Rheometrie, Haake GmbH, Karlsruhe 1995

[14] D.A. Skoog and J.J. Leary, Instrumentelle Analytik, Springer Verlag, Berlin, Heidelberg, New York 1996

[15] H.G. Völz, Industrielle Farbprüfung - Grundlagen und Methoden, VCH Verlagsgesellschaft mbH, Weinheim 1990

[16] U. Zorll (Ed.): Römpp-Lexikon "Lacke und Druckfarben", Thieme Verlag, Stuttgart 1998

[17] A. Zosel, Lack- und Polymerfilme - Viskoelastische Qualitätsmerkmale, Curt R. Vincentz Verlag, Hanover 1996

[18] Further recommendations: see [13] and [14] in Chapter 6

9 Environmental protection and safety at work

The daily activities of coating specialists are nowadays no longer driven purely by chemical, technological and commercial matters. Just as much emphasis is now placed on safety, health and environmental issues, both for operational reasons and in the interests of the general public. Furthermore, the sector is governed - in virtually every industrial country at least - by a whole host of laws, provisions and regulations.

This chapter introduces some of the basic concepts relating to the regulatory side of the industry, with particular reference to the situation in Germany. Two trends must be kept in mind in this respect: firstly we are having to adapt to a continual tightening of the existing regulations, and secondly national legislation is increasingly being shaped by EU directives that apply right across Europe.

9.1 Air pollution control

Existing statutory requirements

Volatile Organic Compounds (VOCs)

The terms solvent and VOC are not uniformly defined either across Europe or throughout the world. According to DIN ISO 11890-1,2 (1998) VOCs are organic liquids (or solids) which evaporate spontaneously under the respective ambient conditions.

The WHO (World Health Organisation) has classified them on the basis of their boiling ranges, as follows:

Volatile Organic Compounds (VOCs)	50 to 100 ... 240 to 260 °C
Very Volatile Organic Compounds (VVOCs)	<0 ... 50 to 100 °C
Semi Volatile Organic Compounds (SVOCs)	240 to 260 ... 0.380 to 400 °C

The abbreviations “SVOC” and “VVOC” are increasingly being used alongside “VOC” to describe solvents for coating materials.

According to the *Solvent Directive* (“Regulation on implementing Council Directive 1999/13/EC on the limitation of emissions of volatile organic compounds due to the use of organic solvents in certain activities and installations”, 31st *BImSchV (federal immission control act)*, organic compounds that have a vapour pressure of 0.01 kPa or more at 293.15 K (20 °C) are covered by the term VOC solvent. Solvents having a lower vapour pressure are not covered by it.

The VOC value is calculated as follows:

$$\text{VOC (g/l)} = \frac{\text{weight of volatile matter (g)} - \text{weight of water (g)}}{\text{volume of coating} - \text{volume of water (l)}}$$

Other calculation methods exist, too. This means that stated VOC values or levels must always be accompanied by the definition and the method of calculation. For coating materials used on wood surfaces:

$$\text{VOC (g/l)} = (100 - \text{nvm} - \text{water fraction}) \cdot \text{density of coating material} \cdot 10$$

nvm: non-volatile matter (%)

Brock, Groteklaes, Mischke: European Coatings Handbook

ISBN 978-3-86630-849-7

As the first level in the hierarchy of legislative air pollution control measures, the Bundesimmissionsschutzgesetz (BImSchG) [federal German pollution control act] contains the principles, procedures and declarations of purpose that state *that* and - in principle - *how* people, animals and plants are to be protected against immissions[1], including noise pollution. Concrete issues of technical planning do not fall within the scope of the BImSchG, however. Such details are covered by the **Bundesimmissionsschutzverordnungen (BImSchV)** [federal German pollution control regulations], of which there are currently 22. Thus, for example, the

- 4th BImSchV contains the regulations concerning the need for official approval under pollution control legislation according to installation type and material throughput, and the
- 12th BImSchV sets out precautions to be taken against accidents involving substances that are covered by the legislation and are above certain threshold values ("accident regulations").

Under the 4th BImSchV, the lower limit for the need for official approval is a solvent use of 25 kg/h. NB: this figure relates to the entire plant and to the maximum conceivable solvent use.

The 31st *BImSchV* specifies emission limits [7] for VOCs classified as carcinogenic, mutagenic or reproduction-toxic (cmr), VOCs that may cause irreversible damage (R 40), Class I substances as defined in the *Technische Anleitung Reinhaltung der Luft* (German clean air act) dated 1986, and other VOCs.

By law, substitutes must be used for cmr substances. The threshold emissions value applies only if a substitute, as defined in accordance with the solvent directive (and the hazardous substances regulation), cannot be found: 1 mg/m^3 from a mass flow of 2.5 g/h for the sum of all corresponding components. The threshold emission value for all volatile compounds subject to risk phrase R 40 and Class I organic compounds, as defined in *TA Luft* 1986, is: 20 mg/m^3 for a mass flow of 100 g/h and above for the sum of all components.

For factories which use the other conventional solvents, there are different ways to be compliant with the regulation's requirements:

- limit value for total emissions in grams of VOCs emitted per m^2 body surface (OEM finishing with solvent consumption exceeding 15 tonnes a year)
- limit value for exhaust gases in milligrams carbon per m^3 treated exhaust gas plus a limit value for fugitive and untreated, contained emissions as a percentage of the solvents employed
- specific reduction scheme with specification of target emission (sum of contained and fugitive emissions) as a ratio of solids consumption
- general (arbitrary) reduction scheme for which equivalence with the reduction scheme of defined target emission can be demonstrated
- simplified reduction scheme (simplified proof of compliance with the target emission)

The emission limit values and the lower limits relating to the granting of permits are characteristic for each sector and depend on the installation size and the best technology available.

To include products and solvent emissions from non-coating installations, the German government passed the *ChemVOCFarbV* law in 2004, which gives force to Directive 2004/42/EU ("Paints Directive"). *ChemVOCFarbV* specifies the VOC reductions to be made in masonry paints inter alia in two stages (by 2007 and 2010).

"Technische Anleitung zur Reinhaltung der Luft (TA Luft)" (German clean air act) is an administrative regulation for defining and monitoring the objectives of the *BImSchG* and contains measures for

1) *Immissions* in the sense of the BImSchG are air pollutants, noise, heat, etc. that have an influence on people and the environment. *Emissions* refer to pollutants and other influences produced by an installation, also including escapes of substances, for example.

preventing damage to the environment by airborne pollutants and is to be the yardstick for approving new installations and major changes and refurbishment of existing ones. The new TA Luft, which came into force on 1 October 2002, now only regulates dust or particulate emissions arising from the manufacture and processing of coatings. However, differences of opinion exist in respect of legislation, jurisdiction and industry. It is not known when the body of laws, which currently exists only in framework form, will be finalised.

Recent amendments to the act include a new *category of requirement* by comparison with earlier practice. Examples include:

- *prevention* instead of damage limitation
- consideration of the *overall burden* (entire firm) rather than evaluation of individual sources
- measures to deal with the *source of emission* instead of the principle of distribution/dilution
- dynamic change instead of fixed procedures (adaptation to the *state of the art*, not the *rules of the art*)
- principle of *proportionality* of cost to benefit and not merely of economic justifiability.

The important *dynamic change clause* also extends to the use of low-solvent or solvent-free coating systems, methods of application, waste-air purification, etc. See also section 6.9.2 in regard to methods of waste-air purification.

An important new instrument in this respect is the Solvent Management Plan (SMP), a kind of *solvent audit*. Firms will be required to publish their internal and external solvent flow at regular internals and to strive for a more efficient use of solvents. Coating shops using from as little as 5 t of solvents per year are required to prepare a SMP.

Solvent reduction in accordance with the EU Directive (VOC Directive)

The Directive is currently at the draft stage, with several different versions still under discussion. It is intended to lay down specific emission limits for the various types of coating manufacturing and processing firms (even those that until now have fallen below the threshold limits) or to bring about a reduction in those limits, which will differ from one sector to the next.

An important new instrument in this respect will be the *Solvent Management Plan* (SMP), a kind of *solvent audit*. Firms will be required to publish their internal and external solvent flow at regular intervals and to strive for a more efficient use of solvents. Coating shops using from as little as 5 t of solvents per year will be required to prepare a SMP.

Solvent audit

The solvent audit was introduced through the solvent directive of the German "Immissionsschutz" law to act as a new planning and reference instrument.

- Its purpose is to determine solvent consumption when adherence to threshold values is being tested
- as proof of compliance with threshold values for fugitive and/or total emissions
- as proof of compliance with the reduction scheme

It returns an overview of all the relevant solvent input and output streams in an installation.

Best available technologies – the EU variant of the state of the art [7]

Standards concerning measures to prevent environmental pollution through the use of the best available techniques (BAT), to protect against severe environmental pollution, to avoid and recycle waste, to use energy efficiently, to prevent and follow-up on accidents are specified in what are called European BAT Reference Notes (BREFs). The BREF document for coating installations, issued in spring 2007, contains standards on limiting emissions. The threshold values are defined in terms of throughput (per unit, per m^2, per kg) and not primarily via specified concentration threshold values. Table 9.1 (page 400) shows the emission values and their classification under EVABAT [7]. The table also reveals the major differences that exist between EU countries.

Table 9.1: Emission limit values and best available techniques (EVABAT)

Industry sector	New installation	Old installation	EVABAT
car; >200 t/a solvent – consumption capacity	45 g/m²	60 g/m²	10 to 35 g/m²
car; >15 t/a in D,A,CR,SR,S,SF,NL	35 g/m²	35 g/m²	
driver's cabin; >200 t/a solvent consumption capacity	55 g/m²	75 g/m²	10 to 55 g/m²
driver's cabin; >15 t/a in D, A	45 g/m²	45 g/m²	
driver's cabin, commercial vehicle, truck; >15 t/a in CR, SR	55 g/m²	60 g/m²	
commercial vehicle, truck; >200 t/a solvent consumption capacity	70 g/m²	90 g/m²	15 to 50 g/m²
commercial vehicle, truck; >15 t/a in D, A	70 g/m²	70 g/m²	
omnibus; >200 t/a LM consumption capacity (incl. degreasing, adhesive processing etc..)	150 g/m²	225 g/m²	92 to 150 g/m²
omnibus; >15 t/a in D, A	150 g/m²	150 g/m²	
omnibus; >15 t/a in CR, SR	70 g/m²	90 g/m²	
rail-car truck; >200 t/a solvent consumption capacity	metal coating	metal coating	70 to 110 g/m²
rail-car truck; >5 t/a in D, A	110 g/m²	110 g/m²	
agricultural and construction machinery; >200 t/a solvent	0.375 kg VOC / kg solids	0.375 kg VOC / kg solids	0.2 to 0.33 kg VOC /kg solids
metal surfaces; >200 t/a solvent consumption capacity	0.375 kg VOC / kg solids	0.375 kg VOC / kg solids	0.1 to 0.3 kg VOC /kg solids
plastic surfaces; >200 t/a solvent consumption capacity	0.375 kg VOC / kg solids	0.375 kg VOC / kg solids	0.25 to 0.35 kg VOC /kg solids
wood surfaces; >200 t/a solvent consumption capacity	1 kg VOC /kg solids	1 kg VOC /kg solids	0.25 kg VOC /kg solids or 3 to 60 g/m²
wood surfaces; >25 t/a in D, A	0.75 kg VOC /kg solids	0.75 kg VOC /kg solids	

9.2 Water pollution control

The Wasserhaushaltsgesetz (WHG) [German water management act] - similarly to air pollution control legislation - regulates every aspect of ground and surface water management in conjunction with regional German water legislation. On the basis of the WHG most of the German Länder (states) have passed “indirect discharger” regulations relating to the discharge of substances that are harmful to water.

Administrative regulations (TA Wasser) [German water pollution act] contain sector-specific limits for certain substances and at the same time describe the state of the art in regard to the corresponding coating installation technology and waste water treatment.

Water hazard classes (WGK) in Germany categorise substances on the basis of their toxicity to fish, daphnia and algae. The significance of the four classes is as follows:

WGK 0: generally not harmful to water
WGK 1: slightly harmful to water
WGK 2: harmful to water
WGK 4: very harmful to water

Water-based coatings are generally only slightly harmful to water; solvent-based coatings or those containing heavy metals are classed in WGK 2 where necessary.

Local statutes often compete with higher-level requirements or may even deviate from them altogether. They control the permissible discharge of non-hazardous substances that affect the chemical oxygen demand (COD) for example, according to the capacity of local sewage treatment plants.

See also section 7.11 in connection with the separation of coating residues, especially from the water from spray booths.

9.3 Waste legislation and waste management

The minimum legal requirements in the areas of waste air control, waste water purification and coating storage can in principle be satisfied by a one-off investment or by choosing a suitable coating material and method of application. The long-term commitment demanded by waste disposal, on the other hand, can only be met with certainty if it is built upon a foundation of accurate and up-to-date information and flexible waste organisation within the company.

Since 1996 the **Kreislaufwirtschafts- und Abfallgesetz (KrW-/AbfG)** [German life-cycle management and waste act] has been managing the classification and handling of "waste for recycling" and "waste for disposal", according to its ability to be recycled. The objective here is environmentally sustainable recycling, and - unlike earlier legislation - the act places emphasis on the following waste hierarchy (already described in section 7.11):

avoidance - reduction - material recycling - energy recovery - landfill

Regional **waste disposal plans** have been drawn up in accordance with §6 of the AbfG, including definition of the various categories and means of disposal of special waste.

The **Technische Anleitung Abfall (TA Abfall)** [German waste management act], like TA Luft, is an administrative regulation that sets out national requirements for the state of the art disposal of waste. It includes detailed regulations on the approval of waste disposal plants, the allocation of waste to disposal routes, the monitoring of waste flows and other organisational regulations.

The **Abfallbestimmungsverordnung** [German waste control regulations] allocates defined classes of waste to particular disposal routes and monitors waste flows. It concerns all of those waste types for which TA Abfall prescribes incineration or disposal as special waste, particularly paint and coating sludge, old coatings and distillation residues. Disposal routes have not been laid down for cured (= solvent-free) coating shop waste (drums, filter mats, backing papers, etc.), which means that, depending on local conditions, it may even be disposed of as domestic-type commercial waste. The trend is generally moving towards treating it as special waste, however.

Local waste statutes are produced by the local bodies responsible for waste disposal, such as local councils, specialist associations, private contractors, etc. They regulate the admission of the different types of waste to the various incineration, recycling and landfill sites according to the local options available.

Assistance is provided in the form of the **waste catalogue** (prepared by the regional working party on waste) which categorises waste by type and by origin and allocates a five-digit waste key number and a disposal route code in each case.

The system is illustrated by the following examples:

55370	Solvent blends not containing halogenated organic components	SAV
55512	Old coatings, old paints, not cured	SAV(SAD)
57118	Plastics packaging	no compulsory requirement
59904	Organic peroxides	CPB/SAV

SAV: Special waste incineration
SAD: Special waste landfill
CPB: Chemico-physical treatment
no further details: no compulsory national requirements

The **European Waste Catalogue (EWC)** has been in force in Germany too since 1.1.1999 in relation to the declaration of waste. It states that waste must be separated as completely as possible - from the raw materials development laboratory through to the coating workshop and coating stripper - according to the possibilities for recycling and disposal. This saves huge unnecessary costs and reduces the amount of waste for which more complex disposal is required.

The **Abfall- und Reststoffüberwachungsverordnung** [waste and refuse monitoring regulations] in essence demands proof of disposal and recycling as a means of initial control: securing the permissible disposal route even before the disposal process begins. It requires appropriate declarations from the waste generator, the disposal firm and the relevant authorities regarding the admissibility of the stated disposal route.

In accordance with **Verpackungsverordnung** [packaging regulations] private and commercial end users have been entitled since 1993 to return transit and sales packaging via the sales route or to send it for material recycling by means of the dual waste management system. Disposable containers bearing labels indicating that their contents are harmful or dangerous to the environment are excluded from these provisions. In cases where small volumes makes it impractical to supply reusable containers, closed recycling systems are currently under construction for smaller tin and plastic containers.

9.4 Safe handling of paints and coatings

The handling of coating materials can in principle give rise to two types of danger:

- risk of fire and explosion (in the case of solvent-based materials) and
- risk to health by intake (absorption through the skin, inhalation, swallowing) of harmful substances.

Incorrect handling can increase the risk of the following conditions:

- skin diseases
- respiratory diseases
- allergies
- damage to immune system
- damage to nervous system
- genetic changes[1)]
- cancer[1)]
- liver and kidney damage
- infertility

The **Gefahrstoffverordnung (GefStoffV)** [German regulations covering hazardous substances] encompasses classifications, labelling, danger symbols and packaging in respect of the handling of hazardous working substances. The fourth amendment of 1.5.1994[2)] introduced substantial changes.

The currently valid *danger symbols* (black against an orange background) are illustrated in Figure 9.1. They include the new hazard symbol **N** "Dangerous to the environment" and a number of new terms. In particular, "Slightly toxic" has been replaced by "Harmful".

Under another provision, the statutory *EU Safety Data Sheet* must be supplied to the purchaser (professional user) no later than at the time of the first delivery of a substance.

1) Nowadays mutagenic and carcinogenic constituents have virtually disappeared from the formulations.
2) An overview is provided in [1]

Figure 9.1: Danger symbols for the labelling of hazardous substances and preparations

The **Explosionsschutzrichtlinien (Ex-RL)** [German guidelines on explosion protection] provide information about the assessment of explosion hazards and appropriate protective measures together with a zoning system for hazardous areas from 0 to 11 (decreasing probability of the development of an explosive atmosphere).

It must contain: Name of product and company, composition with information on ingredients, potential hazards, first aid measures, measures for fire fighting and accidental release, handling or storage, exposure controls and personal protection, physical and chemical properties, stability and reactivity, toxicological and ecological information, disposal information, transport information and regulations. For implementation of the *31st BImSchV*, the primary requirements now are: solids content, solvent content, VOC content, content of particularly critical VOC.

Protective equipment and safety precautions

Some key examples of personal protective equipment to be used in the laboratory and, most importantly, on site are listed below:

- conductive footwear to protect against electrostatic discharges, also conductive protective gloves
- suitable protective clothing, particularly in respect of its fire and melting behaviour
- skin protection - after contact with solvents - by means of appropriate skincare products
- eye protection provided by protective goggles with side protection, or full face mask if necessary
- breathing apparatus if ventilation installations are inadequate, with or without fresh air supply[1)].

Examples of some of the basic safety precautions to be taken in the laboratory and on site (and which must be supported and enforced by safety officers) include the following:

- areas at risk from explosion must be clearly labelled; prohibition of flames, unshielded lights and smoking
- hazardous substances must never be stored in containers that could lead to their being mistaken for foodstuffs
- food and tobacco must never be allowed to come into contact with hazardous substances; eating and smoking should also be prohibited in coating areas

1) Further details are contained for example in the leaflet "Arbeitssicherheit und Gesundheitsschutz in Autoreparatur- und Karosseriebetrieben" published by the technical working party of the automotive refinishing coatings specialist division of the Verband der Lackindustrie e.V.

- fire extinguishers must be provided and kept ready for use
- labels, safety warnings and data sheets must be displayed and observed
- emergency exits must be labelled and kept clear
- First Aid equipment and trained First Aiders must be provided
- important telephone numbers such as emergency/fire/ambulance must be kept to hand
- an emergency plan must be prepared and kept to hand; it should cover appropriate actions to be taken, including designated assembly points for employees in the event of fire.

Definitive guidelines

The **Verordnung über brennbare Flüssigkeiten (VbF)** [German regulations on combustible liquids] covers the classification of solvents and corresponding finished products by flash point and water solubility into hazard classes that must be observed in the storage and handling of coating materials. See also section 8.2.2 (Determining the flash point) in regard to classification into groups and hazard classes.

The **maximale Arbeitsplatzkonzentration (MAC value)** taken from TRGS[1)] 900 is the maximum allowable concentration of gas, vapour or air-borne contaminants in the air at the workplace that according to current data is generally not harmful to health by exposure for eight hours per day over a 40 or 42 hour working week. It therefore does not relate to short-term peak values or cumulative amounts. Measurements are carried out at the workplace to monitor concentrations. Depending on the category of risk, peak limits in respect of the level by which the MAC value can be exceeded and the frequency of such occurrences are also laid down, however.

The **technische Richtkonzentration (TRK)** [technical recommended concentration] of a hazardous substance is the concentration in air that may be reached in accordance with state of the art practices and that should be used as a guide to the safety precautions to be taken and the measurements to be used to monitor the workplace. TRK values are stated only for those hazardous substances for which no MAC values have yet been established. The TRK values are published on an annual basis in the form of TRGS 900.

9.5 Transportation

The **Gefahrgutverordnung Strasse (GGVS)** [German regulations covering the transport of dangerous goods by road] lays down the labelling, choice of container, etc., to be used for road transport according to the viscosity, flash point, toxicity, irritant effect and miscibility with water of a substance.

Equivalent provisions exist for transport by sea (GGVSee) and by rail (GGVE).

9.6 REACH (Registration, Evaluation, Authorisation of Chemicals)

The law was adopted on 1 June 2007 by the European parliament. Its purpose is to address the existing gap since 1982 between two valuation bases for chemical substances: existing substances (produced and sold before 1981) and new substances (produced and sold after 1981).

Key content: No distinction between new and existing substances. All must be registered, and data of the same quality must be provided for all.

Producers and processors share responsibility along the value chain. The uses of the substances must be named in the registration, and other uses either are not permitted or must be post-registered. The use of the substances along the supply chain must be accompanied by a risk assessment.

1) Technische Regeln für Gefahrstoffe [technical rules for hazardous substances]. They set out the latest industrial health, safety, hygiene and technical requirements in respect of the distribution and handling of hazardous substances.

Hazardous substances are only approved for a certain period. A replacement must be found for them, and their use is allowed only if no suitable replacements are available.

This affects manufacturers and importers (into the EU) of substances, downstream users and processors of substances and thus virtually all manufacturing companies, including formulators producing preparations (paints, inks, adhesives ...) as well as companies using these substances and preparations to provide services (craftsmen).

Further effects: Manufacturers, importers and users must ensure that - under reasonably foreseeable conditions - substances are produced, sold or used individually or in preparations in such a way that human health and the environment are not adversely affected.

Specific measures: risk assessment (Chemical Safety Assessment) by manufacturers or importers.

All participants in the product chain: measures to reduce risk.

9.7 Eco-audits: information and limits

Definition

An eco-audit is the most comprehensive possible comparison of the environmental influences of different products, services, systems, methods, etc. Eco-audits - even if they contain errors or are incomplete - are increasingly being used as the basis for decision-making in procurement and purchasing but additionally also to reveal weak points and to improve the environmental properties of products.

Procedure

There is currently no generally recognised or compulsory standard procedure that meets the requirements for a comprehensive, environment-focused product assessment. A series of initiatives are in progress on the part of government bodies and of economic, consumer and environmental

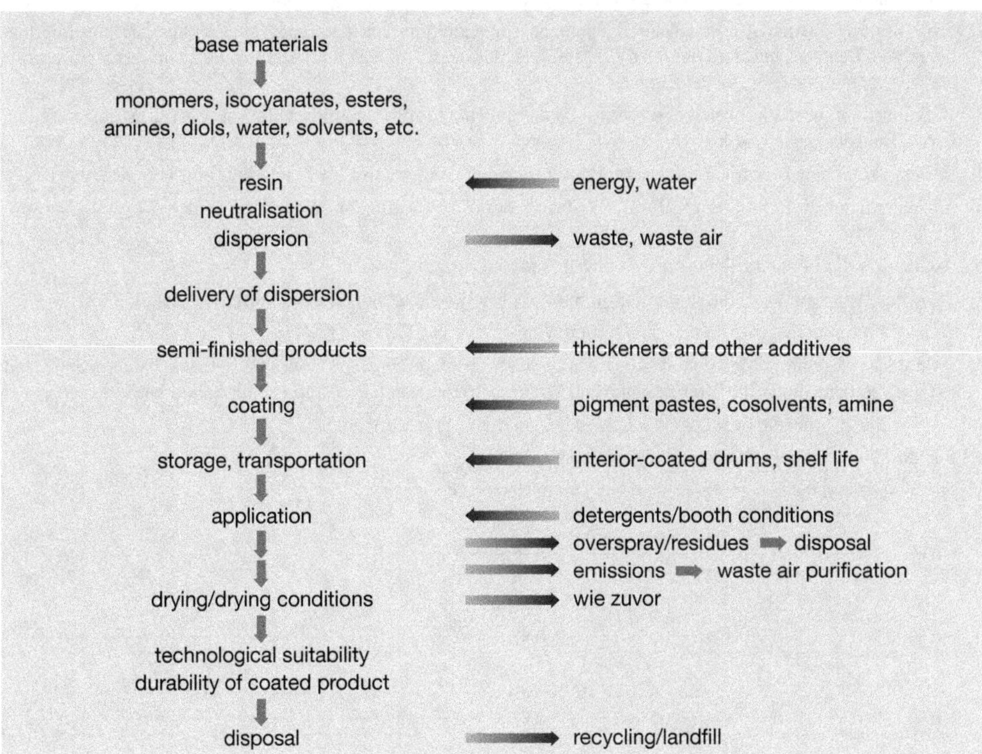

organisations to fill this gap. In attempting to produce an eco-audit, a clear distinction must be made between the

- *material audit* (matrix-type list of all of the environmental loads throughout the entire product life-cycle, compiled systematically without any rating) and the
- *audit evaluation*, in which the items from the material audit are evaluated in terms of their effects on the environment, such as climatic impact (e.g. CO_2 emission). Depletion of stratospheric ozone (key phrase: gap in the ozone layer), photochemical formation of oxidants (keyword: summer smog), eutrophication (over-use of fertilisers in water-bodies and soils), acidification, burden from fossil resources (e.g. raw materials and fossil fuels), burden on nature areas, direct damage to health (through hazardous substances or noise), direct damage to ecosystems

The diagram (page 405) uses the example of the life cycle of a (water-based) coating to describe all of the constituents, stages, handling procedures and marginal conditions that need to be taken into consideration even in the preparation of the material audit.

In the audit evaluation the various criteria mentioned above may have thoroughly contradictory effects on the result of an eco-audit. One example is the replacement of a forced-drying high solid coating with a water-based coating. The saving in solvent is cancelled out by the longer process time and possibly higher energy requirement on the one hand and by the content of amine and cosolvent on the other. Although the content of the latter substances may be lower than that of the solvents in the equivalent product, their ecotoxicity and photochemical ozone-forming potential may cause them to be rated as being more critical than butyl acetate or benzines, for example.

At present the individual author of an eco-audit is thus completely at liberty to build economic or strategic interests into his or her weighting of the factors!

Sources and references for Chapter 9

[1] G. Bachmann, J. Oberflächentechnik, 2 (1995), page 32

[2] Ernst Basler + Partner AG, Staatliche Maßnahmen zur Reduktion von anthropogenen VOC-Emissionen Informationen aus Deutschland und Österreich (Phase 2), Schlussbericht, Zollikon(CH) 2007, www.ebp.ch/files/projekte/efk_voc_lenkungsabgabe_p2_bericht.pdf

[3] W. Baumann u. A. Muth, Farben und Lacke - Daten und Fakten zum Umwelt-schutz, Vol. 1 and 2, Springer Verlag, Berlin, Heidelberg, New York 1997

[4] K. Dören,W. Freitag,D. Stoye, Wasserlacke: Umweltfreundliche Alternative für Beschichtungen, TÜV Rheinland 1992

[5] A. Goldschmidt, H.-J. Streitberger, BASF Handbook: Basics of Coating Technology, 2nd edition, Vincentz Network Hanover 2007

[6] Ökobilanzen für Produkte, Reihe "Texte Umweltbundesamt" Berlin 1992.

[7] Th. May: "Ausgehverbot(e) für Lösemittel" in www.aktuelle-wochenschau.de/2007/woche27/woche27.html

[8] Th.May: Umweltmanagement im Lackierbetrieb, Vincentz Verlag, Hanover 1997

[9] D.Ondratschek (editor): "Taschenbuch Besser Lackieren" 2008, 65th edition, Vincentz Network Hanover 2007, and previous editions (up to 1997: Taschen-buch für Lackierbetriebe, thereafter: Jahrbuch für Lackierbetriebe

[10] VdL-Leitlinien "Umwelt, Gesundheitsschutz und Sicherheit", Frankfurt 1995

[11] RömppOnline, mit updates, siehe www.roempp.com

[12] Further recommended reading: see [13] and [14] in chapter 6

Authors biographies

Thomas Brock, born in 1950, studied chemistry at the universities of Münster and Cologne. After obtaining his doctorate, he spent two-and-a-half years working in development and production at Gebr. Schmidt GmbH, a printing inks manufacturer. He then spent 10 years developing automotive coatings in the Applied Technology department of Herberts GmbH (today DuPont Performance Coatings), where he oversaw the introduction of waterborne coatings. He also researched novel two-pack systems, and optimised the dispersion of pigments in pigmented coatings. Since March 1995, he has been professor of Coatings Technology at the Niederrhein University of Applied Sciences in Krefeld, Germany. His special interests are metrology, testing, rheology, coatings production, coatings processes and application methods. In addition to lecturing, he has produced numerous publications and technical papers down the years, especially in seminars run by technical academies, the German Research Associations and since 2006 he is chairman of division "Paint Chemistry" in German Chemical Society.

Michael Groteklaes, born in 1956, studied chemistry at RWTH University in Aachen, Germany. After obtaining his doctorate, he was appointed head of Plastics Compounding and Processing at Solvay GmbH in Rheinberg, Germany, in 1989. He joined Chemson Polymer-Additive GmbH in Arnoldstein, Austria, three years later to work in the field of formulation of polymer additive compounds. In 1993, he was appointed professor of Coatings Chemistry and Inorganic Chemistry at the University of Applied Sciences in Niederrhein, Krefeld. The focus of his activities is on the dispersion and stabilisation of pigments and fillers, corrosion protection, masonry protection, powder coatings and the formulation of coating materials. His interest is more in the "inorganic aspects" of coatings. Outside his lecturing responsibilities, he is engaged in running advanced technical training courses.

Peter Mischke, born in 1951, studied chemistry at Hamburg University. After obtaining his doctorate, he went into industry where he worked primarily on metal surface pretreatment and then in the development of synthetic resins. In 1991, he was appointed professor of "Paint Chemistry and Process Engineering" at the University of Applied Sciences in Niederrhein, Krefeld. He teaches chemotechnical subjects as well as polymer, binder, interface and colloid chemistry for paint engineers. He also serves as a lecturer at various vocational training events and is the author of the book "Film Formation in Modern Paints".

Brock, Groteklaes, Mischke: European Coatings Handbook

ISBN 978-3-86630-849-7

Index

Brock, Groteklaes, Mischke: European Coatings Handbook

ISBN 978-3-86630-849-7

B

C

D

E

F

I

K

L

M

Q

R

S

T

U

V

W

X

Y

Z

Technical name	Systematic name
acrylic acid	propenoic acid
adipic acid	hexanedioic acid
alkyl (meth)acrylate	alkyl (meth)acrylate
allyl ...	2-propenyl ...
bisphenol A	2,2-bis(4-hydroxyphenyl)propane
bisphenol F	bis(4-hydroxyphenyl)methane
butyl acetate	acetic acid n-butyl ester
butyl diglycol	diethylene glycol monobutyl ether
butyl diglycol acetate	diethylene glycol monobutyl ether acetate
butyl glycol	ethylene glycol monobutyl ether
butyl glycol acetate	2-butoxyethyl acetate
dehydrated castor oil	trans,trans-9,11-octadecadienoic acid
dipropylene triamine	3,3'-iminodi-(1-propylamine)
elaeostearic acid	octadeca-9,11,13-trienoic acid
epichlorohydrin	1-chloro-2,3-epoxypropane
ethyl acetate	ethyl acetic ester
ethyl diglycol	diethylene glycol monoethyl ether
ethyl diglycol acetate	2-(2-ethoxyethoxy)ethyl acetate
ethyl glycol	2-ethoxyethanol
ethylene glycol	ethane-1,2-diol
ethylenediamine	1,2-ethanediamine
formaldehyde	methanal
fumaric acid	trans-butenedioic acid
glycerol	propane-1,2,3-triol
glycidyl ...	2,3-ethoxypropyl ...
hexahydrophthalic acid (anhydride)	cyclohexane-1,2-dicarboxylic acid (anhydride)
isobutanol	2-methylpropan-1-ol
isophorone diisocyanate	3-isocyanatomethyl-3,5,5-trimethylcyclohexyl
isocyanate	
isophthalic acid	1,3-benzenedicarboxylic acid
isoprene	2-methylbuta-1,3-diene
isopropanol	propan-2-ol
lauric acid	dodecanoic acid
linoleic acid	octadeca-9,12-dienoic acid
linolenic acid	octadeca-9-cis,12-cis,15-cis-trienoic acid
maleic acid (anhydride)	cis-butenedioic acid (anhydride)
melamine	triaminotriazine
methacrylic acid	2-methylpropenoic acid
1-methoxypropyl acetate	1,2-propyleneglycol-1-methylether-2-acetate
methylol ...	hydroxymethyl ...
α-methylstyrene	isopropenylbenzene
neopentyl glycol	2,2-dimethyl-1,3-propanediol
oleic acid	cis-octadec-9-enoic acid
pentaerythritol	2,2-bis(hydroxymethyl)-1,3-propanediol
phenol	hydroxybenzene
phthalic acid (anhydride)	benzene-1,2-dicarboxylic acid (anhydride)

Technical name	Systematic name
1,2-propylene glycol	propane-1,2-diol
Technical name	**Systematic name**
propylglycol	ethylene glycol monopropyl ether
ricinoleic acid	cis-12-hydroxyoctadec-9-enoic acid
stearic acid	octadecanoic acid
terephthalic acid	benzene-1,4-dicarboxylic acid
trimellitic acid (anhydride)	benzene-1,2,4-tricarboxylic acid (anhydride)
trimethylol propane	1,1,1-tris(hydroxymethyl)propane
urea	carbamide
vinyl ...	ethenyl ...

www.ingramcontent.com/pod-product-compliance
Lightning Source LLC
LaVergne TN
LVHW080343011026
849161LV00009B/2174